Criminalistics

Criminalistics
Forensic Science, Crime, and Terrorism

James E. Girard, PhD
American University

JONES & BARTLETT
LEARNING

World Headquarters
Jones & Bartlett Learning
40 Tall Pine Drive
Sudbury, MA 01776
978-443-5000
info@jblearning.com
www.jblearning.com

Jones & Bartlett Learning
Canada
6339 Ormindale Way
Mississauga, Ontario L5V 1J2
Canada

Jones & Bartlett Learning
International
Barb House, Barb Mews
London W6 7PA
United Kingdom

Jones & Bartlett Learning books and products are available through most bookstores and online booksellers. To contact Jones & Bartlett Learning directly, call 800-832-0034, fax 978-443-8000, or visit our website, www.jblearning.com.

Substantial discounts on bulk quantities of Jones & Bartlett Learning publications are available to corporations, professional associations, and other qualified organizations. For details and specific discount information, contact the special sales department at Jones & Bartlett Learning via the above contact information or send an email to specialsales@jblearning.com.

Production Credits
Chief Executive Officer: Ty Field
President: James Homer
SVP, Chief Operating Officer: Don Jones, Jr.
SVP, Chief Technology Officer: Dean Fossella
SVP, Chief Marketing Officer: Alison M. Pendergast
SVP, Chief Financial Officer: Ruth Siporin
Publisher, Higher Education: Cathleen Sether
Acquisitions Editor: Sean Connelly
Associate Editor: Megan R. Turner
Production Manager: Julie Champagne Bolduc
Associate Production Editor: Jessica Steele Newfell
Associate Marketing Manager: Jessica Cormier
Manufacturing and Inventory Control Supervisor: Amy Bacus
Assistant Print Buyer: Jessica DeMarco
Composition: Publishers' Design and Production Services, Inc.
Cover Design: Scott Moden
Photo Research and Permissions Manager: Kimberly Potvin
Assistant Photo Researcher: Carolyn Arcabascio
Cover Images: (top) © Ramzi Hachicho/ShutterStock, Inc.; (bottom row, left to right) © Edw/ShutterStock, Inc., Courtesy of Forensic Source © 2010, © Yuri Bathan (yuri10b)/ShutterStock, Inc.
Printing and Binding: Courier Kendallville
Cover Printing: Courier Kendallville
Illustrations: Erik Garcia

Library of Congress Cataloging-in-Publication Data
Girard, James.
    Criminalistics : forensic science, crime, and terrorism / James E. Girard.—2nd ed.
      p. cm.
    Includes index.
    ISBN 978-0-7637-7731-9 (casebound : alk. paper)
    1. Forensic sciences.   2. Criminal investigation.   I. Title.
    HV8073.G564 2011
    363.25'62—dc22
                                                    2009052358
6048

Printed in the United States of America
14 13 12 11 10   10 9 8 7 6 5 4 3 2 1

Dedicated to all my children and grandchildren.

Brief Contents

Contents

SECTION 1
Introduction to Criminalistics. 3

CHAPTER 1
Investigating the Crime Scene 5

CHAPTER 2
Investigating and Processing Physical Evidence 31

CHAPTER 13

Biological Fluids: Blood, Semen, Saliva, and an Introduction to DNA 329

CHAPTER 14

Forensic DNA Typing 357

SECTION 6

Terrorism 389

CHAPTER 15

Computer Forensics, Cybercrime, and Cyberterrorism 391

Resource Preview

Chapter Objectives: Concise learning objectives provide a preview of key chapter concepts and serve as a useful guide for reviewing chapter material.

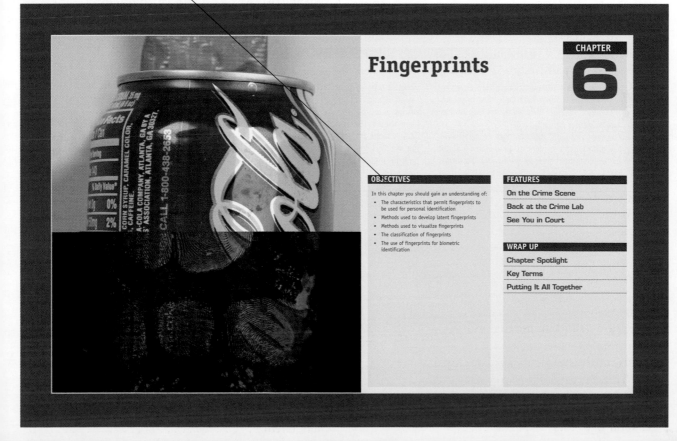

Fingerprints

OBJECTIVES

In this chapter you should gain an understanding of:

- The characteristics that permit fingerprints to be used for personal identification
- Methods used to develop latent fingerprints
- Methods used to visualize fingerprints
- The classification of fingerprints
- The use of fingerprints for biometric identification

FEATURES

On the Crime Scene

Back at the Crime Lab

See You in Court

WRAP UP

Chapter Spotlight

Key Terms

Putting It All Together

YOU ARE THE FORENSIC SCIENTIST

As mentioned in Chapter 1, the smallest objects found as evidence are referred to as *trace evidence*. These objects (i.e., fibers, glass fragments, gunshot residue) are so easily transferred from one individual to another that they may provide evidence of association between a suspect and the victim. Because they are so readily transferred, investigators must take great care to avoid losing or cross-contaminating this evidence.

Usually trace evidence is transferred from one object to another in a process referred to as *direct transfer*. On other occasions, trace evidence is transferred from one object to another by way of an intermediate object, in a process known as *secondary transfer*. It also is possible that two or more intermediate objects may be involved in secondary transfer. It is important that investigators consider the possibility of secondary transfer whenever they examine trace evidence.

Consider the following trace evidence, which was found at the scene of a murder. In the victim's room, where the murder occurred, there is a fabric-covered chair. The suspect's jacket and sweater are seized from his apartment and examined for fiber evidence. The chair has fibers on it that match fibers from the jacket and the sweater. The jacket has fibers from the chair. The sweater does not have fibers from the chair on it.

1. Explain how the fibers might have been transferred.
2. Does this evidence prove that the suspect was in the victim's room more than once?

You are the Forensic Scientist: Realistic case studies and accompanying discussion questions challenge readers to think like a practicing forensic scientist.

BACK AT THE CRIME LAB

Questioned document examination often involves studying obliterated writing. Examiners must take care with this type of evidence to ensure that the sample is not further altered or destroyed.

For example, organic solvents used to clean surfaces may destroy trace evidence that was invisible to the naked eye but might be readily viewed under ultra-violet or infrared light. Infrared photographs of questioned documents are commonly taken, even when writing is visible, to see whether the document has been altered. Looking at questioned documents under various types of light can help examiners identify writing imprints, altered writing, and writing that may have been covered by paint.

Back at the Crime Lab: Summary of scientific principles and procedures.

on the CRIME SCENE — Meth Labs in the Neighborhood

Over the past 20 years, the United States has seen an increase in the number of illegal, underground production sites for methamphetamine ("meth"), a drug that causes a similar reaction as cocaine but is far cheaper to produce and obtain. However, a real laboratory with expensive equipment is not necessary for the production of the drug. Meth can be—and is—produced easily in a kitchen with a slow cooker or in makeshift kitchens using Sterno and cans.

Meth also is manufactured from products that are readily available at the local grocery store. The basic ingredient is pseudoephedrine, the active ingredient in many over-the-counter decongestants. Other materials used to manufacture meth include widely available products such as diet pills, batteries, matches, peroxide, charcoal lighter fluid, paint thinner, rubbing alcohol, drain cleaners, and anhydrous ammonia (which farmers use in fertilizers).

Besides the obvious problems of illegal drug manufacture, distribution, and use/abuse, meth labs pose a hazard to other people living in the vicinity. Meth labs are often set up in kitchens, garages, or other places that may be hidden from view. The production of meth requires the use of corrosive, reactive, toxic, and ignitable substances, so the labs in which it is produced are prone to fire or explosions. Indeed, an estimated 15% of meth labs are discovered because they either blow up or burn down a structure.

In addition, meth production inevitably leaves behind a residue of harmful substances or fumes rising from cookers that can permeate walls, carpeting, and drapery, or even seep between apartments in an apartment or condominium complex. Although there is no infallible method for tracking whether a dwelling has housed a meth lab, the state of Washington requires owners to ensure that the dwelling has less than 5 µg (micrograms) of meth per square foot; in Oregon, the maximum level is 0.5 µg per square foot. Missouri is creating a list of addresses where meth labs have been reported. Antidrug campaigns recommend that home buyers or renters contact local officials to determine if a seizure of chemicals has occurred at the property.

Meth labs pose a particular health risk for children. Often the children of meth producers are neglected or abused by their parents. Their risks from exposure to the toxic and corrosive chemicals used in meth production are equally concerning. Furthermore, the toxic by-products of a meth lab are often disposed of improperly outside, contaminating areas where children play.

Sources: KCI: The Anti-Meth Site. Is there a meth lab cookin' in your neighborhood? http://www.kci.org/meth_info/neighborhood_lab.htm; and U.S. Department of Justice. Dangers to children living at meth labs. http://www.ojp.gov/ovc/publications/bulletins/children/pg5.html

On the Crime Scene: High-profile cases or issues are highlighted relating to relevant theories.

SEE YOU IN COURT

In the late 1970s, a wave of child murders occurred in Atlanta. Wayne Williams was arrested for the offenses, based principally on the carpet fiber evidence found on the bodies, which was similar to carpet fibers found in his vehicle. Williams was convicted and sentenced to life in prison, where he remains today.

However, controversy has dogged this case since the moment the fibers were introduced into evidence. Contrary to the practice today, the jury was never told that the fibers were not unique, but rather heard that they matched carpet fibers that could be found throughout the Atlanta metropolitan area.

Probabilities are difficult to establish in many, if not most, physical evidence cases. Determining the likelihood of a match for new car carpet, for example, would be possible if one knew how many cars with that type and color of carpet were sold within a certain distance of the location of interest. Obviously, as the area included increases, the probabilities drop. Given that cars generally do not stay in confined spaces, one would have to make explicit assumptions to justify the development of a probability statement. No such evidence was employed in the Williams case, which is why this conviction remains tainted in the eyes of many modern forensic scientists.

See You in Court: Relevant court cases explain how forensics played a role in the verdict.

YOU ARE THE FORENSIC SCIENTIST SUMMARY

1. Section 215 of the USA Patriot Act allows the FBI to order any person or entity to turn over "any tangible things," so long as the FBI "specifies" that the order is "for an authorized investigation . . . to protect against international terrorism or clandestine intelligence activities." Section 215 vastly expands the FBI's power to spy on ordinary people living in the United States, including U.S. citizens and permanent residents.
   a. The FBI need not show probable cause, nor even reasonable grounds to believe, that the person whose records it seeks is engaged in criminal activity.
   b. The FBI need not have any suspicion that the subject of the investigation is a foreign power or an agent of a foreign power.
   c. The FBI can investigate U.S. citizens based in part on their exercise of First Amendment rights, and it can investigate non-U.S. citizens based solely on their exercise of First Amendment rights. For example, the FBI could spy on a person because it doesn't like the books she reads, or because it doesn't like the websites she visits. It could spy on her because she wrote a letter to the editor that criticized government policy.
   d. Anyone served with Section 215 orders is prohibited from disclosing that fact to anyone else. The people who are the subjects of the surveillance are never notified that their privacy has been compromised.
2. Yes, the police officer would be able to search the backpack under the emergency exceptions, plain view doctrine, and open fields exceptions.

Chapter Spotlight

- There are three types of WMDs: chemical, biological, and nuclear weapons.
- Chemical weapons, which have the greatest potential for terrorist use, are classified into six categories: choking agents, blister agents, blood agents, irritating agents, incapacitating agents, and nerve agents.
- TIMs, which are commonly used by the chemical industry, can be used as WMDs.
- A point detector is a sensor that samples the air around it.
- The PID, SAW, and colorimetric tubes are all used as point detectors for sensing chemicals present in the air.
- Standoff detectors are used to warn of clouds of CWA or TIM approaching from a distance.
- A "dirty bomb," which is also known as an RDD, uses conventional explosives to spread nuclear contamination.

- Radioactive materials can emit three types of radiation: alpha, beta, and gamma rays.
- The amount of damage caused by radioactive materials depends on the type and penetration power of the radiation, the location of the radiation, the type of tissue exposed, and the amount or frequency of exposure.
- Radioactive contamination is detected by a Geiger counter.
- BWs disseminated through the air pose a threat to the general public.
- Current biological detection systems that use immunoassays or DNA-based technologies are not as reliable as are chemical detection systems.

Key Terms

Alpha particle A radioactive particle that is a helium nucleus.

Beta particle A radioactive particle that has properties similar to those of an electron.

Biological weapon (BW) Disease-producing microorganisms, toxic biological products, or organic biocides used to cause death or injury.

Blister agent A chemical that injures the eyes and lungs and burns or blisters the skin.

Chemical warfare agent (CWA) A chemical used to cause disease or death.

Choking agent A chemical agent that attacks the lung tissue, causing the lung to fill with fluid.

Cholinesterase An enzyme found at nerve terminals.

Curie A unit of radioactivity.

Enzyme-linked immunosorbent assay (ELISA) A sensitive immunoassay that uses an enzyme linked to an antibody or antigen as a marker for the detection of a specific protein, especially an antigen or antibody. It is often used as a diagnostic test to determine exposure to a particular infectious agent.

Gamma ray A high-energy photon emitted by radioactive substances.

Geiger counter An instrument that detects and measures the intensity of radiation.

Gray The new international unit that is intended to replace the rad (l Gy = 100 rad).

Half-life The time required for half of the atoms originally in a radioactive sample to decay.

Immunoassay A test that makes use of the binding between an antigen and its antibody to identify and quantify the specific antigen or antibody in a sample.

Incapacitating agent A chemical that disables but does not kill immediately.

Ionizing radiation Radiation capable of dislodging an electron from an atom, thereby damaging living tissue.

Nerve agent A chemical that incapacitates its target by attacking the nerves.

Neurotransmitter A chemical that carries nerve impulses across the synapse between nerve cells.

Point detector A sensor that samples the environment wherever it is located.

Rad Radiation absorbed dose; the basic unit of measure for expressing absorbed radiant energy per unit mass of material.

Rem Roentgen equivalent for man; a dose of ionizing radiation.

Sarin A nerve gas.

Sievert The new international unit intended to replace the rem (l Sv = 100 rem).

Standoff detector A sensor that reacts to distant events or hazards and can be used to warn of approaching chemicals.

Synapse A narrow gap between nerve cells across which an electrical impulse is carried.

Tabun A nerve gas.

Transmutation Conversion of one kind of atomic nucleus to another.

Weapon of mass destruction (WMD) A weapon that kills or injures civilians as well as military personnel. WMDs include nuclear, chemical, and biological weapons.

Wrap Up: Each chapter concludes with answers to the case study, a chapter summary, key terms, review questions, review problems, and suggestions for further reading specific to the chapter's subject matter.

Resources

Instructor Resources

- **PowerPoint Presentations**, providing you with a powerful way to make presentations that are both educational and engaging. Slides can be modified and edited to meet your needs.
- **Lecture Outlines**, providing you with complete, ready-to-use lesson plans that outline all of the topics covered in the text. Lesson plans can be edited and modified to fit your course.
- **Electronic Test Bank**, containing multiple-choice and scenario-based questions, allows you to originate tailor-made classroom tests and quizzes quicky and easily by selecting, editing, organizing, and printing a test along with an answer key that includes page references to the text.
- **Answers to Review Problems**

The resources found on www.jblearning.com have been formatted so that you can seamlessly integrate them into the most popular course administration tools. Please contact Jones & Bartlett Learning technical support at any time with questions.

Student Resources

Essential components to the teaching and learning system are interactivities and additional resources that help the students grasp key concepts in criminology.

criminaljustice.jblearning.com/criminalistics

Make full use of today's teaching and learning technology with our interactive Web site, which has been specifically designed to complement *Criminalistics: Forensic Science, Crime, and Terrorism, Second Edition*. Some of the resources available include:

- Chapter outcomes
- Interactivities including Web links and chapter spotlight
- Vocabulary explorer including an interactive glossary, flashcards, and crossword puzzles

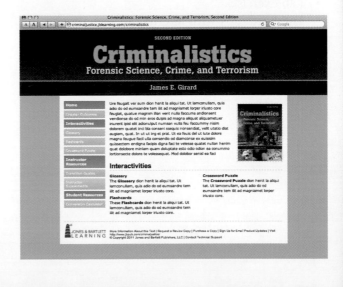

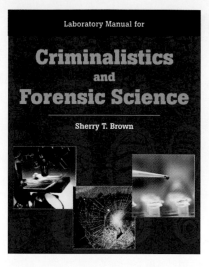

Basic Laboratory Manual for Forensic Science
Ann Wolpert Burgess, Georgia A. Pasqualone,
Michael J. Piatelli—Boston College

http://criminalistics.jbpub.com/basicforensicslab

While most laboratory manuals require sophisticated and expensive equipment, this online manual can be used without this apparatus. Instead any student can apply the labs in any setting.

The manual has 14 online labs and each lab includes:

- Objectives for the laboratory
- PowerPoint lectures and/or video recommendations
- Crime scene cases
- Lab exercises
- Lab write-ups with alternative scenarios to the case to enhance critical thinking

Laboratory Manual for Criminalistics and Forensic Science
Sherry T. Brown, York College of Pennsylvania

http://criminalistics.jbpub.com/forensicslab

This online laboratory manual includes 22 labs that introduce students to the laboratory aspect of crime evidence analysis, such as hairs, fibers, paint, fingerprints, bite marks, and more. The author wrote this manual based on 25 years of instruction with extensive classroom testing.

Additionally, this lab manual includes an instructor manual to assist with the laboratory design, offering substitute options, answers to questions, and hints for success.

Preface

The criminal justice system has learned to rely heavily on the analysis of physical evidence as scientific procedures and methods have become increasingly more reliable and telling than eyewitness testimony. The influence of television programs showing the use of highly sophisticated analytical equipment to solve crimes has caused juries to come to expect scientific evidence to be presented in all criminal cases. Greater stress is now placed on investigators to handle physical evidence in an appropriate scientific manner for later presentation in court. The introduction of DNA typing and database matching have revolutionized how physical evidence from the crime scene is processed. Forensic investigators must possess both a sound understanding of the scientific principles that underlie the measurements they make and a keen knowledge of how to locate physical evidence without disrupting any trace elements at the scene.

In many ways, the attacks of September 11, 2001, expanded the role of criminalistics from traditional examining of crime scenes and physical evidence to assisting the Department of Homeland Security in deterring terrorism. Threats of terrorism from both within and outside of the United States borders widen the scope of those working in the criminal justice system. I have included sections of this book that speak directly to these issues because of the changed nature and role of criminalistics.

New laws passed since 9/11 have placed a precarious balance between the rights and freedoms of individuals and the protection of society as a whole. This tension is evident when we are asked by politicians how much personal freedom we are willing to sacrifice in the name of national security. We now stand in long lines to pass through extensive security monitoring to board airplanes. We are limited in what we can carry with us on these flights. We face the potential of having our telephone conversations recorded. We can even be questioned about the material we check out of public and academic libraries. While these issues are of great importance to the individual, they are of even greater importance to those working in the criminal justice field.

There are no easy answers to these issues, but it is the goal of this textbook to present information to students to help them understand how forensic measurements are made and to find a balance that protects the individual and benefits society as a whole.

Organization

The organization and approach of this text place forensic science with the framework of the basic principles of chemistry, biology, and physics and assumes the reader has little or no scientific background.

The first two chapters introduce the student to the crime scene and physical evidence. In Chapter 1, we learn to secure and document the crime scene. Next, we learn to collect, preserve, package, inventory, and then submit evidence to the crime lab. In Chapter 2, common types of physical evidence are described, and basic scientific principles familiarize students with crime scene reconstruction. This early description of the many types of physical evidence found at crime scenes not only establishes the importance of a careful methodical approach to the crime scene but also gives students a firm foundation for how this evidence will be used to reconstruct the events that transpired during the commission of the crime.

Chapters 3, 4, and 5 offer a solid introduction to the core physical properties that are normally used to examine trace evidence. Chapter 3 shows how the physical properties can be used to characterize evidence. Chapter 4 describes the many types of microscopes used to examine fiber, hair, and paint evidence. The addition of paint evidence is new to the second edition and a response to requests from users of the first edition. Chapter 5 describes optical physical properties, such as color and refractive index, and how they can be used to characterize glass evidence. Wherever possible in these chapters, physical properties are discussed in the context of characterizing physical evidence, building a bridge to understanding how patterns and chemical and biological properties will be used to characterize evidence in the chapters that follow.

Next, students are introduced to pattern evidence. Chapters 6 covers fingerprints—their classification and the methods used to visualize latent fingerprints. A new box in Chapter 6 describes the new hand-held fingerprint scanners that search fingerprint databases while wirelessly connected to the Internet. The focus of Chapter 7 turns to questioned documents, with discussion of handwriting, typed and word processed documents, ink, indented writing, and security printing. Chapter 8 is devoted to firearms and describes handguns, rifles, shotguns, and submachine guns. Techniques used to compare fired bullets and shell casings are described, as well as the methods used to restore obliterated serial numbers.

We then focus on chemical evidence. Chapter 9, which introduces readers to the periodic table and inorganic chemistry, provides a useful introduction to the examination of bullets and gunshot residue. In addition, it provides a foundation for more advanced chemical principles that will be presented in later chapters. Chapter 10 describes the chemistry of fire and introduces the student to organic chemistry through a discussion of hydrocarbon accelerants that are used by arsonists. In Chapter 11, drugs of abuse are arranged by category, and the techniques used to detect them in bulk or in person samples are described.

Chapters 12, 13, and 14 deal with biological evidence. Chapter 12 describes how toxicological measurements are made. Even if the measurements are made after a person has died, they can often be used to reconstruct events that transpired days before. Biological fluids, such as blood, semen, and saliva are the focus of Chapter 13. Techniques used to locate and characterize biological evidence are presented, along with an introduction to DNA. Chapter 14 presents the separation and characterization of short tandem repeats (STRs) by capillary electrophoresis and how this information is used to establish paternity and match offender profiles.

The final section of the text focuses on terrorism. Chapter 15, a new chapter added for the second edition, describes computer forensics, cybercrime, and cyberterrorism. Chapter 16 describes the construction of explosive devices such as improvised explosive devices and the methods used to test for explosive residue. Chapter 17 presents the three major types of weapons of mass destruction—chemical, nuclear, and biological—and the techniques being developed to detect these threats, both point and standoff detectors.

Course Use

Criminalistics: Forensic Science, Crime, and Terrorism, Second Edition, offers the flexibility to tailor a course to suit both instructors' preferences and the needs of particular audiences. The full text may be used for a comprehensive two-semester course, or the book may be broken down in several ways for a one-semester course. The text is arranged in a traditional format, beginning with the crime scene and physical evidence, followed by sections on trace evidence, pattern evidence, and terrorism. Those who have been teaching a one-semester criminalistics course with a different text can use the first 12 chapters of this text in sequence. Other options for a one-semester course are to use the first five chapters, followed by choices for the remaining chapters depending on the teacher's preferences. Those instructors who stress chemical and biological evidence may choose to skip Chapters 6, 7, and 8. Those wanting to stress terrorism may choose to skip the first three chapters.

Acknowledgments

In preparing the second edition of this book I have added topics that were not covered in the first edition. A section on the composition and analysis of paint has been added to Chapter 4. Most importantly, a whole new chapter (15) on computer forensics and cybercrime has been added as a response to instructor requests. Since the publication of the first edition in 2007, there have been many advances in the field of forensic science, and the book has been updated throughout to incorporate these changes.

I would like to express my gratitude and appreciation to everyone who contributed to this book. I extend special gratitude to Erik Garcia, who worked with me for more than four years to create the wonderful graphic drawings in this text; Jonathan Edwardsen, who analyzed samples and produced the chromatograms and spectra contained in the book; Seth Reuter, who worked with me to develop the computer forensics chapter; and Connie Diamant, my wife, who crafted the case studies and put up with me during this project.

I would also like to thank the reviewers of *Criminalistics: Forensic Science, Crime and Terrorism, Second Edition,* for their helpful comments and suggestions.

Russell Carter
Northern Virginia Community College
Manassas, Virginia

Steven Christiansen
Green River Community College
Auburn, Washington

Mark Conrad
Troy University
Dothan, Alabama

Daniel David
Green River Community College
Auburn, Washington

Chris DeLay
University of Louisiana
Lafayette, Louisiana

David Ferster
Edinboro University of Pennsylvania
Edinboro, Pennsylvania

Kimberly Glover
Saint Leo University
Saint Leo, Florida

Don Haley
Tidewater Community College
Norfolk, Virginia

Donald Hanna
Cedarville University
Cedarville, Ohio

Robyn Hanningan
Arkansas State University
Jonesboro, Arkansas

John Kavanaugh
Scottsdale Community College
Scottsdale, Arizona

Michael Meyer
University of North Dakota
Grand Forks, North Dakota

Evaristus Obinyan
Virginia State University
Petersburg, Virginia

Gregory Russell
Arkansas State University
Jonesboro, Arizona

Jill Shelley
Northern Kentucky University
Highland Heights, Kentucky

David Tate
Purdue University
West Lafayette, Indiana

Dean Van Bibber
Fairmont State University
Fairmont, West Virginia

Harrison Watts
Cameron University
Lawton, Oklahoma

Robert Webb
Illinois Central College
East Peoria, Illinois

John Wyant
Illinois Central College
East Peoria, Illinois

Certain chapters were reviewed by experts. My thanks to the content experts who reviewed specific chapters:

Joseph M. Ludas and James W. Gocke—Chapter 6: Fingerprints
Sirchi® Finger Print Laboratories, Inc.
100 Hunter Place
Youngsville, North Carolina 27596

Len Pinaud—Chapter 8: Firearms
State Coordinator for Range3000

Detective Darrel Taber—Chapter 15: Computer Forensics and Cybercrime
Arlington County Police Department
Computer Forensics Unit
Arlington, Virginia

John E. Parmeter, PhD—Chapter 16: Explosives and Chapter 17: Detecting Weapons of Mass Destruction
Sanida National Laboratories
PO Box 5800
Albuquerque, New Mexico 87185-0782

Special thanks to the following people for their contributions to the text:

Maureen Dolan
Arkansas State University
Jonesboro, Arkansas

Carolyn Dowling
Arkansas State University
Jonesboro, Arkansas

Robyn Hannigan
Arkansas State University
Jonesboro, Arkansas

Amy Harrell
Nash Community College
Rocky Mount, North Carolina

Ellen Lemley
Arkansas State University
Jonesboro, Arkansas

Tanja McKay
Arkansas State University
Jonesboro, Arkansas

Kelly Redeker
Arkansas State University
Jonesboro, Arkansas

Introduction to Criminalistics

Investigating the Crime Scene

OBJECTIVES

In this chapter you should gain an understanding of:

- The steps taken to preserve a crime scene
- Documentation of the crime scene
- Ways to systematically search the crime scene
- Methods for collecting, preserving, identifying, packaging, and transporting evidence
- The chain of custody
- The Fourth Amendment and its application to the search and seizure of evidence

FEATURES

On the Crime Scene

Back at the Crime Lab

See You in Court

WRAP UP

Chapter Spotlight

Key Terms

Putting It All Together

In cases that involve capital crimes (such as murder), it is standard procedure to establish a pathway that will be used by all who enter the crime scene. This path begins outside the crime scene at the command center and leads to the focal point of the crime—the body. The location of the path is chosen to allow easy access while minimizing its impact on the crime scene.

1. As the location of the path is being decided, which constraints are placed on the investigators, and how do they minimize these constraints?
2. What are the first steps taken as the path is created?

Introduction

The collection and preservation of evidence are essential for any successful criminal investigation. Indeed, if all of the evidence and information surrounding the crime are not properly collected, preserved, and analyzed, the entire investigation may be jeopardized. In addition, the collection and preservation of evidence are accompanied by another essential element of crime scene investigation—namely, a record of what occurred at a particular time and location and which actions were taken by specific individuals.

Physical evidence is usually collected by the police or civilian crime scene technician and includes any and all relevant materials or objects associated with a crime scene, victim, suspect, or witness. Almost any object can be a piece of physical evidence under the right circumstances. Physical evidence can be collected at the scene of a crime from the body or area (such as a car, home, or workplace) of a victim, suspect, or witness. This chapter describes the methods and procedures followed by crime scene investigators in the processing of crime scenes.

Securing the Crime Scene

The first person to arrive at a crime scene is referred to as the first responder. The first responder's top priority is to offer assistance to any injured persons. By contrast, it is the responsibility of later responders or police officers to secure the crime scene. To safeguard evidence and minimize contamination, access to the scene must be limited, and any persons found at the scene must be identified, documented, and then removed from the scene. As additional officers arrive, they will begin procedures to isolate the area, using barricades and police tape to prevent unauthorized persons from entering the scene.

The scene of a violent crime may be difficult to control. Reporters, friends or family members of the victim, and curious passersby may all want to get as close as possible. Anyone who enters the crime scene has the potential to (unintentionally) contaminate the crime scene and destroy evidence. The first responder must have the authority to exclude individuals from entering the scene and must also identify the family and friends of victims and/or suspects and remove them from the immediate area. All nonessential persons (including law enforcement officers not working on the case, media, politicians, and other parties) must be kept out of the crime scene.

Identifying, Establishing, Protecting, and Securing the Boundaries

The initial boundary established around the crime scene should be larger than the scene. This boundary can easily be shifted inward later but is not as easily enlarged because the surrounding areas may have been contaminated during the ensuing interval. The responding officers must document all actions and observations at the scene as soon as possible. This step is essential for preserving information that might otherwise be forgotten and for providing information that may substantiate or contradict later investigative leads and theories. Documentation should be permanently preserved as a part of the case record.

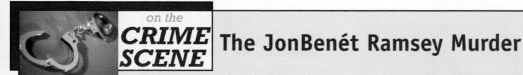

on the CRIME SCENE: The JonBenét Ramsey Murder

On Christmas Day in 1996, John and Patsy Ramsey attended a party a few blocks from their Boulder, Colorado, home along with their 6-year-old daughter, JonBenét. The Ramseys arrived home from the party around 9:30 P.M., carried JonBenét to her room, and put her to bed for the night. Shortly thereafter, the adults also went to bed.

Patsy awoke around 5 A.M. the next morning, went downstairs, and headed into the kitchen to make coffee. On the stairs she found a ransom note, addressed to John, that demanded $118,000 for the return of JonBenét. The letter further stated that the girl was kidnapped by a "small foreign" faction and warned John to comply with the ransom demands or the child would be beheaded. The note also instructed John to expect a call between 8 A.M. and 10 A.M. the next day. Boulder's 9-1-1 center recorded a call from Patsy at 5:25 A.M., and police arrived on the scene 7 minutes later.

Upon entering the Ramsey home, Boulder police focused on the ransom note and its instruction that a call would be forthcoming. The police waited for the call, along with John and Patsy, until the afternoon, but no call ever came. Despite the failure to receive any message from the purported kidnapper, the police did not mount a search of the house at this time.

By late afternoon, the police instructed John and a family friend to search the house for anything "unusual." They were not accompanied by a detective or police officer. Unfortunately, John found the body of his slain child in the basement of the house. JonBenét had been beaten and strangled with a garrote fashioned from a rope and the handle of a paintbrush later found in the Ramsey basement. At this point the case was transformed from a kidnapping to a murder. Investigators who were brought to the crime scene found two different boot prints, fibers, the ligature used for the garrote, and DNA evidence.

Almost 10 years later, in 2006, John Karr—then living in Thailand—confessed to authorities that he had murdered JonBenét. Karr was extradited to Colorado from Thailand, and DNA tests were performed to ascertain the truthfulness of his admission. The DNA results proved that Karr was not implicated in the crime, however.

Patsy died in 2006 from ovarian cancer. This case remains unsolved.

The first responder should document the following items:

- The state of the scene upon arrival (including the location, condition, and appearance of people and any relevant objects).
- Existing conditions upon arrival. (Were the lights on or off? Were doors and windows closed, partially open, or fully open? Was the room ransacked?)
- All personal information concerning witnesses, victims, and suspects.
- Actions and statements of witnesses, victims, suspects, and all other personnel who entered or exited the scene.
- Any items that may have been moved and who moved them. (Moved items should not be restored to their original condition; crime scenes should be documented exactly as found.)

The first responders also must ensure that physical evidence is not lost, contaminated, or moved.

All physical evidence should be preserved for later identification, collection, and submission.

Once the boundaries of the crime scene have been established, a single path into and out of the scene should be created. All personnel are required to use this pathway to exit the scene so as to help preserve physical evidence. If any evidence is moved—even inadvertently—its original location must be documented. This step is especially important if reconstruction of the crime is to be attempted later. All personnel at the scene should be identified and their names recorded in case investigators need to interview these individuals or obtain known samples (such as fingerprints, hairs, or shoeprints) from them at a later time. By identifying all persons who entered or exited the scene and documenting whether they touched or moved any objects, investigators establish and safeguard the **chain of custody**.

Once a lead investigator arrives at the crime scene, it is his or her responsibility to direct the

FIGURE 1-1 The lead investigator directs the investigation.

investigation (**FIGURE 1-1**). The lead investigator's first task is to evaluate the scene to establish whether the scene should be processed for physical evidence. The collection and analysis of physical evidence can be a very timely and costly process, so some jurisdictions collect physical evidence only in instances of more serious crime. If the decision is made to process the scene for collection of physical evidence, then the lead investigator will direct and control the processing, starting with the recognition of physical evidence.

Documenting the Scene and the Evidence

The state of the crime scene must be thoroughly documented to permanently record the condition of the crime scene and its physical evidence. Documentation is the most important and time-consuming activity at the scene. It involves four major tasks:

1. Note taking
2. Photography
3. Sketching
4. Videography

Documentation is critical for a variety of reasons. From a legal perspective, maintaining the chain of custody proves that nothing was altered prior to analysis of the evidence. If any items at the crime scene have been moved, their original locations and the circumstances that led to their relo-

cation must be documented. A photograph of each object should be taken before and after its move. Maintaining the chain of custody also requires documenting who discovered a specific item, when it was discovered, what the item's appearance was, and who took control of the item at each of the stages from the item's recovery to its presentation in court. From a scientific standpoint, documentation later helps the analyst understand how the evidence relates to the overall scene and may suggest which types of evidence analysis to perform.

Even before documentation begins, the lead investigator may have developed a hypothesis about what occurred at the crime scene. The investigator then records facts that may corroborate, refute, or modify this hypothesis. The investigator also must anticipate various questions that may arise and be willing to look for evidence to help answer those questions. As the investigation proceeds, the working hypothesis may be modified, and the documentation and preservation of evidence may help lead to the development of a new scenario.

Note Taking

Notes document the core of the crime scene and physical evidence. These notes must be made in ink in a bound notebook, the pages of which have been previously numbered sequentially. Errors should never be corrected by erasure or by removal of pages but rather by simply crossing out errors with a pen and adding the corrected information. Likewise, no blank spaces should be left for observations or conclusions to be inserted later. All notations must be recorded in a strictly chronological order. Note taking starts with the first responders, who must log the time and place of arrival, appearance of the scene, names and addresses of persons present, and other pertinent information. Even medical personnel must keep logs of their activities at the scene.

All evidence must be documented in the notes. Notes are the principal way of refreshing one's memory months or years later, so they must contain sufficient details for this purpose. The condition of the evidence, the time of its discovery, the name of the discoverer, and the placement, collection, packaging, and labeling of the evidence must all be described, as must the eventual disposition of the evidence. This documentation may be done in a separate evidence log. Additionally, every photograph taken must be documented in the notes, including the film roll (for analog cameras) or file name (for digital cameras), frame numbers, date, time, subject matter, location, camera settings, specialized lighting or film, and other relevant data. Sometimes such information is included in a separate photography log, as discussed later in this chapter.

An audio recording may be used to supplement the written notes. Alternatively, one may simply narrate while videotaping the scene. The advantage of this technique is that a lot of detail can be included without the cumbersome task of having to write it all down right away. However, soon after leaving the crime scene, the audio notes should be transcribed and combined with the original written notes so as to produce a new, complete set of written notes. It is essential that both the original written notes and any audiotape be preserved as evidence.

Written notes tie together all other forms of documentation. During subsequent hours, days, and weeks, they may reflect changes in the working hypotheses, a narrowing of the focus to a single hypothesis, additional forensic observations, and eventually a theory of the crime. During legal proceedings, the investigator's notes are made available to the defendant's attorney before trial. Defense attorneys may point out erasures and misspellings in the notes in an attempt to discredit the investigator—which is why such care must be taken when making the original notes.

Photography

Photographs of the crime scene must be taken without the photographer or anyone else disturbing elements of the scene (**FIGURE 1-2**). Ideally, all items will be in their original, undisturbed state. Any changes made to the items or their environment from the time of their original discovery must be documented in the notes.

A systematic series of photographs is the best way to record the crime scene and any pertinent physical evidence. This photo record should include three types of shots: overall, midrange, and close-up photos (**TABLE 1-1**). The crime scene investigator should take as many shots as practical, using proper guidelines.

Each crime scene should be photographed as thoroughly as possible. Wide-angle photos show where the crime occurred as well as the surrounding area, entrance, and exit. Intermediate distance photos show the location of evidence and its relationship to the entire scene. Close-ups record the

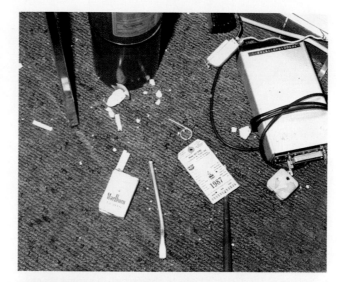

FIGURE 1-2 The items of evidence found at the scene of this safe burglary included a pack of cigarettes, a broken screwdriver, and the safety pin and tag from the fire extinguisher.

appearance of evidence at the crime scene as found by investigators.

If the crime occurred indoors, all walls, ceilings, and floors of the room and of any adjacent rooms should be photographed. If a body is present, it should be photographed from intermediate range to show its relationship to the crime scene (**FIGURE 1-3**). After the medical examiner's (ME's) staff has removed the body, the area under it should be photographed. Additional photographs should be taken of the crime scene from different angles and perspectives.

TABLE 1-1	
Guidelines for Photographing a Crime Scene	
Type of Shot	**Techniques**
Overall	*Interior:* Using the corners of the room as guides, take overlapping shots. Record all doors and windows. Attach camera to tripod so that a slower shutter speed can be used in low light to increase the depth of field and decrease shadows.
	Exterior: Record all structures, roads, paths, addresses, and street signs.
Midrange	Record a sequential series of shots that highlight the physical evidence within the crime scene. Use flash to brighten details.
Close-up	Record using side lighting and documentation placards at first. If shadows are present, activate the flash. Take shots with and without a scale or ruler.

FIGURE 1-3 Crime scene photographs should show the relationship of the victim's body to the rest of the crime scene. The numbered signs in this photo indicate the location of relevant evidence.

A series of close-up photos should be taken to record details about each piece of evidence and any victims, suspects, and witnesses. Close-ups should contain an evidence identifier (code number) and at least one scale (**FIGURE 1-4**), but preferably two scales held at right angles to each other to establish the proper dimensions of the object. Afterward, the same evidence can be photographed again without the scale and identifier. It may be necessary to take several shots of the same object under different lighting conditions.

A 35-mm single-lens reflex (SLR) is the camera of choice. Although investigators have used film SLR cameras for years, digital versions are quickly replacing analog models. The digital image produced is much easier to manipulate. Digital cameras also possess the additional advantage that several still shots can be stitched together electronically to form a three-dimensional panoramic view of the scene. When using such cameras, crime scene photographers must be careful to record the original digital image because digital photos that have been enhanced or altered by computer programs (such as Adobe® Photoshop®) may be deemed inadmissible in court.

As noted earlier, each photograph taken at a crime scene must be recorded in a photo log. This log should record the following data:
- Date and time
- Camera settings (F-stop and shutter speed)
- Film roll number and exposure number (for a film camera)

FIGURE 1-4 Rulers should be used with close-up photos to record the size of the evidence.

- File name and exposure number (for a digital camera)
- Type of shot (overall, midrange, or close-up)
- Distance to the subject
- Brief description

Sketching

While photographs are very useful in recording evidence and scene detail, they may produce a false sense of perspective and do not always present an accurate representation of the dimensions and spatial relationships. Therefore, after taking the necessary photographs, an investigator should make a sketch of the crime scene (**FIGURE 1-5**).

The purpose of a sketch is to accurately record distances between objects at the scene. Sketches allow the investigator to emphasize the most relevant objects and features and to eliminate unnecessary details. The sketches produced at the scene are **rough sketches**. That is, the investigator might not be able to accurately draw the scene to scale, but the sketch must contain all the information necessary to allow a professional to prepare a finished sketch later on. In particular, the sketch must include the following information:

- Case identifier
- Date, time, and location
- Weather and lighting conditions
- Name of the sketch
- Identity and assignments of personnel
- Dimensions and layout
- Measurements and positioning (as indicated by two immovable objects)
- Key or legend (identifying objects that are given designations in the actual sketch)
- Orientation
- Scale

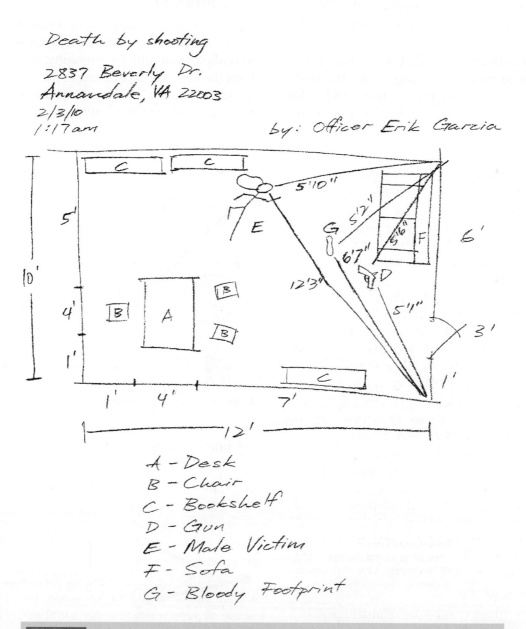

A – Desk
B – Chair
C – Bookshelf
D – Gun
E – Male Victim
F – Sofa
G – Bloody Footprint

FIGURE 1-5 A sketch records important spatial details of the crime scene.

The sketch is usually drawn from an overhead view. Three techniques can be used to record measurements for crime scene sketches: **triangulation method**, **baseline method**, and **polar coordinate method** (FIGURE 1-6). All three techniques require the technician to establish two fixed points. All subsequent measurements of the crime scene are then made relative to these points. Both fixed points must be permanent objects (such as the corner of a room, a tree, or a utility pole) that will not be moved as the crime scene is searched.

- Triangulation method: Measures the location of the evidence (x, y) from fixed points (A, B).
- Baseline method: Draws a line between the fixed points (A, B) and measures the distance to the evidence (x, y) at a right angle from this line.
- Polar coordinate method: Uses a transit or a compass to measure the angle from the north and the distance to the evidence (X). This method is most commonly used in a large-area crime scene (outside or in a warehouse) when

a wall or side of a building is used to establish the fixed points (A, B).

Later, as the case is being prepared for court, a computer professional will use a computer-aided design (CAD) program to prepare **finished sketches** based on the information contained in the investigator's rough sketches (FIGURE 1-7). These finished sketches will be suitable for courtroom presentation. In some instances, the CAD program may produce sketches that zero in on certain rooms or areas, showing bodies, firearms, blood spatters, and other items of interest.

Videography

Just as audio recording can be a valuable addition to written notes, so video recording may be used to complement still photography. A video is perhaps the best way to document the overall view of the scene because it shows the relationships of various pieces of evidence both to one another and to the scene.

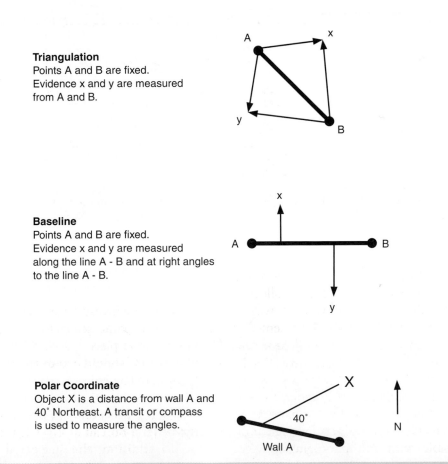

Triangulation
Points A and B are fixed.
Evidence x and y are measured from A and B.

Baseline
Points A and B are fixed.
Evidence x and y are measured along the line A - B and at right angles to the line A - B.

Polar Coordinate
Object X is a distance from wall A and 40˚ Northeast. A transit or compass is used to measure the angles.

FIGURE 1-6 The three techniques used to record measurements at crime scenes are triangulation, baseline, and polar coordinate.

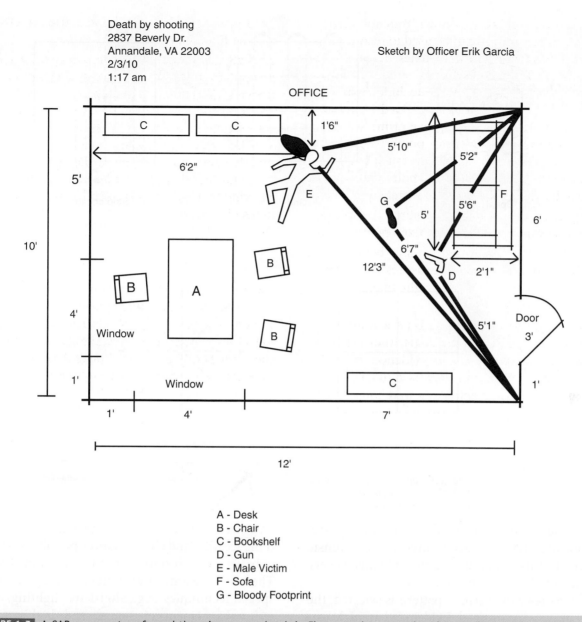

Death by shooting
2837 Beverly Dr.
Annandale, VA 22003
2/3/10
1:17 am

Sketch by Officer Erik Garcia

OFFICE

A - Desk
B - Chair
C - Bookshelf
D - Gun
E - Male Victim
F - Sofa
G - Bloody Footprint

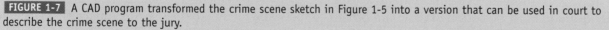

FIGURE 1-7 A CAD program transformed the crime scene sketch in Figure 1-5 into a version that can be used in court to describe the crime scene to the jury.

Using a video camera is also an excellent way to document the investigation and evidence collection process and provides a very convenient way of recording the locations of evidence documented in still photos. As an added benefit, the audio track can simultaneously record a running narrative. Finally, forensic scientists (who frequently are not present at the scene) often find that a video allows them to understand the scene very well at a later time. However, because video resolution is lower than the resolution possible with still photography, video cameras should not replace SLR cameras as the sole means of documenting important information.

Systematic Search for Evidence

Crime scene investigators must methodically search the entire crime scene to ensure the collection of less obvious pieces of physical evidence. The lead investigator should coordinate this search. To search the scene, law enforcement personnel should select and follow a particular search pattern (**FIGURE 1-8**). **TABLE 1-2** lists common patterns recommended by the Federal Bureau of Investigation (FBI).

In addition, the suspected points of ingress and egress must be searched along with any **secondary crime scenes**, which include all sites

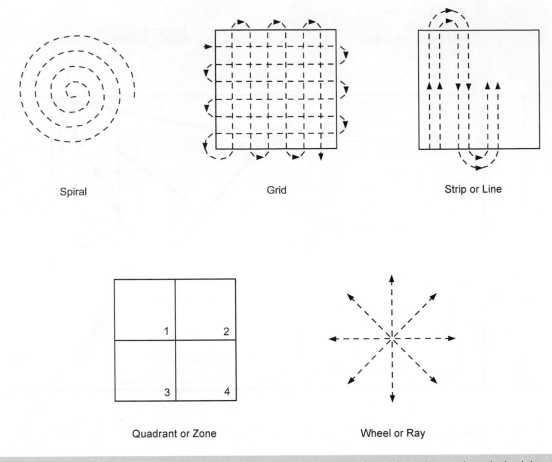

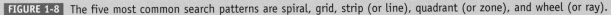

FIGURE 1-8 The five most common search patterns are spiral, grid, strip (or line), quadrant (or zone), and wheel (or ray).

where a subsequent criminal activity took place. It is not usually necessary to involve the forensic scientist in this search unless the evidence is very complex or the crime is a major one.

No matter which search pattern is selected, the search must be prioritized. Evidence that might potentially degrade must be collected first, and other evidence that is found nearby must not be disturbed. In particular, weather or traffic may impose

an order of search that is considered less than ideal. If the weather and circumstances permit, however, it is best to search outside crime scenes in daylight. The lead investigator also must be prepared to use special techniques (e.g., alternative lighting, chemical enhancement of stains) and highly trained personnel (e.g., blood spatter analysts, projectile trajectory analysts), both factors that may further restrict the order in which evidence is documented and collected.

Recognition of Physical Evidence

The first step in processing the scene is the recognition of the gross (obvious) pieces of physical evidence (**FIGURE 1-9**). This process may begin as soon as the first responder appears on the scene and continues throughout the crime scene investigation. By the time the scene is ready to be processed, many of the more obvious pieces of physical evidence will undoubtedly have been discovered already.

TABLE 1-2	
Common Search Patterns	
Name	**When This Technique Is Used**
Spiral	Crime scenes with no physical boundaries (e.g., open field, water)
Grid	Large outdoor crime scenes
Strip (or line)	Outdoor crime scenes where a coordinator arranges many searchers
Zone (or quadrant)	Indoor crime scenes where different teams are assigned small areas to search and one searcher is assigned to several areas of equal size
Wheel (or ray)	Circular crime scenes

TABLE 1-3	
Evidence to Search for in Relation to the Crime	
Crime	**Type of Evidence**
Homicide	Weapons, bullets, cartridges, and cross-transfer evidence (fingerprints, shoeprints, tire tracks, blood, hairs, fibers, or soil)
Arson	Accelerants and ignition devices
Burglary	Tool marks, broken glass, blood, fingerprints, and shoeprints
Pedestrian hit-and-run	Cross-transfer on both the vehicle (blood and tissue, fabric, hair, and fibers) and the pedestrian (paint, plastic, glass, and tire impressions)

The ability to recognize what is and what is not evidence is a skill that is best learned through experience. If every object present at the scene were documented, collected, and sent for analysis, the crime laboratory would quickly be overwhelmed. Conversely, if important evidence is not collected, a case might not be prosecuted successfully. Because recognition of physical evidence is so important, many police departments employ specialized evidence retrieval technicians for this purpose. More commonly, however, police on the scene will be responsible for evidence recognition, documentation, and collection. The investigator must ask, "What am I searching for?" The answer to this question depends on the nature of the particular crime (TABLE 1-3).

Collection, Preservation, Inventory, and Transportation of Physical Evidence

Once the discoverer, time of discovery, location, and appearance of evidence have been thoroughly documented, the evidence must be collected, preserved, inventoried, and packaged in preparation for submission to the crime lab. Exactly how this process occurs depends on the type of evidence and local protocols.

Types of Evidence

A physical evidence sample that is collected at the crime scene is called a **questioned sample** (or an unknown sample) because it is an object with an unknown origin. One way to establish a connection between a questioned sample and a particular person or place is to secure a known sample from the relevant person or place for comparison. **Known samples** also are called control samples, reference samples, or standard samples. Sometimes, the known sample must be created by the forensic scientist. For example, a questioned handwritten note may be compared to a known sample by having the suspect submit samples of his or her handwriting. If a crime involves a gun, the forensic scientist will fire the suspect's gun in a way that allows for recovery of the bullet. This bullet then becomes the known sample. The forensic scientist then can compare the scratches (rifling markings) on the known bullet to the unknown bullet and determine if they came from the same gun.

When forensic scientists begin to test evidence, they first seek to place the unknown sample into a class of objects, which limits the number of known samples to be used for comparison. The next step is **individualization**, the process of proving that a particular unknown sample is unique, even among members of the same class, or proving that a known sample and a questioned sample share a unique common origin. Complete individualization is currently possible for only a few types of evidence, including fingerprints, DNA, and physical matches (or "jigsaw fits").

In the early summer of 1994, Nicole Brown Simpson (ex-wife of former football star O. J. Simpson) and her friend Ronald Goldman were stabbed to death, their bodies found in the front courtyard of Nicole's home. Because of a history of violence between O. J. and his ex-wife, O. J. was an obvious suspect.

At the crime scene, a great deal of blood was found on the ground. To the investigators, the multiple serious cuts made to the victims with a knife suggested that this was a personal and emotional crime. Detectives left footprints in the blood, did not follow proper protocol in collecting blood samples, and then drove around town on a hot day with the samples in their vehicles. Investigators entered O. J.'s home without a warrant and moved evidence before photographing and documenting it properly.

While none of these events was significant enough to compromise the evidence, collectively they suggested to jurors a lack of attention to detail, laziness, and, perhaps, intentional tampering on the part of investigators. Although other mistakes were made during the trial, the errors associated with the collection and preservation of evidence were enough to convince a jury that there was reasonable doubt about O. J.'s involvement in the crime.

Even though the DNA of O. J. was found at the crime scene and records showed that he had purchased a knife consistent with the wounds inflicted on Nicole, the errors made during evidence collection were too great for jurors to overcome. This case serves as an important reminder to investigators that how a crime scene is approached is critical to finding the truth, and how the scene is handled affects the way evidence will be presented to a jury.

Impression Evidence: Tire Tracks, Footprints, Tool Marks, Latent Fingerprints, and Palm Prints

Impression evidence often is developed or enhanced by use of specialized photographic techniques (alternative light sources and filters) and/or by application of chemical developers (e.g., fingerprint powder, ninhydrin, silver nitrate). Collection may be accomplished by photography, by physical lifting (with tape for fingerprints, gel for bite marks, or molding materials for tire impressions), or by actually seizing the entire object that holds the evidence.

Biological Evidence: Blood, Saliva, Semen, and Vaginal Fluids

Biological evidence often is located visually, but also by touch and smell. Often, it needs to be enhanced and/or developed via chemical means before analysis is possible. Swabs and gauze can be used to sample stains. Dried stains can be scraped with a scalpel. In addition, the entire object holding the stain (e.g., clothing, chair, door) may be taken or a portion of the object that is stained cut out of the object using a scalpel.

Firearms and Ammunition Evidence

Firearms and ammunition evidence is usually located by sight. Investigators should consider every firearm to be loaded, so their first priority should be to render the firearm safe (so that it will not discharge accidentally). If the recovering officer is not sure that the firearm has been rendered safe and is unable to do so, the firearms section of the crime laboratory should be contacted so that arrangements can be made to disarm the weapon. Firearms should be placed in paper envelopes, paper bags, or cardboard boxes. Plastic bags collect moisture, so they should not be used because they can cause the firearm to rust. Firearms should not be marked in any permanent way. Instead, a tag should be attached to the trigger guard with the appropriate identifying information.

Arson and Bomb Evidence

Arson and bomb evidence often is difficult to find. Such scenes frequently are filled with a large amount of debris, and potentially useful evidence may be washed away as the fire is extinguished. Arson evidence usually is located by sight and smell. Collect carpet, wood, and other absorbent materials located close to the suspected origin of the fire or blast, place it in a clean paint can, and seal the can tightly. Material from different areas should be packaged separately. Place any suspected flammable liquids in a small glass bottle with a tight-fitting lid. The smooth surfaces of ignition devices, fuses, and exploded bomb components should be packed in separate bags and preserved for fingerprint development in the crime lab. Evidence collectors should not handle unexploded devices but rather should call the bomb squad in such instances.

Chemicals and Controlled Substances

Chemicals and controlled substances are located by visual observation (e.g., drugs, drug paraphernalia), by use of field (spot) tests, or by odor (e.g., humans, dogs). The entire sample found should be packaged, labeled, and sent to the crime lab.

Trace Evidence

Trace evidence includes items that are extremely small—even microscopic. These items are collected in a variety of ways, including with a forceps, tweezers, or gloved hand by scraping (e.g., the undersides of fingernails for blood and tissue evidence), taping (for lifting fingerprints), or vacuuming (for collection of hair and fibers). If necessary, the entire item containing the evidence, such as clothing or cars, can be collected and analyzed later.

When collecting trace evidence, the forensic technician must document and collect not only questioned samples but also known samples. For instance, if a victim is found to have soil on his clothing, it should be sampled. At the same time, it is essential to take several known soil samples from various areas of the scene. Also, hair and fiber samples from victims and suspects must be taken for later comparison with questioned hairs and fibers that may have been retrieved via vacuuming the scene.

To ensure the best outcome, the ME and investigators must work cooperatively to collect evidence. If a crime victim undergoes autopsy, for example, the ME will automatically collect organ and tissue samples for pathological and forensic analysis to establish the cause of death. In addition, blood samples will be taken for toxicological analysis.

The ME will routinely collect samples of the following items from the victim's body:

- Clothing
- Bullets (in case of a shooting victim)
- Hand swabs (to look for gunshot residue in case of a shooting victim)
- Fingernail scrapings
- Head and pubic hairs
- Blood
- Vaginal, anal, and oral swabs (in case of a sex crime)

Packaging Evidence

The type of packaging used depends on the evidence itself. The package must protect and preserve the evidence. Paper envelopes are routinely used for a variety of objects, such as paint chips, glass fragments, soil samples, hairs, and fibers (**FIGURE 1-10**). For liquid samples, it is crucial that the sample not evaporate. For small liquid samples, screw-cap glass bottles are satisfactory (**FIGURE 1-11**); larger samples can be put into a new paint can and then sealed (**FIGURE 1-12**).

FIGURE 1-10 Brown paper bags used for evidence collection may come preprinted with the necessary chain of custody form.

FIGURE 1-11 Wet or liquid samples may be placed in airtight jars. After the top is screwed on, the jar should be sealed with tape to prevent contamination.

FIGURE 1-12 Once the lid is in place, an airtight seal prevents evaporation of liquid evidence. The rubber septum in the lid later may be punctured with a syringe to take a sample for analytical purposes.

Blood is of special concern. While blood is commonly found at crime scenes, the DNA present in blood will degrade if the sample is not properly preserved. Moist blood that is placed in an airtight container readily supports the growth of mold and mildew; unfortunately, a sample that becomes contaminated in this way has no evidentiary value. For this reason, wet blood found at a crime scene should be allowed to air dry in place; once dried, it can be sampled by scraping and placing it in a paper envelope. Alternatively, a wet blood sample may be collected on a swab. The swab should be allowed to dry and then placed in a paper envelope.

Clothing can be put into large paper sacks. No folding of clothing is usually permitted, because this may potentially cause cross-transfer of trace evidence from one area of the garment to another.

Generally, it is best to take the entire piece of evidence as it is found at the scene. This packaging method ensures that other trace evidence in the sample will also be available for the forensic examiner. For example, doors, pieces of drywall, and flooring may be removed from the scene and sent to the lab. In cases of hit-and-run homicide, an entire car may be submitted. Only if the questioned sample is adhering to a huge structure should it be removed and submitted separately from the structure on which it was found.

Submitting Evidence to the Crime Laboratory

When evidence is collected at the scene, it is removed and stored temporarily in a separate evidence collection area, which is constantly guarded. Once all evidence is collected, it is ready to be transported and submitted to the crime lab or to an evidence storage area maintained by the police.

Evidence may be submitted to the crime lab either via mail or by personal delivery. Mail and parcel carriers have special regulations that may apply to shipment of evidence from a crime scene. Unloaded rifles and shotguns can be shipped by any person via the United States Postal Service (USPS), although the shipment must be to an addressee who holds a Federal Firearms License (FFL). Unlicensed individuals cannot ship unloaded handguns via the USPS, but those with an FFL can. Postal regulations prohibit the shipping of live ammunition, but both FedEx and United Parcel Service (UPS) permit such shipments. U.S. postal regulations also prohibit the shipping of certain chemicals, radiological agents, and explosives, but such materials and items often can be transported via either UPS or FedEx. Etiological agents (viable microorganisms and their associated toxins that can cause disease in humans) can be shipped in specially marked and constructed containers via registered mail.

BACK AT THE CRIME LAB

After assisting injured individuals at a crime scene, the responding officer needs to ensure the following steps are executed with the utmost proficiency to avoid compromising the investigation at this initial stage:

- Limit access to the crime scene.
- Identify, document, and remove any person(s) at the crime scene.

- Physically define the perimeter of the crime scene well beyond the actual observed crime site.
- Establish a single path in and out of the crime scene.
- Record all actions and observations as soon as possible.

Each item sent to the laboratory should be packaged separately. In addition, every shipment must include an evidence submission form, which provides essential details about the sample: the submitter's name, a case number, a list of items sent, a list of analyses requested for each piece of evidence, and a brief case history. The forensic scientist is not limited by the investigator's request for particular tests, however. In some cases, the results from the requested list of tests may prompt the forensic scientist to perform additional analyses. In other cases, the forensic scientist may decide that the requested examinations cannot be performed if the sample of evidence submitted is too small.

Chain of Custody

Once the case has moved to court, all evidence will be subject to questions about maintenance of the chain of custody, which is a written chronological record of each person who had an item of evidence in his or her possession. The prosecution must account for the evidence along every step of the way—from its discovery, to its collection, to its analysis, to its storage, to its transfer. Throughout the entire process, including court proceedings and appeals, the prosecution must maintain secure custody of the evidence.

The chain of custody starts with the original discoverer. Evidence should not be moved prior to its documentation and retrieval. If it is moved prior to its documentation and if all of the facts surrounding that movement are not documented completely in the notes, then the chain of custody has been broken, and a court may subsequently rule that the evidence is tainted and inadmissible. The chain of custody also is part of the reason why, when possible, evidence is sent in large pieces with identifying marks directly on it (such as the collector's name or initials, date, and description) rather than by cutting out or removing smaller portions that cannot be labeled individually. The case identifier and other pertinent information listed on packaging materials holding the evidence provide yet another means of trying to document the chain of custody. The chain of custody is preserved by making certain the investigator's notes completely document everything that happens to each piece of evidence at the scene.

A chain of custody form should be attached to each evidence container (**FIGURE 1-13**). Failure at

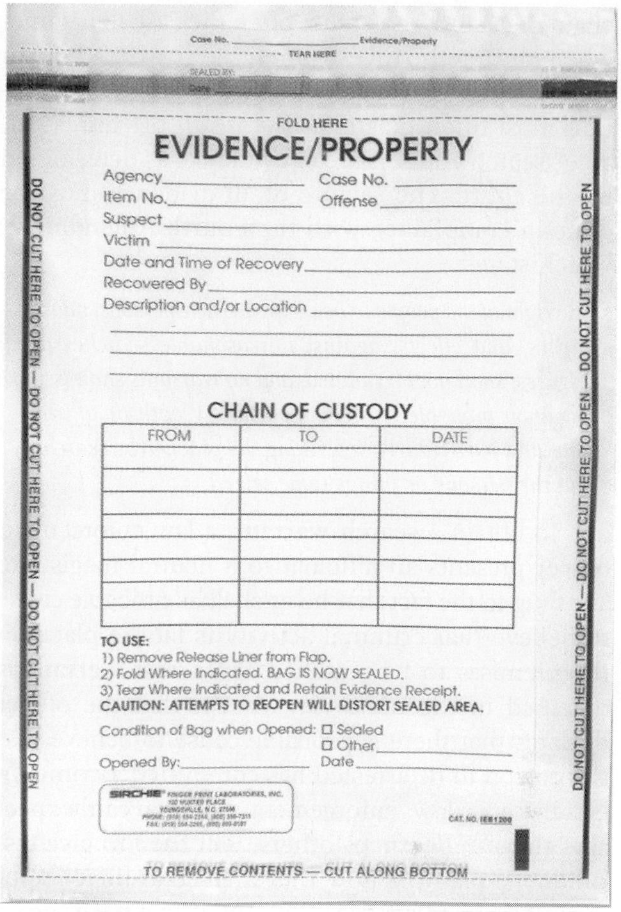

FIGURE 1-13 Each individual who handles a piece of evidence must complete the chain of custody form.

any stage to properly document who has possession of the evidence and what that person did with it (and when) can lead to contamination of the evidence and its subsequent exclusion at trial. Because every person who handles the evidence must be able to show an unbroken chain of custody, it is best practice to keep the number of individuals who come in contact with the evidence to an absolute minimum.

Criminal Evidence and the Fourth Amendment

One of the most frustrating experiences for police is the exclusion of incriminating evidence at trial. After expending all the effort involved with searching, identifying, packaging, shipping, and analyzing physical evidence, having that evidence later excluded from the court proceedings can be demoralizing. Although sometimes evidence may be excluded because of improper or undocumented chain of custody, it is more often the case that somewhere during the investigation the suspect's **Fourth Amendment** privileges were violated. This situation most often occurs as the result of "unreasonable" search and seizure of evidence, as determined by the court. The seizure of all evidence must be done in compliance with the Fourth Amendment, which states:

The right of the people to be secure in their persons, houses, papers, and effects, against unreasonable searches and seizures, shall not be violated, and no warrants shall issue, but upon probable cause, supported by oath or affirmation, and particularly describing the place to be searched, and the persons or things to be seized.

To obtain a **search warrant**, a law enforcement officer presents an affidavit to a neutral magistrate attesting to the fact that he or she has probable cause to believe that criminal activity is taking place on the premises to be searched. An **arrest warrant** is obtained in the same way, except that the officer declares that there is probable cause to believe that the person to be arrested has committed a crime. In either case, law enforcement officers can expect that they, and perhaps others, will have to give testimony explaining how they arrived at the conclusion of probable cause.

Probable cause is not defined in the Fourth Amendment, but rather is purely a judicial concept based on case law. Probable cause is a fairly minimal standard and is to be determined according to the "factual and practical considerations of everyday life on which reasonable and prudent men, not legal technicians, act" (*Brinegar v. United States*, 1949). The courts have ruled that:

In determining what is probable cause ... we are concerned only with the question whether the affiant had reasonable grounds at the time of the affidavit ... for the belief that the law was being violated on the premises to be searched; and if the apparent facts set out in the affidavit are such that a reasonably discreet and prudent man would be led to believe that there was a commission of the offense charged, there is probable cause ... (Dumbra v. United States, *1925*).

In particular, the magistrate will give significant weight to the following sources of information:

- An informant who has previously been proven reliable
- An informant who implicates himself or herself
- An informant whose testimony has been at least partly corroborated by police

- A victim of the crime being investigated
- A witness to the crime being investigated
- Another police officer

The list of what may be seized with a search warrant is extensive. Contraband and the "fruits and instrumentalities of crime" are subject to seizure. The 1967 case of *Warden v. Hayden* also established that such evidentiary items as fingerprints, blood, urine samples, fingernail and skin scrapings, voice and handwriting exemplars, conversations, and other documentary evidence may be obtained via warrants. There are limits to what the magistrate will allow, however. For example, law enforcement personnel are not permitted to administer an agent to induce vomiting or to require a suspect to undergo surgery under anesthesia to retrieve a bullet from the chest. By contrast, an individual may be detained on the reasonable grounds that his or her natural bodily functions (i.e., defecation, urination) will produce evidence that is held internally.

Exceptions to the Fourth Amendment

Even though the Fourth Amendment requires officers to have a warrant before conducting a search, the vast majority of searches and seizures occur without a warrant. Over time, the U.S. Supreme Court has issued a number of exceptions to the rights that are afforded by the Fourth Amendment, including border searches, consent searches, search incident to an arrest, plain view doctrine, emergency exceptions, open fields, stop and frisk procedures, and vehicle inventories.

Border Searches
A customs search at a border was authorized by the First Congress. Such searches and seizures require no warrant, no probable cause, and not even any degree of suspicion.

Consent Searches
An individual can waive his or her Fourth Amendment privileges and consent voluntarily and knowingly to a search of either his or her person or premises by an officer without a warrant.

Search Incident to an Arrest
Arrests can occur in one of two ways. First, an arrest warrant may be obtained and then executed

with an ensuing arrest. Second, a law enforcement officer is allowed to arrest a person in a public place and without a warrant. The person must have, on probable cause, committed a felony or have, in the presence of the officer, committed a misdemeanor. Searches of the arrestee and the area under his or her immediate control are permitted to protect the officer from harm, prevent the destruction of evidence, and seize items that the arrestee might otherwise use to escape.

Plain View Doctrine

If law enforcement officers are legally in a position where they can plainly see contraband or other evidence, they can seize it without a warrant. The reasoning is that a person's expectation of privacy is small for those things that he or she willingly exposes to others, including things exposed in the individual's home.

Emergency Exceptions

When police arrive at the scene of a crime, if they have reasonable suspicion that an injured person is inside a dwelling, they may enter the premises without a warrant to render emergency medical attention. They also may do a sweep of the premises looking for additional victims and perpetrators. During this sweep, they can look only in exposed areas and likely hiding spots. Police also may enter locations without permission to assist at fire scenes and other emergencies.

Open Fields

This exception is similar to the plain view doctrine. It holds that an individual has no expectation of privacy in open fields, which, by their very nature, are open and exposed to public view.

Stop and Frisk Procedures

Law enforcement officers may pat down a suspicious person whom they fear may be armed and dangerous.

Vehicle Inventories

Vehicles present a special law enforcement problem. Because a vehicle may quickly leave the jurisdiction of the investigating officers, whenever a driver or an occupant of a vehicle is arrested, the need to search the vehicle may be immediate.

The Supreme Court and the Fourth Amendment

The Supreme Court has ruled in two cases involving seizure of evidence at a crime scene without a warrant.

Mincey v. Arizona (1978)

On October 28, 1974, one undercover police officer and several plainclothes officers raided an apartment occupied by Rufus Mincey, who was suspected of dealing drugs. A shootout ensued, during which Mincey shot and killed the undercover officer. Three occupants of the apartment were wounded, including Mincey. The only actions taken by the narcotics agents at the scene were a search for additional victims and a call for medical care for the wounded individuals. Within 10 minutes of the shooting, two homicide detectives arrived and took charge of the scene. They conducted a 4-day search of the apartment, without a warrant, including extensive sketching and photographing; opening drawers, cabinets, and cupboards; examining all objects contained therein; ripping up carpets; digging bullet fragments out of floors and walls; emptying clothing pockets; and completing an inventory of the contents of the premises. In total, they confiscated 200 to 300 objects.

Mincey, who was interrogated in the hospital by detectives while he was barely conscious, was indicted for, and later convicted of, murder, assault, and various narcotic offenses. Both during the original trial and during the subsequent appeal, Mincey's attorneys complained that much of the evidence against him was obtained during the warrantless 4-day search and that his statements made in the hospital were not voluntary. The Arizona Supreme Court reversed the homicide and assault convictions on state grounds but let the narcotics convictions stand, holding that the evidence obtained in the search of a homicide scene was proper and that the hospital statements were voluntary. In essence, the Arizona Supreme Court reasoned that a warrantless search of a homicide scene is reasonable, given the seriousness of the crime. Upon appeal to the U.S. Supreme Court in 1978, the court ruled in Mincey's favor, saying:

The search cannot be justified on the ground that no constitutionally protected right of privacy was invaded, it being one thing to say that one who is legally taken into police custody has a lessened right of privacy in his person, and quite another to argue that he also has a lessened

In the early 1980s, in a juvenile court in Canton, Ohio, an attorney appointed to represent two juveniles facing serious felony charges made a startling discovery. The two suspects had been charged with the brutal armed robbery of an elderly couple in their own home. Using a hammer as a weapon, the perpetrators beat the couple and took two pillowcases full of stolen property. Within minutes after police received the call, officers had located the two juveniles, each of whom was carrying a bag. This is where the problem began.

Officers from three jurisdictions—city, county, and township—caught the suspects, and each wanted a part of the case. Hence, the officers divided the stolen goods into three piles, never noting which item came from which pillowcase. The officers returned to the scene of the crime for a "show up," where witnesses can identify the suspects. The officers pooled all of the goods on the hood of a police car and asked the victims to identify the stolen goods, which they did. The officers then redivided the stolen property for "bagging and tagging."

The two juveniles were facing a hearing motion, known as a waiver, to be tried as adults. The state admitted each piece of evidence on the record, employing one officer who obligingly identified each item as well as the person who bagged the evidence and initialed the bag. On cross-examination, a second officer was called to the stand and testified that the first officer was the one who had bagged and tagged the evidence. The trial court judge halted the hearing and ordered counsel into chambers.

Because the chain of custody had been destroyed, the trial could not continue. The suspects pled guilty to juvenile charges and went to juvenile detention for the maximum sentence of 3 years. As these officers discovered, the chain of custody is critical every step of the way, from the crime scene to the courtroom.

The primary error on the part of the officers was the division of evidence. Here is where the chain of custody was destroyed. Evidence only is valuable in context with its location at the scene and other items of evidence. When the officers divided the evidence in what appeared to be a random method, the evidence lost its value. The second mistake was the repooling of evidence at the scene for the witnesses to look over. Evidence was mixed together, with no note of each item's location and context. Third, the evidence was redivided for "bagging and tagging," again with no context. All items of evidence must be collected singularly from the scene. On the chain of custody form, the investigator must note the location of each item relative to other items at the scene (both permanent landmarks and other items of evidence) with measurements and accompany those notes with a photo of the item including an evidence marker.

While the chain of custody command may vary from jurisdiction to jurisdiction, there is no argument about one point: The evidence collection procedures must ensure that the item of evidence is collected at the scene in context with the scene itself.

---

right of privacy in his entire house. Nor can the search be justified on the grounds that a possible homicide inevitably presents an emergency situation, especially since there was no emergency threatening life or limb, all persons in the apartment having been located before the search began.

The court ruled that the seriousness of Mincey's offense was not enough to necessitate a warrantless search, because there was no indication that evidence would be lost, destroyed, or removed while investigators obtained a search warrant, and there was no indication that a warrant would have been difficult to obtain.

Michigan v. Tyler (1978)

In this Supreme Court case, a local fire department responded to a call about a fire in Loren Tyler and Robert Tompkins' furniture store, shortly before midnight on January 21, 1970. Plastic containers holding a suspected flammable liquid were discovered at about 2 A.M., and the chief summoned police investigators. A detective took several photos but stopped his investigation because of extensive smoke and steam. By 4 A.M., the fire had been extinguished, and the chief and detective left, taking the plastic containers with them. At about 8 A.M. the next day, the chief and an assistant reentered the scene to make a cursory inspection. An hour later, the assistant and the detective made another examination and removed evidence. On February 16, 1970, a member of the Michigan State Police arson section visited the site and took photos; the same investigator subsequently made several additional visits. During these visits, various pieces of physical evidence were retrieved, including a portion of a fuse.

Tyler and Tompkins were charged with conspiracy to burn real property and other offenses. Evidence retrieved from the building as well as the arson investigator's testimony were used to obtain their convictions. The defendants' attorneys objected to the testimony and evidence, saying that no warrant was issued, nor had consent been granted for the searches and seizures. The Michigan State Supreme Court reversed the verdicts and remanded the case for a new trial. The State of Michigan appealed to the U.S. Supreme Court, which upheld the ruling of the Michigan Supreme Court, saying that, "A burning building clearly presents an exigency of sufficient proportions to render a warrantless entry 'reasonable,' and, once in the building to extinguish a blaze, and for a reasonable time thereafter, firefighters may seize evidence of arson that is in plain view and investigate the causes of the fire."

The Supreme Court concluded that the later warrantless entries were not part of the initial emergency circumstances and, therefore, violated the Fourth and Fourteenth Amendments. The court excluded evidence obtained during these entries from the respondents' retrial.

Conclusion

The underlying message of the rulings of the Supreme Court is clear. While various Fourth Amendment exceptions permit law enforcement officials and medical personnel to enter a crime scene, render aid, and collect evidence in plain view, officers must obtain a search warrant before they conduct a careful, detailed examination of the crime scene.

WRAP UP

1. The investigators must make a calculated estimate of the path taken by the criminal at the crime scene and, if possible, the path taken by the victim. The pathway the investigators use to access the scene must not coincide with the path taken by either the perpetrator(s) or the victim(s).
2. During its creation, the route of the pathway is carefully photographed. The photos should include long- and medium-range shots to record the pathway's overall appearance before many individuals travel on it. Additionally, images of the focal point(s) of the crime scene should be taken from the pathway using a telephoto lens before it is disturbed in any way. Any items of evidence located on or near the path must be photographed and then packaged.

Chapter Spotlight

- When arriving at the crime scene, the first and foremost responsibility of an investigator is to identify, establish, protect, and secure the boundaries of the crime scene. Upon arrival, the investigator should document the state of the crime scene, any existing conditions, and any personal information concerning witnesses, victims, and suspects.

- A piece of evidence is either a questioned sample or a known sample. A questioned sample (also called an unknown sample) must be compared to evidence with known origins.

- Evidence is first placed within a class of objects, which limits the number of known samples to which it may be compared. After classification, evidence undergoes individualization—the process of identifying the sample as a unique specimen.

- Documenting the scene requires specific and exact notes that allow the investigator to recall—sometimes years later—the process of events and details uncovered.

- Scene documentation should include the following measures:
  - A working hypothesis of the crime

- Notes, either handwritten or tape recorded (to be transcribed later)
- Overall, midrange, and close-up photographs
- Video for setting the scene and providing a better sense of item placement
- Sketches based on three main orientation techniques—triangulation, baseline, and polar coordinate

- The search for evidence may occur in several patterns: spiral, grid, strip, zone, or wheel.

- The treatment of evidence, including its collection, transportation, and preservation, is specific to the type of evidence involved. For example, blood samples need to be treated in a very different manner than hair, alcohol, or glass fragment samples.

- A clear and concise chain of custody is vital to maintain the best prosecutorial case.

- Search and seizure rules are continually undergoing revisions through court challenges.

Arrest warrant A judicial order requiring that a person be arrested and brought before a court to answer a criminal charge.

Baseline method A technique used to record measurements for crime scene sketches that draws a line between the fixed points (A, B) and measures the distance to the evidence (x, y) at a right angle from this line.

Chain of custody The chronological record of each person who had an item of evidence in his or her possession and when they had it.

Finished sketch A drawing made by a professional that shows the crime scene in proper perspective and that can be presented in court.

Fourth Amendment This amendment in the U.S. Constitution gives citizens the right to be secure in their persons, houses, papers, and effects against unreasonable searches and seizures.

Individualization The process of proving that a particular unknown sample is unique, even among members of the same class, or proving that a known sample and a questioned sample share a unique common origin.

Known sample Standard or reference samples from verifiable sources.

Physical evidence Any object that provides a connection between a crime and its victim or between a crime and its perpetrator.

Polar coordinate method A technique used to record measurements for crime scene sketches, using a transit or a compass to measure the angle from the north and the distance to the evidence (X). This method is most commonly used in a large-area crime scene (outside or in a warehouse) when a wall or side of a building is used to establish the fixed points (A, B).

Questioned sample A sample of unknown origin.

Rough sketch A drawing made at the crime scene that indicates accurate dimensions and distances.

Search warrant A court order that gives the police permission to enter private property and search for evidence of a crime.

Secondary crime scenes Sites where subsequent criminal activity took place.

Trace evidence Evidence that is extremely small or is present in limited amounts that results from the transfer from the victim or crime scene to the suspect.

Triangulation method A technique used to record measurements for crime scene sketches, measuring the location of the evidence (x, y) from fixed points (A, B).

Putting It All Together

Fill in the Blank

1. A physical evidence sample that is collected at the crime scene is called a(n) _____ sample.

2. Known samples are also called _____ samples.

3. A known sample that has been produced is sometimes called a(n) _____.

4. _____ has the responsibility for securing the crime scene.

5. The boundary around the crime scene should be _____ than the scene could possibly be.

6. As soon as is possible, the responding officer at a crime scene should _____ all of his or her actions and observations at the scene.

7. The first responders must ensure that _____ is not lost, contaminated, or moved.

8. The _____ has the responsibility of directing the crime scene investigation.

9. Once the boundaries of the crime scene have been established, a(n) _____ path into and out of the scene is created.

10. The taking of _____ documents the core of the crime scene and the physical evidence.

11. Photographs of the crime scene must show all items in their _____, _____ state.

12. _____ photos are used to record details about each piece of evidence.

13. An investigator only needs to draw a(n) _____ sketch of the crime scene.

14. As the case is being prepared for court, a(n) _____ will prepare finished sketches of the crime scene based on information in the investigator's rough sketch.

15. Four common search patterns recommended by the FBI are _____, _____, _____, and _____.

16. Impression evidence is developed or enhanced by the use of specialized _____ techniques and the application of _____ developers.

17. Biological evidence is developed by _____ means.

18. The first priority of an investigator who locates a firearm at a crime scene is to _____.

19. Carpet samples close to suspicious fires are collected and placed in a(n) _____.

20. Trace evidence is collected manually by _____, _____, and _____.

21. List four types of evidence that may be retrieved by the medical examiner from a deceased victim: _____, _____, _____, and _____.

22. Whenever possible, evidence should be _____ (left on or taken off) the object on which it is found.

23. As evidence is collected, it is temporarily stored in a collection area that is constantly _____.

24. Crime laboratories require a(n) _____ to accompany all evidence shipments.

25. From its original finding through court proceedings, the prosecution must maintain the _____ of the evidence.

26. Evidence is most often excluded from trial because of violations of the defendant's _____ rights.

27. To obtain a search warrant, a law enforcement officer must show _____.

28. In determining probable cause, the magistrate may consider _____, _____, _____, and _____.

29. When police arrive at the scene of a crime, if they think an injured person is inside, they can enter the premises without a warrant because of the _____ exception to the Fourth Amendment.

True or False

1. A forensic scientist working in a crime lab is able to perform only those tests that are requested by the investigating officer.

2. If evidence is moved prior to documenting the crime scene, then the chain of custody is broken.

3. Probable cause is clearly defined in the Fourth Amendment.

4. There are strict limits to what a warrant allows to be seized.

5. Searches at U.S. borders require warrants.

6. If illegal contraband is in plain view, a law enforcement officer can seize it without a warrant.

7. In the case of *Mincey v. Arizona*, the Supreme Court ruled that a warrant should have been obtained.

8. In the case of *Michigan v. Tyler*, the Supreme Court ruled that no Fourth Amendment violations were committed by the firefighters' entry to extinguish the fire.

Review Problems

1. One evening about dusk, a man was walking his dog along a local street when the dog started to bark and then ran into the bushes along the road. When the man chased after his dog, he discovered his dog barking at a dead body concealed in the bushes. It was midsummer, the temperature was 78°F, and the day had been sunny. Heavy rain was predicted for the next day. Luckily, the man was carrying his cell phone; he called 9-1-1 to report his findings. He investigated the body from a safe distance and learned that the victim had experienced bleeding from his chest. After 10 minutes, the police responded and began to interview the man. Describe in detail the responsibilities of the first responding officer.

2. The lead investigator arrived on the scene of the crime described in Problem 1 and determined that the scene should be searched for physical evidence. One of the officers drew a rough sketch of the crime scene (**FIGURE 1-14**).

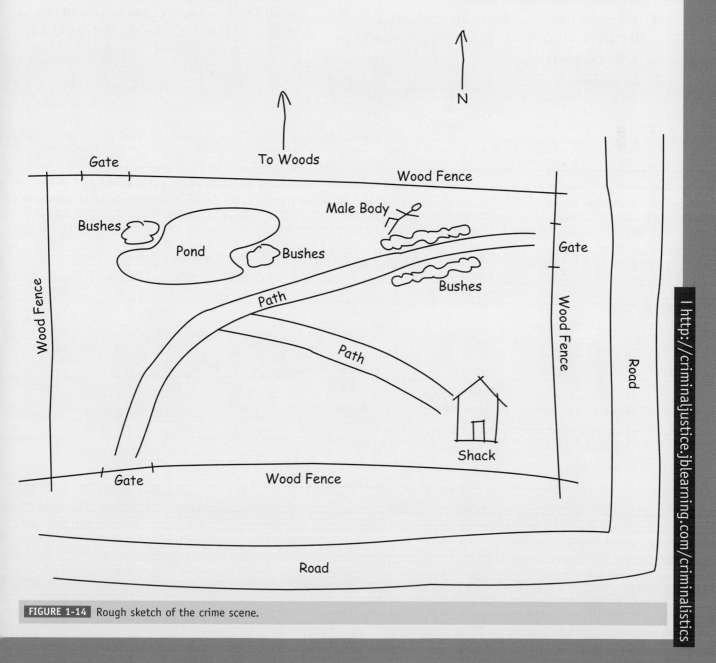

FIGURE 1-14 Rough sketch of the crime scene.

Describe in detail how the lead investigator should now complete the search for physical evidence. What steps will he or she take to ensure that the evidence is not cross-contaminated?

3. One of the officers searching the scene described in Problems 1 and 2 found a bloody knife on the footpath at the entry to the woods. From a distance it appeared to be the murder weapon, and the handle of the knife may have fingerprint impressions left in the blood. How should the knife be handled? Should a swab sample be taken? If so, where on the knife? How can the possible fingerprints be preserved? How should the officer proceed?

4. When the following objects are transported to the crime lab, how should they be packaged?
   a. Blood (wet)
   b. Blood (dry)
   c. Paint chips
   d. Arson liquid accelerant
   e. Carpet from arson scene
   f. Firearms
   g. Impression evidence

5. An officer responding to a homeowner reporting a burglary found a broken window at the rear of the house on the first floor. He determined that the burglary had taken place while the owner was away and that the owner had discovered the robbery when he returned home. The officer called the crime lab, and a forensic examiner arrived on the scene. List in detail which steps the examiner should take to recover all possible evidence that might have been left at the scene of the crime.

Further Reading

Becker, R. *Crime Scene Investigation*, ed 2. Sudbury, MA: Jones and Bartlett, 2005.

Fisher, B. *Techniques of Crime Scene Investigation*, ed 7. Boca Raton, FL: CRC Press, 2003.

Gardner, R. M. *Practical Crime Scene Processing and Investigation*. Boca Raton, FL: CRC Press, 2005.

Lee, H., and Harris, H. *Physical Evidence in Forensic Science*. ed 2. Tucson, AZ: Lawyers and Judges Publishing, 2006.

Lee, H., Palmbach, T., and Miller, M. *Henry Lee's Crime Scene Handbook*. London: Academic Press, 2001.

Jones, P. *Practical Forensic Digital Imaging: Applications and Techniques*. Boca Raton, FL: CRC Press, 2010.

Redsicker, D. *The Practical Methodology of Forensic Photography*, ed 2. Boca Raton, FL: CRC Press, 2001.

SWGIT, Scientific Working Group on Imaging Technology, International Association for Identification, http://www .theiai.org/guidelines/swgit.

Technical Working Group for Bombing Scene Investigation. *A Guide for Explosion and Bombing Scene Investigation*. Washington, DC: National Institutes of Justice and U.S. Department of Justice, 2000.

Technical Working Group for Crime Scene Investigation. *Crime Scene Investigation: A Guide for Law Enforcement*. Washington, DC: National Institutes of Justice and U.S. Department of Justice, 2000.

Technical Working Group for Fire/Arson Scene Investigation. *Fire and Arson Scene Evidence: A Guide for Public Safety Personnel*. Washington, DC: National Institutes of Justice and U.S. Department of Justice, 2000.

Vince, J. J., Jr., and Sherlock, W. E. *Evidence Collection*. Sudbury, MA: Jones and Bartlett, 2005.

http://criminaljustice.jblearning.com/criminalistics

Answers to Review Problems

Interactive Questions

Key Term Explorer

Web Links

http://criminaljustice.jblearning.com/criminalistics

Investigating and Processing Physical Evidence

OBJECTIVES

In this chapter you should gain an understanding of:

- Common objects found at a crime scene that qualify as physical evidence
- Different types of crime labs and their organization
- The functions performed by a forensic scientist
- Class and individual characteristics of physical evidence
- Reconstruction of a crime scene
- The admissibility of physical evidence and the role of an expert in court

FEATURES

On the Crime Scene

Back at the Crime Lab

See You in Court

WRAP UP

Chapter Spotlight

Key Terms

Putting It All Together

YOU ARE THE FORENSIC SCIENTIST

As mentioned in Chapter 1, the smallest objects found as evidence are referred to as *trace evidence*. These objects (i.e., fibers, glass fragments, gunshot residue) are so easily transferred from one individual to another that they may provide evidence of association between a suspect and the victim. Because they are so readily transferred, investigators must take great care to avoid losing or cross-contaminating this evidence.

Usually trace evidence is transferred from one object to another in a process referred to as *direct transfer*. On other occasions, trace evidence is transferred from one object to another by way of an intermediate object, in a process known as *secondary transfer*. It also is possible that two or more intermediate objects may be involved in secondary transfer. It is important that investigators consider the possibility of secondary transfer whenever they examine trace evidence.

Consider the following trace evidence, which was found at the scene of a murder. In the victim's room, where the murder occurred, there is a fabric-covered chair. The suspect's jacket and sweater are seized from his apartment and examined for fiber evidence. The chair has fibers on it that match fibers from the jacket and the sweater. The jacket has fibers from the chair. The sweater does not have fibers from the chair on it.

1. Explain how the fibers might have been transferred.
2. Does this evidence prove that the suspect was in the victim's room more than once?

Introduction

Physical evidence is merely one piece of the puzzle when investigators are trying to solve a case. In some types of crimes (e.g., homicide, sexual assault), it may be the most important factor in proving the link between the suspect and the victim. Physical evidence may also be essential to prove that the same suspect is linked to a series of incidents. In other cases, the implications of the physical evidence must be confirmed by the testimony of witnesses and/or the confession of the suspect to warrant a conviction. This chapter describes how physical evidence is identified, classified, and then presented to a court of law.

Types of Evidence

Four types of **evidence** are distinguished: testimony, physical, documentary, and demonstrative (**TABLE 2-1**). The most common types of physical evidence are listed in **TABLE 2-2** . Because a crime scene tends to include so many physical items, it is impractical to treat each and every object that is encountered as evidence. Nevertheless, it is extremely important to identify those items that might provide significant probative information related to the crime. To do so, experienced investigators

TABLE 2-1

Types of Evidence

Type	Definition	Example
Physical evidence	Tangible objects—that is, items that are real, direct, and not circumstantial	A weapon used to commit a crime; trace evidence found at the crime scene (e.g., blood, hair, fibers)
		Property recovered after a crime is committed; fingerprints shoeprints tire tracks handwriting
Documentary evidence	Any kind of writing, sound, or video recording; its validity is usually authenticated by expert testimony	A transcript of a recorded telephone conversation
Demonstrative evidence	Real evidence used to illustrate, demonstrate, or recreate a prior event	A cardboard model of the crime scene
Testimony	Evidence in the form of witnesses speaking under oath in court	Eyewitnesses; hearsay witnesses character witnesses

TABLE 2-2
Common Types of Physical Evidence

Drugs: Any drugs, either licit prescription drugs or illicit substances.

Blood, semen, or saliva: Either dried or liquid blood, semen, or saliva that may be useful in identifying unknown persons or in establishing a connection between objects or persons.

Fibers: Any synthetic or natural fibers that may be useful in establishing a connection between objects or persons.

Fingerprints: Both visible and latent (invisible) fingerprints.

Firearms or ammunition: Any firearm, ammunition, shell casing, or bullet.

Glass: Holes in glass, glass fragments, cracks in glass.

Hair: Any human or animal hair.

Impressions: Impressions left by a wide variety of objects, such as shoeprints, tire treads, palm prints, and bite marks.

Organs or body fluids: Organs and body fluids undergo toxicological analysis for drugs, alcohol, and poisons.

Explosives: Objects containing explosive chemicals or objects covered with residues from an explosion.

Paint: Dried or liquid paint.

Petroleum products: Grease or oil stains, gasoline, paint thinner, or kerosene.

Plastic bags: Common household garbage bags.

Plastic, rubber, or other polymers: Common plastics found in the home.

Powder residue: Residue from the discharge of a firearm.

Serial numbers: Firearm identification numbers, vehicle identification numbers, serial numbers of computers and electronic devices.

Documents: Handwriting samples, typewritten (printer or typewriter) samples, paper, ink, erasures, and heat treatment.

Soils or minerals: Soil, gravel, or sand from various locations.

Tool marks: Objects that leave an impression, such as crowbars, screwdrivers, and hammers.

Parts from vehicles: Objects broken from an automobile.

TABLE 2-3
U.S. Federal Rules of Evidence

Article I
General Provisions

Rule 104(b): Relevancy Conditioned on Fact
When the relevance of evidence depends on the fulfillment of a condition of fact, the court shall admit it upon, or subject to, the introduction of evidence to support a finding of the fulfillment of the condition.

Article IV
Relevancy and Its Limits

Rule 401: Definition of Relevant Evidence
"Relevant evidence" means evidence having any tendency to make the existence of any fact that is of consequence to the determination of the action more probable or less probable than it would be without the evidence.

Rule 402: Relevant Evidence Generally Admissible; Irrelevant Evidence Inadmissible
All relevant evidence is admissible except as otherwise provided by the Constitution of the United States, by act of Congress, by these rules, or by other rules prescribed by the Supreme Court pursuant to statutory authority. Evidence which is not relevant is not admissible.

Rule 403: Exclusion of Relevant Evidence on Grounds of Prejudice, Confusion, or Waste of Time
Although relevant, evidence may be excluded if its probative value is substantially outweighed by the danger of unfair prejudice, confusion of the issues, or misleading the jury, or by considerations of undue delay, waste of time, or needless presentation of cumulative evidence.

(**TABLE 2-3**). In court proceedings, the judge is responsible for determining what is relevant and what is not. Relevant evidence is deemed admissible; irrelevant evidence is deemed inadmissible.

The Modern Crime Lab

In the mid-1960s, there were roughly 100 crime labs in the United States. Today, there are roughly 350 crime laboratories in this country, more than 80% of which are affiliated with police agencies. Clearly, the number of crime labs in the United States has increased dramatically in recent times, for two major reasons.

First, there has been an incredible increase in the crime rate in the United States. As the crime rate has increased, so has the percentage of crime that is drug related—and, in conjunction, the number of drug samples sent to crime labs. That is because all drugs that are seized by law enforcement authorities must be sent to a forensic lab for chemical analysis (to prove they are actually illicit drugs) before they can be used as evidence in court.

who are familiar with the circumstances of the crime scenes they examine must make logical decisions about precisely which items will be examined in more detail. In making this decision, an investigator does not rely on a list of what to take and what to leave behind. Instead, the investigator learns to focus on those objects whose scientific analysis is likely to yield important clues and that have provided useful forensic evidence in the past.

Once an object is collected at a crime scene, it is analyzed in the forensic laboratory. Based on the results of the scientific investigation, the prosecutor then decides whether the item will be presented to the court. Whether this object is considered evidence or not is solely determined by its relevance to the crime being investigated and the legality of its collection. In the United States, evidence is defined by the U.S. **Federal Rules of Evidence (FRE)**

on the CRIME SCENE — Physical Evidence and the Innocence Project

On November 16, 1983, as a 28-year-old woman was walking from work to home in Lowell, Massachusetts, an unknown man came up to her and tried to engage her in casual conversation. The woman didn't take up her end of the conversation, but the man forced her into a nearby yard and sexually assaulted her.

The following evening, within 100 yards of the first attack, a 23-year-old woman, who also was walking home from work, was pushed to the ground by a man wielding a knife. After struggling with her assailant, the second victim escaped her attacker and called the police. She described the assailant as a man wearing a red, hooded sweatshirt and a khaki-colored military-style jacket.

On the night of the second attack, the police stopped a suspect, Dennis Maher, who was wearing clothes that matched the description given by the second victim. A search of his car turned up an army field jacket, a military-issue knife, and a rain slicker. Maher, a U.S. Army sergeant, was arrested and charged with both attacks in Lowell plus an unsolved rape case that had occurred the previous summer in Ayer, Massachusetts. All three victims identified Maher from photographic lineups, even though their original descriptions of their attackers varied. Maher, however, insisted that he was innocent.

The Lowell attacks were tried together. Relying on the identifications made by the victims and no other physical evidence, Maher was convicted of the assaults. A month later, he was convicted of the Ayer rape, where physical evidence existed but was never tested.

In 1993, the Innocence Project—part of the Benjamin N. Cardozo School of Law established by Barry C. Scheck and Peter J. Neufeld—took up Maher's case. Members of the project repeatedly tried to gain access to the physical evidence from the victims that was collected at the time of the incidents, but they were told that the evidence couldn't be located. In 2001, a law student scrounging around the basement of the Middlesex County Courthouse found the box of evidence containing the clothing and underwear of one of the victims. The Massachusetts State Police Crime Laboratory found seminal fluid stains on the underwear as well as possible bloodstains on the clothing. Finding biological material on these items allowed for DNA testing by Forensic Science Associates.

Although the evidence found on the clothing was deemed inconclusive, the test results on the underwear produced a genetic profile of the assailant that excluded Maher as the donor of the sample. Prosecutors soon afterward located the evidence from the Ayer case, and testing by Orchid Cellmark again revealed that Maher was not the source of the biological material found on the victim.

On April 3, 2003, after 19 years in prison proclaiming his innocence, Dennis Maher was exonerated and released from prison.

The second reason the number of crime labs has increased is the 1966 Supreme Court decision in the case *Miranda v. Arizona*. In its ruling, the Supreme Court established the need for the so-called Miranda warning, which requires arresting officers to advise criminal suspects of their constitutional rights and right to counsel. As a consequence, fewer defendant confessions are now made, which has forced prosecutors to seek more thorough police investigation and to use more physical forensic evidence as part of their cases.

A few crime labs are owned by private companies and provide a particular specialty to law enforcement. Orchard Cellmark, for example, is well known for its work on forensic DNA. Battelle Corporation has expertise in arson cases, and Sirchie Corporation is known for its work on fingerprinting and trace evidence collection.

National Laboratories

The U.S. federal government does not have a single federal forensic laboratory with unlimited jurisdiction. Instead, four crime laboratories have been established to deal with evidence from suspected violations of federal (rather than state or local) laws.

TABLE 2-4	
Services Provided by the FBI and the ATF Crime Labs	
FBI Crime Lab	Units: Chemistry, DNA, Explosives, Firearms and Tool Marks, Forensic Audio, Video and Image Analysis, Latent Prints, Questioned Documents, Materials Analysis, Special Photographic Analysis, Structural Design, Trace Evidence, Investigative and Prosecutive Graphics, and Hazardous Materials Response
ATF Crime Lab	Investigates crimes relating to firearms, explosives, tobacco, and alcohol

These laboratories are operated by the following government agencies:

- Federal Bureau of Investigation (FBI; part of the Department of Justice)
- Drug Enforcement Administration (DEA; part of the Department of Justice)
- Bureau of Alcohol, Tobacco, Firearms, and Explosives (ATF; part of the Department of the Treasury)
- U.S. Postal Service (USPS; Inspection Service)

Among the federal crime laboratories, those operated by two organizations stand out above the rest—the FBI and ATF laboratories. **TABLE 2-4** lists the services provided by these federal labs.

State and Municipal Laboratories

Every state has established its own crime lab; these state-based facilities serve both statewide and local law enforcement agencies (if the local jurisdiction does not have its own lab). In addition, some states (e.g., California, New York, Illinois, Michigan, Texas, Virginia, Florida) have developed a statewide system of regional laboratories whose activities are coordinated by the state government in an effort to minimize duplication of services and to maximize inter-laboratory cooperation.

By contrast, local crime labs serve city and county governments. For example, some larger cities operate their own labs, which are independent of their respective state laboratories.

Divisions of the Crime Lab

A crime lab typically includes six divisions that report to the director's office.

1. *Biological/Serological Division:* Deals with anything pertaining to fluids.
2. *Chemistry or Toxicology Division:* Deals with unknown substances, drugs, or poisons.
3. *Trace Evidence or Microscopy Division:* Deals with anything small enough to require a microscope for viewing, such as hairs or fibers.
4. *Ballistics, Firearms, and Tool Marks Division:* Deals with guns or weapons.
5. *Latent Fingerprints Division:* Locates, photographs, processes, and compares latent fingerprints to known candidates and to fingerprints in the Automated Fingerprint Identification System.
6. *Questioned Document Division:* Examines documents to identify the writer and to detect a forgery or alteration.

For their findings to be widely accepted, crime labs must establish their credentials as forensic laboratories. Accreditation by professional organizations such as the American Society of Crime Laboratory Directors, the National Forensic Science Technology Center, and the College of American Pathologists serves this function. In addition, labs may perform—and be certified for—specialized services. For example, a lab that specializes in the analysis of teeth and bite marks might apply to the National Board of Forensic Odontology for accreditation in this specialty area. Individual lab workers may also enhance their own credentials by joining associations such as the American Academy of Forensic Sciences and the American Board of Criminalistics.

To warrant accreditation, a crime lab must meet minimum requirements established by the certifying authority. Among other things, it must develop the following documents and programs:

- A quality control manual. *Quality control* measures ensure that test results (e.g., the results of DNA analysis) meet a specified standard of quality.
- A *quality assurance* (QA) manual. QA serves as a check on quality control. That is, a laboratory's QA measures are intended to monitor, verify, and document the lab's performance.
- A lab testing protocol. *Protocols* are the procedures and processes followed by the laboratory to ensure that it performs tests correctly and accurately. For example, a lab may perform

validation studies to confirm that it is performing specific types of tests properly.

- A program for proficiency testing. *Proficiency testing* determines whether lab workers as individuals and the laboratory as an institution are performing up to the standards established by the profession. The laboratory or worker is given a sample, for which the results of the analysis are already known; if the lab's or worker's results do not match the known results, clearly there is a problem. Proficiency tests may be either blind (the worker is unaware that he or she is being tested) or known (the worker is aware of the test and can consult any resources necessary).

Functions of a Forensic Scientist

The forensic scientist performs the following steps as he or she processes physical evidence:

1. Recognize physical evidence.
2. Document the crime scene and the evidence.
3. Collect, preserve, inventory, package, and transport physical evidence.
4. Analyze the physical evidence.
5. Interpret the results of the analysis.
6. Report the results of the analysis.
7. Present expert testimony.

The first step, recognition of physical evidence, begins at the crime scene. Because all subsequent steps involve working with evidence retrieved from the actual scene, processing the crime scene is not only one of the first events to occur following commission of a crime but also one of the most important. If the case is to proceed smoothly, the collection and processing of physical evidence must be done both thoroughly and correctly. Yet another important aspect of forensic training is learning how to choose the appropriate analysis for the evidence that has been gathered.

Additional Information

The primary goal in analyzing physical evidence is to make the facts of a case clear. Through the analysis and interpretation of physical evidence, the expert can provide additional information that ties together the facts of the case.

Information on the *Corpus Delicti*

Facts dealing with the **corpus delicti** ("body of the crime") prove that a crime has actually taken place and that what happened was not an accident. Examples of such evidence for the crime of burglary might include tool marks on a broken door, which strongly indicate a forced entry, or a ransacked room from which jewelry is missing. For an assault, relevant evidence might include the blood of the victim on a suspect's clothes and a bloody knife, both of which indicate foul play.

Information on the *Modus Operandi*

The **modus operandi** (MO; the "method of operation") is the characteristic way in which career criminals commit a particular type of crime. In bur-

BACK AT THE CRIME LAB

A forensic scientist must be a "jack-of-all-trades" when it comes to the sciences. In particular, the forensic scientist needs basic knowledge in the following areas:

- Physics: Ballistics, explosion dynamics, fluid viscosity, and dust impression lifting
- Chemistry: Arson investigation, chemical decomposition of matter, and soil analysis
- Biology: Genetic fingerprinting, biological decomposition, and DNA analysis

- Geology: Soil samples
- Statistics: Statistical relevance of comparison tests that can stand up to cross examination in court

While many of these tasks will be carried out by trained specialists, having a basic understanding of each discipline ensures that the forensic investigator will not make mistakes in handling, preparing, and analyzing evidence.

glary cases, the type of tools used and the tool marks they leave, methods of ingress and egress, and types of items taken are all important clues. In arson cases, the type of accelerant used, its position in the building, and the technique that was used to ignite the fire often turn out to be a particular arsonist's "signature." Indeed, comparing the MO for a specific case to closed arson cases can sometimes lead to the identification of the arsonist.

Linking a Suspect and a Victim

Identifying a link between a suspect and a victim is extremely important, particularly in cases of violent crime. Blood, hairs, fibers, and cosmetics can all be transferred between victim and perpetrator, which are examples of **Locard's exchange principle**. This principle states that whenever two objects come into contact with one each other, there is an exchange of materials between them. For this reason, every victim and every suspect must be thoroughly searched for trace evidence.

Linking a Person to a Crime Scene

Perpetrators as well as victims often leave fingerprints, shoeprints, footprints, tire tracks, blood, semen, fibers, hair, bullets, cartridge cases, or tool marks at the scene of a crime—another example of the Locard exchange principle. Conversely, victims, perpetrators, and even witnesses may carry glass, soil, stolen property, blood, and fibers away from the scene of the crime, and this evidence can be used to prove their presence at the scene.

Disproving or Supporting a Suspect's or Witness's Testimony

Suppose a person is accused of a hit-and-run accident. Examination of the undercarriage of the car reveals blood and tissue, but the vehicle's owner claims he ran over a dog. A species test on the blood would reveal whether it came from a human source, thereby supporting or disproving the investigator's hypothesis of the crime.

Identification of a Specific Suspect

Fingerprints and DNA left at the scene of a crime are the most conclusive ways of identifying a suspect. The probability of finding a fingerprint at a crime scene is more likely than the likelihood of finding a DNA sample.

Providing Investigative Leads

Physical evidence can be used to direct the course of an investigation. For example, a paint chip left at the scene of a hit-and-run accident can be analyzed and used to narrow the search for the type of car that might have been involved in the accident.

Eliminating a Suspect

Physical evidence has exonerated many more suspects than it has convicted.

State of the Evidence

The crime scene and all of the evidence in it are subject to the effects of time. For example, sunlight or other environmental factors such as rain, snow, or wind may all alter the crime scene and destroy evidence. The moment an object is considered to be physical evidence, it becomes a mute witness to the crime. Investigators must move quickly to identify and protect evidence before environmental effects begin to alter its appearance and composition.

Biological evidence is most susceptible to change. A bloodstain found on the wall of a crime scene shortly after a shooting initially will be wet and red. As it is exposed to air, it will clot, dry, and eventually turn brown. If the crime scene is outside, blood also may be exposed to direct sunlight, which can change it from a red drop to a black dry spot in a relatively short period of time. In addition, rain can quickly wash it away.

Other physical evidence also may undergo changes due to time and physical influences. For example, a bullet may pass through a body of a suspect, picking up blood that could later be used to identify this person. But if the bullet then smashes through a wall, where the blood is scraped off its surface, its shape will likely be altered. The blood evidence on the bullet is scattered as it passes through the wall and might even be totally lost.

Likewise, part of a torn document left at a crime scene may be altered by exposure to the sun or water. The rays of the sun might bleach out the color, and water might change its texture. If a forensic scientist was trying to match the exposed piece to an unexposed piece of the same document, the two pieces might appear to be quite different.

Objects of evidence whose appearance changes with time can test the wits of crime scene investigators. These changes in appearance also challenge the forensic scientist, who must try to compare the evidence from the crime scene to a reference material, known as an **exemplar**. The forensic scientist must be able to show that the evidence (questioned sample) and the known sample (exemplar) have a common source. This leads the forensic scientist to use one of the most basic tenets of forensic science: **explainable differences**. That is, to show a common source for the two samples, the forensic expert must have a sound scientific explanation for any differences between the evidence (questioned sample) and the reference material (exemplar).

Why Examine Physical Evidence?

A forensic scientist examines physical evidence for one of two purposes: identification or comparison. **Identification** is the process of elucidating the physical or chemical identity of a substance with as much certainty as possible. Comparison is the process of subjecting both the evidence (questioned sample) and the reference material (exemplar) to the same tests to prove whether they share a common origin.

The comparison of evidence to reference material is an aspect of forensic science that differentiates it from all other applications of science. An object becomes evidence only when it contributes information to the case; otherwise, it is excluded from consideration. A bloodstain on the jeans of a homicide victim, for example, might appear to be an important piece of evidence until it is determined that the stain came from the victim. We already know the victim was present at the crime scene, so such a stain cannot be used to identify the perpetrator. Nevertheless, the location of the stain might provide valuable clues about the manner in which the victim was assaulted.

Characteristics of Physical Evidence

Identification

When a forensic scientist attempts to identify an object, he or she takes measurements that describe the physical and chemical properties of that object with as near-absolute certainty as scientific techniques will allow. For example, the crime laboratory might use chemical tests to determine if a white powder found at a crime scene is an illicit drug,

such as cocaine or heroin, or the residue from bomb making, such as TNT. In cases involving biological material, such as blood, semen, and saliva, the crime lab might use molecular biological tests to determine the identity of the sample.

Many forensic analyses involve comparison of the questioned sample to some standard sample. A test is considered valid if it is reproducible, sensitive, and specific. To be reproducible, the test's analysis of the standard sample must always yield the same, correct results. To be sensitive, it must be able to accurately identify the unique characteristics of the substance. To be specific, the test must give a definitive result for a particular substance.

For example, if the forensic chemist is testing a white powder sample found at a crime section to determine whether it is cocaine, the test results must narrow the possibilities down so that cocaine is the only substance that would produce a positive result. That is, no other substance should give the same results as cocaine. Sometimes a battery of several different tests must be carried out to reach this conclusion with certainty.

One of the forensic scientist's most prized skills is the ability to pick the appropriate test for each questioned sample. In making this choice, the scientist must take a variety of issues into account, including the quantity and quality of the evidence (questioned sample). Some standard tests are easy to perform with large samples of material but do not have enough sensitivity to analyze trace quantities of evidence.

Once these testing procedures have been used repeatedly and shown to give reproducible, accurate results when used by several different laboratories, the test protocols are permanently recorded and become protocols that are accepted by the court. The FBI has established several scientific working groups (SWGs) that develop testing protocols in conjunction with crime laboratories as new standard tests emerge. For instance, in 2005 the Scientific Working Group for Materials Analysis issued protocols for the elemental analysis of glass to crime laboratories. Other FBI SWGs are focusing on gunshot residue, DNA analysis, the analysis of human hair, and other forensic evidence. Professional organizations, such as the American Society for Testing Materials, also establish standards for test methods that have been adopted by crime laboratories for specific analysis.

Forensic scientists must rely on their experience to know when they have performed enough tests on the questioned sample. At this point, they should have enough data to draw a conclusion about the questioned sample to a reasonable degree of scientific certainty—that is, beyond any reasonable doubt. The result is then ready to be presented to a court of law.

Associative Evidence

Physical evidence located at a crime scene can be used either to associate a suspect with a crime or to rule out that person as a suspect. Indeed, physical evidence excludes or exonerates people from suspicion more often than it implicates them.

Blood and other body fluids, fingerprints, hairs, bullets, firearms, and imprint evidence can all be **associative evidence**. These items are considered to be of uncertain origin until they are compared to a known standard (exemplar) that may be collected from suspects, victims, or witnesses. Two types of associative evidence are identified: that with **class characteristics** and that with **individual characteristics**. Associative evidence that has class characteristics can be classified only as belonging to a certain class of objects; such an item may be excluded as belonging to other classes of objects. When an object is examined and placed in a class, multiple sources remain as possibilities. By contrast, associative evidence that has individual characteristics can be associated with only a single source. When an object is individualized, the number of possible sources is reduced to just one. In some cases, it is even possible to state that a questioned object is unique.

For manufactured objects, class characteristics include, but are not limited to, the size, shape, style, and pattern of an object when it is made. These characteristics originate as a result of repetitive, mechanical steps that are repeated as copy after copy of the object is made. The distinctive tread pattern on the sole of a new shoe is a good example (FIGURE 2-1). The size, shape, style, and pattern are distinctive to the class of shoes that are produced by one manufacturer for one style. A tread pattern found at a crime scene would allow the forensic scientist to determine the size of the shoe, the manufacturer, and the style, which would allow the investigator to determine that the print was made by someone who owns a shoe from this particular class of shoes. Although such a shoeprint can be associated with a class of shoes with high probability, the shoe pattern alone is not enough to definitively identify which owner within the class of all owners of these shoes committed the crime. By contrast, a worn shoe with wear patterns unique to one individual would allow the forensic scientist to link this one shoe to the tread pattern found at the crime scene.

Analysis of class characteristics allows the forensic scientist to identify the object as being a member of a class of objects with high probability. Samples of drugs, fibers, hair, glass, blood, and soil all are examples of evidence that can be associated with a class. Unfortunately, without using sound, scientific information, the forensic scientist will be unable to associate the object with a single source (owner) with very high probability.

Individual characteristics may also include scratches and imprints that are left on an object that transform the object from being simply a member of a class to a unique object. For example, tool marks and impressions may turn an object with class characteristics into an object that has individual characteristics.

Tool Marks

A **tool mark** is created when some kind of tool creates an impression, cut, scratch, or abrasion in another surface. Even when the suspected tool is not recovered, marks left at one crime scene might match those found at other crime scenes. This information helps investigators link separate crimes and often leads to new investigative leads as evidence from the multiple scenes is pooled together.

Tool marks can be made by a variety of tools and are often left during the course of a burglary. Screwdrivers or crowbars, for example, may be used as levers to pry open windows and doors. Pliers and wire cutters may be used to cut through screens, vinyl windows, and padlocks.

Careful examination of the impression that a new tool leaves in a softer surface will provide class evidence, perhaps indicating the size and shape of the tool. A worn tool, by contrast, may have unique characteristics, such as a chip out of one side or a surface that is more worn on the right than on the left. If the edge of the worn tool is scraped against a softer surface, it will cut a striated line that is a mirror image of the pattern on the tool's edge.

Markings left on an object at a crime scene can be compared to an object that has been scratched in the crime laboratory with the same tool. To duplicate the tool mark, the forensic examiner makes

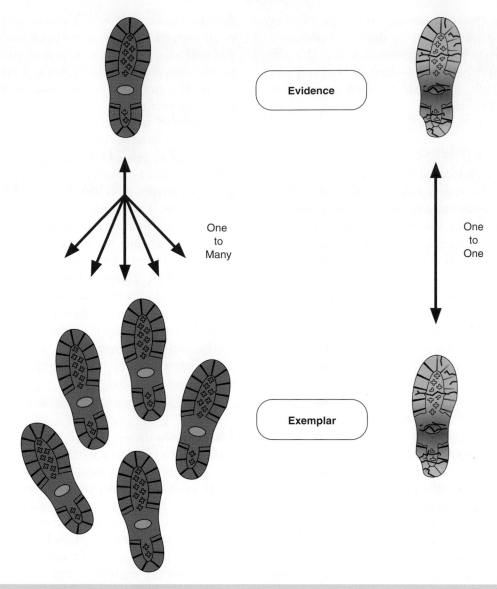

Evidence

One
to
Many

One
to
One

Exemplar

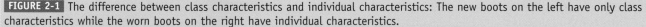

FIGURE 2-1 The difference between class characteristics and individual characteristics: The new boots on the left have only class characteristics while the worn boots on the right have individual characteristics.

several test marks. For example, he or she may apply different levels of pressure or change the angle at which the tool makes contact with the surface. In addition, the forensic scientist may make tool marks against a variety of surfaces. A lead or rubber-based sheet, which is soft, is often pressed against the scored area to record the tool mark. The resulting tool mark is then compared with the tool mark observed at the crime scene using a comparison microscope.

When the impressions left at the crime scene are sufficiently unique and match the impressions made in the crime lab (**FIGURE 2-2**), the evidence has been individualized. A forensic examiner then can testify that the impressions could be made only by using the questioned tool.

At the crime scene, photographing the area around the impression is usually the first step in the recovery of tool mark evidence. Photos from a distance serve to show the impression's context in relation to the overall crime scene. Close-up photos taken at an optimal angle, with proper lighting, will record the unique features of the impression. Unfortunately, photography is not the best recovery method for tool marks because the finer details of the striations produced are not captured well by

FIGURE 2-2 A cast of a tool mark.

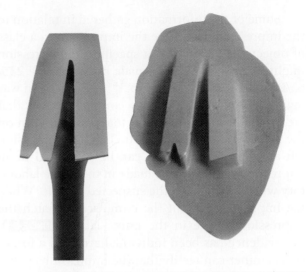

FIGURE 2-3 A cast of a hammer claw.

a photograph. Nevertheless, photography is not destructive, so additional tests can be carried out after the impression is photographed.

If possible, the object that was marked by the tool should be removed from the crime scene for later examination at the laboratory. Care must be taken in removing, packaging, and shipping the marked object to prevent any damage to or contamination of the evidence. If the marked object cannot be removed and shipped to the laboratory, the forensic examiner should make a cast of the mark. During this process, casting material that is made of silicone rubber or other filled plastics is applied to the impression and allowed to dry. The resulting cast can then be compared with the suspect tool (**FIGURE 2-3**). If the tool suspected of making the tool mark is recovered, it is critical that the investigator *not* try to fit it into the actual tool mark, because that may alter the impression.

Impressions

Impressions of one type or another often are left at crime scenes, such as impressions left by shoes or tires. These impressions may be deep (e.g., ruts in mud) or ultra-thin (e.g., barely visible marks in dust).

The first step in recording an impression is to photograph it. As described in Chapter 1, this piece of evidence should be photographed both alone and with a scale inserted into the scene to give a sense of the context and scale (**FIGURE 2-4**). Photographs taken from a distance serve to show the impression in relation to the overall crime scene. Close-up photos taken at an optimal angle, with proper lighting, will record the features of the impression.

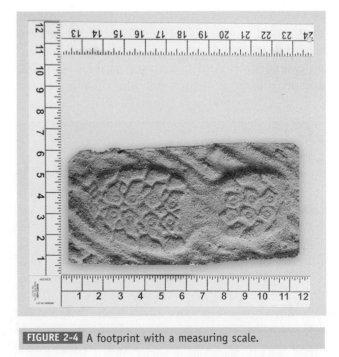

FIGURE 2-4 A footprint with a measuring scale.

Some of the information gathered in relation to the impression will place the impression in a class of objects. For example, a specific tire impression might suggest that it was made by one of the 235/65/R16 tires made by Dunlop. If the impression was made by a worn tire, it may contain features that are unique, such as a surface that is more worn on the right side than on the left. The tire impression left at a crime scene then can be compared to an impression that has been made in the crime laboratory with the tire from the suspected vehicle. When the impressions left at the crime scene match the impressions made in the crime lab (**FIGURE 2-5**), the evidence has been individualized, and a forensic examiner can testify that the impressions could be made only by the questioned tire.

Ideally, the entire object containing the impression will be collected and sent to the laboratory for subsequent testing. This step may be possible with items such as paper or floor tiles. In other circumstances, the investigator may not be able to physically remove the entire object, such as when a shoeprint is left in the earth. He or she must then "lift" the impression, so that the forensic examiner can subsequently compare this information with the suspect shoe or tire.

A variety of techniques are used to lift impressions. For example, clear tape may be applied to the impression and then pressed to eliminate any air pockets. A more elaborate but reliable lifting technique uses electrostatics to lift the impression (such as from dust) onto a plastic sheet. The electrostatic dust print lifter uses a plastic-coated metal sheet to which is applied a large negative charge. The plastic coated sheet is placed on top of the dust print, and the power is turned on. Any dust under the

plate will take on a positive charge and will be attracted to the negatively charged plastic plate. The dust print that is transferred to the plastic-coated sheet will appear as a precise mirror image of the original print (**FIGURE 2-6**). This technique works well on rough-surfaced floor tiles or flooring with an irregular surface from which it would be difficult to lift impressions with tape.

To preserve tire tracks or shoeprints made in dirt or snow, the investigator should make a cast of the impression. First, the investigator photographs the impression to record the image. Then, the investigator places a casting frame around the impression. Next, he or she pours casting material made from silicone rubber or dental stone into the impression (**FIGURE 2-7**). (Plaster of Paris is no longer used for this purpose because it crumbles too easily.) The cast needs a minimum of 30 minutes to set, and it should not be examined further for 24 hours. Because most gypsum casting materials generate heat during curing, an insulating medium must be applied to a snow impression before casting is attempted with this medium. For example, specialty waxes may be sprayed onto the snow impression to lock in the impression before the casting material is applied.

The laboratory procedures for comparing a recovered impression to an impression made by the

FIGURE 2-6 The electrostatic plate will lift dust from horizontal and vertical surfaces, thereby recording the impression.

FIGURE 2-5 An impression of a worn tire.

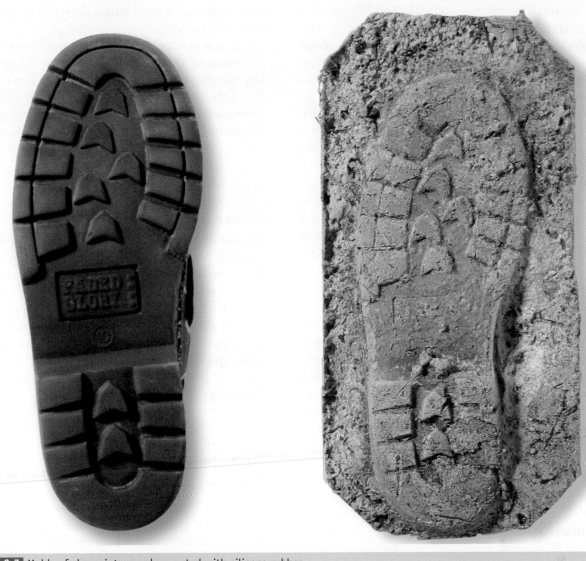

FIGURE 2-7 Molds of shoeprints may be created with silicone rubber.

suspect are possible only if the actual shoe or tire suspected of leaving the impression has been recovered. Test impressions may be necessary to compare the characteristics of a suspect's shoe to those of the impression found at the crime scene. As the examiner compares the two impressions, he or she will determine which class and individual characteristics the two impressions share. If the two impressions were made by shoes of the same size, width, shape, and tread design, for example, the examiner concludes that the suspect's shoe cannot be excluded from the class of shoes being considered. If the two impressions also share a sufficient number of uniquely individual characteristics such as uneven wear, gouges, cracks, or broken tread, then the evidence supports a finding that both the evidence and

the test impression were made by this particular shoe and only this shoe.

Crime Scene Reconstruction

The physical evidence left at the scene of the crime may also be used to establish some of the events that occurred before, during, or immediately after the crime. That is, the evidence may help investigators determine the order in which the events took place. The reconstruction of some part of the crime scene may prove to be very important in corroborating or refuting a description of events that have been reported by witnesses or suspects.

Reconstruction of a crime scene begins with an examination of the crime scene. This step is followed by the collection and analysis of physical evidence and other independent sources of information, such as witness descriptions, photographs, sketches, autopsy findings, and written reports. Reconstruction is a complex process that involves the use of inductive and deductive reasoning, probability, statistics, and pattern analysis to analyze the data provided by the physical evidence. Often the team that is attempting to reconstruct the crime scene must solicit input from specially trained experts as part of this process. Reconstruction is used often in criminal cases in which there is no eyewitness evidence or where eyewitness evidence is considered unreliable.

Pattern Evidence

The physical evidence that is left at a crime scene provides the foundation for reconstructing the events that took place and—ideally—the sequence in which they occurred. Although the evidence alone does not describe all the events of the crime, it can be considered a "mute witness" that either supports or contradicts statements given by suspects or eyewitnesses. In addition, physical evidence can generate investigative leads. The collection and analysis of physical evidence form the foundation of crime scene reconstruction.

Explosion Patterns

An analysis of damage patterns at the scene of an explosion can provide investigators with information that will allow them to form a hypothesis about how the detonation took place and then begin to reconstruct the event. Some of the features that can be observed in the damage can be of great value:

- The direction in which the blast traveled
- The location of maximum damage
- Analysis of the debris field

When investigators mark the scene with indicators that show the direction of the blast, they will be able to establish the site of the detonation. Fragments and debris will be blown in all directions from this location, but can eventually be traced back to that location. Given that what is left of that location will hold the largest amount of residue and possibly bomb fragments, investigators will have the best chance of finding chemical residue there that may help identify whether an explosive was used or whether the explosion resulted from a

malfunction in a mechanical device (e.g., gas stove, gas tank, furnace).

If a powerful explosive is used, investigators can estimate the weight of the bomb by determining the size of the crater left behind. The diameter (d) of the crater, measured in meters, is used to determine the weight (w) of the bomb, measured in kilograms, by using the following equation:

$$w \cong \frac{d^3}{16}$$

Firearm Ballistics

Reconstruction of crime scenes involving firearms often is necessary to determine the cause of death—whether homicide, suicide, or accidental death. Reconstruction can also provide information that places the shooter and the victim at precise locations within the crime scene. The reconstruction of the trajectory of the bullet may also prove or disprove the testimony of a witness.

Entry and Exit Hole Geometry

In cases where the bullet passed through an object, the shapes of the entry and exit holes will indicate where the bullet entered the object. If the bullet remained intact, the entry hole is most often smaller than the exit hole. Most bullet holes are elliptically shaped, so trigonometry can be used to estimate the angle of entry. The following equation is used to estimate the angle of entry (θ) from measurements of an elliptical bullet hole:

$$\text{Cos } \theta = \frac{\text{Shorter dimension}}{\text{Longer dimension}}$$

Bullet Trajectory

Crime scene investigators use two methods to determine the trajectory of a bullet after they locate any bullet holes in walls, floors, ceilings, or other objects at the scene. The older method uses physical objects, such as rods and strings, to find trajectory. The rods are inserted into the holes, and string is used to estimate the trajectory. The newer method uses a laser to visualize the bullet's trajectory (FIGURE 2-8). The laser can be mounted on a tripod and its beam shot through a hollow plastic tube (trajectory rod) that is inserted into the bullet hole. The laser finder has a protractor attached to the laser; once the laser has been sighted through the trajectory rod, the angle of entry is then read

FIGURE 2-8 A laser can be used to determine a bullet's trajectory.

from the protractor. The laser device is more expensive but easier to use over longer distances.

Bullet Ricochet

When examining a bullet's trajectory, investigators must take into account the possibility of **ricochet**. Ricochet is the deviation in the flight path of a bullet as a consequence of impact with another object. That is, the bullet hits a hard surface and rebounds to hit someone or something at the scene. If the investigator finds that the victim was shot because of an accidental ricochet rather than an intentional shot aimed at the individual, then the investigation changes to an accident investigation.

Not all projectiles have the same tendency to ricochet. Low-velocity, heavy bullets are more likely to ricochet. By contrast, high-velocity, lightweight bullets tend to expand on contact and are more likely to break up on impact with a hard surface. Careful examination of the scene will indicate ricochets as damage to the walls, floor, and ceiling. Ricochet is more likely in scenes involving concrete or brick walls. Bullets recovered from a victim may have markings that indicated that they struck a very hard surface. Furthermore, particles of paint, plaster, or soil that were picked up at the ricochet impact site may still be found attached to the bullet.

Shell Casings

Shell casings from automatic and semiautomatic weapons found at a crime scene can provide useful information. Shell casing patterns at crime scenes are difficult to interpret, however. Although most semiautomatic weapons eject casings to the right, many factors influence the ejection pattern—for example, how the shooter's body is positioned, how the shooter holds the gun, whether the shooter is stationary or in motion, and how hard the ground is. Once the suspect's weapon has been located, forensic firearms experts can experiment by varying these factors as they try to reconstruct the events involving the gun.

Bloodstain Patterns

In violent crimes, interpretation of bloodstain patterns may provide vital information about what actually happened at a crime scene. Often, bloodstain patterns open a window to the events that occurred during the commission of the crime. The size, shape, and pattern formed by bloodstains found at a crime scene all can be used to reconstruct events.

The bodies of adult males hold approximately 5 to 6 quarts of blood; the bodies of adult females contain 4 to 5 quarts of blood. Thus it is not surprising that in violent crimes, large amounts of blood are often found at the scene. If the crime is committed indoors, the walls, the floors, and even the ceiling may all be spattered with bloodstains. Bloodstain evidence may even be present in rooms other than the one where the crime was committed.

Active Bloodstains

Active bloodstains are caused by blood that travels because of force, rather than because of gravity. Active bloodstains could result from an impact to the victim's body by a weapon, such as a knife, hammer, or bullet. Bloodstains caused by impact usually form a spatter pattern in which numerous small droplets of blood are dispersed. Active bloodstains are produced by pumping pressurized blood onto a surface when an artery is cut and the heart continues to pump. Depending on the nature of the wound, the volume of blood may be large (gushing) or relatively small (spurts). The overall pattern of the projected bloodstains may reflect the oscillation of pressure produced as the heart pumps. An object that is covered with blood, such as a knife, also may produce an active bloodstain: Blood can be thrown from the knife as it is moved or if it is stopped suddenly.

Bloodstains produced in this way are known as cast-off stains.

By observing the shape of a bloodstain, an experienced investigator may be able to determine the direction in which the droplets were traveling when they hit a surface. If the bloodstain is elliptically shaped with a tail, the direction of travel may be easily determined. The tail of the bloodstain points to the direction of travel of the blood droplet (FIGURE 2-9). Conversely, a round spatter pattern indicates that the angle of impact is perpendicular to the object or at 90°. An investigator must be careful with these stains because one exception is possible: If the large drop of blood throws off a smaller droplet on impact, the tail of the smaller "satellite" stain will point toward the "parent" drop (FIGURE 2-10). Once the investigator determines that a small stain is a satellite, however, the information presented by the bloodstain tails will be consistent.

The angle of impact of a bloodstain can be determined from its dimensions. The trigonometric calculations that follow are based on the fact that a drop of blood that is moving through the air will assume a spherical shape. When it hits a surface, the blood will produce a stain that is longer than it is wide (oval or elliptical) if it is falling at any angle other than 90° (FIGURE 2-11).

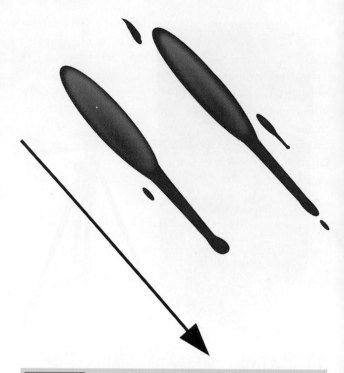

FIGURE 2-9 The tails of these bloodstains indicate their direction of travel.

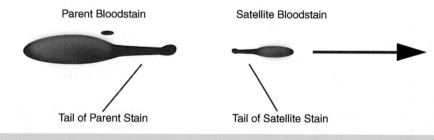

Parent Bloodstain | Satellite Bloodstain

Tail of Parent Stain | Tail of Satellite Stain

FIGURE 2-10 This larger bloodstain created a satellite bloodstain, as can be seen by the satellite's tail.

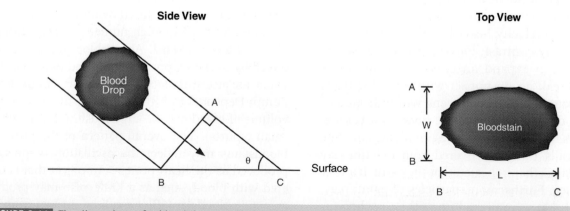

Side View | **Top View**

Blood Drop

A

θ

Surface

B | C

A | W | B | L | B | C

Bloodstain

FIGURE 2-11 The dimensions of a blood drop reveal the angle of impact.

In this example, the width of the stain is the same as the diameter of the drop before impact. The length of the stain, however, is related to not only the diameter of the drop, but also the angle of impact (θ). The width of the drop is equal to AB, which is one side in a right triangle that lies opposite to angle θ. The length of the drop is equal to the hypotenuse (BC) of the triangle. The ratio of the width (short side) to the length (long side) of the bloodstain is equal to the sin of the impact angle:

$$\text{Sin } \theta = \frac{\text{Width}}{\text{Length}}$$

Example

Determine the angle of impact of a bloodstain that is 10 mm wide and 20 mm long.

Solution
Use the equation

$$\text{Sin } \theta = \frac{\text{Width}}{\text{Length}}$$

to determine the ratio.

$$\text{Sin } \theta = \frac{10 \text{ mm}}{20 \text{ mm}} = 0.5$$

Take the arcsine of 0.50 to find the angle.
The arcsine of 0.50 is 30°.
The angle of impact is 30°.

Passive Bloodstains

Passive bloodstains are formed from the force of gravity. They may take the form of drops, a pool, or a blood flow. Passive bloodstains might cover a victim, the area under the victim, and objects near the victim. Examining the passive bloodstain may suggest how much time has passed since the blood was deposited. The drying times of drops and pools can be estimated from experiments carried out in the laboratory on the same surface material and at the same temperature and humidity.

The pattern produced by dropping blood is caused by the surface tension of the blood drop and the high viscosity of blood. Blood has a viscosity (thickness) that is four times greater than the viscosity of water. Vertically dropping blood generally produces a circular pattern. As the distance between the source of blood and the floor increases (i.e., the distance the blood falls increases), the size of the circular pattern increases (FIGURE 2-12). Once the distance the drop must fall exceeds 48 inches, the drop sizes remain the same.

The texture of the surface on which the drop falls also affects its size and shape. Hard, nonporous surfaces produce a circular stain with smooth edges. Softer, porous surfaces produce stains that have scalloped edges.

Transfer Bloodstains

Transfer bloodstains are deposited on surfaces as a result of direct contact with an object that has wet blood on it. Examination of transfer bloodstains may indicate points of contact between suspects and objects present during a crime. Transfer bloodstains also are used to establish the movement of individuals or objects at the crime scene.

Transfer stains are left behind when an object that is covered with wet blood, such as a knife or an assailant's hand, contacts another object, such as a dishcloth used to wipe off the blood. The pattern produced by the transfer might be detailed enough that it will establish the type of object that left it. If a particular pattern (such as a shoeprint) is found repeatedly at the scene, the movement of the object (i.e., the person wearing the shoes) can be determined. After the initial transfer, the amount of blood deposited decreases with each successive step until no more blood is left and the trail disappears. Even when the trail of transfer stains becomes too faint to be seen by the naked eye, however, it may be possible to visualize the print by using a chemical treatment.

Point of Convergence

When multiple bloodstains appear to have originated in the same place, investigators may be able to use trigonometry to determine their **point of convergence**. The point of convergence is the most likely point of origin of the blood that made the stains (FIGURE 2-13).

Example

Suppose that stain 1 is 10 mm long and 5 mm wide, and stain 2 is 8 mm long and 6 mm wide. Using the earlier equation, we can determine that the incident angle is 30° for stain 1 and 48.5° for stain 2. If the blood that caused these stains originated from the same place, we can then identify that origin.

Solution
To do so, we draw a line from the center of each stain to a point where each line intersects. At the point of intersection (C), we draw a line at a right angle to the floor and high enough that it will show the origin of the blood. At this point, we draw two right triangles that have a common point of intersection on the

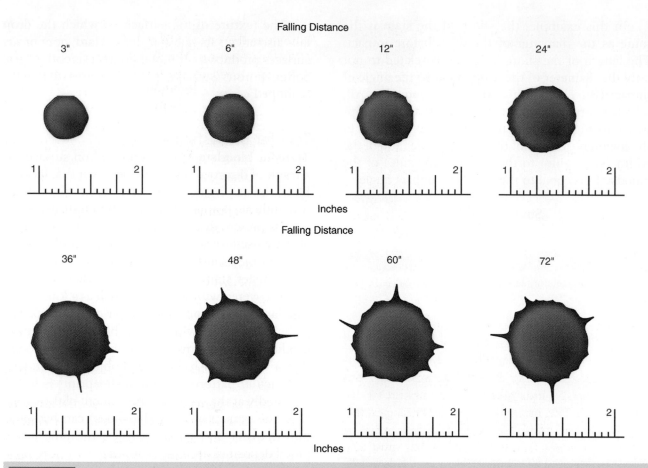

Falling Distance

3"　　　　6"　　　　12"　　　　24"

Inches

Falling Distance

36"　　　　48"　　　　60"　　　　72"

Inches

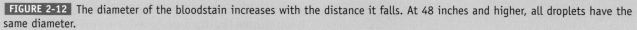

FIGURE 2-12 The diameter of the bloodstain increases with the distance it falls. At 48 inches and higher, all droplets have the same diameter.

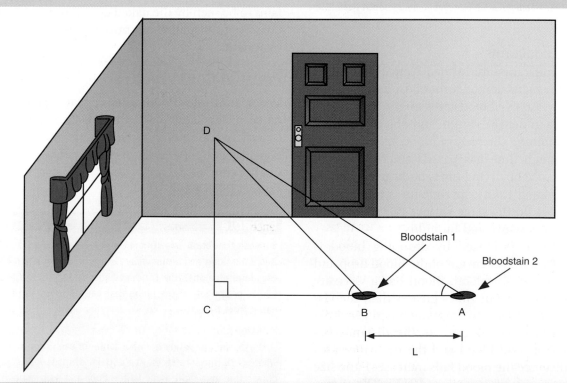

FIGURE 2-13 Investigators may use bloodstains found at a crime scene to calculate the position from which the blood originated, known as the point of origin (D).

line just drawn (D) and that share a common line (CD). We can use either of these triangles to calculate the length of CD and the height of point D, the origin of the blood. Suppose we take triangle ACD, where the side of the triangle (AC), which is adjacent to the point of impact, is 0.75 yd from point C. The length of CD is calculated by the following equation:

$$(\tan 30°)(0.75 \text{ yd}) = 0.433 \text{ yd}$$

This calculation tells us that the blood originated 0.433 yd from the floor at point D in the diagram. Of course, this assumes that both stains originated at the same moment and were not deposited at two different times. Such complications illustrate the complexity and the limitations of blood pattern analysis.

Physical Evidence in Court

Physical evidence has value in court proceedings only when the forensic scientist who testifies about it understands—and can explain to a jury—how the evidence was analyzed and how the results of this analysis may be interpreted in the context of the crime scene. The procedures and technologies that are used in the crime laboratory not only must be based on sound scientific principles, but must also satisfy the criteria of admissibility previously established by the courts. Consequently, the forensic expert needs to understand the judicial system and the standards of admissibility for scientific expertise, which may vary from state to state and sometimes even from court to court within states.

During the early part of the twentieth century, the guidelines for determining the admissibility of scientific information were governed by what is known as the Frye standard (*Frye v. United States*, 1923). At that time the judge decided whether the techniques used by the forensic scientist to examine evidence could be admitted to court as evidence. Such a "general acceptance" test required the scientific test to be generally accepted by the scientific community. Today, courts expect experts to present scientific papers that have been reviewed by other respected scientists and books that have been written about the test procedures. In addition, if other courts have admitted the procedure in prior judicial proceedings, then that precedent is taken into consideration. As a specific test has become more commonly used in courts, some jurisdictions have issued judicial notices that the test is generally accepted and that experts do not have to prove the test valid every time it is used in an individual court. More recently, the Frye standard has generated debate about its inability to deal with new and innovative scientific tests that may not be generally accepted.

In federal courts, the FRE govern the admissibility of all evidence, including expert testimony. FRE rule 401, for instance, allows anything that materially assists the finding of fact (by the jury) and is deemed relevant by the law (by the judge). FRE rule 702 regulates the admissibility of testimony by experts:

If scientific, technical, or other specialized knowledge will assist the trier of fact to understand the evidence or to determine a fact in issue, a witness qualified as an expert by knowledge, skill, experience, training, or education may

SEE YOU IN COURT

Defense attorneys often question whether critical evidence was not collected because the crime scene technicians favored some evidence over other evidence because of their operating theory of the crime. Of course, the task of selecting which prosecutorial evidence to present in court actually belongs to investigators working with prosecutors; this decision is not made by crime scene technicians. Crime scene and laboratory technicians should know as little as possible about the testimonial evidence. As a result, crime scene technicians may collect a great deal of physical material that in the end proves worthless. That is part of the job.

At the time a crime scene is examined, some evidence will be obvious; other evidence will not. Scene technicians must be able to indicate with certainty that nothing that could have been analyzed remains uncollected. Similarly, lab technicians must be able to justify the testing protocols employed and the failure to perform other tests. Where samples are too small to preserve any material for a defense testing lab, justifications must be clear why that is so. Technicians must be familiar with the use of their testimony, evidence, and findings in a courtroom and the limits of that use.

testify thereto in the form of an opinion or otherwise, if (1) the testimony is based upon sufficient facts of data, (2) the testimony is the product of reliable principles and methods, and (3) the witness has applied the principles and methods reliably to the facts of the case.

Coppolino v. State (1968) reinforced the wide discretion and flexibility that a judge has when deciding admissibility of scientific evidence. In this case, the medical examiner testified that the victim died of an overdose of succinylcholine chloride, a drug prescribed as a muscle relaxant. A toxicology report had found an abnormally high concentration of succinic acid (a by-product of the body's metabolic breakdown of succinylcholine chloride) in the victim's body, and the medical examiner relied on these results in determining the cause of death. The defense argued that the toxicology data were gathered by a new test that had not gained wide acceptance in the scientific community. The court ruled in favor of the prosecution, however, and admitted the evidence. In its decision, the court noted that scientific progress is inevitable—new tests are constantly evolving to solve forensics-related problems. This ruling led to the Coppolino standard, under which a court is allowed to admit a novel test or a controversial scientific theory on a particular issue if an adequate foundation proving its validity can be laid even if the scientific community as a whole isn't familiar with it.

A landmark ruling in the early 1990s led to the Daubert standard (*Daubert v. Merrell Dow Pharmaceuticals, Inc.*, 1993), the test applied today in federal courts to determine admissibility of scientific evidence. The Daubert test requires special pretrial hearings for scientific evidence and special procedures on discovery. This rather strict test requires the trial judge to assume the responsibility of "gatekeeper" in ruling on the admissibility of scientific evidence presented in his or her court. Judges can use the following guidelines to help them gauge the validity of the scientific evidence:

1. Has the scientific technique been tested before?

2. Has the technique or theory been subject to peer review and publication?

3. What is the technique's potential rate of error?

4. Do standards exist that can verify the technique's results?

5. Has the technique or theory gained widespread acceptance within the scientific community?

Although the Daubert standard was considered overly restrictive when it was first proposed, it was affirmed in 1999 (*Kumho Tire Co. Ltd. v. Carmichael*). In this case, the court ruled that the trial judge should play the "gatekeeper" role for expert testimony as well as scientific evidence:

We conclude that Daubert's general holding—setting forth the trial judge's general "housekeeping" obligation—applied not only to testimony based on "scientific" knowledge, but also to testimony based on "technical" and "other specialized" knowledge. . . . We also conclude that a trial court may consider one or more of the more specific factors that Daubert mentioned when doing so will help determine that testimony's reliability. But, as the court stated in Daubert, the test of reliability is "flexible" and Daubert's list of specific factors neither necessarily nor exclusively applies to all experts in every case.

Expert Testimony

Factual evidence often is given by a lay witness who can present facts and observations. That is, the layperson's testimony should be factual and should never contain personal opinions. The expert witness, by comparison, must evaluate the evidence and provide an analysis that goes beyond the expertise of the layperson. The expert's opinion must rest on a reasonable scientific certainty that is based on his or her training and experience. That opinion may be attacked by the opposing counsel, so the expert needs to be prepared to defend the conclusions reached. At the same time, the expert should be willing to concede limitations to the tests that are being presented. The forensic scientist should present a truthful, persuasive opinion and not serve as an advocate for one side or the other.

Because a forensic expert's results may be an important factor in determining the guilt or innocence of a suspect, experts are often required to present to the court the results of their tests and their conclusions. Judges are given the responsibility of accepting a particular individual as an expert witness. The expert must establish that he or she possesses particular knowledge or skill or has experience in a trade or profession that will help the jury in determining the truth of the issues presented.

The court usually limits the subject area in which the expert is allowed to present testimony. Before testifying, the prosecution qualifies the expert by presenting his or her education, experience, and prior testimony to the court in a procedure known as *voir dire* (French, but originally from the Latin *verum dicere*, meaning "to speak the truth"). This interrogation of the expert may be quite intense, because acceptance of an unqualified expert (or, conversely, exclusion of a qualified expert) is considered grounds for overturning the verdict in a higher court.

During the *voir dire* process, counsel often cites information that emphasizes the expert's ability and proficiency relative to the information to be presented, such as degrees, licenses, awards, and membership in professional organizations. If the expert teaches, has authored books, or has written scientific papers on the scientific issues, this information also helps to establish his or her competency. Generally, courts rely on the training and experience of experts to assess their knowledge and experience. In addition, judges are more likely to qualify an expert if the expert has been qualified previously by other courts.

Also during the *voir dire* process, the opposing attorney is given the opportunity to cross-examine the expert and to highlight any weakness in his or her education or experience. The question of precisely which credentials are suitable for expert qualification is subjective and is an issue that courts tend to avoid. Although the court may not necessarily disqualify an expert because of a perceived weakness in education or experience, many attorneys nevertheless want the jury to be aware of the expert's weaknesses.

Once an expert is deemed qualified, the judge will require the expert's testimony to stay within the limits previously set. Just how much weight a judge or jury will place on an expert's testimony depends on the information presented by the expert. Although experience and education are important factors, the expert's ability to explain complicated scientific data in clear, down-to-earth language may be even more important in establishing his or her credibility. The prosecution and defense counsel will try to lead the jury to accept their opposing theories of what happened, but it is up to the expert to present evidence and results to the jury in a clear and objective manner.

During cross-examination, the opposing counsel often challenges the accuracy and interpretation of test results obtained during the forensic examination of the evidence. When challenging the accuracy of the test results, the defense may raise the specter of cross-contamination during collection, transportation, storage, and testing of the evidence. The likelihood of cross-contamination of evidence has decreased significantly in recent years as the training for crime scene investigators has improved and clear chain of custody documentation for all physical evidence has been established.

The forensic scientist's interpretation of the scientific test results is likely to be the most significant area of disagreement during cross-examination. The forensic scientist forms an opinion based on what he or she considers the most probable explanation of the results. During his or her testimony, this individual must be careful to express an expert opinion that has a solid foundation based on test results.

YOU ARE THE FORENSIC SCIENTIST SUMMARY

1. Fibers from the chair were found on the suspect's jacket, and fibers from his jacket were found on the chair. This evidence strongly supports the hypothesis that the suspect sat in the chair wearing his jacket. At the same time, fibers from the suspect's sweater were found on the chair, but no chair fibers were found on his sweater. This evidence suggests that the fibers from his sweater were transferred to the chair by a secondary transfer. The suspect got sweater fibers on his jacket earlier, and, when he sat in the chair, the sweater fibers on the jacket were transferred to the chair.

2. No, the fiber evidence indicates that the suspect sat in the chair. The evidence doesn't indicate how many times the suspect sat in the chair.

Chapter Spotlight

- Four types of evidence are distinguished: testimony, physical, documentary, and demonstrative evidence.

- There are approximately 350 crime labs in the United States. The FBI and the ATF laboratories are the two primary federal crime labs.

- Crime labs typically have six different divisions, dealing with the following areas: biological/serological analysis; chemistry/toxicology; trace evidence or microscopy; ballistics, firearms, and tool marks; latent fingerprints; and questioned documents.

- A variety of professional boards and societies are employed to accredit laboratories in regard to specialized kinds of evidence. Accreditation generally requires the establishment in the lab of quality control and quality assurance (QA) manuals, testing protocols, and proficiency testing procedures.

- The forensic scientist performs seven major steps as part of evidence analysis: recognition, documentation, collection and preservation, analysis, interpretation, reporting, and presentation of testimony.

- A forensic scientist's examination of the evidence can serve many purposes, ranging from determining the MO employed to linking a suspect to the crime scene. Nevertheless, the overall goal is to establish the unique status of the evidence, crime, and suspect.

- The conditions of the crime scene can affect the evidence available for collection.

- Evidence collected at a crime scene can be used to recreate or reconstruct the scene and determine how the crime occurred.

- A forensic scientist must be familiar with *Changes in the Standards for Admitting Expert Evidence in Federal Civil Cases Since the Daubert Decision*, and the use of evidence in a courtroom.

- The Daubert test requires special pretrial hearings for scientific evidence and special procedures on discovery.

- Expert witnesses are usually challenged on one of two bases: accuracy of testing procedures and interpretation of the evidence.

Key Terms

Active bloodstain A bloodstain caused by blood that traveled by application of force, not gravity.

Associative evidence Evidence that associates individuals with a crime scene.

Class characteristic A feature that is common to a group of items.

Corpus delicti The "body of the crime."

Evidence Information about a crime that meets the state or federal rules of evidence.

Exemplar Representative (standard) item to which evidence can be compared.

Explainable differences A sound scientific explanation for any differences between the evidence and the reference material.

Federal Rules of Evidence (FRE) Rules that govern the admissibility of all evidence, including expert testimony.

Identification The process of matching a set of qualities or characteristics that uniquely identifies an object.

Individual characteristic A feature that is unique to one specific item.

Locard's exchange principle Whenever two objects come into contact with each other, there is an exchange of materials between them.

Modus operandi The "method of operation," also known as the MO.

Passive bloodstain A pattern of blood formed by the force of gravity.

Point of convergence The most likely point of origin of the blood that produced the bloodstains.

Ricochet Deviation of a bullet's trajectory because of collision with another object.

Tool mark Any impression, cut, scratch, or abrasion caused when a tool comes in contact with another surface.

Transfer bloodstain A bloodstain deposited on a surface as a result of direct contact with an object with wet blood on it.

Putting It All Together

Fill in the Blank

1. Whether an object is considered evidence is determined by its _____ to the crime.

2. In court, it is the responsibility of the _____ to determine what is relevant.

3. Investigators must move quickly to identify and protect evidence before _____ effects begin to alter its appearance.

4. _____ (Biological/chemical/physical) evidence is most susceptible to change.

5. Another name for physical evidence is the _____ sample.

6. Another name for the known sample is the _____.

7. If a piece of paper that is evidence is exposed to the sun and changes color, a forensic expert may match it to a known sample of paper that is a lighter color. The court must then be told about these _____ (unexplainable/explainable) differences.

8. _____ is the determination of the physical and chemical identity of a substance with as much certainty as existing analytical techniques will permit.

9. Before a test method can be adopted as a routine test in a crime laboratory, it must be shown to give reproducible, accurate results for a(n) _____ sample.

10. The FBI has established _____ that developed testing protocols and standard tests.

11. When an expert has drawn a conclusion "to a reasonable degree of scientific certainty," it means the conclusion has been substantiated beyond any _____.

12. Size, shape, color, style, and pattern are all considered _____ characteristics.

13. When an object is classified as belonging to a certain class, it is _____ from other classes.

14. Evidence that can be associated with a single source with extremely high probability possesses _____ characteristics.

15. A(n) _____ is any impression, cut, scratch, or abrasion caused when a tool comes in contact with another surface.

16. Tool marks are encountered most often in cases of _____.

17. _____ is the first step in the recovery of tool mark evidence.

18. Impressions can be lifted from rough surfaces by the use of _____.

19. Impressions left in snow can be cast by spraying _____ into the impression before adding casting materials.

20. In reconstructing a bombing, investigators mark the scene with indicators to show the direction of the blast so that they can establish the _____.

21. The weight of explosives used in a bombing can be estimated by the _____.

22. Two methods are commonly used to determine bullet trajectory; the older method uses rods and strings, whereas the newer method uses _____.

23. A(n) _____ is the deviation in the flight path of a bullet as a result of its impact with another object.

24. Low-velocity, heavy bullets are _____ (more/less) likely to ricochet.

25. A(n) _____ bloodstain is formed when blood travels because of force, not gravity.

26. A(n) _____ bloodstain is formed from the force of gravity.

27. Drops of blood falling on hard, nonporous surfaces will have a(n) _____ shape.

28. Drops of blood falling on soft, porous surfaces have _____ edges.

29. The "general acceptance" test that requires the scientific test to be generally accepted before it can be admitted as evidence in court is known as the _____ standard.

30. The ruling that allows the court to admit novel or new tests is the _____ standard.

31. The _____ standard requires that the trial judge act as a gatekeeper for the admission of scientific evidence.

32. Before the judge qualifies an expert witness, the opposing attorney is given the opportunity to _____ the expert.

True or False

1. The forensic examiner should try to fit the suspected tool into the tool mark at the crime scene.

2. The tail of an elliptically shaped bloodstain points to the direction of travel of an active bloodstain.

3. The angle of impact of an active bloodstain can be determined by its dimensions.

4. Transfer bloodstains can be used to establish the movement of individuals at the crime scene.

5. The court never limits the subject area in which an expert witness is allowed to present testimony.

6. Testimony by an expert witness will include that person's opinion about the results of the tests that he or she has observed.

Review Problems

1. A crater left in concrete after a bombing in a parking garage measures 2 m. Estimate the weight of the bomb in kilograms.

2. A crater from a bomb that was left on the sidewalk outside a business measures 1.5 yd. Estimate the weight of the bomb in pounds.

3. A bullet hole in a door at a crime scene indicates that the bullet was shot from inside the room, then passed through the door and into the hall. The bullet hole is elliptically shaped.

The shorter dimension is 11 mm, and the longer dimension is 15 mm. What is the angle of entry?

4. A bullet hole in a wall at a crime scene indicates that the bullet was shot from outside the room, then passed through the wall and into the room. The bullet hole is elliptically shaped. The longer dimension is 20 mm, and the shorter dimension is 14 mm. What is the angle of entry?

5. If a bloodstain is 5 mm long and 3 mm wide, what is its angle of impact (θ)?

6. If a bloodstain is 10 mm long and 3 mm wide, what is its angle of impact (θ)?

7. A circular drop of blood at a crime scene was measured to have a diameter of 0.8 in. What distance did it fall? What was the angle of impact?

8. A circular drop of blood at a crime scene was measured to have a diameter of 0.5 in. What distance did it fall? What was the angle of impact?

9. If bloodstain 1 in Figure 2-13 is 5 mm long and 3 mm wide, and distance BC is 3 ft 6 in., how far off the floor did the blood originate?

Further Reading

Becker, Ronald. *Crime Scene Investigation*, ed 2. Sudbury, MA: Jones and Bartlett, 2005.

Bevel, T., and Gardner, R. M. *Bloodstain Pattern Analysis with an Introduction to Crime Scene Reconstruction*, ed 2. Boca Raton, FL: CRC Press, 2002.

Bodziak, W. J. *Footwear Impression Evidence*, ed 3. Boca Raton, FL: CRC Press, 2008.

DuPasquier, E., et al. Evaluation and comparison of casting materials in forensic sciences: application to tool marks and foot/shoe impressions. *Forensic Science International.* 1996; 82, 33.

Fisher, B. *Techniques of Crime Scene Investigation*, ed 7. Boca Raton, FL: CRC Press, 2003.

Handbook of Forensic Services. Washington, DC: U.S. Department of Justice, http://www.fbi.gov/hq/lab/handbook/forensics.pdf.

Reference Manual on Scientific Evidence, ed 2. Washington, DC: Federal Judicial Center, 2000.

Scientific Working Group on Bloodstain Pattern Analysis (SWGSTAIN). Recommended terminology. *Forensic Science Communications.* 2009; 11(2). http://www.fbi.gov/hq/lab/fsc/current/standards/2009_04_standards01.htm.

Vince, J. J., Jr., and Sherlock, W. E. *Evidence Collection.* Sudbury, MA: Jones and Bartlett, 2005.

Wonder, Y. A. *Bloodstain Pattern Evidence-Objective Approaches and Case Applications.* New York: Elsevier, 2006.

http://criminaljustice.jblearning.com/criminalistics

Answers to Review Problems

Interactive Questions

Key Term Explorer

Web Links

Trace Evidence

SECTION

2

CHAPTER 3

Physical Properties: Forensic Characterization of Soil

CHAPTER 4

The Microscope and Forensic Identification of Hair, Fibers, and Paint

CHAPTER 5

Forensic Analysis of Glass

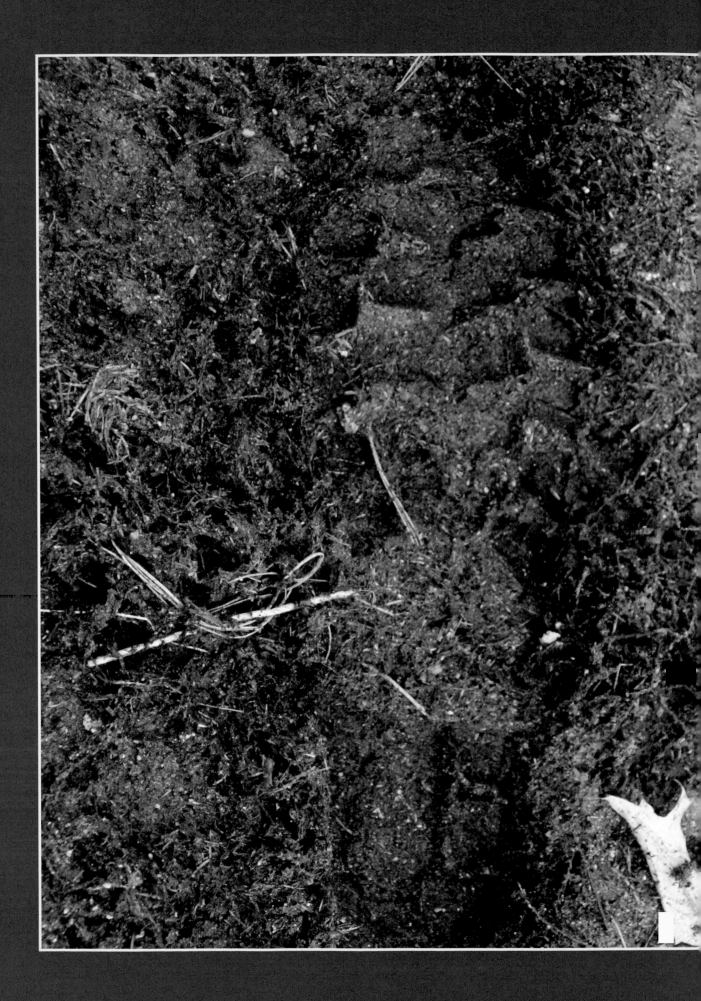

Physical Properties: Forensic Characterization of Soil

OBJECTIVES

In this chapter you should gain an understanding of:

- The difference between physical and chemical properties
- Conversions between the English system of measurements to the metric system
- The forensic characteristics of soil
- Ways to collect and preserve soil evidence

FEATURES

On the Crime Scene

Back at the Crime Lab

See You in Court

WRAP UP

Chapter Spotlight

Key Terms

Putting It All Together

Soil provides useful evidence because it is either sticky or **friable**. A friable object is easily crushed and made into a powder. Whether sticky or friable, however, soil is easily transferred as trace evidence from the crime scene onto objects and people present at the scene. This trace evidence then may be transferred again by secondary transfer onto carpets or into vehicles.

Soil also has properties similar to those of soft, plastic objects such as Play-Doh; in particular, it can retain an impression that has been pressed into it. This property is especially useful in providing evidence of association. For example, a worn boot might leave an impression at the crime scene, and the footwear later can be found to be owned by the suspect.

1. A suspect in a burglary is found to be wearing brand-new size 12 Wolverine boots. Outside the broken window, in the garden of the home that was robbed, is a boot impression is found that matches one made by a size 12 Wolverine boot. Does the evidence prove that this suspect was the burglar?
2. The burglar knocked over a flowerpot in the home, which fell to the ground and broke. A soil sample found in the suspect's car contains the same vermiculite material (a common component of potting soil) that was found in the soil of the flowerpot. Is this class or individual evidence?

Introduction

When common objects are found at crime scenes, forensic laboratories first examine those items in an attempt to identify properties that may serve to place the object within a particular class. The search for distinguishing properties then continues, as the forensic scientist tries to discover additional characteristics that will allow the object to be individualized. This chapter describes the properties that are most useful for characterizing soil, glass, fibers, and other physical evidence. It also discusses how to collect evidence and measure its distinguishing properties.

Physical and Chemical Properties

The distinguishing characteristics that are used to identify different objects are called properties. A person can easily recognize his or her own car among hundreds of other cars in a parking lot by characteristics such as make, model, color, dents in the body, and belongings left on the back seat. In much the same way, we can recognize different substances by their characteristics or properties. Properties of substances can be grouped into two main categories: physical and chemical.

The **physical properties** of a substance are those properties that can be observed and recorded without referring to any other substance. Color, odor, taste, hardness, density, solubility, melting point (the temperature at which a substance melts), and boiling point (the temperature at which a substance boils or vaporizes) are all physical properties. For example, if you were to describe the physical properties of pure copper, you might report that it is a bright, shiny metal; is malleable (can be beaten into thin sheets) and ductile (can be drawn into fine wire); and melts at 1083°C (1981°F) and boils at 2567°C (4653°F). No matter what its source, pure copper always has the same properties. These **intensive physical properties** are independent of amount of substance measured and depend only on the identity of the substance. For example, to melt a substance (e.g., when ice turns into water), heat is applied to change it from a solid to a liquid, but this change in state does not change the composition of the substance. Thus the appearance of the substance may change but there is no change in composition. By contrast, **extensive physical properties**—such as mass, volume, and length—depend on the amount of substance present.

The **chemical properties** of a substance are those properties that can be observed when the substance reacts or combines with another substance to change its chemical composition. For example,

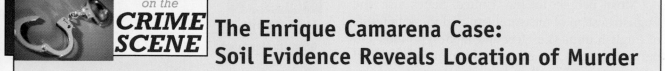

CRIME SCENE

The Enrique Camarena Case: Soil Evidence Reveals Location of Murder

In 1985, the bodies of Enrique "Kiki" Camarena Salazar, a U.S. Drug Enforcement Agency (DEA) agent, and Alfredo Zavala Aguilar, the Mexican reconnaissance pilot with whom Salazar worked, were found in the Mexican state of Michoacan. Camarena was an undercover agent assigned to clandestine surveillance of Mexican drug lords. The Mexican Judicial Police reported that both men died in a shootout between police and the drug dealers at El Mareno ranch in Michoacan. Close examination of the bodies, however, revealed that they had been tortured and that Camarena's head had been crushed. Furthermore, the Mexican police showed little or no interest in finding the perpetrators of this crime—unusual behavior for a crime in which a law enforcement agent had died.

DEA officials sent evidence taken from Camarena's body to the FBI Crime Laboratories for further testing to conclusively determine what happened to the two men. FBI Special Agent Ron Rawalt, a forensic geologist, requested soil samples from both Camarena's body and the location the body was found. Soil had mixed with body fluids at wound sites throughout the slain agent's body, a process that takes some time. This rocky soil contained tan and brown volcanic ash and fragments of rhyolite, a fine-grained igneous rock that is rich in silica. By contrast, soil samples taken from the ranch in Michoacan contained a coarse, greenish-black basaltic glass but no rhyolite.

Further analysis of the material found in the body's wounds led FBI geologists, with the help of Smithsonian Institute scientists, to determine that the material was consistent with that found in the El Tequilla ash flow in the Guadalajara Basin, nearly 100 km north of the site in Michoacan. Investigators now knew that the agent was not murdered in Michoacan—but they needed still more information to solve the case.

Extensive field studies narrowed the location of the original grave site in the Guadalajara Basin. The investigators also knew that grass had burned nearby because charred material was present in Camarena's body. Furthermore, they needed to find an area where the ash flow carrying the rhyolite was at least 6 to 8 ft. deep. Using these criteria, a location was determined and cadaver dogs were used to find the exact grave site.

The evidence showing that the body was originally buried in Guadalajara allowed the investigators to solve the mystery. The Mexican police, facing mounting pressure from the U.S. government to find Camarena (before his body had been discovered), had conspired with the drug lords to move the FBI agent's body from Guadalajara (where the crime occurred) to Michoacan. The Mexican police subsequently raided the farm in Michoacan, where they killed the occupants, placed the dead agent's remains in a prominent position, and declared that they had found Camarena's body.

when gasoline burns, it undergoes a chemical change: It combines with oxygen in the air to form carbon dioxide and water. Likewise, copper turns green when it is exposed to the atmosphere for a long time; the surface reacts with moisture, oxygen, and pollutants in the atmosphere to form a new substance. If a substance does not react with another substance, its lack of reactivity also qualifies as a chemical property. For example, one chemical property of gold is that, unlike copper, it does not tarnish or change when exposed to the atmosphere.

Whether a forensic scientist measures a physical property or a chemical property of an object that is evidence, that measurement must conform to a standard system of measurements that is accepted throughout the world.

The Metric System

In forensics and all other scientific fields, it is very important to take accurate measurements. For example, medical treatment relies on accurate measurements of factors such as temperature, blood pressure, and blood sugar concentration, just as baking requires careful measurement of flour, water, and sugar combined with the regulation of baking time and temperature.

All measurements are made relative to some **reference standard**. For example, if you measure your height using a ruler marked in meters, you are comparing your height to an internationally recognized reference standard of length called the meter.

Most people in the United States use the English system of measurement and think in terms of English units (i.e., feet and inches, pounds and ounces, gallons and quarts). If a man is described as being 6 ft 4 in. tall and weighing 300 lb, we immediately think of a large man. Similarly, we know how much to expect if we buy a half-gallon of milk at the grocery store. Most Americans, however, have a far less clear idea of the meaning or size of a meter, liter, or kilogram. These common metric units of measurement are used by the scientific community and in most countries other than the United States. Although the United States is committed to changing to the metric system, the pace of change, so far, has been extremely slow.

The International System of Units

The **International System of Units (SI)** was adopted by the International Bureau of Weights and Measures in 1960. The SI is an updated version of the metric system that was developed in France in the 1790s, following the French Revolution.

The standard unit of length in the SI is the meter. Originally, the meter was defined as one ten-millionth of the distance from the North Pole to the equator measured along a meridian. Not surprisingly, this distance proved difficult to measure accurately. Thus, for many years, the meter was defined as the distance between two lines etched on a platinum iridium bar that was kept at the International Bureau of Weights and Measures in Sevres, France. Today, the meter is defined even more precisely, as being equal to 1,650,763.73 times the wavelength of the orange-red spectrograph line of the krypton isotope, $^{86}_{36}$Kr.

The standard unit of mass in the SI is the kilogram. It is defined as the mass of a platinum iridium alloy bar that, like the original meter standard, is kept at the International Bureau of Weights and Measures.

SI Base Units

There are seven base units of measurement in the SI system (TABLE 3-1). Because these base units are often inconveniently large (or small) for many measurements, smaller (or larger) units—as defined by the use of prefixes—often are used instead. TABLE 3-2 lists the prefixes used in the SI, which includes the measurements that are most commonly used in forensic science.

TABLE 3-1

SI Base Units

Quantity Measured	Name of Unit	SI Symbol
Length	meter	m
Mass	kilogram	kg
Time	second	s
Electric current	ampere	A
Thermodynamic temperature	kelvin	K
Amount of substance	mole	mol
Luminous intensity	candela	cd

TABLE 3-2

SI Prefixes

Prefix	Symbol	Exponential Form	Decimal Form
mega	m	10^6	1,000,000
kilo	k	10^3	1,000
hecto	h	10^2	100
deka	da	10^1	10
deci	d	10^{-1}	0.1
centi	c	10^{-2}	0.01
milli	m	10^{-3}	0.001
micro	μ	10^{-6}	0.000001
nano	n	10^{-9}	0.000000001
pico	p	10^{-12}	0.000000000001

In addition to the seven SI base units, many other units are needed to describe other physical quantities. All of these units are derived from the seven base SI units. For example, volume is measured in cubic meters (m^3), and area is measured in square meters (m^2). TABLE 3-3 lists commonly used SI derived units.

Because the SI is based on the decimal system, conversions within it are much easier than conver-

TABLE 3-3

SI Derived Units

Physical Quantity	Unit	SI Symbol
Area	square meter	m^2
Volume	cubic meter	m^3
Density	kilogram per cubic meter	kg/m^3
Force	newton	N
Pressure	pascal	Pa
Energy or quantity of heat	joule	J
Quantity of electricity	coulomb	C
Power	watt	W
Electric potential difference	volt	V

sions within the English system. Units within the SI always differ by factors of 10, so conversions from one SI unit to another are made by moving the decimal point the appropriate number of places.

Example

A search of a crime scene turns up 0.0583 kg of cocaine. How many grams of cocaine have been found?

Solution

1. Use Table 3-2 to find the relationship between kilograms and grams: 1 kg = 1000 g.
2. From this relationship, determine the conversion factor by which the given quantity (0.0583 kg) must be multiplied to obtain the answer in the required unit (g).

$$\text{conversion factor} = \frac{1000 \text{ g}}{1 \text{ kg}}$$

3. Multiply 0.0583 kg by the conversion factor to obtain the answer.

$$0.0583 \text{ kg} \left(\frac{1000 \text{ g}}{1 \text{ kg}} \right) = 58.3 \text{ g of cocaine}$$

When the multiplication is complete, the kilograms cancel out of the equation.

Example

A bullet hole is found in a wall 4236 mm from the floor. How many centimeters is the bullet hole from the floor?

Solution

1. 1 cm = 10 mm

$$\text{conversion factor} = \frac{1 \text{ cm}}{10 \text{ min}}$$

2. Multiply 4236 mm by the conversion factor to obtain the answer.

$$4236 \text{ mm} \left(\frac{1 \text{ cm}}{10 \text{ mm}} \right) = 423.6 \text{ cm}$$

Conversion from the SI to the English System (and Vice Versa)

TABLE 3-4 lists units commonly used in the English system for length, mass, and volume. **TABLE 3-5** gives the relationships that must be used to convert from SI units to English units and vice versa.

Example

A bloodstain is found 25 in. from the victim's body. How many centimeters is the body from the bloodstain?

Solution

1. Use Table 3-5 to find the relationship between inches (in.) and centimeters (cm).

$$1 \text{ in.} = 2.54 \text{ cm}$$

2. The answer needs to be expressed in centimeters. Therefore the conversion factor is:

$$\frac{2.54 \text{ cm}}{1 \text{ in.}} = \text{conversion factor}$$

3. Calculate the answer.

$$25. \text{ in.} \left(\frac{0.454 \text{ kg}}{1 \text{ lb}} \right) = 27.24 \text{ kg of marijuana}$$

Example

The Border Patrol finds 60 lb of marijuana in the trunk of a car. How many kilograms of marijuana have been found?

Solution

1. Use Table 3-5 to find the relationship between pounds and kilograms.

$$1 \text{ lb} = 0.454 \text{ kg}$$

2. The answer needs to be expressed in kilograms. Therefore the conversion factor is:

$$\frac{0.454 \text{ kg}}{1 \text{ lb}} = \text{conversion factor}$$

3. Calculate the answer.

$$60 \text{ lb} \left(\frac{0.454 \text{ kg}}{1 \text{ lb}} \right) = 27.24 \text{ kg of marijuana}$$

TABLE 3-4

Units of Measurement in the English System

Length	Mass	Volume
12 inches (in.) = 1 foot (ft)	16 ounces (oz) = 1 pound (lb)	16 fluid ounces = 1 pint (pt)
3 feet = 1 yard (yd)	2,000 pounds = 1 ton	2 pints = 1 quart (qt)
1760 yards = 1 mile (mi)		4 quarts = 1 gallon (gal)

TABLE 3-5

Conversion Factors for Common SI and English Units

Length	Mass	Volume
1 inch (in.) = 2.54 centimeters (cm)	1 ounce (oz) = 28.4 grams (g)	1 fluid ounce (fl oz) = 29.6 milliliters (mL)
1 yard (yd) = 0.914 meter (m)	1 pound (lb) = 454 grams (g)	1 U.S. pint (pt) = 0.473 liter (L)
1 mile (mi) = 1.61 kilometers (km)	1 pound (lb) = 0.454 kilogram (kg)	1 U.S. quart (qt) = 0.946 liter (L)

Mass and Weight

Matter is the stuff that makes up all things. Matter occupies space and has **mass**. Iron, sand, air, water, and people, for example, all have mass and occupy space. Mass is a measure of the quantity of matter that an object contains. The mass of an object does not vary with location. Thus a person has the same mass on Earth and on the Moon. **Weight**, by contrast, measures the force exerted on an object by the pull of gravity. On Earth, it measures the force of attraction between our planet and the object being weighed. On the Moon, which has one sixth the gravity of Earth, an astronaut weighs only one sixth as much as he or she does on Earth. Thus weight varies with location.

The mass of an object is determined by weighing it on a balance. In the older two-pan balance, a balance beam is suspended across a knife edge and identical pans are hung from each side (**FIGURE 3-1A**). The object to be weighed is placed on the left pan, and a series of standard weights is placed on the right pan. When the mass of the standard weights in the right pan is identical to the mass of the object in the left pan (i.e., they balance), the indicator (located between the two pans) points to the center mark.

A modern electronic balance has only one pan and does not routinely use reference weights (**FIGURE 3-1B**). A typical electronic analytical balance has the capacity to weigh a 100-g to 200-g object with a sensitivity of 0.01 mg. When an object is placed on the pan, it pushes the pan down with a force equal to its mass times the acceleration of gravity (m × g). The electronic balance then uses an electromagnet to return the pan to its original position. The electric current required to generate this amount of force is proportional to the mass, which is displayed on a digital readout.

Temperature

For certain measurements, including those involving temperature and energy, forensic scientists continue to use units that are not SI units. The SI unit for temperature is the kelvin (K), but forensic scientists use the **Celsius scale** for many temperature measurements. On this scale, the unit of temperature is the degree Celsius (°C); this scale was devised so as to set the freezing point of water at 0°C and its boiling point at 100°C, both at atmospheric pressure. The 100 degrees between these two reference points are of equal size.

(A)

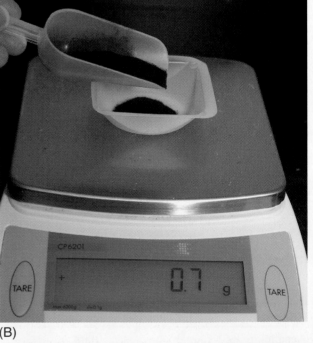

(B)

FIGURE 3-1 (A) A traditional two-pan balance. The object to be weighed is placed on the left pan, and standard weights are placed on the right pan until the two pans balance. (B) A modern electronic laboratory balance has only one pan and a direct digital readout.

In the United States, most temperatures, including those given in weather reports and cooking recipes, are measured in Fahrenheit. On this scale, water's freezing point is 32°F, and its boiling point is 212°F. Thus there are 180 degrees between the freezing and boiling points. Given that the Celsius temperature scale has only 100 degrees between these same two points, 5 degrees on the Celsius scale is equivalent to 9 degrees on the **Fahrenheit scale**.

To convert from Celsius to Fahrenheit, use the following equation:

$$°F = \frac{9}{5}(°C) + 32 \text{ or } °F = 1.8(°C) + 32$$

To convert from Fahrenheit to Celsius, use the following equation:

$$°C = \frac{5}{9}(°F - 32) \text{ or } °C = 0.56(°F - 32)$$

The measurement of temperature is one of the physical properties commonly recorded at crime scenes and in forensic labs. Indeed, one simple way to determine the identity of an unknown liquid is to measure its boiling point. In the past, temperature was measured with a glass thermometer that contained either mercury or alcohol colored with a red dye. When the thermometer came into contact with a hotter object, heat was transferred to the thermometer and the mercury (or alcohol) expanded and moved up the glass column. If the object was cooler, the mercury (or alcohol) contracted and moved down the glass column. Because mercury is poisonous and glass breaks easily, this type of thermometer is rarely used today.

Currently, temperature is measured with either electronic or optical thermometers. The electronic thermometer uses a thermocouple to measure temperature. Inside the probe, two different metals are joined together. Electrons have a lower free energy in one of the metals than the other, so they flow from one metal to the other. The flow of electrons depends on temperature and is measured by a small circuit that converts the electrical signal into temperature. A switch on these thermometers changes the output from Fahrenheit to Celsius.

Optical thermometers (pyrometers) measure temperature without making contact with the object. They are available as hand-held point-and-shoot guns (**FIGURE 3-2**). All objects release infrared radiation (IR), and the wavelength of the IR emitted is directly proportional to the temperature of the

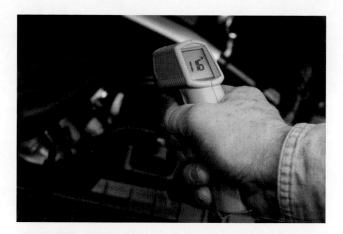

FIGURE 3-2 An optical thermometer measures the temperature without contacting the object being measured.

object. After using an infrared detector to convert the IR energy to an electrical signal, the optical thermometer reports temperature in either Fahrenheit or Celsius. It is important to remember that this device records the temperature of the surface of the object. An optical thermometer should not be used to measure the temperature of an object that may be warmer in the interior than on the surface, such as a dead body.

Algor mortis is the natural drop in body temperature that occurs after death. Normal body temperature is 98.6°F. If the temperature of the area surrounding the body is in the range of 18°C (64°F) to 20°C (68°F), then the temperature of a naked body will drop 1.5°C per hour for the first 8 hours after death. To identify the time of death, the rectal temperature of the body is measured with a thermometer and the **postmortem interval** determined.

Example

Normal body temperature is 98.6°F. What is this temperature in degrees Celsius?

Solution

$$°C = \frac{5}{9}(°F - 32) = \frac{5}{9}(98.6 - 32) = 37°C$$

Example

The medical examiner determines that a body found at a crime scene wearing only underwear has a rectal temperature of 87.8°F. Estimate the postmortem interval.

Solution
1. Convert the temperature of the dead body from Fahrenheit to Celsius:

$$°C = \frac{5}{9}(°F - 32) = \frac{5}{9}(87.8 - 32) = 37°C$$

2. Determine how many degrees the body cooled. The living body was at 37°C, and it dropped to 31°C:

$$37°C - 31°C = 6°C \text{ drop}$$

3. Given that body temperature decreases 1.5°C per hour after death, then:

$$6°C = (1.5°C/hr) \, (x \text{ hours})$$

$$x \text{ hours} = \frac{6°C}{1.5°C/hr} = 4 \text{ hr}$$

The victim has been dead for 4 hours (postmortem interval).

Density

When examining an object of evidence, density may be useful to help establish its composition. Density often is confused with weight, as in the following inaccurate examples: "mercury is heavier than water" and "iron is heavier than aluminum." What is actually being compared is the densities of these materials, not their weights. The **density** (d) of a substance is defined as the mass (m) of the substance per unit volume (V):

$$d = \text{sample mass/sample volume} = m/V$$

Density is an intensive property of matter. An intensive property does not depend on how much matter you have; thus the density of a material is the same regardless of how big a sample of that material you are investigating. Gases are less dense than liquids, and liquids are less dense than solids. The common units of density for solids and liquids are g/cm^3 or g/mL and for gases, g/L. A good fact to memorize is that the density of water is 1 g/mL. Knowing this density allows you to estimate the weight of any volume of water. The densities of some common substances are listed in TABLE 3-6 .

The procedure for determining the density of a solid object is simple. First, the object is weighed on an electronic balance and its mass determined. Next, a graduated cylinder is partially filled with water and the volume recorded (V_1). The solid object is then carefully lowered into the graduated cylinder. Once the object is completely submerged, the total volume of the object and the water in the graduated cylinder (V_2) is measured (FIGURE 3-3). The difference between the original volume and the final volume ($V_2 - V_1$) is the volume of the object.

TABLE 3-6

Densities of Some Common Substances at 20°C

Substance	Density (g/mL)
Gases	
Air	0.00124
Carbon dioxide	0.00198
Hydrogen	0.00009
Liquids	
Mercury	13.6
Ethyl alcohol	0.79
Gasoline	0.69
Water	0.998
Solids	
Gold	19.3
Lead	11.5
Iron	7.8
Aluminum	2.7
Glass	2.5
Ice	0.92

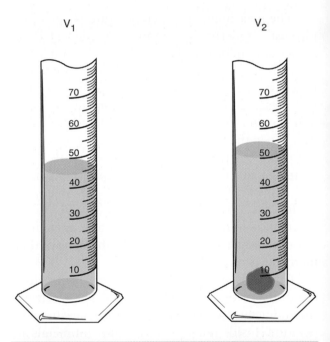

FIGURE 3-3 The volume of an object can be determined by measuring the volume of water it displaces ($V_2 - V_1$).

Example

The district attorney wants to know if a metal object recovered from a crime scene is iron or aluminum. When you weigh the object, you find that it weighs 39 g. You take a graduated cylinder and fill it with water until it reads 45 mL. Next, you carefully submerge the metal object into the graduated cylinder; the level of water raises to 50 mL. What is the density of the object? Is it iron or aluminum?

Solution

1. To determine the volume of the object:

$$V_{object} = V_2 - V_1$$
$$V_{object} = 50 \text{ mL} - 45 \text{ mL} = 5 \text{ mL}$$

2. To determine the density:

$$m = 39 \text{ g}$$
$$V = 5 \text{ mL}$$
$$d = m/V = 39 \text{ g}/5 \text{ mL} = 7.8 \text{ g/mL}$$

3. Compare the density of the metal object to the densities in Table 3-6. Using Table 3-6, we learn that the density of iron is 7.8 g/mL. Thus this object is iron.

Density exhibits an inverse relationship with temperature. In general, the density of a substance decreases as temperature increases. The amount of change is greatest for gases: Their density decreases on average by 0.33% per 1°C increase. The density decrease for liquids and solids with increases in temperature depends on the specific substance. The density of water decreases by 0.025% per 1°C increase, the density of alcohol decreases by 0.011% per 1°C increase, and the density of mercury decreases by 0.018% per 1°C increase. It is important to make sure the temperature of an object is constant (not changing) before making density measurements. Ideally, these measurements will be made in a constant-temperature chamber.

To make a rough estimate of the density of an object, simply immerse it in water. If the object sinks, it has a density greater than 1 g/mL. If it floats, it has a density less than 1 g/mL. This flotation technique can be used to characterize soils.

Soils

Soil is a complex mixture of inorganic and organic materials. The inorganic part comprises remnants of rock fragments that were formed over thousands of years by the weathering of bedrock. The organic part is derived from the decayed remains of plants. The mineral (inorganic) particles of soil are composed primarily of silicates; depending on the location, however, soil also may contain phosphates and limestone. The size of the particles in a soil determines its texture. Relatively coarse particles (diameter of 0.10 to 2.0 mm) form sand, slightly finer particles form silt, and the finest particles (0.002 mm or less) form clays. Typical soil contains all three types of particles.

A typical soil is made up of several main layers, also known as horizons (**FIGURE 3-4**). The uppermost, or O horizon, contains heavily decomposed organic matter. The A horizon, which includes the **topsoil**, is dark in color. It contains most of the humus—decayed plant and animal materials—in the soil. The A horizon is rich in microorganisms and other creatures, including earthworms, insects, and small burrowing animals that keep the soil aerated. Below the topsoil is the lighter-colored, humus-poor, and more compacted B horizon, also known as the **subsoil**. This layer, which may be several feet thick, consists of inorganic particles of the parent rock and contains minerals and organic materials leached from the A horizon. The C horizon consists of a layer of fragmented bedrock mixed with clay that rests on solid bedrock.

Because they are mixtures of materials, soils have highly individual characteristics. The diversity of materials and the rapid rate at which they change both horizontally and vertically provide a forensic scientist with a powerful tool with which to distinguish one soil from another. The kinds of igneous rock that could be present in soils are almost unlimited, and they are both identifiable and recognizable because they differ in color, mineralogy, and texture. It is this diversity in earth materials, combined with our ability to measure and observe the

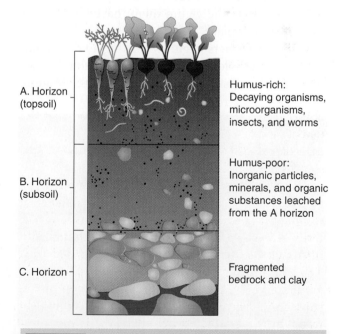

A. Horizon (topsoil) — Humus-rich: Decaying organisms, microorganisms, insects, and worms

B. Horizon (subsoil) — Humus-poor: Inorganic particles, minerals, and organic substances leached from the A horizon

C. Horizon — Fragmented bedrock and clay

FIGURE 3-4 Soil contains three major layers, or horizons, in which the thickness varies considerably depending on the type of soil.

different kinds of materials, that gives forensic scientists the power to discriminate among various soil samples.

Forensic Characteristics of Soils

Most forensic soil examinations involve comparison of many soil samples taken from different locations. Matching a crime scene sample to a reference establishes a high probability that two samples have a common source. In comparison studies of soils, not only are the natural components of the soils observed, but any artifacts in the soil—such as human-made objects like paint, fiber fragments, or chemicals—are considered as well.

Soil evidence often is found at a crime scene and transferred at the scene of the crime onto the criminal. Mud or soil in the treads of an athletic shoe or tire, for example, may link a suspect to the crime scene. As is the case with most physical evidence, forensic soil analysis is a comparative technique. That is, if soil is found on the shoes of the suspect, it must be carefully collected and compared to soil samples from the crime scene. Soil evidence that is not from the crime scene also is often very helpful to investigators. Careful examination of soil found on a victim's shoes by an expert in forensic geology, for instance, may allow investigators to determine specific sites where the victim might have been before the crime took place.

Most examiners begin by making a visual comparison of the color of the soil sample with reference materials. Forensic experts estimate there are more than 1000 different soil colors that can be distinguished from one another. Side-by-side comparison of the color and texture of soil samples is a simple test that eliminates a large percentage of samples as not being matches (FIGURE 3-5). Because the color of soil becomes darker when it is wet, it is important that all specimens be dried in an oven before their comparison.

Soil color has been standardized according to the Munsell soil color notation. The Munsell soil color book is used to help describe the color of the soil in question (FIGURE 3-6). The color of each soil sample in question is compared to a color chart, where each color is given a Munsell code.

Examination of the soil under a low-power microscope may reveal small soil components, such as roots of plants or human-made fibers or plastic fragments. Soils often are passed through sieves to separate soil components by size. After the larger components are removed, examination with a high-power microscope (higher magnification) will help in identifying small rock or mineral fragments in the soil. The size distribution of particles also can be determined if the sample is passed through a nest of sieves (FIGURE 3-7). In such a setup, the sieve on the top has a screen with the largest openings, and each lower screen has smaller holes than the screen above it. After the soil is added through the top, the entire nest of sieves is shaken to agitate the soil and cause components to drop through the screens. As the components fall through the sieves, each subsequent screen catches smaller and smaller components. When the separation is complete, the fractional weight of each screen is determined. The

(A)

(B)

FIGURE 3-5 (A) Samples found on the suspect's shoes are compared with soil samples collected from all possible locations where the soil could have originated. (B) Samples also are needed from areas where the suspect claimed the soil originated (alibi samples).

FIGURE 3-6 Soil sample color is described with a Munsell soil color code; the soil is matched to color chips in the manual.

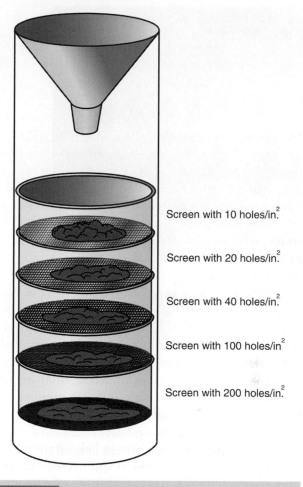

Screen with 10 holes/in.^{2}

Screen with 20 holes/in.^{2}

Screen with 40 holes/in.^{2}

Screen with 100 holes/in^2

Screen with 200 holes/in.^{2}

FIGURE 3-7 In a nest of sieves, the top sieve has the coarsest screen. Each subsequent screen has an increasingly finer screen that traps ever-smaller particles.

results of the sieve nest can then be used to compare soil samples.

Soils are extremely heterogeneous, and their composition can vary greatly over even short distances, both horizontally and vertically. A surface sample offers little possibility for comparison with soil from a grave that has a depth of 5 ft. Soil sampling in many cases is the search for the one sample that is a match. Surrounding samples may merely serve to demonstrate the range of local differences. Screening techniques during sampling that eliminate samples that do not match (i.e., elimination samples) are, therefore, extremely important.

Forensic soil examiners must have training in numerous aspects of geology, such as recognition of minerals and rocks. A **mineral** is a solid inorganic substance that occurs naturally. Most are crystals and have physical properties that are unique to their crystal structure. In particular, they have characteristic colors, densities, geometric shapes, and refractive indexes that can be measured to identify the mineral's presence in a sample.

More than 2000 minerals naturally occur on Earth, but many are very rare. A forensic geologist usually encounters only about 50 minerals on a routine basis. In addition to being present in soil, minerals also are used in the manufacture of many human-made products.

Rocks are composed of a combination of minerals. As the natural environmental process of weathering takes place, fragments of large rocks are reduced in size and become components of nearby soil.

BACK AT THE CRIME LAB

Many types of tests can be performed on standard soils and soil evidence in the forensic crime laboratory:

- Visual comparison of soil color
- Low-power microscopy

- High-power microscopy
- Sieves
- Density-gradient tube
- Chemical tests

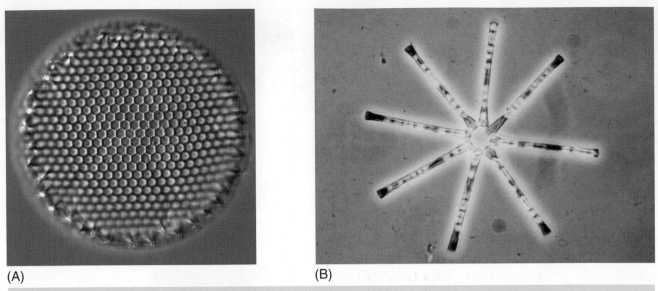

(A)   (B)

FIGURE 3-8 (A) Coscinodiscus, a salt water species. (B) Asterionella, a colony of freshwater pennate diatoms.

The combination of minerals and human-made components in soil creates a unique "signature" that the forensic geologist can use to compare soil samples. If two samples share enough similarities, the examiner may be able to determine that they share a common origin.

How can soil analysis benefit law enforcement? Consider the following case: A police officer in a Midwestern city who was arresting a man on a minor crime noticed what looked like a bad case of dandruff on the suspect's shoulders. Later examination showed it was not dandruff but rather diatomaceous earth, a white powder composed of fossilized skeletons of one-celled organisms called diatoms, whose coral-like skeletons are porous and have unique, specific shapes (FIGURE 3-8). The diatomaceous earth on the suspect matched the diatomaceous earth insulating material that was in a safe that had been burglarized the previous day.

Gradient Tube Separation of Soil: Separation by Density

In the past, forensic laboratories used a **density gradient column** to compare soil samples. To find the density distribution of particles, a density gradient column is used to separate known weight samples that have been previously sieved.

The density gradient column consists of two glass tubes, each of which is filled with layers of two liquids, where every layer has a different density (FIGURE 3-9). Often, the liquid at the bottom of the tube is pure tetrabromoethane (density = 2.96

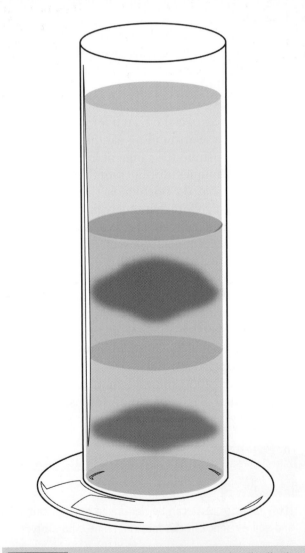

FIGURE 3-9 In a gradient tube separation, the distribution of the components is based on density.

g/mL). The liquid at the top of the tube is typically pure ethanol (density = 0.789 g/mL). In between are layers consisting of various mixtures of these chemicals, which have densities ranging between the two extremes of the pure substances.

To compare two soil samples on the basis of density, each sample is carefully placed on the top of the liquid in one of the tubes in the density gradient column. The various components fall through the layers, with each component being held up—that is, floating—at the point where its density is less than the density of the layer beneath it. The distributions of particles within the tubes can then be compared, revealing whether the two samples have a common origin.

To see the value of this type of soil assessment, consider the following example: A murder was committed on a college campus. The victim was found to have soil adhering to the cleats on his athletic shoes. Soil samples were collected at several locations on campus and compared by the density gradient column method. As shown in FIGURE 3-10, the crime scene sample was most similar to the garden sample, strongly indicating that the victim was in the garden before his death.

The gradient tube separation method is not considered a definitive test because soils collected from different locations can give similar density distribution patterns. Instead, the density gradient column is most useful when forensic geologists are comparing soils. More sophisticated testing of soil specimens, such as atomic absorption spectroscopy and X-ray fluorescence spectroscopy, should be used to confirm the results of these tests.

Collection and Preservation of Soil Evidence

Crime scene investigators who collect soil samples must be certain to sample the exact, original location of the crime. Soil samples should be collected as soon as possible because the soil at the crime scene can change dramatically over time. To establish the variation in the soil at the crime scene, the crime scene investigator should collect specimens at the scene of the crime and reference specimens with a 100-yard radius of the crime scene. But the investigator should not stop there: He or she should also take soil samples from all paths into and out of the site. If the suspect claims to have been elsewhere

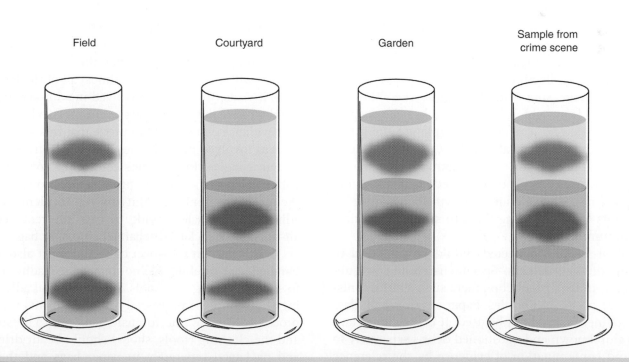

FIGURE 3-10 Comparison of reference samples from the courtyard, college gardens, and baseball field to the sample taken from the shoes of the victim (crime scene). The crime scene sample matches the soil from the garden.

On May 9, 1991, the body of 9-year-old Rebecca O'Connell was discovered in a secluded area in Sioux Falls, South Dakota. Rebecca had gone missing the previous day after visiting a local store. She had been raped, both vaginally and anally, and stabbed in the neck, shoulder, chest, hip, and hands. The forensic pathologist determined that her death was the result of a cut to her jugular vein.

Witnesses noticed a light blue pickup truck leaving the area where Rebecca's body was found, and they gave a description of the driver to police. Three days later, a police detective spoke to Donald Moeller, who owned a blue pickup truck, matched the general description of the driver, and was at a local store at the same time as the victim. Moeller denied any involvement and gave blood and hair samples for analysis. The next day, Moeller fled the state, leaving behind his light blue pickup truck. The police obtained a search warrant for his vehicle and his home.

Soil samples were taken from the wheel wells of the truck, the crime scene, and several unpaved roads near Sioux Falls. Retired forensic geologist Dr. Wehrenberg found numerous consistencies between the soils at the crime scene and the truck, including an unusual blue mineral called gahnite found in the left wheel well of the truck and at the crime scene. Based on the characteristics of the samples, Dr. Wehrenberg was able to rule out the locations where Moeller said he drove his truck.

The soil evidence strengthened the case against Moeller. He was convicted and sentenced to death for the rape and murder of Rebecca in 1992. In 1996, his conviction was overturned on procedural grounds. Moeller was retried, reconvicted, and resentenced to death in 1997. This time, the conviction was upheld. Moeller has continued his appeals.

at the time of the crime, the investigator should take soil samples at that location as well, so as to prove (or disprove) the alibi. Sometimes the investigator may notice discrepancies in soils' color, texture, and composition at the site, suggesting that a soil might have been dug up from a greater depth or transferred from a different location.

When collecting samples, the investigator should make sure that the depths (i.e., the soil horizon) from which he or she takes the suspect soil sample and a reference sample are the same. Given that only a small sample (less than a tablespoon) is needed for testing, taking a specimen of the top surface is all that is needed. Investigators should also draw a map identifying soil sample locations and correlating them to sample identification numbers.

All samples collected should be packaged in individual containers. Specimens should be submitted in leak-proof containers such as film canisters or plastic pill bottles. Paper envelopes or glass containers should not be used. It is very important to ship known and questioned debris separately to avoid contamination of evidence by the reference materials.

When a lump of soil is found, it is important that it be packed and subsequently preserved intact, because such a sample may have substantial evidentiary value. To see why, consider that an automobile collects and builds up layers of debris on its undercarriage. These layers represent a composite sample that is unique to that car. During the normal use of a car, it will pick up soil from almost all of the locations it visits. If the car is involved in a hit-and-run accident and the impact jars a lump of soil loose, this evidence can be used to prove that this specific car was involved in the crime. That is, soil from the undercarriage of the suspect's car may be compared with soil left at the scene to prove (or disprove) that the car was present at the hit-and-run incident. Soil adhering to vehicles should be carefully removed, air-dried, and packaged separately in paper bags.

Soil found on a suspect or victim must also be handled very carefully. Do not remove soil adhering to shoes, clothing, and tools. Tools with soil adhering should not be processed for latent prints until processed first by the forensic geologist. The soil and the clothing (tools, shoes) should be air-dried and packaged separately in paper bags and then shipped to the laboratory.

Soils often are considered to be physical evidence. The high diversity of minerals, rocks, fossils, organic material, and other items in soils allows forensic geologists to discriminate between samples. A forensic geologist will characterize and compare soil samples in the lab, and many samples can be excluded based on these comparisons. Specimens are considered to come from a single source only if there are a sufficient number of correlations between soil samples.

Usually, the forensic geologist looks for the unusual in a soil sample, such as an uncommon mineral, a microfossil, or human-made materials. Sometimes, however, simply matching soil, rocks, minerals, or fossils found on a suspect to a particular location will provide all the evidence that is needed. Hans Gross, one of the earliest forensic scientists, noted in his 1893 handbook, "Dirt on shoes can often tell us more about where the wearer of those shoes had last been than toilsome inquiries."

1. The plastic properties of soil allowed the impression of the boots to be captured. Unfortunately, the suspect is wearing new boots. The impression left by these boots will fall into the category of class evidence (all size 12 Wolverine boots). The new boots will carry no individualizing wear marks that might allow investigators to conclude that the boot prints were unique. The boot print evidence is class evidence and does not individualize the evidence to this one suspect.

2. Because the soil in the pot is friable, when the pot hit the floor, the solid clump of soil was broken apart and the components dispersed as a powder. This powder could have stuck to the suspect's shoes. When the suspect drove his car, the soil components dropped onto the carpet of the car. Because soil in our yards does not usually contain vermiculite, the soil in the car did not come from there. The vermiculite present does narrow the soil to a smaller class of soil samples, although it does not individualize the sample.

Chapter Spotlight

- Soil can be identified by physical and chemical properties.

- Physical properties are properties that can be observed directly, including color, odor, taste, hardness, density, and melting and boiling points.

- Chemical properties are properties that can be observed when the sample reacts with another substance. For instance, if the sample neutralizes a basic solution, then the original sample is acidic in nature.

- Scientists should be comfortable using the metric system and the SI.

- Mass measures how much of a material is present.

- Weight measures how much force is exerted on a mass by gravity.

- Density measures mass per volume. It is affected by temperature, so warmer objects are often slightly less dense than colder objects.

- Temperature can be measured in Fahrenheit or Celsius. The conversion from one to the other can be calculated using the following equations: $°F = (9/5)°C + 32$ or $°C = 5/9(°F − 32)$.

- Algor mortis is the drop in body temperature that occurs after death. It generally follows the pattern of a $−1.5°C$ per hour decrease if the surrounding temperature is between 18 and 20°C.

- Soil has four major horizons:
  - The O horizon is the organic layer made of decomposing litter with very little to no base soil.
  - The A horizon is located below the O horizon and is a mixture of organics and soil.
  - The B horizon includes primarily mineral soils and is heavily influenced by the leachate from the upper layers. It often is quite thick in comparison to the other horizons.
  - The C horizon comprises mostly fragmented bedrock and clay materials.

- Soil analysis includes visual identification, including determination of the soil's color and identification of specific components such as roots and minerals.

- Sieve analysis is used to separate components of a soil sample according to size.

- Gradient tube separation is used to separate components of a sample according to density.

- Soils are characterized by significant horizontal and vertical variability.

- Elimination samples are very important to limit the number of noncomparable samples.

Key Terms

Celsius scale A temperature scale on which water freezes at 0°C and boils at 100°C at sea level.

Chemical property A characteristic of a substance that describes how it reacts with another substance.

Density The mass of an object per unit volume.

Density gradient column A glass tube filled (from bottom to top) with liquids of sequentially lighter densities.

Extensive physical property A property that is dependent on the amount of material present.

Fahrenheit scale A temperature scale on which water freezes at 32°F and boils at 212°F at sea level.

Friable Easily broken into small particles or dust.

Intensive physical property A property that is independent of the amount of material present.

International System of Units (SI) The system of measurement (metric units) used by most scientists.

Mass A measure of the quantity of matter.

Mineral A naturally occurring inorganic substance found in the Earth's crust as a solid.

Physical property A property that can be measured without changing the composition of a substance.

Postmortem interval Time elapsed since death.

Reference standard Physical evidence whose origin is known and that can be compared to evidence collected at a crime scene.

Subsoil Soil lying beneath the topsoil, which is compacted and contains little or no humus.

Topsoil The surface layer of soil, which is rich in humus.

Weight A measure of the force of attraction of the Earth for an object.

Putting It All Together

Fill in the Blank

1. The _____ properties of a substance are those properties that can be observed when the substance reacts or combines with another substance so as to change its chemical composition.

2. The _____ properties of a substance are those properties that can be observed and recorded without referring to any other substance.

3. The International System of Units is known as the _____ units.

4. The basic SI unit for length is the _____.

5. The basic SI unit for mass is the _____.

6. The basic SI unit for volume is the _____.

7. Distance on a map is expressed in _____ (kilometers or millimeters).

8. A kilogram is approximately _____ pounds.

9. A meter is approximately _____ yards.

10. _____ is a measure of the quantity of matter in an object.

11. The Celsius temperature scale defines the freezing point of water to be _____°C.

12. The Fahrenheit temperature scale defines the boiling point of water to be _____°F.

13. A(n) _____ is immersed in a liquid to make a temperature measurement.

14. A(n) _____ measures temperature without touching the object.

15. The density of water is _____ g/mL.

16. The top layer of soil, the O horizon, is mostly _____ matter.

17. The B horizon, which is located below the topsoil, is mostly _____ matter.

18. The igneous rock found in soils differs in _____, _____, and _____.

19. When a nest of sieves is used to separate soil components by size, the bottom-most sieve will contain the _____ (smallest/largest) size particles.

20. More than _____ minerals naturally occur on Earth, but a forensic geologist routinely encounters only about _____ minerals.

21. A separation of soil components can be done by using a(n) _____ tube.

22. A density gradient column contains two liquids of different density that are added to give a gradient from _____ (most dense/least dense) on the bottom of the tube to _____ (most dense/least dense) on the top.

23. When a soil sample is placed in a density gradient column, the inorganic minerals in the soil should _____ (float near the top/sink to the bottom).

24. To establish the variation in soil at the crime scene, the investigator should collect reference soil samples within a(n) _____ radius of the scene.

True or False

1. When a substance melts, its composition changes.

2. An optical pyrometer should be used to measure the temperature of a dead body.

3. The density of an iron bar depends on the length of the bar.

4. The density of water increases as it is heated.

5. Soil samples should be collected from the crime scene as soon as possible.

6. Soil samples should be collected at all alibi locations claimed by the suspect.

Review Problems

1. Canadian police investigating an arson call to tell you they confiscated 825 mL of an accelerant. Express this amount in:
   a. Liters
   b. Quarts
   c. Gallons

2. Mexican police have confiscated 153,000 g of cocaine. Express this amount in:
   a. Kilograms
   b. Pounds

3. A bag containing 3 lb 8 oz of cocaine is found in a suitcase by a customs inspector at Dulles Airport. How many kilograms have been found?

4. In the state of New York, the minimum prison time imposed for unlawful possession of cocaine depends on the amount of cocaine a person is found to possess. Penal law 220-21 states that possession of 4 oz or more is a Class A1 felony with a penalty of 15 years to life in prison. Joe Bradshaw has been arrested and found to have a package in his pocket containing 150 g of cocaine. How many ounces of cocaine are in the package? Has Joe committed a Class A1 felony?

5. In the state of New York, the minimum prison time imposed for unlawful possession of cocaine depends on the amount of cocaine a person is found to possess. Penal law 220-18 states that possession of 2 oz or more is a Class A2 felony with a penalty of 3 years to life in prison. Mary Gordon has been arrested and found to have a package in her purse containing 10 g of cocaine. How many ounces of cocaine are in the package? Has Mary committed a Class A2 felony?

6. In Illinois, unlawful possession of 900 g of heroin is a Class 1 felony with a penalty of 10 to 50 years in prison. Harvey Nelson was arrested and found to have 50 oz of heroin in his car. How many grams of heroin did he have? Is he guilty of a Class 1 felony?

7. In Illinois, unlawful possession of 500 g of marijuana is a Class 3 felony with a penalty of

1 to 5 years in prison. Judy Whitman was arrested and found to have 10 oz of marijuana in her car. How many grams of marijuana did she have? Is she guilty of a Class 3 felony?

8. The state of Alabama imposes a $0.52 tax on beer per gallon. If 5,250,000 liters of beer were sold in Alabama in 2009, how much money was raised by this tax?

9. You are investigating a murder. The coroner measures the body of the cadaver to be 34°C.

    a. What is the average temperature of a living person in degrees Fahrenheit and degrees Celsius?

    b. By how many degrees has the body temperature of the victim dropped since death?

    c. What is the postmortem interval?

10. You are working in a forensic lab. The microscope manufacturer recommends that your microscope be kept at a constant 75°F temperature. The temperature of the lab is 15°C. Should you turn on the heat or the air conditioning?

11. The district attorney wants to know if a metal object recovered from a crime scene is iron or aluminum. You weigh the object and find that it weighs 540 g. You take a 500-mL graduated cylinder and fill it with water until it reads 100 mL. When you carefully submerge the metal object into the graduated cylinder, the level of water rises to 300 mL. What is the density of the object? Is it iron or aluminum?

12. Thieves broke into a jewelry store and stole a large amount of gold that was later used to make expensive rings and other jewelry. The police arrest a suspect and impound a large amount of yellow metal. The suspect claims it is brass. You select a piece of the yellow metal and find it weighs 482 g. You take a 200-mL graduated cylinder and fill it with 50 mL of water. When you carefully submerge the metal object into the graduated cylinder, the level of water rises to 75 mL. What is the density of the yellow metal? Is it gold?

Further Reading

Girard, James E. *Principles of Environmental Chemistry*, ed. 2. Sudbury, MA: Jones and Bartlett, 2009.

Munsell Soil Color Charts. Baltimore, MD: MacBeth, a division of Kollmorgan Instruments Corporation, 1994.

Murray, R. C., and Solebello, L. P. Forensic examination of soil. In R. Saferstein (ed.), *Forensic Science Handbook,* vol. 1 (ed 2). Upper Saddle River, NJ: Prentice-Hall, 2002.

Murray, R. C., and Tedrow, J. C. F. *Forensic Geology*. Upper Saddle River, NJ: Prentice-Hall, 1992.

Petraco, N., and Kubic, T. *Color Atlas and Manual of Microscopy for Criminalists, Chemists, and Conservators*. Boca Raton, FL: CRC Press, 2004.

http://criminaljustice.jblearning.com/criminalistics

Answers to Review Problems

Interactive Questions

Key Term Explorer

Web Links

http://criminaljustice.jblearning.com/criminalistics

The Microscope and Forensic Identification of Hair, Fibers, and Paint

OBJECTIVES

In this chapter you should gain an understanding of:

- The parts of a compound microscope and how it works
- The use of a comparison microscope to compare two objects
- The large working distance and the larger depth of field afforded by the stereomicroscope
- Differentiation of amorphous and crystalline materials by use of a polarized light microscope
- The structure of hair and the microscopy techniques used to identify human hair
- The characteristics of natural fibers, human-made fibers, and the fabrics made with both types of fibers
- The composition of different types of paint and how paint samples are characterized
- The use of microspectrophotometers and scanning electron microscopes in the forensic lab

FEATURES

On the Crime Scene

Back at the Crime Lab

See You in Court

WRAP UP

Chapter Spotlight

Key Terms

Putting It All Together

The microscopic techniques used to investigate trace fiber and hair evidence focus on their morphology. *Morphology* is a term that is used to describe the size, shape, and color of the evidence. The methods employed for morphological comparisons of questioned and control samples often are used to show that two samples do not match. Indeed, more suspects are exonerated by morphological comparison than are implicated by it. Only very rarely will the examiner be able to confirm that the individual who left behind the questioned trace sample was the same person who is the source of the control sample.

The utility of morphological comparison methods is limited because morphology often is not capable of individualizing hair and fibers. When two samples are similar enough to have originated from the same person, additional tests must be performed to establish their origin.

1. A bank was robbed by a man wearing a black ski mask. In a subsequent search of the bank parking lot, police recovered a black ski mask. What trace evidence might be present? How should that evidence be investigated?
2. The examiners at the forensic lab found a short, brown beard hair inside the ski mask. Should the investigators focus their search on men with brown beards?

Introduction

During the search of a crime scene, investigators will inevitably encounter many types of physical evidence. Although the search may initially focus on large objects that may prove to be evidence, trace evidence must not be overlooked. *Trace evidence* is a generic term used to describe small, often microscopic, objects that are readily transferred between people and places. The range of objects falling into the category of trace evidence is enormous and can include hair, fibers, glass, soil, feathers, pollen, dust, and paint.

In 1932, with a borrowed microscope and a few other pieces of basic equipment, the Federal Bureau of Investigation (FBI) established its technical laboratory in Washington, D.C. The microscopic comparison of fibers and hairs was among the first examinations performed by this laboratory. Since then, forensic laboratories (including the FBI laboratory) have greatly expanded their capacity to handle trace evidence thanks to the development of modern analytical instruments. Even with much more sophisticated instruments available, however, forensic scientists today usually begin their examination of trace evidence with a microscope. This chapter describes the various types of microscopes available in forensic laboratories that are used to examine hairs and fibers.

Magnifying Small Details

Forensic scientists often are confronted with the need to analyze many different types of materials obtained during criminal investigations. The identification or comparison of miniscule traces of a wide range of materials is a common occurrence in a modern crime lab. The earliest detectives—including the fictional Sherlock Holmes—used magnifying glasses to carefully examine evidence in the field and microscopes to study objects brought to their laboratories.

Early crime labs relied almost entirely on the light microscope for examining minute details of evidence that were not visible to the naked eye. This kind of a microscope offered less than 10 times (10×) magnification capabilities but was still considered a wonder.

Refraction

A magnifying glass is a lens that is thicker in the middle than at the edge. It makes objects appear larger than they are by refracting (bending) light rays as they repeatedly pass through the air and back through the lens. When light passes at an angle through the interface between two transparent media (e.g., air and glass) that have different densities, an abrupt change in direction—that is, **refraction**—of the beam is observed as a consequence of the difference in the velocity of the light in the two media.

The phenomenon of refraction is commonly observed with an object that is immersed in a glass of water. For example, when a glass of water contains a straw (**FIGURE 4-1**), refraction makes it appear that the straw is broken at the surface of the water. As the rays of light leave the water and enter the air, their velocity increases, causing them to be refracted. How much light bends (and thus the focal length) depends on the change in the **refractive index** (η) as the light enters and leaves the magnifying glass. The refractive index is a ratio of the velocity of light in a vacuum to its velocity in any other medium:

$$\text{Refractive Index} = \eta = \frac{\text{Velocity of light in a vacuum}}{\text{Velocity of light in a denser media}}$$

The refractive index is 1.0 for air and 1.33 for water at 25°C. A typical glass has a refractive index of about 1.4 (a difference of 0.4 for light passing from air to glass).

To use a magnifying glass to examine an object, you look through the lens (**FIGURE 4-2**). The magnified image is called the **virtual image**. As the lens is moved closer to the eye, the virtual image increases in size by 5 to 10 times. Higher

magnification requires two lenses that are at fixed distances from each other in a hollow tube, a setup known as a compound microscope.

Types of Microscopes

A microscope has at least two lenses: an objective lens and an ocular lens or eyepiece (**FIGURE 4-3**). The **objective lens**, which is the lower lens in the hollow tube, produces a "**real**" image—a magnified and inverted version of the object being examined. The real image is then viewed through the **ocular lens** (the smaller lens at the other end of the hollow tube in the eyepiece), producing a virtual image in the brain of the viewer. The virtual image is a magnified image produced in the viewer's mind of the real image; it has the same orientation as the real image, but its magnification exceeds the magnification of the real image by the ocular lens (**FIGURE 4-4**). The magnification power of a microscope is greater than that of a magnifying glass because the former device has two lenses. The magnifying power of a microscope is determined by multiplying the power of the objective lens by the power of the ocular lens. A **compound microscope** can magnify objects up to 1500 times.

A compound microscope—even one with perfect lenses using a tungsten light bulb to illuminate the object—has limitations, however. In particular,

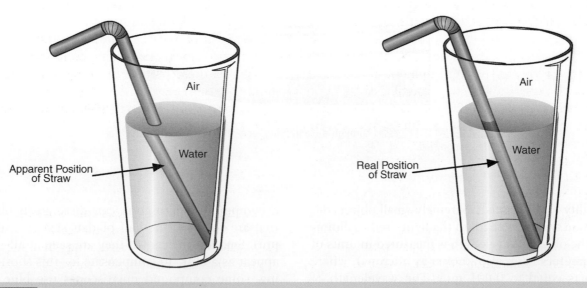

Air

Water

Apparent Position of Straw

Air

Water

Real Position of Straw

FIGURE 4-1 The glass on the left shows the apparent position of a straw in water as seen by the eye. Refraction makes it appear that the straw is broken at the surface of the water.

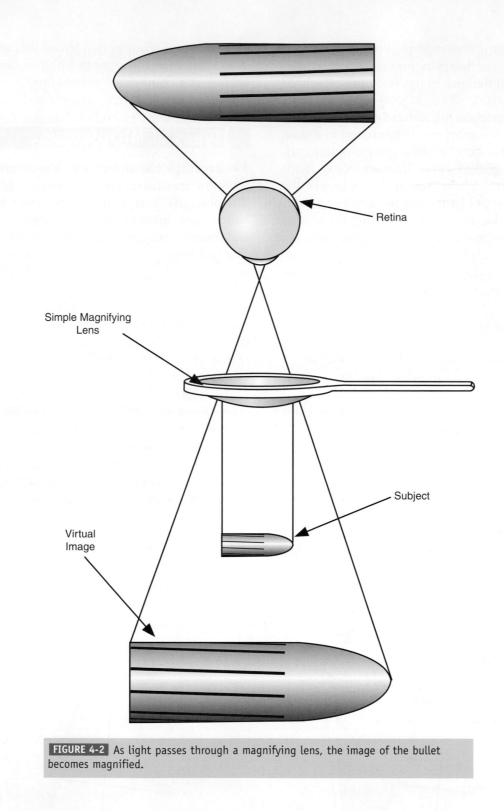

Retina

Simple Magnifying
Lens

Subject

Virtual
Image

FIGURE 4-2 As light passes through a magnifying lens, the image of the bullet becomes magnified.

its ability to distinguish extremely small objects depends on the wavelength of the light used to illuminate the object. Wavelength is measured in units of **micrometers** (μm; also known as microns), where 1 μm is equal to 0.001 mm. The wavelength of white light (which is typically used to illuminate objects in compound microscopes) is 0.55 μm, and a compound microscope can distinguish objects that are roughly one-half of that size (i.e., 0.275 μm). Smaller objects—if they are seen at all—may appear as a blur. To compensate for this shortcoming, some compound microscopes use blue light (which has shorter wavelength than white light) to illuminate the object of interest.

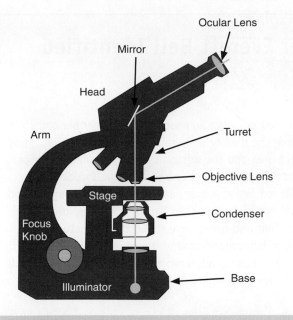

FIGURE 4-3 A typical compound microscope uses a light bulb to illuminate the specimen (light microscope).

Five types of microscopes are typically used for forensic examinations:

1. Compound microscope
2. Comparison microscope
3. Stereoscopic microscope
4. Polarizing microscope
5. Microspectrophotometer

Each of these microscopes has a specific function that makes it very effective for obtaining certain types of information. The **scanning electron microscope (SEM)**, which can provide magnification of more than 100,000 times, also is occasionally used for forensic examination of evidence.

Compound Microscopes

The mechanical system of a compound microscope has six important parts:

- Base: The stand on which the microscope sits.
- Arm: A metal piece that supports the tube body of the microscope. It is also used as a handle to pick up the microscope.
- Body tube: A hollow tube that holds the objective lens at the lower end and the eyepiece lens at the upper end. Light passes from the objective lens through the body tube to the eyepiece, where it is observed by the eye.
- Stage: The platform that supports the specimen. The specimen is placed on a glass slide that is clipped into place on the stage. The x- and

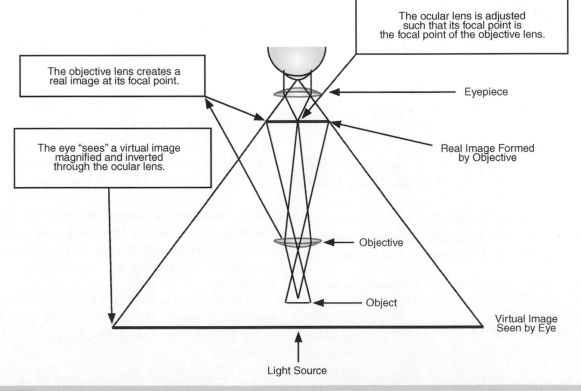

FIGURE 4-4 The passage of light through two lenses forms a virtual image of the object as seen by the eye.

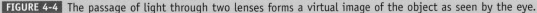

On August 22, 1995, the body of 8-year-old Sandra was discovered by police in a wooded area near her church in southern Michigan. The child had been brutally assaulted before she died of strangulation. Her bicycle was found near her partially hidden body. Police cordoned off the wooded area and the adjacent church parking lot to perform a detailed search of the area. Fingerprints were taken from the victim's body and from the bicycle.

The coroner discovered material under Sandra's fingernails, on her abdomen, and on her hands. Vaginal and rectal swabs and analysis of her clothing revealed no biological material other than her own. Furthermore, the bicycle provided no incriminating fingerprint evidence. However, hair and fibers were found on the victim.

The police launched a major investigation, canvassing the neighborhood around the church. They found that Everett Bell, a local resident and an employee of a Hardee's® restaurant, was seen in the area around nightfall riding his bicycle. Bell's employer revealed that Bell had been sent home early from work that day. The Hardee's manager gave police Bell's work pants, shirt, and cap.

The FBI laboratories found blue cotton fibers under Sandra's fingernails. When they were examined under a microscope, it was clear that these fibers matched the blue cotton fibers from the suspect's shirt. Blue and green fibers found on Sandra's clothes also matched those from the shirt. In addition, a dark blue-gray polyester fiber matching the fibers of the suspect's pants matched fibers found on the rectal swab. Of particular interest, the sample from the victim contained "nubs;" the end of the fiber was melted during manufacture into bulbs that are called nubs. The nubs that were lifted with tape from Bell's shirt matched the nubs lifted from the surface of the victim's shirt.

Hardee's employees were given particular shirts manufactured by WestPoint Stevens Alamac Knits. Bell's shirt was a short-sleeve polo shirt that Hardee's had recently replaced with a striped shirt. Therefore, few of the solid shirts were still being worn by employees. Tape applied to the surface of Bell's shirt removed a significant number of the nubs that proved to be identical to nubs found on the victim's clothes, establishing physical contact between the victim and the suspect.

Bell later confessed to Sandra's murder as well as to five other rapes and at least nine other murders.

---

y-axis stage control knobs are used to move the slide back and forth under the objective lens.

- Coarse adjustment: A knob that focuses the microscope on the specimen by raising and lowering the body tube of the microscope.
- Fine adjustment: A knob that adjusts the height of the body tube in smaller increments so the specimen can be brought into fine focus.

The optical system has four fundamental parts: the illuminator, the condenser, the eyepiece, and the objective.

There are several types of electric lighting (**illuminators**) for microscopes. Tungsten is the least expensive electric illumination; it is also hotter and less bright than the other kinds of illuminators. Fluorescent provides cooler and brighter light than tungsten. This is beneficial when you are viewing slides for long periods of time or when you are observing live specimens, such as protozoa. Halogen provides the very brightest illumination.

In the **condenser**, the light rays from the illuminator are focused through a lens in the center of the stage and onto the specimen. The condenser has an iris diaphragm that can be opened or closed to regulate the amount of light passing into the condenser.

The *eyepiece* is the part of the microscope that you look through. The *objective* is the second lens of the microscope. Compound microscopes have a "nosepiece" with a rotating objective turret, which allows you to change the objective lens and amplification power. The viewed image is magnified by both the objective and the eyepiece lens. Thus the total magnification is equal to the product of the magnifying power of the two lenses. For example, if the eyepiece lens has a magnification of 10× and the objective lens has a magnification of 40×, the total magnification will be the product of the two lenses' magnifications: 10 × 40 = 400×. Forensic examiners typically use a 10× eyepiece with nosepiece that contains a 4×, 10×, 25×, or 40× objective lens, providing magnification of 40×, 100×, 250×, or 400×, respectively.

Each objective lens is stamped with a numerical aperture (NA). As the NA doubles, the lens is able to resolve details that are half as distant from each other (twice as close to each other). The NA is effectively an index of the ability of the objective lens to resolve fine detail. The higher the NA (up to about 1000× the NA), the more detail the microscope can discern. Anything beyond 1000×, however, is considered "empty magnification" that doesn't add anything to the clarity of the image.

The forensic examiner must balance the microscope's magnifying power with both the field of view and the depth of focus. The **field of view** is the area that the forensic examiner sees when looking through the eyepiece. As the magnifying power increases, the field of view shrinks. The examiner should start with a low magnification. After taking in this (relatively) "big picture," the examiner should switch to a higher-power objective lens to scrutinize particular parts of the specimen more closely. The examiner should also adjust the **depth of focus** (the thickness of the region in focus, which affects the clarity of the image) as necessary. As with the field of view, when the magnifying power increases, the depth of focus decreases.

Thus both field of view and depth of focus decrease as magnifying power increases. As a consequence, high powers of magnification sacrifice both the size of the area in view and the thickness of the region in focus. Low powers of magnification sacrifice detail within the region under examination and in focus. For this reason, a trade-off is necessary when using a compound microscope.

Comparison Microscopes

To compare two specimens, such as evidence taken from a crime scene with a reference sample taken from a suspect, the forensic examiner often uses a **comparison microscope**. A comparison microscope is essentially two compound microscopes that are connected by an optical bridge—that is, a series of

mirrors and lenses that link the two microscopes' objective lenses. The comparison microscope has a single eyepiece through which the forensic examiner sees the images from both compound microscopes placed side by side.

Comparison microscopes that compare semitransparent hairs and fibers are lighted from below the stage. Other comparison microscopes are designed to compare bullets, bullet cartridges, and other opaque objects (FIGURE 4-5). Because these specimens do not transmit light, illumination from below the stage is not possible; instead, optical microscopes for viewing these items are equipped with a vertical or reflected illumination system (FIGURE 4-6).

An experienced examiner will mount the evidence on the same side for all comparisons. For instance, an examiner who is right-handed should always place the evidence on the right microscope stage and the reference material on the left stage. When this is done without fail, the evidence and the standard reference will not be confused.

Stereoscopic Microscopes

The **stereoscopic microscope** (FIGURE 4-7) is the most commonly used microscope in crime laboratories. Although it is able to magnify images by only 10× to 125×, many types of physical evidence are routinely examined under low magnification.

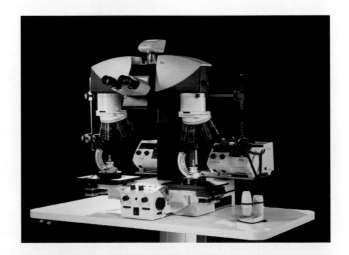

FIGURE 4-5 This comparison microscope is used to compare the marks or scratches on a bullet recovered from a crime scene to a test bullet fired from a gun suspected of being a murder weapon.

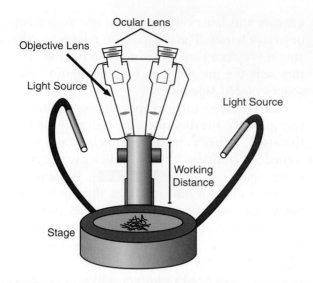

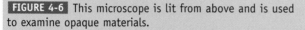

FIGURE 4-6 This microscope is lit from above and is used to examine opaque materials.

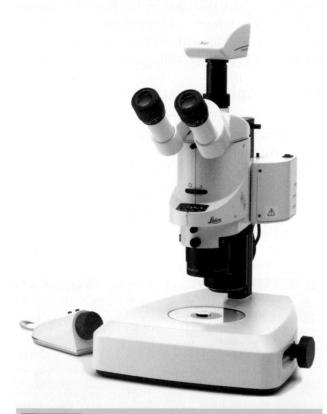

FIGURE 4-7 Although it will magnify only 125×, the stereoscopic microscope is the most commonly used microscope in forensics.

The stereoscopic microscope has several features that make it particularly easy to use. This device has two eyepieces, which make it more comfortable for the viewer. It also produces a three-dimensional image that is presented in a right-side-up, frontward orientation (unlike the upside-down, backward orientation of the image produced by a compound microscope). Its large **working distance**, which is the distance between the objective lens and the specimen, means that the forensic examiner can use this device to view large items. Finally, the stereoscopic microscope can be lighted either from below, to illuminate semitransparent objects, or vertically from above, to illuminate opaque objects.

Polarizing Microscopes

The compound microscope has been used to determine the type of minerals present in geological samples for more than 100 years. Minerals whose atoms are arranged in random order are classified as **amorphous materials**. Minerals whose atoms are arranged in a distinct order are classified as crystalline materials.

By observing how light interacts with thin sections of the minerals, the forensic examiner can classify minerals—and classification is the first step in forensic identification. The classification techniques described here rely on the fact that light traveling through the minerals in different directions will behave differently, but predictably. These differences can be detected with a polarizing microscope. To examine and classify a mineral, the sample must be dry and thin enough to be placed on a microscope slide under a cover slip. If the material is too thick, a portion can be shaved off with a scalpel.

Isotropic materials—such as gases, liquids, and cubic crystals—have the same optical properties when observed from any direction. Because isotropic materials have only one refractive index, they appear the same when light passes through them from different directions. By contrast, **anisotropic** (or **birefringent**) **materials**—such as quartz, calcite, and asbestos—are crystalline. In anisotropic minerals, the arrangement of atoms is not the same in all directions. Thus, when an anisotropic material is examined, the arrangement of atoms in the substance appears to change as the direction of observation changes. Photons of light passing through asbestos fibers from different directions will encounter different electrical neighborhoods, which in turn will affect the path and time of travel of the light beam in different ways.

Besides providing information on the shape, color, and size of different minerals, polarized light microscopy can distinguish between isotropic and anisotropic materials. This technique is used to identify the presence of certain minerals and their anisotropic properties.

Plane-Polarized Light

According to the wave theory of light, light waves oscillate (vibrate) at right angles to the direction in which the light is traveling through space. The oscillations occur in all planes that are perpendicular to the path traveled by the light. These oscillations may be produced by a light bulb and viewed by an observer looking directly into the beam (FIGURE 4-8A). If the beam of light is passed through a sheet of Polaroid® material, the light that is transmitted through the Polaroid sheet oscillates in one plane only. Such light is called **plane-polarized light**. If a second Polaroid lens is inserted before your eye, that lens must be aligned in a parallel plane so that plane-polarized light is passed. If the second Polaroid lens is rotated, less plane-polarized light is transmitted through the second lens, and, at a rotation of 90°, no light is transmitted (FIGURE 4-8B).

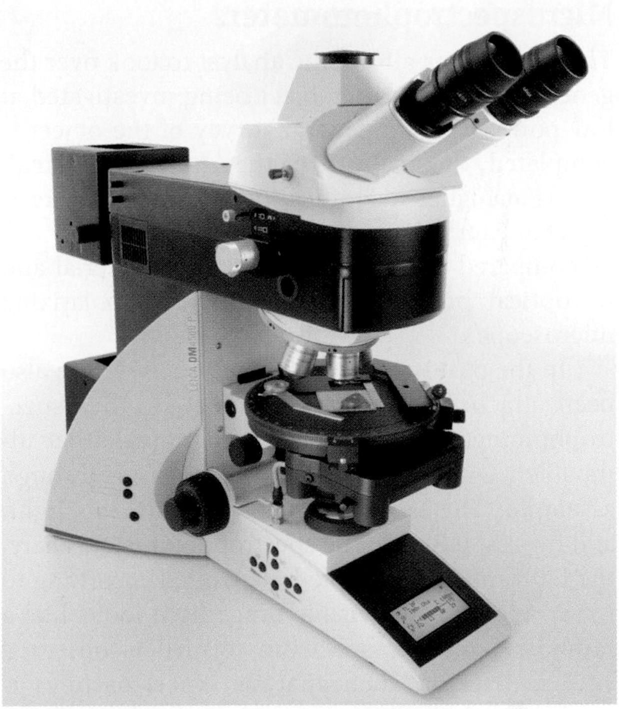

FIGURE 4-9 A polarizing microscope includes two polarizing filters: the polarizer lens and the analyzer lens.

Polarized Light Microscopy

A **polarizing microscope** includes two polarizing filters: the **polarizer** lens, which is fixed in place below the specimen, and the analyzer lens, which is located above the specimen (FIGURE 4-9). The sample stage, which is placed between the two filters, can be easily rotated.

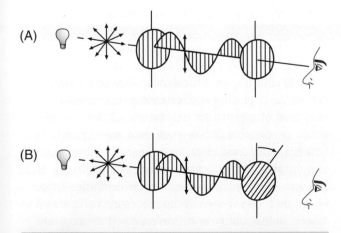

FIGURE 4-8 (A) Light from a light bulb is polarized in one plane as it passes through the first polarizer, and it will reach the eye only if the second polarizer is in the same position. (B) When the second polarizer is rotated, it cuts off the light and stops its transmission.

Besides information about the gross fiber morphology and color of the sample, analysis of plane-polarized light can determine whether the sample exhibits **pleochroism**. Pleochroism is the property that causes a substance to show different absorption colors when it is exposed to polarized light coming from different directions. The observed colors change with the orientation of the crystal and can be seen only with plane-polarized light. As the sample stage is rotated, these substances change color.

Polarized light microscopy also is used to identify human-made fibers and paint. If a fiber is collected from a victim and placed on a microscope slide, it may be difficult to tell one type of fiber from another. If it is a glass fiber, however, its color will not change as the microscope stage is rotated. That is, because glass fibers are isotropic, they are unaffected by rotation under polarized light. Other human-made fibers are anisotropic and change colors as the stage is rotated. It is an easy matter to compare the pleochroism of a fiber found at a crime scene to a fiber taken from the suspect, thereby proving (or disproving) a link between the suspect and the scene.

Microspectrophotometers

The microscope allows the analyst to look over the general features of the object being investigated at low power. Once a general survey of the object is completed, the analyst can then hone in on small details at higher magnification. When an interesting characteristic is identified, its morphology can be compared with that of a reference material and its optical properties identified via a polarizing microscope.

In the past two decades, microscopes have also been attached to spectrophotometers. In a spectrophotometer, a lamp emits radiation—either ultraviolet (UV) or infrared (IR)—that passes through a sample. The sample absorbs some of this light, and the rest passes to the spectrophotometer. There, the light from the sample is separated according to its wavelength (the visible range light looks like a rainbow), and the spectrum formed is observed with a detector. Such analysis (**spectroscopy**) is used to elucidate the composition of unknown materials in many different fields. Spectroscopy can also be used to match compounds, measure concentrations, and determine physical structures.

A **microspectrophotometer** combines the capabilities of a spectrophotometer with those of a microscope. The microscope magnifies the image of the specimen for the spectrometer and allows the forensic examiner to analyze extremely small samples or a small region of a sample. The spectrophotometer then measures the intensity of light at each wavelength. Microspectrophotometers (also called microspectrometers, microscope spectrophotome-

ters, and microphotometers) are extremely useful. In addition to measuring the light absorbed or transmitted by the sample, a microspectrophotometer can measure the intensity of light reflected from a sample, the intensity of light emitted when a sample fluoresces, or the intensity of polarized light after it has interacted with a sample.

Microspectrophotometers have numerous advantages over conventional spectrometers. For instance, they can easily take spectra of samples as small as 2.5 µm wide while viewing the sample directly. The latter feature allows for more precise measurements of the sample while eliminating interference from the surrounding material. Also, because many human-made substances—even colorless ones—absorb more in the UV region than in the visible light and IR regions, microspectrophotometers are very useful for analysis of synthetic fibers (discussed later in this chapter).

To see how this works, suppose that comparison of two fibers—one found at a crime scene and the other taken from a suspect—using a polarized light microscope indicates that the appearance and morphology of the two fibers are identical. When the fibers are analyzed in an IR microspectrophotometer, each produces an IR spectrum. At first glance, the two spectra appear very similar and indicate that both fibers are nylon. With a careful comparison of the two spectra, however, small differences become apparent. When these IR spectra are compared with the IR spectra of known reference fibers, one fiber is found to be nylon-6 and

SEE YOU IN COURT

In the late 1970s, a wave of child murders occurred in Atlanta. Wayne Williams was arrested for the offenses, based principally on the carpet fiber evidence found on the bodies, which was similar to carpet fibers found in his vehicle. Williams was convicted and sentenced to life in prison, where he remains today.

However, controversy has dogged this case since the moment the fibers were introduced into evidence. Contrary to the practice today, the jury was never told that the fibers were not unique, but rather heard that they matched carpet fibers that could be found throughout the Atlanta metropolitan area.

Probabilities are difficult to establish in many, if not most, physical evidence cases. Determining the likelihood of a match for new car carpet, for example, would be possible if one knew how many cars with that type and color of carpet were sold within a certain distance of the location of interest. Obviously, as the area included increases, the probabilities drop. Given that cars generally do not stay in confined spaces, one would have to make explicit assumptions to justify the development of a probability statement. No such evidence was employed in the Williams case, which is why this conviction remains tainted in the eyes of many modern forensic scientists.

the other nylon-6,6. Although the word *nylon* is used to describe an entire class of fibers, there are actually five common nylons sold commercially in the United States: nylon-6,6; nylon-6,12; nylon-4,6; nylon-6; and nylon-12. All nylons share some physical properties (such as solubility in solvent), but some subgroups differ in appearance and absorption of dye due to differences in molecular structure. As this example shows, the microspec-trophotometer has the ability to differentiate fibers that fall into the same generic class (nylon) but are structurally different (nylon-6 versus nylon-6,6).

Scanning Electron Microscopes

When the evidence available is very small and more magnification is needed, an SEM is often used as part of the forensic analysis (FIGURE 4-10A). An SEM

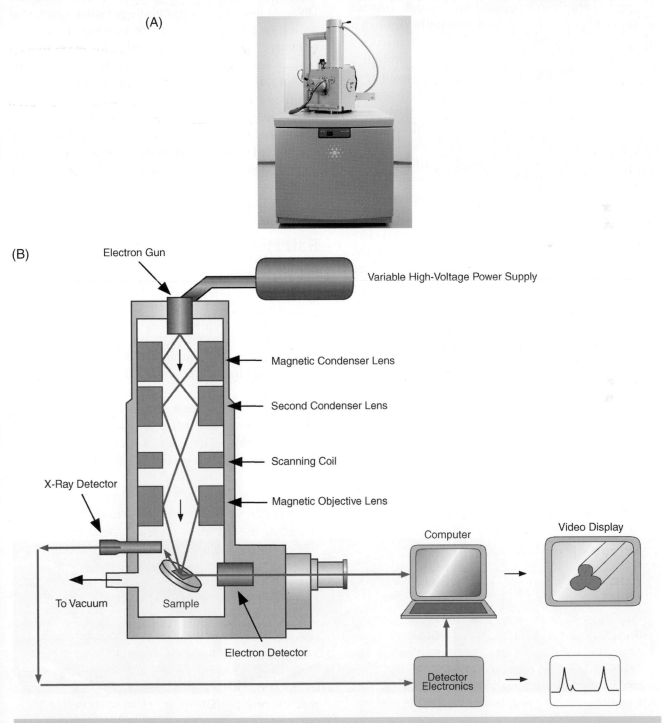

(A)

(B)

Electron Gun

Variable High-Voltage Power Supply

Magnetic Condenser Lens

Second Condenser Lens

Scanning Coil

Magnetic Objective Lens

X-Ray Detector

Computer

Video Display

To Vacuum

Sample

Electron Detector

Detector Electronics

FIGURE 4-10 (A) A scanning electron microscope (SEM). (B) A schematic diagram of an SEM.

can magnify an image 100,000 times and has a depth of focus that is more than 300 times that available with an optical microscope.

An SEM operates on the same basic principle as the polarizing microscope but uses electrons rather than light to identify properties of the object. Electrons, which are produced in the electron source found in the SEM, travel through a specimen in a way that is similar to a beam of light passing through a sample in an optical microscope. Instead of glass lenses directing light wavelengths through a specimen, however, the electron microscope's electromagnetic lenses direct electrons through a specimen (FIGURE 4-10B). Because the wavelength of electrons is much smaller than the wavelength of light, the resolution achieved by the SEM is many times greater than the resolution possible with the light microscope. Thus the SEM can reveal the finest details of structure and the complex surface of synthetic fibers (FIGURE 4-11).

Some of the electrons striking the surface of the sample are immediately reflected back toward the electron source; these are called backscattered electrons. An electron detector can be placed above the sample to scan the surface and produce an image from the backscattered electrons. Other electrons that strike the surface of the sample penetrate it; they can be measured by an electron detector that is placed beneath the sample. In addition, some of the electrons striking the surface cause X-rays to be emitted. By measuring the energy of the emitted X-ray, the elemental composition of the surface can be determined through a technique called energy dispersive X-ray spectroscopy (EDX). If a fiber is coated with a dulling agent, such as titanium dioxide, the EDX will detect its presence.

Forensic Applications of Microscopy: Hair

Hair is frequently the subject of forensic examination. This analysis has some limitations, however—namely, forensic examination of an individual hair will not result in definitive identification of the person from whom the hair was shed, unless the hair has a tag attached that can be analyzed for DNA. At best, a hair's morphology (when compared with a reference hair) may be consistent with the reference hair; it cannot be said definitively to be a perfect match. Hair does not possess a sufficient number of unique, individual characteristics to be positively associated with a particular person to the exclusion of all others. While hair samples are commonly used to exclude a suspect, they can be considered only as contributing evidence—to connect the suspect to the crime scene, for example, or to connect multiple crime scene areas to each other through transfer of evidence.

Hair Morphology

Hair is composed primarily of **keratin**, a very strong protein that is resistant to chemical decomposition and can retain its structural features for a long time. In fact, even hairs of ancient Egyptian mummies have been found to retain most of their structural features. This durability makes hair resistant to physical change and an excellent class of physical evidence.

Each strand of hair grows out of a tube-like structure known as the hair follicle, which is located within the skin. The shaft of each hair consists of three layers: the cuticle (the outer layer), the cortex

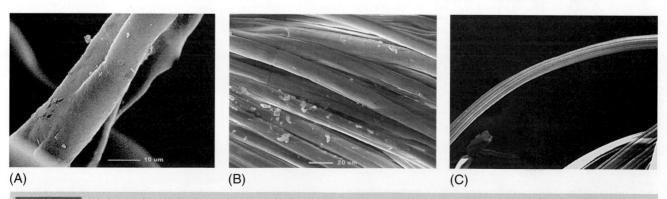

(A) (B) (C)

FIGURE 4-11 (A) Cotton fibers, (B) rayon fibers, and (C) polyester fibers, as seen through a scanning electron microscope.

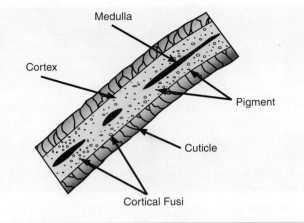

FIGURE 4-12 The shaft of a hair, showing the cuticle, cortex, medulla, pigment, and cortical fusi.

(the middle layer), and the medulla (the innermost layer) (**FIGURE 4-12**).

Cuticle

The **cuticle** consists of scales of hardened, flattened, keratinized tissue that overlie the cortex. These scales have a shape and pattern that are unique to the animal species from which they came—for example, the cuticle of a cat hair differs from the cuticle of a human hair. In addition, the acidity of modern shampoos can be adjusted to make these scales stand up or lay down. When the scales lay flat, hair reflects light and has an attractive luster—think of the attractive model's hair in a shampoo commercial.

One way to observe the scale pattern is to make a cast of the surface. To do so, the forensic examiner presses the hair into a soft material, such as clear nail polish or soft vinyl. Later, when the material hardens and the hair is removed, the impression of the cuticle is left behind in the clear plastic. This plastic will transmit light, so the impression of the cuticle can be examined on the plastic's surface with a compound microscope under high magnification.

Cortex

The **cortex** consists of an orderly array of spindle-shaped cortical cells, which is aligned parallel to the length of the hair. The cortex contains pigment bodies, which contain the natural dye melanin, and cortical fusi, which are irregular-shaped air spaces of varying sizes. Cortical fusi are commonly found near the root of a mature human hair, although they may be present throughout the length of the hair. As part of a side-by-side comparison of evidentiary hair samples and reference hair samples, the forensic examiner may compare the samples on the basis of the color, shape, and distribution of their pigment bodies and cortical fusi.

If hair has been dyed, dye may sometimes be observed in the cuticle and the cortex. If the hair has been bleached, the cortex will be devoid of pigment granules (though it may have a yellow tint).

The cortex can be observed with a compound microscope. The hair is usually mounted in a liquid medium that has about the same index of refraction as the hair. This choice minimizes the amount of reflected light and optimizes the amount of light that penetrates the hair, facilitating the observation of the cortex.

Medulla

The **medulla** consists of one or more rows of dark-colored cells that run lengthwise through the center of the hair shaft. The shape of the medulla may be cylindrical (human) or patterned (animals) (**FIGURE 4-13**). The pattern observed is specific to the particular species from which the hair came.

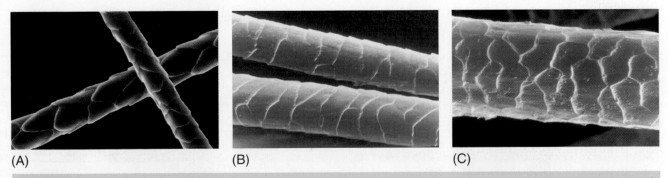

(A) (B) (C)

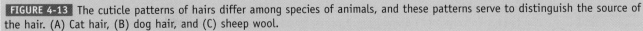

FIGURE 4-13 The cuticle patterns of hairs differ among species of animals, and these patterns serve to distinguish the source of the hair. (A) Cat hair, (B) dog hair, and (C) sheep wool.

The medulla may be a continuous, fragmented, or interrupted line of dark cells. The medullar ratio (the ratio of the medullar diameter to the hair diameter) for animal hair is generally greater than 0.5; the medullar ratio for human head hair is generally less than 0.33.

Hair Growth

The root of the hair (refer to Figure 4-12) is surrounded by cells that provide all the necessary ingredients to generate hair and to sustain its growth. Human hair goes through three developmental stages: the **anagenic phase**, **catagenic phase**, and **telogenic phase**. The size and shape of the hair root are different in each of these phases (**FIGURE 4-14**).

The anagenic phase is the initial growth phase, during which the hair follicle is actively producing the hair. During this phase, which may last as long as 6 years, the follicle is attached to the root by the dermal papilla. During the anagenic phase, the hair must be "pulled" to be lost, in which case it may have a **follicular tag**. The follicular tag includes cells from the root that contain DNA (**FIGURE 4-15**). Although a human hair itself has no genomic DNA, it does contain mitochondrial DNA. The follicular tag can be used for DNA typing for either genomic or mitochondrial DNA.

FIGURE 4-15 The follicular tissue may potentially include DNA, such as with this forcibly removed head hair with follicular tissue attached.

The catagenic phase is a short period of transition between the anagenic and telogenic phases. It lasts only 2 to 3 weeks. During this phase, hair continues to grow, but the root bulb shrinks, takes on a club shape, and is pushed out of the follicle.

During the telogenic (final) phase, hair naturally becomes loose and falls out. Over a 2- to 6-month period, the hair is pushed out of the follicle, causing the hair to be naturally shed. The former connection, the dermal papilla, is no longer attached to the hair but is assimilated by the follicle.

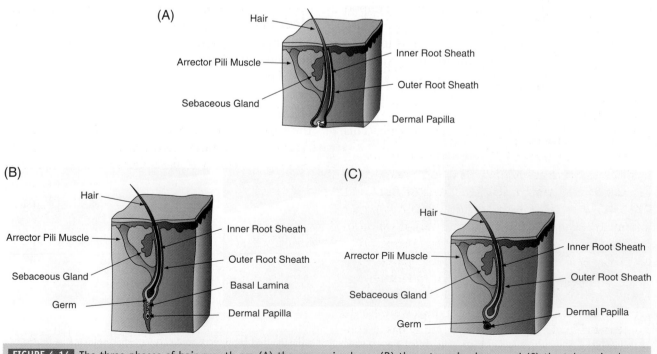

FIGURE 4-14 The three phases of hair growth are (A) the anagenic phase, (B) the catagenic phase, and (C) the telogenic phase.

Hair grows at a constant rate of approximately 1 cm per month. If the shed hair has grown after bleaching or coloring, an estimate of the time since chemical treatment can be established by assessing the presence (or absence) of color pigment or dye in the cuticle and cortex.

Comparison of Hair by Microscopy

When hair evidence is found at the scene of a crime, the forensic examiner must first answer two questions: (1) Is the hair human? and (2) does it match the hair of the suspect?

If the victim is covered with pet hair, a trained forensic examiner should be able to distinguish an animal hair from a human hair with little difficulty. To do so, the examiner will look for matches in published reference photos and use microscopy to reveal details of the sample's cuticle (i.e., scales), medullar ratio, and medullar pattern. As mentioned earlier, these characteristics are often unique to a particular species.

Human hair comparison is a much more difficult task. An examiner typically uses a comparison microscope to view the evidence hair and the known hair side by side. First, the examiner assesses the samples' color, length, and diameter. Next, the examiner compares features such as the shape of the medulla (if present) and the distribution, color, and shape of the pigment granules (including whether the hair has been colored or bleached). Other abnormalities due to disease, fungal infections, or physical damage can also be used as a basis for comparison if found in a specimen.

The microscopic comparison of human hair has been long accepted by the forensic community as a valid way to include and exclude suspect hairs against reference hairs—but only if the examiner has the proper training to make this determination. In 2002, a study by the FBI reported that significant error rates were associated with the microscopic comparison of human hairs (Houck & Budlowe, 2002). As part of this study, human hair evidence submitted to the FBI over a 4-year period was subjected to both standard microscopic examination and DNA analysis. Of the 80 hairs compared, only 9 (approximately 11%) for which the FBI examiners found a positive microscopic match between the questioned sample and a reference hair were found not to match by DNA analysis. This study has led police and the courts to regard microscopic hair comparisons as subjective; thus positive microscopic hair examinations must be confirmed by DNA analysis before they will be accepted as evidence in court. Microscopic examination of hair is, however, a fast and reliable technique for excluding questioned hairs from further scrutiny.

Collecting Hair Evidence

Hair evidence should be collected by hand, using a bright light to illuminate the area being searched. Hair evidence can be collected using a wide, transparent sticky tape, lint roller, or a special evidence vacuum cleaner (FIGURE 4-16). Strands of hair found as evidence should be individually packaged in paper packets. If the hairs have been collected on tape, the entire lint roller or sticky tape should be packaged, complete with attached hair evidence, in a polyethylene storage bag.

Control (reference) samples should be collected from the victim, the suspect, and any other individual who could have left hair evidence of his or her presence at the crime scene. Reference hairs should be taken from all pertinent regions of the body. Fifty head hairs, taken from different regions of the head, should be plucked (hair with roots is to be preferred) or cut close to the root. In sexual assault cases, 24 pubic hairs also should be plucked or cut close to the root.

The search for hair evidence should begin as soon as it is possible. Hair evidence is easily transferred to and from the crime scene, perpetrator, and victim. Waiting too long to begin the search will bring into question the actual relationship of the evidence to the crime.

Information Obtained by Microscopic Comparison of Hair

A microscopic examination of the hair sample may enable the examiner to determine the area of the body from which the hair originated. Scalp hairs, for

FIGURE 4-16 Hairs can be collected using tape or a special vacuum.

example, tend to have a consistent diameter and a relatively uniform distribution of pigment; in some cases, they may also be dyed, bleached, or damaged. By contrast, beard hairs may be relatively coarse due to constant trimming and tend to have blunt tips from cutting or shaving; an examination of a cross section of such a hair may reveal that it is triangular in nature. Pubic hairs, by comparison, have widely variable cross sections and shaft diameters; these short, curly hairs also may feature continuous medullae.

Microscopic examination cannot reveal the age or sex of the individual who was the source of a particular hair. The exception is infant hair; it tends to be short, with fine pigment granules. Sometimes, microscopic analysis may be able to suggest the race of the hair's owner. TABLE 4-1 lists the characteristics of human hair that are associated with particular racial groups. Head hairs are usually the best samples for determining race, although hairs from other body areas can prove useful as well. Racial determination from the microscopic examination of head hairs from infants can be difficult, however, given the rudimentary nature of hair in young children. Hairs from individuals of mixed racial ancestry may possess microscopic characteristics attributed to more than one racial group. DNA studies have shown that many people have DNA from multiple racial groups; thus race cannot be divided into clearly defined categories, as many people falsely believe. The identification of race is most useful as an investigative tool, but it also can be an associative tool when an individual's hairs exhibit unusual racial characteristics.

Forensic Applications of Microscopy: Fibers

The term *fiber* is used to designate any long, thin, solid object. In scientific terms, a fiber is said to have a high aspect ratio; that is, the ratio of its length to its cross-section diameter is large. Fibers of forensic interest are classified based on their origin and composition (FIGURE 4-17). Most fibers of forensic interest are very common and are encountered daily in homes and workplaces.

Investigators do not have to worry about the composition of fibers being altered over time because the vast majority of fibers do not undergo physical, biological, or chemical degradation while at a crime scene. If one part of the fiber in question has been damaged, microscopic measurements can usually be taken at many other points along the fiber. Because fibers are so small and light, they are easily transferred from one object to another. As a consequence, they often provide evidence of association between a suspect and a crime scene. Of course, if a fiber was transferred once during a crime, it may be easily transferred a second time—such as during the investigation of the crime scene. An investigator must, therefore, be extremely careful to secure all fiber evidence so it won't be accidentally lost and make certain to avoid any cross-contamination.

Most fiber evidence allows the examiner only to place it within a class. Not until a suspect has been identified and reference fibers collected from this individual will fiber evidence be individualized. The forensic examiner should begin the examination of fiber evidence with a low-power stereoscopic microscope. This technique allows the specific fibers in question to be selected and isolated for more detailed analysis.

Natural Fibers

Items of clothing, remnants of cloth, thread or yarn, and pieces of rope are all commonly found at crime scenes, and many are made from natural fibers. Natural fibers, which are derived from plant or animal sources, include materials such as cotton, flax, silk, wool, and kapok. In most textile materials, the natural fibers are twisted or spun together to form

TABLE 4-1

Characteristics of Human Hair

Race	Diameter	Cross Section	Pigmentation	Cuticle	Undulation
Negroid	69–90 μm	Flat	Dense and clumped	—	Prevalent
Caucasian	70–100 μm	Oval	Evenly distributed	Medium	Uncommon
Mongolian	90–120 μm	Round	Dense auburn	Thick	Never

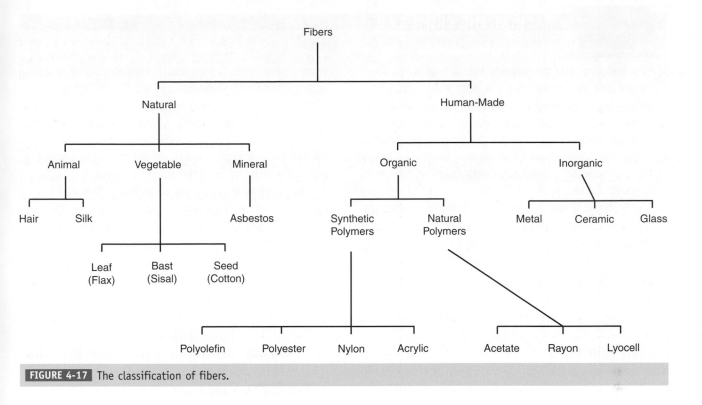

FIGURE 4-17 The classification of fibers.

threads or yarns that are woven or knitted into fabric. When examined under magnification, a thread made from natural fibers appears to be a collection of many small twisted hairs (FIGURE 4-18).

Cotton (a plant fiber) is the most widely used natural fiber. In fact, undyed white cotton is so widely used in clothing and other fabrics that it has almost no evidentiary value. Animal fibers include hair from sheep (wool), goats (cashmere), llamas, and alpacas as well as fur from animals such as rabbits and mink. Identification of animal fur follows the same procedure as the microscopic examination of animal hair described previously.

Yarn

Yarns are categorized into two primary groups: filament and spun. Filament yarns are made of a continuous length of a human-made fiber. Spun yarns are made of short lengths of fibers that are twisted or spun together so they adhere to each other to form a thread or yarn. Whether the thread is a filament or a spun yarn can be determined simply by untwisting the yarn. Filament yarns unravel into long strands of fibers, whereas spun yarns unravel into short lengths of fiber that can be easily pulled apart. Because yarn is made by twisting parallel fibers together, the twist direction is another factor that allows for comparison. Other physical properties used to compare yarn samples include their texture, the number of twists per inch, the number of fibers making up each strand, the blend of fibers (if more than one is present), the color, and pilling characteristics (i.e., tufts of fiber sticking out of the main bundle).

The examination of threads of yarn is normally done under a stereomicroscope using fine tweezers. The specimen is unraveled slowly and its features noted while being viewed under low power.

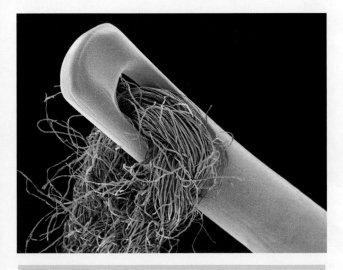

FIGURE 4-18 Threads made of natural fibers appear to be a collection of many small fibers.

Hair and fiber analysis provide a challenge when questioned and known samples are similar but cannot be proved identical through current conventional methods. A new set of methods using stable isotopes is currently being examined for its ability to provide quantitative comparison data even in situations in which today's tools provide only qualitative data (such as hair follicles without tags).

Isotopic forensics depends on the idea that the keratinized components of the body (nails and hair) are formed from the by-products of the food and water that the person has consumed. Regional differences in the isotopic composition of hair fibers can be used to determine whether a hair fiber found at a crime scene belongs to a person who has spent a significant period of time in the region or if it belongs to someone who was merely passing through. Likewise, clothing fibers will show specific isotopic ratios depending on either the location of plant growth or the chemical process through which the artificial fibers were made.

A caveat to this new set of methods: The variability of isotopic values, even within a homogenous population, makes specific identification nearly impossible. Like many of the tests described in this book, isotopic analysis will be used more often to exonerate suspects than to prosecute them.

Woven Fabrics

Woven fabrics are made by intertwining two sets of yarns that are placed at a 90° angle to each other on a loom. The lengthwise strand, which is known as the **warp**, runs the length of the fabric. The **weft** strands are interlaced at right angles to the warp strands. The device used to weave fabrics is called a loom (FIGURE 4-19). A loom can produce a variety of distinctive patterns, known as weaves.

The basic weaves are plain, twill, and satin. The plain weave is the most common weave (FIGURE 4-20); it produces a "checkerboard" pattern with the warp and weft threads. The twill weave produces a pattern where the warp thread travels across the fabric at a fixed angle (FIGURE 4-21). Denim fabric is one of the most common twill weaves. In the satin weave, the front of the fabric is composed almost entirely of warp threads produced

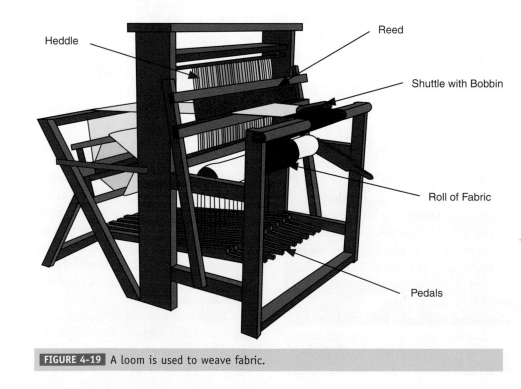

Heddle

Reed

Shuttle with Bobbin

Roll of Fabric

Pedals

FIGURE 4-19 A loom is used to weave fabric.

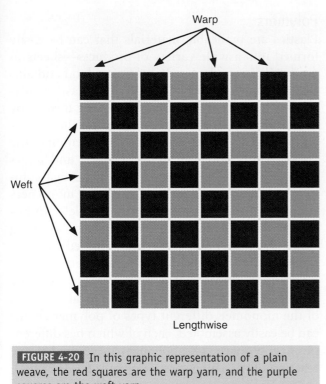

Warp

Weft

Lengthwise

FIGURE 4-20 In this graphic representation of a plain weave, the red squares are the warp yarn, and the purple squares are the weft yarn.

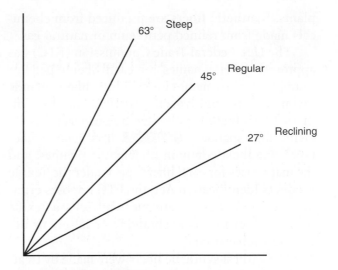

Steep 63°

Regular 45°

Reclining 27°

by repeating the weave (**FIGURE 4-22**). Satin fabrics appear smooth and shiny.

Fabric analysis is usually performed under a stereomicroscope in conjunction with tweezers, a ruler for counting threads, and a protractor for measuring angles. The forensic examiner records the colors of the threads and number of threads per inch, and determines the type of weave present in the cloth. The questioned fabric then is compared to known textile samples.

Synthetic Fibers

Rayon, the first human-made fiber, was introduced in the United States in 1910. The "wash and wear" revolution in the 1950s found Americans replacing their clotheslines with electric clothes dryers. The new wash-and-wear garments were made of acrylic and polyester fibers, which ensured that they dried wrinkle free. Since then, a wide variety of synthetic fibers have replaced natural fibers in many fabrics, garments, rugs, and other domestic products. Today, the U.S. fiber industry produces more than 9 billion pounds of fiber each year and has annual domestic sales that exceed $10 billion.

Modern manufactured fibers are classified into two categories: cellulosic fibers and synthetic fibers. **Cellulosic fibers** are produced from cellulose-containing raw materials, such as trees and other

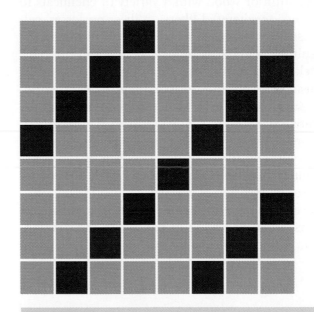

FIGURE 4-21 The twill pattern can be at different angles, such as in this 3-1 twill pattern.

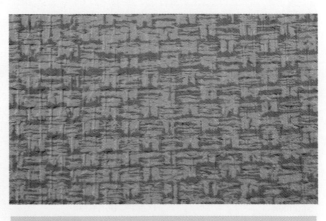

FIGURE 4-22 A typical satin weave.

plants. **Synthetic fibers** are produced from chemicals made from refined petroleum or natural gas.

The U.S. Federal Trade Commission (FTC) has approved generic names for each fiber type. For example, generic names for cellulosic fibers include rayon, acetate, and lyocell; generic names for synthetic fibers include polyester, nylon, acrylic, polyolefin, and spandex. **TABLE 4-2** lists generic fiber types, the most common trademarked names, and the major uses for each fiber type. Under the Textile Products Identification Act, the FTC imposes criminal sanctions against anyone who sells a textile fiber product that is misbranded or falsely or deceptively advertised.

The earliest synthetic fibers were made by treating cotton or wood with a variety of chemicals to produce **regenerated fibers**. In these processes, cellulose is extracted from the cellulosic fiber, and the resulting material is immersed in a solvent. This thick liquid is then forced through a small hole, causing the solvent to evaporate and a fiber to form.

In 1940, only 10% of the fibers used in the United States were synthetic. After World War II, however, chemists at DuPont learned how to create synthetic fibers using chemicals that are by-products of petroleum refinement. Their discoveries led to an explosion of new development in this area. The evolution of the technology to make these new fibers proceeded hand in hand with the development of the plastics industry.

Polymers

Plastics are malleable materials that can be easily formed into a wide variety of products—sheets of sandwich wrap, door panels of cars, food containers, or flexible threads. Chemists call plastics **polymers** because these huge molecules are formed by chemically linking together many smaller molecules. *Polymer* is derived from a combination of the Greek words *poly* (meaning "many") and *meros* (meaning "parts"). The individual parts that combine to form the polymer are known as monomers (from the Greek *monos*, meaning "single"). In homopolymers (from the Greek *homos*, meaning "same"), one type of molecule is joined to other molecules of the same type to form an enormously long linear thread; an example is polyethylene (**FIGURE 4-23A**). By changing the chemical structure of the monomer, different types of polymer chains can be easily assembled, each of which has different physical properties. Copolymers are formed by the reaction of two different monomers (**FIGURE 4-23B**).

Forming Synthetic Fibers

Most synthetic fibers are produced by the melt spinning process. First, the polymer is melted. The molten material is then forced (extruded) through a spinneret, a mold containing a variable number of holes (**FIGURE 4-24**). As it cools, the polymer hardens into a solid.

TABLE 4-2		
Synthetic Fibers		
Generic Fiber Type	**Common Trademark Names**	**Major Uses**
Acetate	Celanese Acetate®	Sportswear, lingerie, draperies, and upholstery
Acrylic	Cresloft®, Creslan®, Dralon®, Acrilan®, Duraspun®, Wear-Dated®	Sportswear, sweaters, socks, blankets, carpets, draperies, and upholstery
Lyocell	Tencel®, Lyocell by Lenzing®	Men's and women's dress clothing
Nylon	Condura®, Supplex®, Tactel®, Antron® Zeftron®, Vivana®, Caprolan®, Capima®	Ski apparel, windbreakers, luggage, rope, tents, and thread
Polyester	Dacron®, CoolMax®, Thermax®, Hollofil®, Thermoloft®, Fortrel®, ESP®, Polarguard®	Permanent-press garments, ties, dress apparel, lingerie, draperies, sheets, and pillow cases
Polyolefin	Herculon®, Innova®, Marvess®, Salus®, Telar®, Alpha Olefin®, Essera®, Kermel®	Pantyhose, knitted sportswear, carpets, upholstery, furniture, and slipcovers
Rayon	Modal®	Coat linings, lingerie, ties, bedspreads, and tablecloths
Spandex	Lycra®, Dorlastan®	Bathing suits, lingerie, and stretch pants

Source: Data from American Fiber Manufacturers Association, Fabric University (www.fabriclink.com).

(A)

Homopolymer

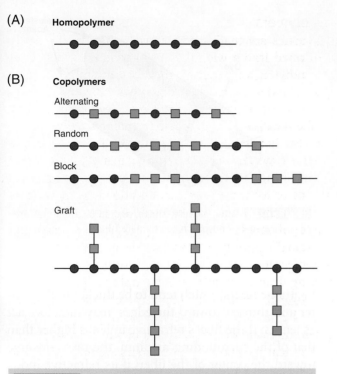

(B) **Copolymers**

Alternating

Random

Block

Graft

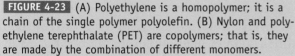

FIGURE 4-23 (A) Polyethylene is a homopolymer; it is a chain of the single polymer polyolefin. (B) Nylon and poly-ethylene terephthalate (PET) are copolymers; that is, they are made by the combination of different monomers.

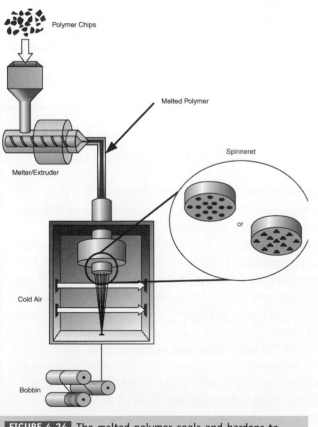

Polymer Chips

Melted Polymer

Melter/Extruder

Spinneret

or

Cold Air

Bobbin

FIGURE 4-24 The melted polymer cools and hardens to form a solid after it is forced through the spinneret.

The shapes of the holes in the spinneret determine the cross-sectional shape of the polymer—for example, round, trilobal, flat, or dumbbell (FIGURE 4-25). Round (hollow) fibers trap air, creating insulation (e.g., in synthetic down comforters). Tri-lobed fibers reflect more light, so they are often used to give an attractive sparkle to textiles. Pentagonal-shaped fibers, when used in carpet, show less soil and dirt. Octagonal-shaped fibers offer glitter-free effects. Not surprisingly, the forensic examiner may be able to use these characteristics to place a synthetic fiber within a particular class.

Companies may use a variety of techniques to dye fibers—sometimes even using different dyeing techniques for the same fiber. If pigments are used to color the fiber, they will be attached to the surface as granules. In contrast, dyes will penetrate farther into the fiber. Many fibers have treated surfaces. For example, shiny fibers may be treated with dulling agents, such as titanium dioxide, to make the fibers appear less lustrous.

Comparison and Identification of Synthetic Fibers

The first step in the comparison of synthetic fibers is an examination of the fibers with a comparison microscope, paying special attention to features such as color, diameter, cross-section shape, pitting or striations, and the presence of dulling agents

Man-Made Fibers

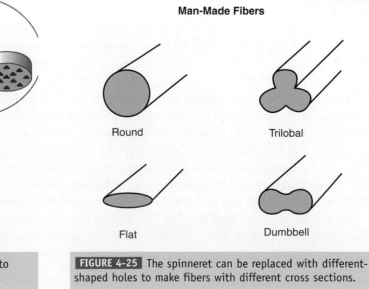

Round

Trilobal

Flat

Dumbbell

FIGURE 4-25 The spinneret can be replaced with different-shaped holes to make fibers with different cross sections.

(**FIGURE 4-26**). The main advantages of microscopic comparison are as follows:

- This technique does not destroy the fiber, unlike other analysis techniques.
- Microscopic comparison is not limited by the sample size; a fiber 1 mm in length can be easily examined.
- Every forensic laboratory has a comparison microscope readily available.

Once the microscopic comparison has shown that the gross morphology of the two fibers is the same, the chemical composition of the samples is compared. Additional tests should be performed to determine whether the fibers belong to the same generic class. Additionally, the analyst should try to place the fiber into a specific polymer subclass. (Recall the earlier example, in which a fiber was first classified as a nylon, and then further classified as nylon-6 versus nylon-6,6.)

The refractive index is another useful physical property that can be used for identification of synthetic fibers. As mentioned earlier, synthetic fibers are manufactured by drawing a molten polymeric substance through a spinneret to create a fiber. The force exerted during this extrusion process causes the molecules in the polymer to align along the length of the fiber, which gives the fiber not only strength but also the ability to refract light.

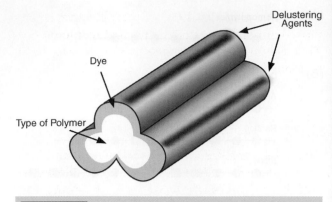

Dye
Type of Polymer
Delustering Agents

FIGURE 4-26 Color, texture, shape, and surface features are all used as bases on which to compare fibers.

Synthetic fibers, which tend to be thicker in the center and thinner toward the edges, may therefore act as lenses. If the fiber's refractive index is higher than that of the surrounding medium, the rays converge toward the center of the fiber; if its refractive index is lower than that of the surrounding medium, the rays diverge toward the edge of the fiber. The internal reflection of light within the fiber is due to the presence of minute differences in the solid state of the fiber. The forensic examiner can identify the refractive index of a fiber by immersing the fiber in a fluid with a comparable refractive index and observing the disappearance of the **Becke line** (the

on the CRIME SCENE Fiber Evidence

Fiber evidence, unlike fingerprints or DNA, cannot pinpoint an offender in any definitive manner. However, this forensic tool has been critical in connecting pieces in crime puzzles. For example, fibers were used to link two Ohio murders in the 1980s that otherwise might have never been connected or solved.

In 1982, Kristen Lea Harrison was abducted from a ball field in Ohio. She was found 6 days later some 30 miles away; she had been raped and strangled to death. Investigators suspected that the orange fibers found in her hair were similar to those found on a 12-year-old female murder victim from 8 months earlier in the same county. These oddly shaped (tri-lobed) polyester fibers led forensic scientists to postulate that they were carpet fiber. Additional evidence near Kristen's body included a box for a special kind of van seat, and plastic wrap around her ankles was linked to the unknown killer. Unfortunately, the leads ended there.

Several days later, however, police noticed a van similar to the one Kristen had been forced to enter. A 28-year-old woman who was abducted, held prisoner, and tortured managed to escape from this van. It proved to have orange carpeting that matched the fibers in Kristen's hair. The unique color enabled scientists to trace it to a manufacturer that supplied information about its limited distribution. Because only 74 yards of this carpet had been shipped to that area of Ohio, investigators were able to quickly narrow down the possibilities. Other evidence established a more solid link, and Robert Anthony Buell was eventually convicted.

bright line that develops as the objective lens is moved out of focus) under a polarizing microscope. This technique is not limited by sample size, and it is nondestructive.

Paint

Crime labs analyze a large number of paint chips. Paint is transfer evidence, and it can provide important clues in investigations, especially in burglaries and vehicular crimes. Paints are applied to many surfaces in our modern society, either because they protect the underlying material or for aesthetic purposes where a beautiful surface is desired. Water-based and oil-based paints are composed of three major ingredients: (1) a pigment, (2) a solvent, and (3) a binder.

The **pigment**, either organic or inorganic, provides the desired color. The primary pigment in white paint is inorganic, usually titanium oxide (TiO_2) or zinc oxide (ZnO). These compounds are white or opaque, and they serve to cover any previous color on the surface to be painted. White lead [$2 PbCO_3 \bullet Pb(OH)_2$] was used extensively as a primary pigment until 1977, when it was banned for interior use because of its toxicity. Pigments—such as carbon black, iron oxide (Fe_2O_3, brown and red), cadmium sulfide (CdS, orange), chromium oxide (Cr_2O_3, green), and various organic compounds—are mixed with the primary pigment to obtain the desired color. Blue and green colors are made by adding organic pigments.

The **solvent** in a paint is the liquid in which the pigment, binder, and other ingredients are suspended, usually in the form of an emulsion. Once the paint has been spread, the solvent evaporates and the paint dries. In oil-based paints, the solvent is usually a mixture of hydrocarbons obtained from petroleum. A compatible organic solvent must be used for thinning the paint and cleaning up after painting. This is usually turpentine, which, like most other suitable solvents, gives off hazardous fumes. In water-based paints, the solvent is water, which can be used both for thinning and for cleaning up.

The **binder** is a material that polymerizes and hardens as the paint dries, forming a continuous film that holds the paint to the painted surface. The pigment becomes trapped within the polymer network. In older oil-based paints, the binder was usually linseed oil. As the solvent evaporates and the paint hardens, the linseed oil chemically reacts with atmospheric oxygen to form a cross-linked polymeric material.

The latex binder in water-based paint is dispersed into the paint and starts to coalesce and form a solid film as water evaporates. The slightly rubbery nature of the partially polymerized material explains why water-based paints are called latex paints. As the water evaporates, further polymerization occurs, and the paint hardens.

Several synthetic polymers are used as binders in latex and acrylic paints. Paints with poly (vinyl acetate) binders now account for about 50% of the paint used for interior work. Those with polystyrene-butadiene binders are less expensive but do not adhere quite as well and have a tendency to yellow. Acrylic paints, made with acrylonitrile binders, are considerably more expensive than other types of paints, but they are excellent for exterior work. They adhere well and are washable, very durable, and resistant to damage by sunlight. In most respects, they are superior to oil-based paints, which they have now largely replaced.

Automobile Paint

Automobile paint is much more complicated. During the manufacturing of automobiles, several different layers of paint are applied (FIGURE 4-27). Each layer has its own unique function. The **electrocoat primer** is an epoxy resin that is applied to the steel body of the car for corrosion resistance. This primer is black or gray in color. Next, the **primer surfacer** is applied to hide any minor imperfections in the surface. This primer is an epoxy—modified polyester or polyurethane. Pigment is added to this primer to cover and hide the electrocoat primer. The next layer of paint is called the

FIGURE 4-27 Several layers of paint are applied to automobiles during the production process.

basecoat. This layer, which is an acrylic-based polymer, gives the automobile its beautiful color. The basecoat might also contain aluminum or metal oxide powders to give the paint a metallic look. Finally, an unpigmented acrylic **clearcoat** is applied to provide resistance to scratches, solar radiation, and acid rain. Not only is each layer a different color, but the organic resin and pigment in each layer are different. A careful analysis that reveals the layer structure will allow an examiner to individualize the paint chip.

Collection of Paint Evidence

Paint evidence is most likely to be found in crimes involving burglary or involving a vehicle. When a paint chip is located, the investigator must be extremely careful to keep the paint chip intact. Paint chips can be handled with forceps or picked up by sliding a thin piece of paper under the chip. It is best not to attempt to remove paint that is smeared on clothing or other objects. The object should be carefully packaged and sent to the laboratory.

A reference sample must be collected for comparison. This sample should always be collected from an undamaged area for comparison in the laboratory. When collecting a sample from a vehicle involved in an accident, it is very important that the collected paint be as close to the point of impact as possible. Remember that not all surfaces of an automobile fade at the same rate, and parts of the car may have been repainted. When collecting an automotive standard the investigator must be careful to remove all the layers of paint (including the undercoat layers) down to the metal (or plastic) body surface. This is easily accomplished by using a sharp scalpel to remove the paint reference sample. If there may have been a cross transfer of paint between two vehicles, all layers of paint need to be removed from each vehicle.

In cases of burglary, tools used by the burglar may have trace paint evidence attached. If such a tool is found, the investigator should package it and send it to the laboratory for examination. A reference paint sample should be collected from an area that has been in contact with the tool, such as a window sash or a doorjamb. Care must be taken to remove all the layers of paint down to base material (e.g., wood, plastic, aluminum). If there are many coats of paint on the window or door that was the

FIGURE 4-28 A jigsaw match is conclusive evidence.

point of entry for the burglar, these paint chips may have unique layers.

An investigator needs to be very careful when handling paint evidence. A physical "jigsaw" match (**FIGURE 4-28**) of a questioned paint chip and an area where the reference paint is being collected is conclusive evidence. The fit of the edges of the two (questioned and reference chips), any surface markings, and the layer structure will establish the uniqueness of the match and will individualize the match.

Forensic Analysis of Paint

Forensic paint analysis begins with a comparison of the paint chip in question and the paint chip from a known source. Considering that there are thousands of paint colors, it is not surprising that color is the most important forensic characteristic. The color, texture, and layer sequence can be easily observed by placing the questioned and known samples under a stereomicroscope (**FIGURE 4-29**) or an illuminated desk magnifier. The layer structure order, color, thickness, and other details should be recorded. It may be necessary to cut the paint chip so that necessary information can be collected. Using a scalpel to cut through the chip can help reveal the layer structure. Sometimes the chip is cast into a block of epoxy, and microtome is used to cut the chip and reveal its layer structure. Samples embedded in epoxy make it possible to grind and polish the edge so that physical details such as pigment size and distribution can be determined.

If the paint chips are big enough, destructive chemical tests on a part of the chip can provide ad-

FIGURE 4-29 Cross section (edge) of a paint chip magnified by a polarized light microscope (PLM), showing four layers of paint.

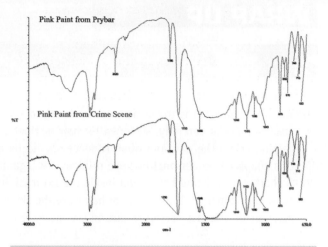

Pink Paint from Prybar

Pink Paint from Crime Scene

FIGURE 4-30 Infrared spectroscopy shows that the paint on a suspect's prybar has an IR spectrum that is identical to paint found at the crime scene.

ditional data. The application of solvent to a paint chip will cause swelling or the generation of colors. This information may help to identify resins and pigments in the chip. These tests can be performed in a porcelain spot plate with very small paint chips under a stereomicroscope. By comparing the reaction of the questioned and known samples to different solvents, trained microscopists can determine if the two came from a common source. Infrared microspectrophotometry is routinely used for paint analysis. Any infrared beam is reflected from the surface of the paint chip to produce an infrared spectrum of the surface. The infrared spectra is then interpreted and compared with the infrared spectra of known organic materials. In this way, the identity of the major organic components present in the paint chip can be determined (FIGURE 4-30).

An SEM with an energy-dispersive X-ray (EDX) can be used to identify the inorganic pigments in a paint sample. The layer structure can be studied in more detail by using the high magnifying power of the SEM, and the identity of the pigment can be determined by the EDX. The chip is cast into an epoxy block and a side is cut away with a microtome. FIGURE 4-31 shows a chip standing on end, with its layer structure exposed, giving a picture of not only the layers but also their elemental composition.

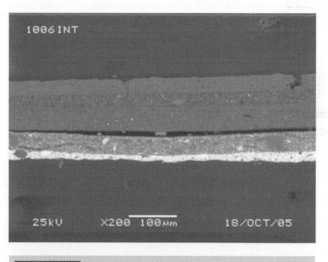

1006 INT

25kV   X200 100μm   18/OCT/05

FIGURE 4-31 SEM-EDX analysis of the same paint chip as Figure 4-29. This technique shows that there is a thin fifth layer of paint (on the bottom) that was not visible by PLM.

WRAP UP

YOU ARE THE FORENSIC SCIENTIST SUMMARY

1. The black ski mask is the same color as the ski mask worn by the robber and probably has been used by him. This mask should be carefully examined for hair or fiber evidence as well as any personal hygiene (e.g., cologne, hair gel) residues. The hair and fiber evidence should be characterized using the microscopic techniques described in this chapter. The personal hygiene residues are organic material; we will learn how to analyze them in Chapter 10.

2. A short, brown beard hair found inside the ski mask indicates the transfer of hair evidence. It is reasonable to assume that a man with this color of hair wore the ski mask at some point. Although the person with the brown beard may not be the last person to put on the mask, it is reasonable that the police should follow this lead.

Chapter Spotlight

- The microscopic techniques used to investigate trace fiber and hair evidence focus on their morphology. The methods employed for morphological comparisons of questioned and control samples often are used to show that two samples do not match, rather than that the individual who left behind the questioned trace sample was the same person who was the source of the control sample.

- The common types of microscopes used in processing trace evidence are the compound microscope, the comparison microscope, the polarizing (optical) microscope, the microspectrophotometer, and the SEM.

- In addition to identifying the refractive index, which provides information about fiber composition, microspectrophotometry can be used to analyze the composition of unknown materials, particularly dyes on fibers.

- SEMs are capable of magnifying an object 10 to 100,000 times. They are used to compare fiber surface morphologies between samples and, when combined with EDX, provide compositional information.

- The portions of a hair that are vital to forensic examination are the cuticle, cortex, and medulla. These layers of the hair shaft may reveal important information about the source of the hair (e.g., if it is human, whether it belongs to the suspect).

- At a crime scene, collection of hair and fiber evidence can be done by using a vacuum with specialized filters and/or using lift tape. In the lab, collection of hair and fiber evidence can be done by gently shaking the evidence over white paper and by using a vacuum or lift tape. Collection of hair and fiber evidence from a body generally involves the use of lift tape.

- Once a sufficient number of hairs have been collected (50 head hairs), these hairs are compared under a microscope, and the morphology is noted and matched to standards and unknown samples from other sources.

- Fiber evidence can be classified as natural (e.g., cotton, flax, wool) or synthetic (e.g., polyester).

- Fibers are compared with standards provided by manufacturers when available.

Amorphous material A solid without order in the arrangement of its atoms.

Anagenic phase The initial phase of hair growth, when the hair follicle is producing hair.

Anisotropic material Material that appears different when the direction of observation is changed.

Basecoat The layer of automotive paint that contains the colored pigment.

Becke line A bright line that develops as the objective lens of a microscope is moved out of focus.

Binder The material that hardens as the paint dries, forming a continuous film.

Birefringent material An anisotropic material.

Catagenic phase The intermediate stage of hair growth, which occurs between the anagenic and telogenic phases.

Cellulosic fibers Fibers that are produced from cellulose-containing raw materials, such as trees or other plants.

Clearcoat Outermost layer of automobile paint that contains no pigment.

Comparison microscope Two microscopes linked by an optical bridge.

Compound microscope A microscope with one body tube that is used for magnification in the range of 25 to 1200 times.

Condenser A lens under the microscope stage that focuses light on the specimen.

Cortex The body of the hair shaft.

Cuticle A scale structure covering the exterior of the hair.

Depth of focus The depth of the area of the specimen that is in focus.

Electrocoat primer First layer of paint applied to the steel body of an automobile.

Field of view The part of the specimen that can be seen through the microscope lenses.

Follicular tag Tissue surrounding the hair shaft that adheres to hair when it is pulled out.

Illuminator The part of a microscope that illuminates the specimen for viewing.

Isotropic materials Materials that have the same optical properties when observed from any direction.

Keratin The primary protein that forms hair and nails.

Medulla A column of cells running down the center of the hair.

Micrometer One-millionth of a meter (μm).

Microspectrophotometer A microscope that measures the interaction of infrared or ultraviolet radiation with a sample.

Objective lens The lower lens of a microscope; the lens closest to the specimen.

Ocular lens The upper lens of a microscope; the lens nearest to the eye.

Pigment Added to paint to give it color.

Plane-polarized light Light that oscillates in only one plane.

Pleochroism A property of a substance in which it shows different colors when exposed to polarized light from different directions.

Polarizer A lens that passes light waves that are oscillating only in one plane.

Polarizing microscope A microscope that illuminates the specimen with polarized light.

Polymer A large organic molecule made up of repeating units of smaller molecules (monomers).

Primer surfacer A layer of automobile paint that slows corrosion of the underlying steel.

"Real" image The actual nonmagnified image.

Refraction The bending of light waves.

Refractive index A ratio of the velocity of light in a vacuum to its velocity in any other medium.

Regenerated fibers Fibers made by treating cotton or wood with a variety of chemicals.

Scanning electron microscope (SEM) A microscope that illuminates the specimen with a beam of electrons.

Solvent The liquid in which the components of paint are suspended.

Spectroscopy Measurement of the absorption of light by different materials.

Stereoscopic microscope A microscope with two separate body tubes that allow both eyes to observe the specimen at low or medium magnification.

Synthetic fibers Fibers produced from chemicals made from refined petroleum or natural gas.

Telogenic phase The final phase of hair growth, during which hair falls out of the follicle.

Virtual image An image that is seen only by looking through a lens.

Warp Lengthwise strand of yarn on a loom.

Weft Crosswise strands of yarn on a loom.

Working distance The distance between the object being investigated and the objective lens.

Putting It All Together

Fill in the Blank

1. A magnifying glass is a(n) _____ lens that is thicker in the middle than at the edge.

2. How much light bends depends on the change in _____ as the light enters and leaves the magnifying glass.

3. When using a magnifying glass, the magnified image is known as the _____ image.

4. A(n) _____ microscope has two lenses that are at fixed distances from one another in a hollow tube.

5. The lower lens in a compound microscope is the _____ lens.

6. The upper lens in the eyepiece of the compound microscope is the _____ lens.

7. A compound microscope can magnify objects up to _____ times.

8. As the numerical aperture (NA) of a microscope lens doubles, it is able to resolve details that are _____ as close to one another.

9. As the magnifying power increases, the field of view _____ (increases/decreases).

10. The thickness of the region that is in focus when using a compound microscope is called the _____.

11. The light rays from the illuminator are condensed and focused through the _____ lens.

12. Side-by-side comparisons of specimens are best performed by using a(n) _____ microscope.

13. The _____ microscope has two eyepieces.

14. The distance between the objective lens and the specimen is the _____.

15. _____ materials have the same optical properties when observed from all directions.

16. As the direction of observation is changed, _____ materials will change their appearance.

17. _____ is the property of a substance in which it shows different colors when exposed to polarized light coming from different directions.

18. The microspectrophotometer is an instrument that attaches a spectrophotometer to a(n) _____.

19. With a microspectrophotometer, the UV, visible, and _____ spectrum of the sample can be measured.

20. A spectrophotometer measures the light intensity as a function of _____ after the light has interacted with the sample.

21. The electrons that are immediately reflected back toward the electron source in a scanning electron microscope are called _____ electrons.

22. Hair is composed primarily of the protein _____.

23. A human hair has three layers: the _____, the _____, and the _____.

24. The medulla of human hair is _____ (cylindrical/patterned).

25. The initial phase of hair growth is known as the _____ (catagenic/anagenic/telogenic) phase.

26. During the final phase of hair growth, known as the _____ phase, hair becomes loose and falls out.

27. The three most basic weaves in fabrics are _____, _____, and _____.

28. Cellulosic fibers are produced from raw materials from trees or plants that contain _____.

29. Synthetic fibers are produced from chemicals made from refined _____.

30. Synthetic fibers made from cellulose fibers are also known as _____ fibers.

31. _____-shaped synthetic fibers reflect more light and give an attractive sparkle to textiles.

32. The first step in comparing two synthetic fibers is to examine the fibers with a(n) _____ microscope.

33. Synthetic fibers can be identified by comparing their _____ _____.

34. Paints have three major components. They are _____, _____, and _____.

35. Paint evidence is especially important in _____ and _____ crimes.

36. Other than color, what is the most important physical property of a paint chip when making a comparison? _____

37. Paint chips can be picked up by sliding a piece of _____ under the chip.

38. A reference paint sample can be obtained from an automobile by using a(n) _____.

39. A unique fit of the edges of two paint chips is called a(n) _____ match.

40. Sometimes a paint chip is cast into a block of epoxy and cut with a microtome to reveal its _____ structure.

41. Infrared microspectrophotometry will reveal the _____ components of a paint chip.

42. Destructive chemical tests can be carried out on paint chips by applying _____ to the paint chips.

True or False

1. Microspectrophotometers can be used to determine fiber subgroups.

2. The race of the suspect may be indicated by microscopic hair analysis.

3. The age of the individual can be determined by microscopic hair analysis.

4. Hair grows at a constant rate of approximately 2 cm per month.

5. The first step in the comparison of synthetic fibers is an examination of the fibers with a comparison microscope.

6. When obtaining a reference paint sample from an automobile, only the top layer of paint should be removed.

7. Blue and green paints are made with organic pigments.

8. A reference paint sample can be obtained from any surface of the automobile being examined.

9. Using an SEM-EDX to analyze the layer structure of a paint chip will give the identity of the pigment in each layer.

10. The color, texture, and layer sequence of a paint chip can be easily observed by placing it under a stereomicroscope.

Review Problems

1. The medical examiner (ME) is examining a corpse from a crime scene. Describe in detail how the ME should collect and preserve hair evidence from the victim's head and clothes. Which of the hairs should be sent to the DNA unit for analysis?

2. You learn from a witness that the corpse from which the ME took samples in Problem 1 lived with a cat. Describe in detail how cat hair differs from human hair. Draw a diagram of a human hair and a cat hair. Show all parts and label the diagram.

3. You have been asked to examine a woven fabric. Describe in detail how you would perform your analysis. Which details about the fabric would you record?

4. A human body was recovered from a remote wooded site. The individual, who had naturally brown hair, regularly bleached his hair, and he had brown hair roots showing. His roommate remembered the day he last bleached his hair. The brown roots of the hair measured 0.25 cm. How much time elapsed between this date and the day he died?

Further Reading

Bell, S. *Forensic Chemistry.* Upper Saddle River, NJ: Prentice-Hall, 2006.

Bisbing, R. E. The forensic identification and association of human hair. In R. Saferstein (Ed.), *Forensic Science Handbook, vol. 1* (ed 2). Upper Saddle River, NJ: Prentice-Hall, 2002.

Caddy, B. *Forensic Examination of Glass and Paint.* New York: Taylor & Francis, 2001.

Deedrick, D., and Koch, S. L. Microscopy of hair—part 1: A practical guide and manual for human hairs. *Forensic Science Communications.* 2000; 6(1). http://www.fbi.gov/hq/lab/fsc/backissu/jan2004/research/2004_01_research01b.htm.

Eyring, M. B., and Gaudette, B. D. An introduction to the forensic aspects of textile fiber examination. In R. Saferstein (Ed.), *Forensic Science Handbook, vol. 2* (ed 2). Upper Saddle River, NJ: Prentice-Hall, 2005.

Forensic Fiber Examination Guidelines Scientific Working Group on Materials (SWGMAT). *Forensic Science Communications.* 1999; 1(1). http://www.fbi.gov/hq/lab/fsc/backissu/april1999/houcktoc.htm.

Forensic Human Hair Examination Guidelines. Scientific Working Group on Materials Analysis SWGMAT. *Forensic Science Communications.* 2005; 7(2). www.fbi.gov/hq/lab/fsc/current/standards/2005_04_standards02.htm.

Houck, M. M., and Budlowe, B. Correlation of microscopic and mitochondrial DNA hair comparisons. *Journal of Forensic Sciences.* 2002; 47:964.

Oien, C. T. Forensic hair comparison: Background information for comparison. *Forensic Science Communications,* 2009; 11(2). http://www.fbi.gov/hq/lab/fsc/current/review/2009_04_review02.htm.

Petraco, N., and Kubic, T. *Color Atlas and Manual of Microscopy for Criminalists, Chemists, and Conservators.* Boca Raton, FL: CRC Press, 2004.

Standard Guide for Microscopic Examination of Textile Fibers in ASTM Standard E 2228-02. ASTM International, 2004.

Thornton, J. L. Forensic paint examination. In R. Saferstein (Ed.), *Forensic Science Handbook, vol. 2* (ed 2). Upper Saddle River, NJ: Prentice-Hall, 2002.

Standard Working Group for Materials Analysis (SWGMAT). Standard guide for microspectrophotometry and color measurement in forensic paint analysis. *Forensic Science Communications,* 2007; 9(4). http://www.fbi.gov/hq/lab/fsc/backissu/oct2007/standards/2007_10_standards01.htm.

http://criminaljustice.jblearning.com/criminalistics

Answers to Review Problems

Interactive Questions

Key Term Explorer

Web Links

http://criminaljustice.jblearning.com/criminalistics

Forensic Analysis of Glass

When a pane of glass shatters, small, sharp pieces called shards are thrown over a wide area. Larger pieces travel in the direction of the blow and are usually found close to the original location of the glass pane. Smaller shards can be propelled up to 10 ft from the pane, also in the direction of the blow. If a pane of glass is shattered by a violent blow, hundreds of tiny backscattered shards will inevitably become caught in the hair or clothing of the person who broke the pane. Because these shards are so small (less than 1 mm long), they are easily dislodged. The speed at which they fall off the perpetrator depends on the type of clothing worn by the individual and his or her subsequent activities. Most shards are lost in the first hour after the event, and the probability of finding glass evidence on a suspect decreases over time.

Investigators collect these tiny shards from a suspect by combing the suspect's hair and shaking his or her clothing over a clean piece of paper. If two or more glass shards from a suspect's hair or clothes are found to be indistinguishable from a control sample of glass from the scene, they can be considered significant associative evidence.

1. A car has its driver's window smashed during an attempted robbery. A suspect who is running down the street is stopped by police. He claims to have nothing to do with the car. What should be the officers' next step?
2. The lab finds glass shards on the suspect's sweatshirt. Which tests should now be done on the glass fragments?

Introduction

Glass has been shown to be very useful evidence because it is often encountered in criminal investigations. For example, when a burglar breaks a pane of window, small fragments of glass are often showered onto his or her hair, clothing, or shoes, and these fragments can later be found on the suspect as transfer evidence. This chapter describes the many different types of glass commonly found at crime scenes and explains how glass fragments can be placed into specific classes through the use of optical and non-optical analysis methods. In addition, the chapter describes how to individualize a glass fragment by making a **fracture match**.

Types of Glass

Glass is a solid that is not crystalline but rather has an amorphous structure. The atoms of an **amorphous solid** have a random, disordered arrangement, unlike the regular, orderly arrangement that is characteristic of **crystalline solids**. Another characteristic property of glass is that it softens over a wide temperature range rather than melting sharply at a well-defined temperature.

Soda-lime glass is the glass commonly used in most windows and bottles. It consists of 70% silicon dioxide (SiO_2), 15% sodium oxide (Na_2O), 10%

calcium oxide (CaO), and 5% other oxides. This type of glass is made by heating together sodium carbonate (baking powder), calcium oxide (lime) or calcium carbonate (limestone), and silicon dioxide (sand). Soda-lime glass has a green to yellow tint, which is most easily seen by looking at the edge of the pane. This color is caused by an iron impurity that is present in the sand. Soda-lime glass starts to soften when it is heated to a temperature of more than 650°C, a fact that can prove useful when investigating fires. For example, if the windows of a burned building are found to be deformed (melted), the temperature of the fire must have exceeded 650°C. Common window glass fractures when its surfaces or edges are placed under tension, and an edge fissure may propagate into visible cracks.

A variety of metal oxides can be added to this basic recipe to give glass a special appearance. For example, the addition of lead oxide (PbO) will give the glass a high brilliance because of its greater internal reflection of light; for this reason, lead glass is used for expensive crystal dinnerware. The addition of cobalt oxides will make the glass blue, manganese oxide will make it purple, chromium oxide will make it green, and copper oxide will make it red or blue-green.

In 1912, the Corning Glass Company found that the addition of 10% to 15% boron oxide (B_2O_3) to glass made the resulting product more shock-

Glass Fragments Solve Hit-and-Run

On June 24, 1995, on the island of Providenciales in the Turks and Caicos Islands, a passerby reported to police that a woman's body was lying on the side of the Leeward Highway. The police found the 42-year-old woman lying face down. Upon investigation, police concluded that the victim had been struck by a car sometime after midnight while walking home from her job as a waitress.

The local constable carefully documented the area surrounding the body. His report of items scattered around the body included earrings, a watch, a pendant and chain, eyeglasses, debris from the undercoating of a vehicle, and nine large glass fragments. The constable photographed the items in their original positions, measured distances from the body to the found objects, and collected soil samples from the surrounding area. These items were packaged and sent to the Miami-Dade Police Crime Laboratory in Florida for analysis.

Eleven days later, a suspect was identified when neighbors reported that his car was missing a headlight. The suspect denied being involved in the accident and requested that his attorney be present for any further questioning. Upon gaining access to the suspect's car, the police found considerable damage to the driver's-side front fender as well as a missing headlight on that side of the vehicle. Since this was an older car, each side had two headlights, each of which contained glass lenses. Because the car had been washed, a careful examination of the vehicle did not reveal any biological material. However, glass fragments were found lodged in the bumper and inside the lamp assembly of the missing light. The constable collected these fragments and samples of debris from the undercarriage of the car for further analysis.

The Miami-Dade Crime Laboratory analyzed the glass fragments in particular to determine whether an association existed between the glass fragments found on the crime scene and the glass fragments found in the suspect's vehicle. At the lab, investigators visually inspected the fragments for fracture matches but did not find one. Later, glass fragments stamped with the markings "e-a-l-e-d" found at the crime scene were matched to those taken from the suspect's car. Equipped with a GRIM2 refractive index measurement apparatus (which is discussed in this chapter), the police lab found that nine of the crime scene fragments had similar qualities to those of the suspected car—enough to be statistically significant. Furthermore, the lab established that all the fragments came from a common source by using elemental analysis with an inductively coupled plasma-optical emission spectrometer (ICP-OES).

and heat-resistant. This borosilicate glass was given the trade name Pyrex™ and was subsequently found to resist attack from virtually all chemicals except hydrofluoric acid (HF), which etches its surface.

Tempered Glass

Tempered glass (also known as safety glass) is more than four times stronger than window glass. During its manufacture, the sand, lime, and sodium carbonate are heated together, and the hot glass that is formed is rolled into sheets. Its upper and lower surfaces are then cooled rapidly with jets of air. This process leaves the center of the glass relatively hot compared to the surfaces and forces the surfaces and edges to compress. Tempered glass is stronger because wind pressure or impact must first overcome this compression before there is any possibility of fracture.

When tempered glass breaks, it does not shatter into pieces with sharp edges, but rather breaks into "dices" (i.e., small pieces without sharp edges). Tempered glass is used in the side and rear windows of automobiles, in large commercial windows, in doors, and even in shower doors and home windows where the window is less than 1 ft from the floor.

Windshield Glass

Automobile windshields are made from **laminated glass** (FIGURE 5-1). Today, windshields are made with two layers of glass, with a high-strength vinyl plastic film such as polyvinyl butyral (PVB) being sandwiched in between the layers. The three pieces

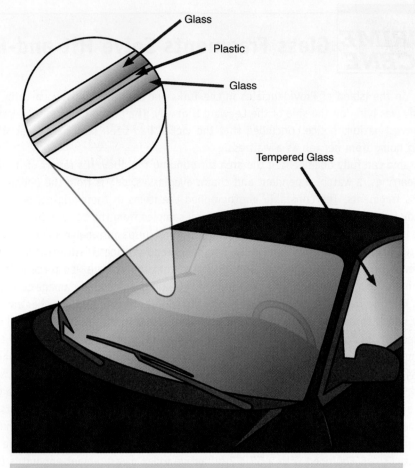

Glass

Plastic

Glass

Tempered Glass

FIGURE 5-1 Automobile windshields are made from laminated glass whereas the side and rear windows consist of tempered glass.

are laminated together by applying heat and pressure in a special oven called an autoclave.

This type of glass is ideal for automobile windshields because of its strength and shatter resistance. The plastic film holds the glass in place when the glass breaks, helping to reduce injuries from flying glass. The film also can stretch, yet the glass still sticks to it. Laminated safety glass is very difficult to penetrate as compared to normal windowpane glass. The glass sandwich construction allows the windshield to expand in an accident without tearing, which helps hold the occupants in a vehicle. Banks use a similar bullet-proof glass that has multiple layers of laminated glass.

Forensic Examination of Glass Evidence: An Overview

For the forensic scientist, the goals in examining glass evidence are twofold:

- To determine the broader class to which the glass belongs, thereby linking one piece of glass with another
- To individualize the glass to one source—a particularly difficult challenge given that glass is so ubiquitous in modern society

To pinpoint the source of the glass evidence, the forensic examiner needs the two usual samples: glass fragments collected from the crime scene and glass fragments taken from some item belonging to the suspect. The examiner must then compare these samples (often side by side via a stereomicroscope) by identifying their characteristics—for example, their color, fracture pattern, scratches and **striations** (irregularities) from manufacturing, unevenness of thickness, surface wear (outside versus inside surfaces), surface film or dirt, and weathering patterns. In particular, the examiner tries to fit the "pieces of the puzzle" together by matching the irregular edges of the broken glass samples and finding any

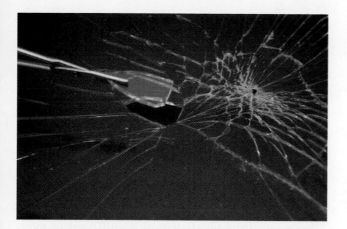

Matching broken pieces of glass. Finding a perfect match is tantamount to individualizing the glass to a single source with complete certainty.

corresponding irregularities between the two fragments (FIGURE 5-2). Finding a perfect match is tantamount to individualizing the glass to a single source with complete certainty.

Non-Optical Physical Properties of Glass

Many non-optical physical properties can be used to compare a questioned specimen of glass to a known sample. These non-optical physical properties include surface curvature, texture, and special treatments. Clearly, frosted glass cannot be a match to a clear window glass. Similarly, a curved piece (such as a fragment from a bottle) cannot come from the same source as a flat piece (such as from a window). And finally, laminated glass would not compare to wire-reinforced glass. Thus these sorts of comparisons are most useful in proving the two pieces *cannot* be associated.

Surface Striations and Markings

When sheet glass is rolled, the rollers leave parallel striation marks, called ream marks, on the surface. Even polishing does not completely remove these marks, and their presence can be enhanced by low-angle illumination and photography. These ream marks may hint at how various pieces should be oriented in the case of an indirect physical match where an intervening piece may be missing. The relative spacing might also be useful as a means of individualization. Surface scratches, etchings, and other markings might be employed in a similar way

as the forensic examiner tries to piece together the puzzle.

Surface Contaminants

The presence of such impurities as paint and putty is useful in two ways. First, the patterns of the adhering materials might suggest how the pieces fit together. Second, chemical analysis of the adhering materials might further individualize the pieces and prove their association.

Thickness

Thickness can be measured to a high degree of accuracy with a micrometer. One must be careful, however, in assuming that the thickness is constant—it is not, particularly in curved pieces of glass. For this reason, the forensic examiner must take several representative measurements of both the known and the questioned samples. Determination of curvature can distinguish flat glass from container, decorative, or ophthalmic glass. Thickness is a very useful way of proving that two pieces of glass, which are otherwise extremely similar, are *not* actually from the same source.

Hardness

A number of scales are used to describe the hardness of substances. Geologists and mineralogists often employ the **Mohs scale**, which indicates a substance's hardness relative to other substances. On the Mohs scale, the softest common mineral—talc—is assigned a relative value of 1, and the hardest common mineral—diamond—is assigned a relative value of 10. Each of the remaining values is assigned to another appropriate common mineral. For example, quartz is assigned the Mohs value 7 and topaz is assigned the Mohs value 8.

The relative positions of the minerals on the Mohs scale reflect their scratching power: A harder substance will scratch a softer one. Thus diamond will scratch everything else on the list; topaz will scratch quartz and everything lower on the Mohs scale, down to talc. Talc, by contrast, will not scratch anything else on the list. For an unknown mineral or substance, its relative hardness is determined by using it to try to scratch the benchmark minerals. Its position on the scale is between the benchmark mineral, which it scratches, and the next mineral on the list, which scratches it. For instance, an unknown that scratched talc and quartz but was itself scratched by topaz would be assigned a relative position between 7 and 8. In this same fashion, all other materials can be ordered appropriately.

The Mohs scale is not very useful for glass samples, however, because all glasses tend to fall in the same range, between 5 and 6. Thus the Mohs scale is too insensitive for forensic work, as are all of the other standard hardness scales. Generally, the forensic lab establishes relative hardness by referring to glass samples in its collection. The relative scratching power of the known and questioned samples is established by trying to scratch these samples with glass in the lab's collection. Either the scratching powers of the known or unknown samples are similar or they are not.

Glass Fractures

Elasticity is the ability of a material to return to its previous shape after a force is exerted on it. For example, when a force is exerted on a pane of glass, it stretches (this bending may not be visible to the naked eye). If the force is not too high, the glass will then return to its original state and no damage occurs. If the force exceeds the glass's elasticity, however, the glass fractures.

The forensic examiner may be able to analyze fractured window panes and determine the direction of an impact and the amount of force applied to them, suggesting what actually happened at the scene. For example, it is often important to establish whether a window was broken from the inside or the outside. At the scene of a homicide, a broken window near the door latch may be an attempt to disguise the crime as a burglary. In the case of a burglary, the window would have been broken from the outside. However, if the homicide was deliberate, the perpetrator may have broken the window from the inside in an attempt to mislead investigators into thinking burglary was the intruder's primary goal.

Characteristics of Glass Fractures

Glass may be subjected to three types of forces (strains):

- **Compressive force** squeezes the material.
- **Tensile force** expands the material.
- **Shear force** slides one part of the material in one direction and another part in a different direction.

Each of these forces causes a deformation, which is resisted by the internal cohesion (stress) of the material. Glass breaks when a tensile strain is applied that is sufficient to overcome the natural tensile stress limit of the material.

If a person places a weight on a horizontal sheet of glass, the pane will experience compressive strain where the load meets the pane. The side holding the weight is called the loaded side, designated as side L, and the unloaded side is designated as side U. The deformation induced by the load will cause side U to expand, so side U will experience a tensile strain. If the tensile strain is sufficient to overcome the tensile strength of the pane, the pane will develop cracks on the unloaded side. Several of these cracks may appear, and they will grow or travel in two directions simultaneously. First, they will grow from the unloaded to the loaded side. Second, they will radiate outward, away from the load point; they are therefore called **radial cracks**. The radial cracks form several pie-shaped (or triangular) sectors radiating from the point of loading. If the load is suddenly removed, these sectors will stay in place because the third side of each of the triangular sections is still solid glass.

If the load persists, however, each sector will continue to be forced outward. This movement causes compressive strains on side U and concurrent tensile strains on side L. These strains will cause new cracks to develop on the loaded side. As before, these cracks grow in two ways: first from the loaded to the unloaded side, and second until they connect two radial cracks. These new cracks are called **tangential cracks** or **concentric cracks**, and the resulting pattern has a spider web appearance (FIGURE 5-3).

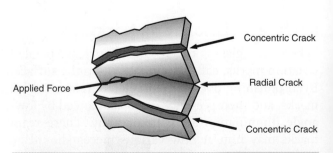

FIGURE 5-3 Radial cracks grow from the loaded point outward and from the unloaded side to the loaded side. Tangential (also know as concentric) cracks grow from one radial crack to another and from the loaded side to the unloaded side.

Note that radial cracks grow from the load point outward and from the unloaded side to the loaded side. In contrast, tangential cracks grow from one radial crack to another and from the loaded side to the unloaded side. This is the case if the weight was placed statically on a pane of glass.

By contrast, when a bullet is shot at the pane of glass, the load is a **projectile**. The load side is known as the **entrance side**, and the unloaded side is called the **exit side**. The same cracking occurs, and the same hole formation happens as when a static load is applied. As the initial velocity of the projectile increases, however, the central hole becomes smaller, the cracking patterns become simpler, and the central hole develops a pattern wherein the exit hole is invariably wider than the entrance hole (FIGURE 5-4).

Examination of the edges of broken pieces of glass will reveal a set of curved lines known as **rib marks** (or "stress" marks). These arcs are always nearly perpendicular to the surface at the side on which the break started and curve until they are nearly parallel to the surface on the opposite side (e.g., the side to which the break grew). In a radial crack, the rib marks will be nearly perpendicular to the unloaded (or exit) side and nearly parallel to the loaded (or entrance) side. Things will be exactly reversed for a tangential crack, which grows in the opposite way. The 3R rule helps in remembering this pattern:

- **R**adial cracks give rib marks, which make
- **R**ight angles on the
- **R**everse side from where the force was applied.

The direction of lateral propagation of the crack is always from the concave sides of the rib marks toward their convex sides. Thus, in a radial fracture, the rib marks will be oriented with their concave sides "cupped" toward the load (or entrance) point.

Forensic Examination of Glass Fractures

If all the glass pieces are present, the first thing to check for is the hole made by the load or projectile (e.g., bullet, hammer), which will be wider on the exit side. This test works best when the hole is made by a high-speed projectile. In the event the hole was made by a low-speed projectile (such as a hammer), this test will not be very meaningful. In this case it is usually best to examine the rib marks. Of course, to make this examination meaningful, each edge must be determined to be either a radial or a tangential crack (which is why it is so important that all pieces be collected), and inside and outside sides of the pieces must be identified (which is why it is so important that the investigators mark the proper orientation of each piece directly on the item, as well as documenting all orientations in their notes and photos).

Thus, if the forensic scientist is examining the edge of a radial fracture, whichever side shows nearly perpendicular rib marks will be the unloaded (or exit) side, meaning the side away from the force that caused the break. Alternatively, if the scientist is examining a tangential fracture, the side showing the nearly perpendicular rib marks will be the loaded (or entrance) side, meaning the side from which the original breaking force was applied.

In the event that the investigator or evidence technician neglected to mark which side of the glass was inside and which side was outside, it is sometimes possible to figure out this information in the lab. Traces of window putty would indicate an outside side, for example, and paint traces of different colors might also be used to distinguish between the two sides.

Of course, the preceding discussion assumes that the glass is not tempered. When tempered glass breaks, it produces small pieces; the fractures cannot be categorized as radial or tangential, so the kind of analysis mentioned above is not applicable.

When there are several bullet holes, analysis can determine the sequence of the impacts. The first shot will cause fractures that simply "run out" (terminate) wherever the original strains have been

FIGURE 5-4 The bullet entered from the backside (entrance side) making a smaller hole, and passed through the glass pane leaving a wider hole at the front surface (exit side).

sufficiently relieved in the material. The radial fractures associated with a second shot will run out when they meet a fracture from the first shot, and so on for all subsequent shots (**FIGURE 5-5**).

The majority of fragments recovered from a suspect's clothing or hair will likely be very small (0.25 to 1 mm). Most glass evidence adhering to a suspect is lost fairly rapidly, depending on the suspect's subsequent activities and the texture of his or her clothing. For example, wool sweaters will retain glass fragments longer than a leather jacket. The size of a fragment may be so small that individual characteristics cannot be found. In these cases the forensic examiner turns to measurements of density and refractive index to characterize glass evidence.

Glass Density Tests

Density tests are often performed on glass fragments. When a forensic scientist measures the density of a glass fragment, he or she is measuring one of its physical properties. Density is a class characteristic, so it cannot serve as the sole criterion used for individualizing the glass evidence to a single source. Such measurements can, however, give the forensic

scientist enough data to warrant further testing of other evidence or to provide enough evidence to exclude the glass fragments as having originated somewhere other than the crime scene. In addition, if a sufficient amount of separate class characteristic evidence can be gathered against a suspect, collectively the evidence may make a strong circumstantial case, which may result in conviction.

To see how this works, consider decorative glass. This type of glass is made by adding different minerals to the glass recipe as the basic ingredients—sand, lime, and sodium carbonate—are being heated. The density of the resulting glass will vary with the type and amount of minerals added. If a recovered glass fragment is placed in a liquid that has a higher density than the glass, the glass fragment will float. If the liquid is less dense than the glass fragment, the glass will sink. When the density gradient column method is used to determine the density of glass, the forensic scientist uses a **density gradient tube** filled with a liquid that has been especially prepared to have a density gradient.

The gradient is prepared such that the density at any level is less than that of any level lower in the tube and greater than that of any level higher in the tube. The gradient is prepared by mixing bromoform and bromobenzene, two dense organic liquids,

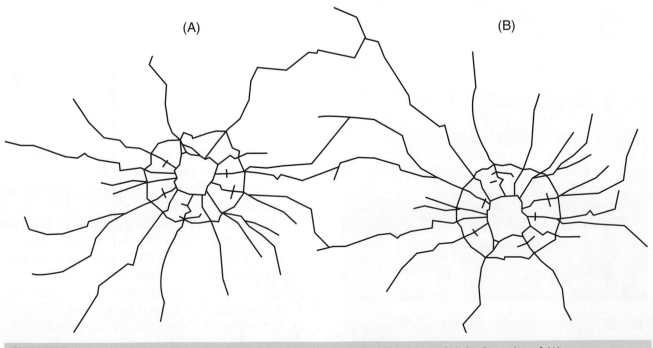

(A)   (B)

FIGURE 5-5 In these two bullet holes in one piece of glass, the formation of (B) preceded the formation of (A).

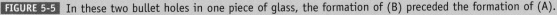

in different proportions. When glass fragments are poured in the top of the column, they fall through the liquid until they become suspended in the liquid at the level that is the same density as the particular glass fragment. Fragments of different densities will, therefore, settle at different levels in the column. The questioned glass fragment's density may then be compared with a glass sample from the crime scene to prove (or disprove) that it is a match.

Density measurements should not be performed on fragments of glass that are cracked or contain an inclusion, because these flaws will make the glass seem less dense than it really is. (An inclusion is a defect that forms when a particle or bubble becomes embedded in the main body of the glass.) Window glass, in particular, does not have a uniform density. For this reason, the variation in density of the known sample should be determined with samples taken from different locations in the window or door whenever possible. Likewise, because the surface or edge of tempered glass is denser than at its interior, care must be taken with tempered glass to measure several known samples. Density comparisons between known and questioned specimens should be made using fragments of approximately equal size.

The Federal Bureau of Investigation (FBI) has reported density results for 1400 glass samples received from 1964 to 1997. From this information, it is known that the range of densities for flat glass, container glass, and tableware glass all overlap.

When the density tests are concluded, any evidence that does not match the known specimen can be excluded. If questioned and known samples are found to have comparable densities, however, further testing is still required. A refractive index test is usually performed to support the comparison. If the density measurement indicates that the specimen from the crime scene matches the reference material, a refractive index test that also indicates a match will improve the discrimination capability by approximately twofold.

Optical Physical Properties of Glass

Color

Comparing the color of a suspect piece of glass with the color of a reference sample can distinguish whether the two samples share a common source. As a consequence, significant color differences between glass fragments can be used as the basis for exclusion of a suspect.

Given that sample size may affect the apparent color, side-by-side comparisons should be made with fragments of approximately the same size. These fragments should be visually compared by placing them on edge over a white surface using natural light. Viewing the glass in this way allows for the optimal observation of color. It also allows the examiner to distinguish between the true color of the glass and any coatings or films that might be present on the glass's surface. In addition, observing the glass using both fluorescent and incandescent light is often helpful in distinguishing colors.

Refractive Index

Light has wave properties. That is, a beam of light traveling from a gas (such as air) into a solid (such as glass) undergoes a decrease in its velocity, such that the beam bends downward as it passes from the air into the glass. The application of this phenomenon allows the determination of the glass's **refractive index**, a measure of how much the light is bent (refracted) as it enters the glass.

The bending of a light beam as it passes from one medium to another is known as refraction. The refractive index, η, is the ratio of the velocity of light in the air to the velocity of light in the glass being measured. The velocities of light in both media are related to the angles that the incident and refracted beams make with a theoretical line drawn vertically to the glass surface (**FIGURE 5-6**).

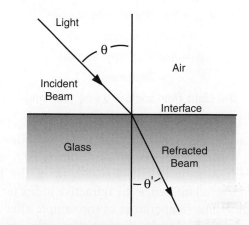

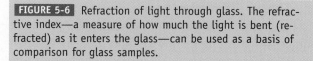

FIGURE 5-6 Refraction of light through glass. The refractive index—a measure of how much the light is bent (refracted) as it enters the glass—can be used as a basis of comparison for glass samples.

on the CRIME SCENE Windshield Fragments

Twenty-one-year-old Iowa State University student Danny Peterson died on June 8, 2002, six days after being hit by a vehicle. Peterson and two friends were walking along the shoulder of a road when a vehicle veered off the pavement, fatally injured Peterson, and sped away. After Danny was transported to the hospital, Detective Jack Talbot of the South Lake Minnetonka Police Department investigated the scene of the accident. He recovered small glass slivers from a windshield on the road and fragments from a headlight in a nearby culvert. Fragments of windshield glass also were collected from Danny's clothes.

There was no straightforward match for the windshield fragments, but the glass from the headlight matched a specific make and model of car—the 2001–2002 Mercury Cougar. Even though the headlight glass seemed important, it was not probative. In fact, the headlight turned out to belong to a 2001–2002 Mercury Cougar that was in the repair shop on the day of the accident.

All investigative leads were followed but to no avail. Three months after the accident, however, a young woman called the police with a promising lead. She suggested that her husband, Guido Vivar-Rivera, might have been involved in the accident. On the night of the accident, Rivera was drunk and upset, saying that someone broke his Pontiac's windshield with a rock. The windshield was repaired soon after the accident.

The police inspected the outside of Rivera's 1996 Pontiac Grand Am and saw stress fractures on the bumper, possibly caused from the impact of hitting Danny. Detective Talbot questioned employees at the repair shop who replaced the windshield. The employees said it was unlikely the windshield damage was caused by a rock.

Detective Talbot and his team arrested Rivera and got a search warrant for the Pontiac Grand Am. In addition to the stress fractures on the bumper, they found a dent on the hood and pieces of broken window glass on the inside of the car. This glass was a match with the windshield fragments found at the scene of the accident and on Danny's clothes.

Rivera was convicted of a felony hit-and-run and served his time at the Hennepin County Adult Correctional Facility.

$$\eta^x_D = V_{air}/V_{glass} = \sin\theta_{air}/\sin\theta'_{glass}$$

where

V_{air} = Velocity of light in air

V_{glass} = Velocity of light in glass

θ = Angle of light in air

θ' = Angle of light in glass

x = Temperature

D = Light from sodium D line (589 nm)

The velocity of light in a liquid sample is always less than that of light in air, so refractive index values for solids are always greater than 1.

The temperature and wavelength of the light being refracted influence the refractive index for any substance. The temperature of the sample affects its density, and the density change affects the velocity of the light beam as it passes though the sample. Therefore, the temperature at which the refractive index is determined is always specified by a superscript in the notation of η. Likewise, the wavelength

of the light used affects the refractive index because light of differing wavelengths bends at different angles. The bright yellow light from a lamp containing sodium, which produces a beam with a wavelength of 589 nm (Figure 5-6), is commonly called the sodium D line. This lamp provides the standard wavelength of light, denoted η_D. Thus the refractive index of a liquid measured at 20°C using a sodium lamp that gave a reading of 1.3850 would be reported as $\eta^{20}_D = 1.3850$.

Single sheets of plate glass, such as those commonly used for making windows, do not usually have a uniform refractive index value across the entire pane. Because the index of refraction can vary as much as 0.0002 from one side to another, the difference in the refractive indices of the questioned plate glass fragment and the reference sample must be smaller than 0.0002 if the forensic scientist is to be able to distinguish the normal variations in a pane of glass from variations that would rule out a match altogether.

The refractive index is one of the most commonly measured physical properties in the forensic laboratory, because it gives an indication of the composition and the thermal history of the glass. Two methods are used to measure the refractive index of glass: the oil immersion method and the Emmons procedure.

Oil Immersion Method

When using the oil immersion method, the forensic examiner places the questioned glass fragments in specialized silicone oils whose refractive indices have been well studied. The refractive index of the oil is temperature dependent: As its temperature increases, its refractive index decreases. Silicone oils are chosen for this task because they are resistant to decomposition at high temperatures. The refractive index of virtually all window glass and most bottles can be compared by using silicone oil as the comparison liquid and by varying its temperature between 35°C and 100°C.

An easy way to vary the refractive index of the immersion oil is to heat it. The suspected glass fragments and immersion oil are placed on a microscope slide, which is then inserted into a hot stage microscope (FIGURE 5-7). The stage of such a microscope is fitted with a heater that can warm the sample slowly while accurately reporting the temperature to ±2°C. A filter inserted between the lamp and the sample allows light with a constant 589-nm wavelength to reach the sample. Increasing the temperature has little effect on the refractive index of the glass but decreases the refractive index of the oil by about 0.004 per 1°C.

When the glass fragments are initially observed through the microscope, they will produce a bright halo around each fragment, known as the Becke line (FIGURE 5-8). As the temperature increases, the refractive index of the oil decreases until the Becke line—and the glass fragments—disappear. At this point (called the match point), the refractive indices of the oil and the glass fragment are the same. The examiner can compare suspect and known samples in this way to determine whether they have the same match point; alternatively, he or she can estimate the refractive index of the glass from graphs that report the refractive index of the oil as a function of temperature.

Automated systems are also available for making refractive index measurements using the immersion method. The Glass Refractive Index Measurement (GRIM) system, for example, combines a hot stage microscope with a video camera that records the behavior of the glass fragments as they are being heated (FIGURE 5-9). That is, the camera shows the contrast between the edge of the glass fragment and the immersion oil as the temperature increases, until it reaches the match point. The GRIM system's computer then converts this temperature to a refractive index using reference information stored in a database.

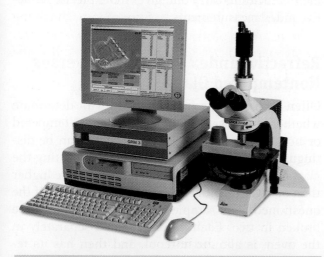

FIGURE 5-7 A hot stage microscope is a key instrument in the forensic examination of glass. The temperature of the sample affects its density, and the density change affects the velocity of the light beam as it passes through the sample—and hence its refractive index.

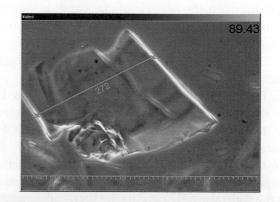

89.43

272

FIGURE 5-8 Oil immersion is one technique used to determine the refractive index of glass. The Becke line appears as a bright halo around the glass fragment.

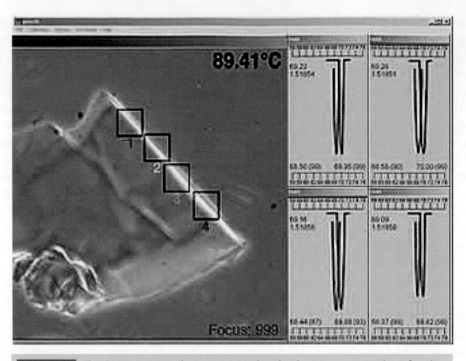

FIGURE 5-9 The GRIM 3 system is an automated technology to measure the refractive index of glass.

Emmons Procedure

The Emmons procedure, which was developed by the Association of Official Analytical Chemists, uses a hot stage microscope in conjunction with different source lamps. It measures the index of refraction at a variety of wavelengths. Most often the refractive index measurements are recorded by first taking a measurement with a sodium lamp (the sodium D line at 589 nm) and then by using a hydrogen lamp (which produces two lines, the C line at 656 nm and the F line at 486 nm). The microscope converts the difference in the refractive indices between the particle of glass and the silicone oil to a difference in brightness contrast, and it enhances the Becke line. This procedure increases the precision of the refractive index measurements taken on the glass particles.

The questioned glass is crushed and placed in the silicone oil on the hot stage. As the temperature of the hot stage increases, measurements are taken at the three different wavelengths (486, 589, and 656 nm). Lines representing the refractive index of the glass as a function of wavelength are recorded for each temperature. These data are then superimposed on a complex graph, known as the Hartmann net. The Hartmann net contains the correlation between the refractive index and the wavelength at fixed temperatures for the silicone oil. The point at which the dispersion lines for the glass samples intersect the dispersion lines for the silicone oil is where the refractive index of the glass sample is determined. Three separate indices of refraction are recorded: η_C, η_D, and η_F. Because three separate measurements are taken on each sample, this method, although more difficult to carry out, gives more precise refractive index measurements.

Refractive Index of Tempered versus Nontempered Glass

Often a forensic examiner needs to determine whether the questioned glass sample is tempered or nontempered glass. Tempered glass can be distinguished from nontempered glass by heating the glass fragments in a furnace at a temperature higher than 600°C in a process known as **annealing**. If the questioned glass sample is large enough, it can be broken in two. Each piece is heated separately in the oven, is allowed to cool, and then has its refractive index measured. Since annealing alters the optical properties of the glass, the change in refractive index between the two annealed pieces can be used to determine if it is tempered or nontempered glass. After annealing, the change in refractive index for tempered glass is much greater than the change observed for nontempered glass.

Variations in Density and Refractive Index

As with other types of evidence, the properties of glass are more often used to exonerate suspects than to individualize samples and definitively prove a connection between a suspect and a crime scene. Indeed, if either the densities or the refractive indices of a questioned glass specimen and a reference glass sample do not match, then the forensic scientist can easily prove that they did not share a common origin. However, glass is so ubiquitous, and so many manufacturers use the same processes to produce each type (e.g., rolling molten glass into flat sheets to make windows), that sometimes even fragments from different sources may have similar indices of refraction or similar densities. Thus individualizing glass samples *accurately* is particularly challenging.

To assist crime labs in making such distinctions, the FBI has compiled density and refractive index data about glass from around the world. These data indicate how widespread the use of a glass with a specific refractive index is. For example, a glass fragment having a refractive index of 1.5278 was found in only 1 out of 2337 specimens in the FBI database, while glass with a refractive index of 1.5184 was found in more than 100 of the 2337 specimens. The forensic scientist can access this FBI database whenever he or she needs to compare the refractive index of a questioned glass fragment to refractive index information and, thereby, calculate the probability that two such samples might be matches as a result of sheer chance (FIGURE 5-10).

The FBI also has correlated the relationship between their refractive indices and densities for 1400 glass specimens (FIGURE 5-11). The results show that once the refractive index of a glass specimen is known, the subsequent measurement of its density will improve the discrimination capability of the measurements by approximately twofold. Most forensic examiners prefer to measure refractive index simply because refractive index measurements are faster and easier to make than density measurements, and often the glass fragment size is too small to get an accurate density measurement. If the glass fragment is large enough, both the refractive index

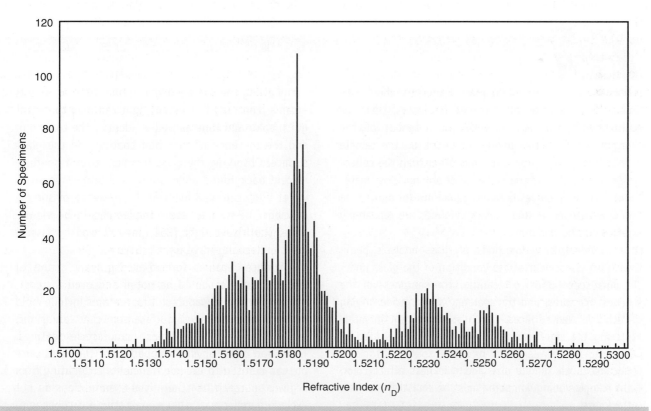

FIGURE 5-10 The frequency of occurrence of refractive indices of glass specimens has been determined by the FBI and is available to forensic examiners in an FBI database.

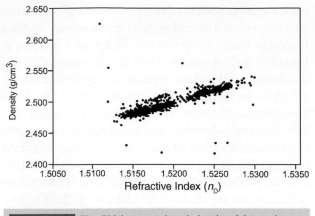

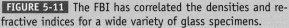

FIGURE 5-11 The FBI has correlated the densities and refractive indices for a wide variety of glass specimens.

and the density should be determined unless other discriminating measurements such as elemental analysis are performed.

Elemental Analysis of Glass

The physical and optical methods for forensic comparison of glass fragments are well established in crime labs and widely accepted in courts throughout the world. These analytical methodologies have two other advantages: (1) These tests are nonde-structive, so the evidence is preserved for additional testing, and (2) the tests are performed using inexpensive instruments. These advantages ensure that these tests will remain the principal methods for the comparison of glass. Methods of elemental analysis—particularly those in which the specimen is consumed during the analysis—should be used only after all nondestructive methods of examination have been completed and in cases in which additional discrimination is necessary.

The elemental composition of glass can be measured by surface techniques such as use of a scanning electron microscope (SEM) or X-ray fluorescence (XRF). The SEM has several disadvantages that limit its value in the analysis of glass fragments. Primary among these is that, because of the irregular shape of the glass fragments, precise quantitative determination of element concentration is not possible.

The XRF, by contrast, is routinely used for elemental analysis of glass. For instance, the glass industry uses XRF as an accurate, precise method of enforcing quality control during glass manufacturing. The XRF instrument focuses a beam of X-rays on the surface of the glass and then measures the energy of the X-rays that are emitted from the glass.

BACK AT THE CRIME LAB

Documentation related to glass fragments should include the condition of recovered fragments, their approximate size, the presence (or absence) of the original surface, the amount of debris on the sample (which may come from locations other than the collection location), and the presence of any nonglass material. The FBI provides detailed guidelines for processing of glass evidence, and these guidelines are repeatedly reviewed and updated.

Nondestructive methods of glass analysis begin with an assessment of the condition of the glass under a microscope prior to cleaning. The sharpness of the edges, fractures, and transparency assist in identifying freshly broken surfaces, which are important in interpreting the significance of evidence. Color is used to distinguish between two or more sources of glass. Because both sample size and thickness affect color, the samples being compared must be similar in size and thickness.

Viewing the sample under natural light as well as incandescent and fluorescent light can help to distinguish color and tone as well as identify the true color and the presence of thin film coatings on the glass surface. Observing the glass fragment over a nonfluorescent background under short- and long-wave ultraviolet light can also identify the presence of surface coatings. In such a case, samples should be viewed under short-wave light (254 nm) followed by assessment under long-wave light (350 nm).

Surface features formed during manufacture or later during use should be noted and used for comparison purposes. Manufacturing features include mold marks and polish lines. Surface scratches, abrasions, and pitting are usually post-manufacture features. These and other physical properties of the glass sample can be used to exclude fragments originating from a given source. When the initial examinations do not exclude fragments, further examination (usually chemical analysis) is required.

The energy of the emitted X-rays can be correlated to the presence of specific elements. In one study, XRF was used to measure the ratios of 10 elements in window glass samples that had virtually identical indices of refraction. When the elemental ratios determined by XRF were compared, the source of 49 of the 50 glass specimens could be correctly determined. One major advantage of XRF is that it does not destroy the sample.

The elemental composition of glass also can be measured by flameless atomic absorption spectrophotometry (FAAS) or inductively coupled plasma (ICP) methods. There are two major disadvantages of using these methods for the analysis of glass fragments. First, the glass fragment must be dissolved in acid and small samples of the resulting solution then injected into the instrument, so the original sample is destroyed. Second, these methods entail use of hazardous chemicals, such as hydrofluoric acid.

Despite these disadvantages, the ICP method, when coupled with an optical emission spectrometer (ICP-OES), has been shown by the FBI to be a dependable method for the determination of 10 elements in glass: aluminum, barium, calcium, iron, magnesium, manganese, sodium, strontium, titanium, and zirconium. The FBI studies also demonstrated that the determination of the concentrations of these 10 elements provides a great degree of discrimination capability. An ICP-OES study of the elemental distribution of automobile side-window glass, for example, found that the probability of two glass samples from different cars being indistinguishable was 1 in 1080, compared to 1 in 5 when just the refractive indices were used as the basis of comparison.

The X-ray fluorescence, flameless atomic absorption spectrophotometry, and inductively coupled plasma are described in more detail in Chapter 9.

YOU ARE THE FORENSIC SCIENTIST SUMMARY

1. Police should detain the suspect and get a warrant to examine his or her hair and clothing for glass shards. If the suspect was near the car when the window was broken, he or she should be covered with glass fragments. The police also should examine the suspect's skin for any cuts or scratches from the broken window. In addition, the police should gather eyewitness testimony. If glass fragments are found, they should be analyzed by the methods described in this chapter.

2. The number of glass fragments found on the suspect is important: The more shards found, the closer his proximity to the event. The police should determine whether the glass is tempered glass, which is used in car side windows.

Chapter Spotlight

- Glass is a solid that is not crystalline; rather, it has an amorphous structure.

- Soda-lime glass is the glass commonly used in windows and bottles. A variety of metal oxides can be added to this glass to give it a special appearance.

- Tempered (safety) glass is more than four times stronger than window glass.

- Automobile windshields are made from laminated glass. The plastic film holds the glass in place when the glass breaks, helping to reduce injuries from flying glass.

- Many nonoptical physical properties—such as surface curvature, texture, special treatments, surface striations, markings, surface contaminants, and thickness—can be used to compare a questioned specimen of glass to a known sample.

- Glass fractures when it is subjected to compressive, tensile, and shear forces that exceed its elasticity. The excessive force produces radial and concentric cracks in the glass.

- The 3R rule: Radial cracks create rib marks at right angles on the reverse side from where the force was applied.

- Glass density tests are performed using the density gradient column method. The density of a glass fragment taken from a suspect may be compared with the density of a glass sample

from the crime scene in this way, either proving or disproving a link between the two.

- Optical physical properties of glass include color and refractive index.

- Comparing the color of a suspect piece of glass can distinguish between glass from different sources.

- To determine the refractive index, forensic examiners often use the oil immersion method. It involves placing glass pieces into specialized silicone oils. The temperature then is varied until the match point is reached and the Becke line disappears.

- Tempered glass can be distinguished from nontempered glass by heating the glass fragments.

- The FBI maintains a database of density and refractive index data that forensic scientists can use as the basis of comparison when analyzing their own sample.

- The elemental composition of glass can be measured by flameless atomic absorption spectrophotometry (FAAS) or inductively coupled plasma (ICP; sometimes with use of optical emission spectrometer, known as ICP-OES) methods or by use of surface techniques such as a scanning electron microscope (SEM) or X-ray fluorescence (XRF).

Key Terms

Amorphous solid A solid in which the atoms have a random, disordered arrangement.

Annealing Heat treatment that produces tempered glass.

Compressive force Force that squeezes glass.

Concentric cracks Cracks that appear as an imperfect circle around the point of fracture.

Crystalline solid A solid in which the atoms are arranged in a regular order.

Density gradient tube A tube filled with liquids with successively higher density.

Entrance side The load side of a projectile.

Exit side The unloaded side of a projectile.

Fracture match The alignment of the edges of two or more pieces of glass, indicating that at one time the pieces were part of one sheet of glass.

Laminated glass Two sheets of glass bonded together with a plastic sheet between them.

Mohs scale A scale that measures the hardness of minerals and other solids.

Projectile The load of a bullet shot at a pane of glass.

Radial cracks Cracks that radiate in many directions away from the point of fracture.

Refractive index Ratio of the speed of light in air to the speed of light in another material (such as glass).

Rib marks The set of curved lines that are visible on the edges of broken glass.

Shear force Force that moves one part of the material in one direction while another part is moving in a different direction.

Striations Fine scratches left on bullets, formed from contact of the bullet with imperfections inside the gun barrel.

Tangential cracks Cracks that appear as an imperfect circle around the point of fracture.

Tempered glass Glass that has been heat treated to give it strength.

Tensile force Force that expands the material.

Putting It All Together

Fill in the Blanks

1. Glass is often found on burglary suspects as _____ evidence.

2. Glass is a solid that is not crystalline, but rather a(n) _____ solid.

3. The atoms of an amorphous solid have a(n) _____, _____ arrangement.

4. Tempered glass is rapidly _____, which makes the surface and edges compress.

5. Automobile windshield glass is laminated with a(n) _____ layer between two layers of glass.

6. When sheet glass is rolled, the rollers leave parallel _____ marks in the surface.

7. The thickness of a questioned glass sample can be measured with a(n) _____.

8. The accepted scale of glass hardness is the _____ scale.

9. On the Mohs scale, the softest material, _____, is given a value of 1 and the hardest material, diamond, is given a value of _____.

10. A force that squeezes glass is called a(n) _____ force.

11. A(n) _____ force expands glass.

12. A force that slides one part of glass in one direction and another part in a different direction is called a(n) _____ force.

13. A(n) _____ crack radiates outward, away from the load point.

14. A(n) _____ crack grows from one radial crack to another and from the loaded side to the unloaded side.

15. When comparing glass fragments for the purpose of matching their color, the fragments should be viewed on _____ over a white surface.

16. The refractive index is a measure of how much light is _____ as it enters a material.

17. The bending of a light beam as it passes from air into glass is known as _____.

18. The _____ and _____ of the light being refracted influence the refractive index for any substance.

19. The D line, which is used to designate the refractive index, indicates _____.

20. The oil immersion method of refractive index measurement uses _____ oils.

21. For the oil immersion method of refractive index measurement, the refractive index of the immersion oil is varied by raising its _____.

22. The halo that is observed around the glass fragment in the oil immersion method is known as the _____ line.

23. The Emmons procedure for measuring the refractive index of glass makes measurements at three different _____ of light.

24. By using a(n) _____ microscope, the Emmons procedure increases the precision of the refractive index measurements.

25. Heating glass in a furnace at temperatures above 600°C is called _____.

26. The scanning electron microscope cannot take precise measurements of elemental concentrations in glass fragments because of the _____ of the glass fragments.

True or False

1. Glass fragments have a sharp melting point.

2. When tempered glass breaks, it shatters into pieces with sharp edges.

3. Density measurements on cracked glass are not to be trusted.

4. Plate glass has a uniform refractive index across the entire pane.

5. One big advantage of the XRF measurement is that it does not destroy the sample.

Review Problems

1. Refer to FIGURE 5-12. Determine the order in which these bullet holes were made. Justify your answer.

2. Refer to FIGURE 5-13. Determine the order in which these bullet holes were made. Justify your answer.

3. Compare the bullet holes in Figures 5-12 and 5-13, and indicate which (if any) were made by a higher-velocity bullet. Justify your answer.

4. Using the equation on page 120, determine the index of refraction for the following glass fragment. The incident angle of the D line light is 45°. The light passing through the glass is refracted at an angle of 28°.

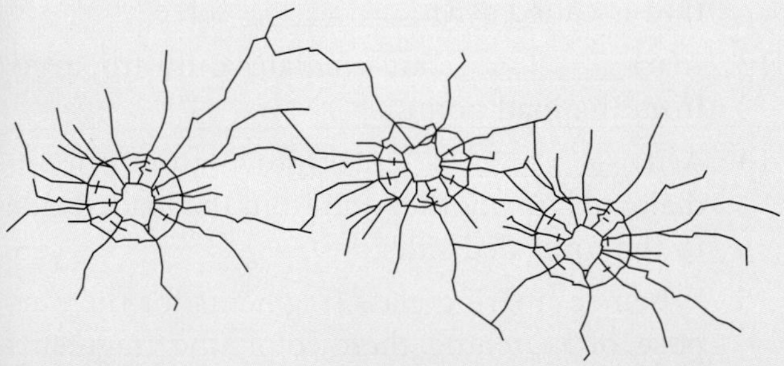

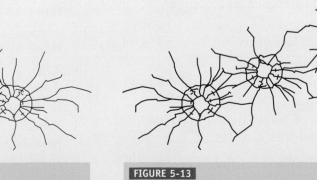

FIGURE 5-12

FIGURE 5-13

5. Using the equation on page 120, determine the index of refraction for the following glass fragment. The incident angle of the D line light is 45°. The light passing through the glass is refracted at an angle of 27°.

Further Reading

Bottrell, Maureen C. Forensic glass comparison: Background information used in data interpretation. *Forensic Science Communications*, 2009; 11(2). http://www.fbi.gov/hq/lab/fsc/current/review/2009_04_review01.htm.

Houck, M. M. *Mute Witnesses Trace Evidence Analysis*. San Diego, CA: Academic Press, 2001.

Koons, R. D., Buscaglis, J., Bottrell, M., and Miller, E. T. Forensic glass comparisons. In R. Saferstein (Ed.), *Forensic Science Handbook, vol. 1* (ed 2). Upper Saddle River, NJ: Prentice-Hall, 2002.

Petraco, N., and Kubic, T. *Color Atlas and Manual of Microscopy for Criminalists, Chemists, and Conservators*. Boca Raton, FL: CRC Press, 2004.

Scientific Working Group for Materials Analysis (SWGMAT). Elemental analysis of glass. 2006. 8(1). www.fbi.gov/hq/lab/fsc/current/standards/2005/standards10.htm.

Scientific Working Group for Materials Analysis (SWGMAT). Glass density determination. 2006; 8(1). http://www.fbi.gov/hq/lab/fsc/current/standards/2005/standards8.htm.

Scientific Working Group for Materials Analysis (SWGMAT). Glass fractures. www.fbi.gov/hq/lab/fsc/current/standards/2005/standards7.htm.

Scientific Working Group for Materials Analysis (SWGMAT). Glass refractive index determination. www.fbi.gov/hq/lab/fsc/current/standards/2005standards9.htm.

Scientific Working Group for Materials Analysis (SWGMAT). Introduction to forensic glass examination. www.fbi.gov/hq/lab/fsc/current/standards/2005/standards4.htm.

http://criminaljustice.jblearning.com/criminalistics

Answers to Review Problems

Interactive Questions

Key Term Explorer

Web Links

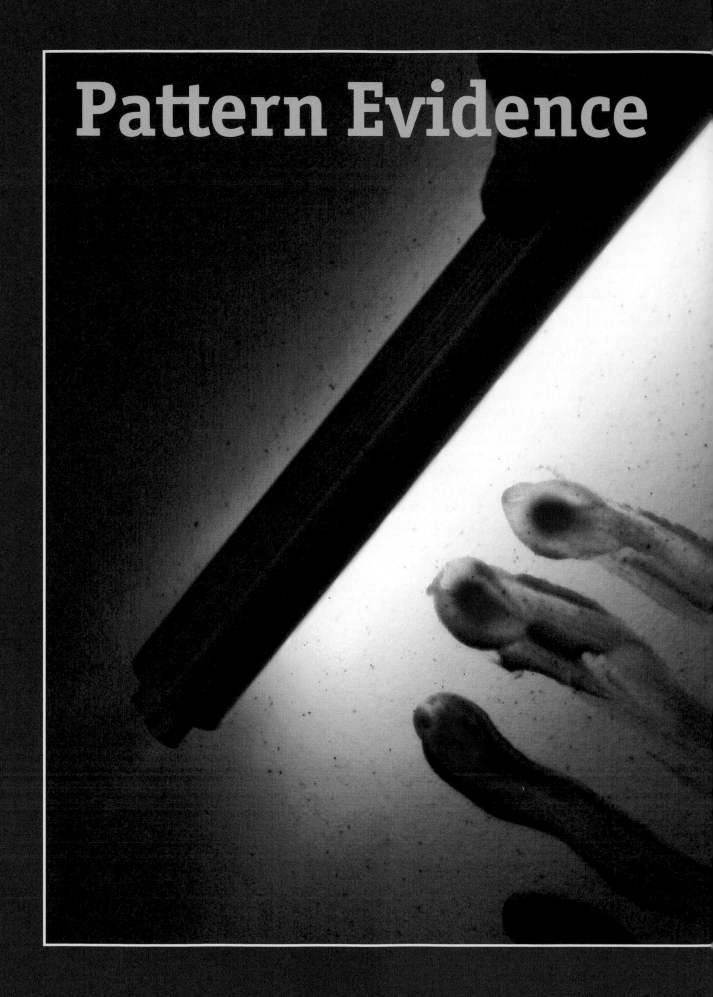

Pattern Evidence

Fingerprints

OBJECTIVES

In this chapter you should gain an understanding of:

- The characteristics that permit fingerprints to be used for personal identification
- Methods used to develop latent fingerprints
- Methods used to visualize fingerprints
- The classification of fingerprints
- The use of fingerprints for biometric identification

FEATURES

On the Crime Scene

Back at the Crime Lab

See You in Court

WRAP UP

Chapter Spotlight

Key Terms

Putting It All Together

Fingerprints can be extremely valuable evidence, so crime scene investigators should make extraordinary efforts to recover any prints present at the scene. All surfaces should be thoroughly investigated. Shining a flashlight at an oblique angle to a surface often helps to illuminate prints. Even if the perpetrator wore gloves, he or she may have removed them to handle small or oddly shaped objects. It is often useful to approach the crime scene as the perpetrator did. Searching at the same time of day may indicate certain surfaces that are more likely to have been touched. A person familiar with the crime scene may be able to suggest places to look for latent prints, as he or she may be able to identify items that are out of place or that the perpetrator may have brought to the crime scene.

As you learn more about fingerprints in this chapter, consider the following questions:

1. A fight in a bar results in the death of a young man. A witness identifies an individual as the culprit. When police question this individual, he says that he was at work during the incident and that co-workers can verify his story. Where in the bar should crime scene investigators focus their search for fingerprint evidence? Which techniques should they use to develop and retain any latent prints?

2. An automobile that has been abandoned in public parking at the airport is suspected of being used in a bank robbery earlier in the day. Which areas of the car should investigators search for fingerprints? Would fingerprint evidence on these surfaces still be present or would it be lost from the surface? Which techniques should investigators use to develop and retain any latent prints?

Introduction

This section of the book explores pattern evidence—that is, those items left at a crime scene that may have arisen from different events but that share an identifiable pattern. Chapters 6, 7, and 8 cover pattern evidence contained in fingerprints, questioned documents (forgeries), and firearms, respectively. As these chapters emphasize, when pattern evidence is found, it often can be individualized and the pattern shown to be so unique that it could be generated by only one source. This chapter focuses specifically on fingerprints, the telltale prints left by the ridges and valleys on our fingers.

History of Fingerprinting

Police have long sought a way to identify the perpetrators of crimes. The first personal identification methodology was devised by a French police expert, Alphonse Bertillon, and introduced in 1883. Bertillon's system of personal identification—which became known as **anthropometry**—was divided into three integrated parts:

- Measurements of the bony parts of the body, conducted with the utmost precision and under carefully prescribed conditions

- The morphological description of the appearance and shape of the body and its measured parts as they related to movement

- A description of marks observed on the surface of the body (e.g., moles, warts, scars), or those resulting from disease, accident, deformity, or artificial disfigurement (tattoos)

Under Bertillon's system, matching at least 11 of the measurements was essential to positively identify a person.

Part of Bertillon's anthropometry system focused on the study of the human ear. According to Bertillon, "It is, in fact, almost impossible to meet with two ears which are identical in all their parts." Bertillon's system required that the ear be described and measured in exquisite detail. In fact, he devoted 15 pages of his book on anthropometry to the human ear and its use as a means of identifying individuals.

Bertillon's system was used until 1903, when Will West was convicted and sentenced to the U.S. penitentiary at Leavenworth, Kansas. A few days after arriving, Will West was brought to Bertillon to be measured and photographed. He denied ever being in the penitentiary before, even though the measurements Bertillon took were almost identical to those already obtained from a prisoner named

Madrid Terrorist Bombing: Fingerprint Evidence Points to Wrong Suspect

On March 13, 2004, less than three days before the Spanish presidential election, 13 backpacks loaded with dynamite and cell phone detonators were detonated on crowded commuter trains in Madrid, Spain. The attacks were carried out by a group of veteran Al Qaeda soldiers, criminals, and religious extremists. In total, 191 people died and more than 2000 were injured.

The ruling conservative Popular Party of Spain had what seemed to be an insurmountable lead at the time of the bombings. Sixty hours after the attacks, however, the Socialist Party won the election over the incumbent Popular Party. The Socialist Party subsequently withdrew Spain's troops from Iraq. Thus Al Qaeda showed that terrorism could be used to effect political change and achieved its objective of "scaring" the Spaniards out of Iraq.

Spanish authorities were able to track the perpetrators based on a mobile phone found in one of the backpacks that failed to detonate. When surrounded by police, the terrorists chose to commit suicide rather than surrender to authorities. Spanish police also found a shopping bag containing detonators in the back of a van at a train station hours after the bombings. The bag held smudged fingerprints. Because the Spanish police were unable to find a match to the prints, they consulted the Federal Bureau of Investigation (FBI).

FBI fingerprint experts in the United States matched one of the Spanish prints to Brendon Mayfield, a U.S. citizen who was a lawyer in Oregon. Mayfield's fingerprints were on file because he had once served in the U.S. Army. Authorities immediately placed Mayfield under surveillance but soon took him into custody when word of the investigation leaked to the media. At the time, investigators said that they weren't sure how Mayfield was involved in the train bombings.

Spanish authorities, however, expressed doubts from the start about the FBI's fingerprint match. Shortly after Mayfield's arrest, officials in Spain released a statement saying the fingerprints belonged to an Algerian, Ouhnane Daoud. Daoud had a residency permit to live in Spain and had a police record. The Spanish police's own forensic scientists determined that the fingerprints were made by the middle and thumb fingers of the Algerian's right hand.

This case prompted a critical review of the FBI's fingerprint analysis protocols. The Inspector General issued a report acknowledging that, although there was an "unusual similarity" between Mayfield's and Daoud's fingerprints, which confused three FBI examiners and a court-appointed expert, the FBI examiners failed to adhere to the bureau's own rules for identifying latent fingerprints. The report also noted that the FBI's "overconfidence" in its own skills prevented it from taking the Spanish police seriously.

"William West." Will West, the new prisoner, continued to deny he had ever been incarcerated, even though his photograph and that of William West appeared to be identical. Eventually, the Leavenworth authorities realized that they had two prisoners who looked almost identical. After fingerprinting the two prisoners, it was discovered that although the two prisoners appeared to be identical twins, there was no resemblance between their fingerprints: The fingerprints of Will West were different from those of William West.

This incident led authorities to reach two conclusions: (1) The Bertillon measurements, and comparison of facial features in general, were an unreliable method of identification, and (2) two people who appear to have identical measurements and facial features nevertheless have unique fingerprints. Indeed, Bertillon's system required the investigator to make 11 individual measurements and was so cumbersome that two different investigators measuring the same person frequently would produce different measurements. The added difficulty of administering the system in a uniform manner also led to its abandonment when fingerprint identification came onto the scene.

Although the Chinese had used fingerprints as signatures on official documents more than 3000 years ago, it was not until 1858 that the first practical application of fingerprinting was discovered. Sir William Herschel, an English government official who worked in India, was vexed at the problem of how to have the many illiterate Indians sign legal documents, while still preventing forgery (a "signature" consisting of an X or some other non-unique mark was clearly open to abuse). To resolve these problems, Herschel began placing inked palm impressions and, later, thumb impressions on contracts as an individualized identifier. He also began fingerprinting all prisoners in jail, many of whom underwent this process multiple times over the years as they drifted in and out of prison. When Herschel examined the fingerprints of the prisoners, he realized that they did not change as the person aged. In fact, he noted that no change had occurred in his own fingerprints in more than 50 years. (Later, scientists would discover that the fingerprints start forming when a fetus is in the ninth or tenth week of development and become permanently fixed by about the seventeenth week of gestation.)

A few years later, Henry Faulds, a Scottish missionary doctor, became interested in fingerprints while working at a hospital in Japan in the 1870s. To record these impressions, he covered his subjects' hands in printer's ink and then pressed them against a hard surface. Like Herschel, Faulds confirmed that fingerprint patterns do not change over time. Even when a person's fingers suffer a superficial injury, they heal in such a way as to return to the exact same fingerprint patterns. In addition, Faulds developed a classification for fingerprint patterns, using terms such as "loops" and "whorls" to describe the patterns.

In 1892 Sir Francis Galton, a half-cousin of Charles Darwin, published an in-depth study of fingerprinting science. His book, which was titled *Fingerprints*, included the first classification system for fingerprints. In it, Galton identified the characteristics by which fingerprints can be identified. These same characteristics (minutia) are still in use today and are often referred to as Galton's Details.

Although Galton's work became the foundation of modern fingerprint science and technology, his approach to classification was too cumbersome to have practical applications in criminal investigations. In 1897 Sir Edward Henry, who had been trained by Galton, devised a more workable classification system independently, which he implemented in India. In 1901 Henry was appointed Assistant Commissioner of Police at Scotland Yard, where he also introduced his fingerprint system. Within 10 years, Henry's classification system had been adopted by police forces and prison authorities throughout the English-speaking world, and it remains in use today (as discussed later in this chapter).

Juan Vucetich, an Argentinean police official and a correspondent of Galton's, devised his own system of fingerprint classification, which he called "icnofalagometrico." Vucetich's system soon found a practical application, which inspired its adoption in many Spanish-speaking countries. In June 1892, Francisca Rojas claimed that she had been brutally attacked and her two children murdered by a ranch worker named Velasquez. Velasquez was arrested but refused to confess to the murder of the two children. A search of the crime scene yielded a number of bloody fingerprints on a doorpost of the Rojas's cabin. These prints did not match the inked impressions that police had made of Velasquez's fingerprints, but rather Rojas's fingerprints. When con-

BACK AT THE CRIME LAB

The formation of fingerprints and footprints is affected by genetics, primarily in terms of the types of whorl, arch, and loop patterns and their respective positions. There is a tendency for fingerprints to run in similar patterns within families. These similarities arise because of the volar pad structure found in the feet and hands, which is genetically transmissible. High pads on fingers and toes tend to form whorls, low pads tend to form arches, and intermediate pads tend to form loops.

Fingerprint formation occurs because of buckling skin. In the fetus, the basal layer of skin grows faster than the epidermal layer, leading to pressure at the skin surface. The skin then buckles, relieving the stress by pushing back toward the softer tissue of the dermis. This buckling creates the ridges that are associated with fingerprints. The shape and style of the fingerprint created are then determined by the minimal energy rule: Nature takes the path of least resistance.

fronted with this evidence, Rojas confessed to the murder of her children; she was later sentenced to life imprisonment.

Fingerprints were first used in the United States by the New York City Civil Service Commission starting in 1901. The commission used fingerprints to prevent impersonations during examinations for civil service positions. Also, in 1903 fingerprinting was instituted by the New York State Prison System and at Leavenworth Penitentiary.

Police from the United States and Canada who attended the 1904 World's Fair in St. Louis, Missouri, were impressed by a presentation by Detective John Ferrier of Scotland Yard on fingerprint identification. In the years following the fair, fingerprinting began to be used in earnest in all major cities in North America. In 1911 the FBI opened an office in Ottawa, Canada, to which Canadian police services sent their complete fingerprint files. In 1924 the records of the FBI Identification Division and the National Bureau of Criminal Investigation housed at Leavenworth Penitentiary were merged and moved to Washington, D.C. Among the records was a core collection of 810,000 fingerprint cards. Today, the FBI has the largest collection of fingerprint records in the world, including fingerprints from more than 64 million people.

Characteristics of Fingerprints

As the previous discussion has emphasized, fingerprints have some important characteristics that make them invaluable evidence in crime scene investigations:

1. A fingerprint is unique to a particular individual, and no two fingerprints possess exactly the same set of characteristics.

2. Fingerprints do not change over the course of a person's lifetime (even after superficial injury to the fingers).

3. Fingerprint patterns can be classified, and those classifications then used to narrow the range of suspects.

The following sections discuss these principles in more detail.

Uniqueness of Fingerprints

In his book *Fingerprints*, Sir Francis Galton described his work that proved the individuality (and permanence) of fingerprints. Galton used fingerprints as a tool in pursuing his primary interest—determining heredity and racial background. Unfortunately for him, Galton discovered that fingerprints do not provide any proof of an individual's genetic history or racial background. Nevertheless, he was able to prove scientifically what Herschel and Faulds already suspected: Fingerprints do not change over the course of an individual's lifetime, and no two fingerprints are exactly the same. According to Galton's calculations, the odds of two individual fingerprints being the same are 1 in 64 billion.

When Edmond Locard later examined Galton's classification system for fingerprints (Galton's Details, now called **friction ridge characteristics**), he determined that if 12 of these details were the same between two fingerprints, it would suffice as a positive identification. Locard chose 12 points as the crucial number in part to improve upon the 11 body measurements (e.g., arm length, height) required to positively identify criminals using Bertillon's anthropometry system.

Over the years, experts have debated how many details in two fingerprints must match before both can be identified as coming from the same individual. Some have suggested that as few as 8 detail matches are sufficient; others have argued that as many as 16 detail matches are required. To date, no comprehensive statistical study has ever been completed to determine the frequency of occurrence of different details and their relative locations. Until such a study is published, it is impossible to establish guidelines for certifying fingerprint matches.

Following its 1973 meeting, the International Association for Identification stated that "No valid basis exists at this time for requiring that a predetermined minimum of friction ridge characteristics must be present in two impressions in order to establish positive identification. The foregoing reference to friction ridge characteristics applies equally to fingerprints, palm prints, toe prints, and sole prints of the human body." This position was affirmed by an international symposium on fingerprint detection and identification held in Ne'urim, Israel, in 1995. The participants in this symposium issued a statement stating "No scientific basis exists for requiring that a predetermined minimum number of friction ridge features must be present in two impressions in order to establish a positive identification."

In the United States, courts rely on a certified fingerprint expert to establish beyond a reasonable doubt, through a point-to-point comparison, that there are sufficient matching details to ascertain that the two prints came from the same individual. The subjectivity of this position is troubling.

In 1999 the first challenge to fingerprint evidence was brought in Philadelphia in the case *United States v. Byron Mitchell*, which involved fingerprints found on the gearshift and door of a getaway car used in the robbery of an armored truck. Defense attorneys argued that fingerprints could not be proven to be unique using the Daubert guidelines (i.e., the Supreme Court's decision in *Daubert v. Merrell Dow Pharmaceuticals*, discussed in Chapter 2). The defense petitioned the courts for a Daubert hearing to determine the admissibility of the fingerprint identification as scientific evidence. The defense asked the question, "Is there a scientific basis for a fingerprint examiner to make an identification, of absolute certainty, from a small distorted latent fingerprint fragment, revealing only a small number of basic ridge characteristics such as the nine characteristics identified by the FBI examiner...."

In its decision on this case, the U.S. District Court for the eastern district of Pennsylvania upheld the admissibility of fingerprint evidence. In particular, the court found that fingerprint evidence is admissible under Rule 702 of the Rules of Evidence and the Supreme Court's decisions in *Daubert v. Merrell Dow Pharmaceuticals*. The court also made the following statements:

1. Human friction ridges are unique and permanent throughout the area of the friction ridge skin, including small friction ridge areas.

2. Human friction ridge skin arrangements are unique and permanent.

Permanence of Fingerprints

Generally, the skin that covers most of the human body is relatively smooth. It consists of several layers. The layer of skin nearest the surface is known as the **epidermis**. The underlying skin, which is called the **dermis**, is 15 to 40 times thicker than the epidermis and contains nerves, blood vessels, oil (sebaceous) and sweat glands, and hair follicles (FIGURE 6-1).

The epidermis contains several tiers of cells. At its bottom, resting on the dermis, is the dermal papillae, which is continually pushing up. As the

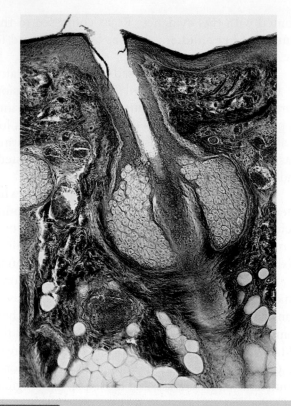

FIGURE 6-1 A cross-section of human skin from the scalp.

cells move toward the surface of the skin, driven along by new cells coming up from beneath, they eventually die. By the time they become the outermost layer of skin, the stratum corneum, they have been transformed into keratin (the same protein that forms hair).

By contrast, skin that is found on the fingers, palms, toes, and soles is not smooth, but rather contains tiny raised lines—friction ridges—that allow the extremities to grasp and hold on to objects (FIGURE 6-2). These series of lines form both ridges (hills) and grooves (valleys) that take on a distinct—and individually unique—pattern. Friction ridges do not run evenly across our fingertips, but rather form identification point **minutiae** that can be easily categorized (as discussed later in this chapter).

Minute pores found on the tops of the friction ridges are constantly emitting perspiration and oil that clings to the surface of the ridges (recall that the dermis contains sweat and sebaceous glands). When the fingertips come into contact with a surface, perspiration (and other material the fingers have recently touched, such as grease and dirt) may be transferred to that surface. The resulting copy of the friction ridges is termed a fingerprint. Note that

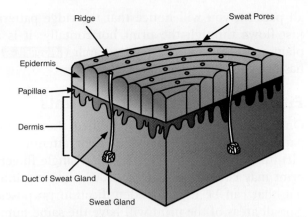

FIGURE 6-2 The surface of the finger contains friction ridges and sweat pores. The ridges form part of the fingerprint, which may be recorded on an object when the finger touches the object and leaves behind perspiration.

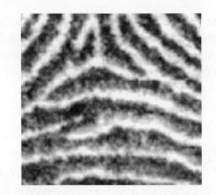

FIGURE 6-3 A delta forms a triangle.

similar impressions may be left behind when a person's bare palm, toes, or soles come into contact with a surface, and are equally unique to the particular individual who made them.

Although it is impossible to change one's fingerprints, criminals have experimented with many painful techniques in attempts to do just that, including use of corrosive acids and burning. For example, the famous Depression-era gangster John Dillinger tried to obliterate his fingerprints with sulfuric acid. Examination of his corpse after he died in a shootout with the FBI, however, revealed that Dillinger's fingerprints could still be matched to the prints taken from him after an earlier arrest. Other criminals have attempted to mutilate their fingerpads by cutting deep into the dermal papillae,

forming a permanent scar upon healing. This mutilation is self-defeating because the permanent scarring provides new features that can be used for identification.

Fingerprint Patterns

A delta is a location in a fingerprint where two lines diverge and form a triangle (**FIGURE 6-3**). Locating a delta is important because classification of fingerprint patterns requires identification of the delta (or the lack thereof). Three major classes of fingerprint patterns are distinguished based on the presence (or absence) of deltas (**FIGURE 6-4**):

- A fingerprint that lacks any deltas is considered to be an **arch**.
- A fingerprint that contains one (and only one) delta is considered to be a **loop**.
- A fingerprint that contains two or more deltas is said to be a **whorl**.

(A)   (B)   (C)

FIGURE 6-4 The three major fingerprint patterns are (A) loop, which contains one (and only one) delta, (B) whorl, which contains two or more deltas, and (C) arch, which does not have any deltas.

In the general population, 65% of fingerprints are loops (the most common pattern) and 35% are whorls. Arches are the least common pattern, being found in only 5% of the population.

Loops

In a loop pattern, friction ridges enter from one side of the print, curve around, and then exit from the same side as they entered. Two types of loop patterns are distinguished based on the hand on which they appear. In an **ulnar loop** (shown in Figure 6-4A), the pattern area comes in from the left and goes back out the left. To see how this works, put your left hand on a book (palm down). Your little finger will be on the left, which is where the loop enters and exits; that is, the loop points toward your left little finger (and toward the ulnar bone in your arm). The same print from the right hand is a **radial loop**. That is, if you place your right hand on the book (palm down) and make the same comparison, you will find that the loop pattern area now enters and exits toward your right thumb (and the radial bone on your thumb side). Obviously, to distinguish between these two types of loops, you must know which hand made the print.

As noted earlier, all loops have a delta. In addition, all loops have a core that is the center of the loop pattern.

Whorls

Sometimes, the inner area of the pattern—that is, the area that tends to form a circle—consists of one or more ridges that make a complete circuit between the two deltas. If the examiner draws a line between these two deltas and the line touches the ridge lines, then the pattern is considered to be a plain whorl. If the ridge lines are not touched, the pattern is termed a central pocket loop. A double loop whorl consists of two separate and distinct loop formations with two separate and distinct shoulders and two deltas.

A fingerprint that contains more than two deltas is considered to be an accidental whorl. A print that is classified as accidental often contains several patterns such as a combination of a plain whorl or loop or an arch.

Arches

Arches have no deltas and no significant core. If you study the image in Figure 6-4C and look at the over-

all pattern, you will notice that the ridge pattern just flows through the print horizontally; it is a plain arch. By contrast, a tented arch (FIGURE 6-5) has ridge lines that rise and fall rapidly.

Fingerprint Identification Points

Once the examiners have identified the general pattern of the print, they turn their attention to identification points (FIGURE 6-6). A single fingerprint may contain as many as 100 or more minutiae that can be used for identification purposes. Not all areas of the print will have the same number of points, however. For instance, the area immediately surrounding a delta usually contains more identification points than the area near the tip of the finger. FIGURE 6-7 shows seven identification points that are clearly visible on the questioned fingerprint.

Methods for Developing Fingerprints

The term *latent* is overused by the popular press and is commonly applied to all chance or unintentional fingerprint impressions that are of evidentiary

FIGURE 6-5 A tented arch has ridge lines that rise and fall rapidly.

Minutiae	Ridge Path Deviations
Ridge ending	
Bifurcation	
Island (short ridge) and point (or dot)	
Lake	
Opposed bifurcations and cross	
Bridge (crossover)	
Double bifurcation and trifurcation	
Hook (or spur)	

FIGURE 6-6 A single fingerprint may contain as many as 100 or more identification points.

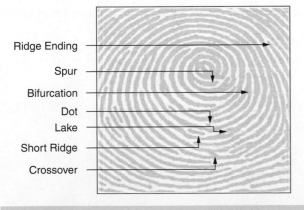

Ridge Ending
Spur
Bifurcation
Dot
Lake
Short Ridge
Crossover

FIGURE 6-7 Seven different minutiae in a fingerprint.

value. In fact, there are three general types of fingerprint impressions:

- A **latent fingerprint** is a friction ridge impression that is not readily visible to the naked eye. This kind of impression is caused by the transfer of body perspiration or oils present on finger ridges to the surface of an object.
- A **patent fingerprint** (also known as a **visible fingerprint**) is a friction ridge impression that is visible to the naked eye, such as a fingerprint deposited on a surface by a bloody hand.
- A **plastic fingerprint** impression is a friction ridge impression that is left on the surface of a soft material such as clay, tar, or wet paint.

Patent and plastic impressions are readily visible, so they usually need no enhancement to visualize the print. By contrast, finding latent (invisible) prints is much more challenging and requires the utilization of techniques that will help visualize the print. Several methods for enhancing a latent print are available; the one of choice depends on the surface that is to be examined. Powders should be selected when the surface is smooth, while chemicals should be used for soft and porous surfaces.

Recovering Fingerprints from Hard and Nonabsorbent Surfaces

To visualize fingerprints made on hard and nonabsorbent surfaces (such as mirror, tile, glass, and painted wood), the crime scene investigator typically applies a specialized fingerprint powder. A variety of such powders is available, so the choice of a particular powder is based on its contrast with the surface being examined.

Black powder, which is composed of black carbon or charcoal, is typically applied to white or light-colored surfaces. Gray powder, which consists largely of finely ground aluminum, is used on dark-colored surfaces; it is also applied to mirrors and polished metal surfaces because these surfaces will appear as black when photographed. Magnetic-sensitive powders are also available in black and gray. Finally, fluorescent powders can be used to develop latent prints; these powders will fluoresce when viewed under ultraviolet light.

The technique used to apply the fingerprint powder depends on the type of powder used. Carbon/charcoal and aluminum powders are applied lightly to a nonabsorbent surface with a fiberglass or camel's hair brush, where they will stick to perspiration residues and/or deposits of body oils left on the surface (FIGURE 6-8). By contrast, magnetic-sensitive powders are spread over a surface with a magnet, known as a Magna Brush (FIGURE 6-9). The Magna Brush is less likely to damage or destroy the print than the previously mentioned brushes, because it does not have any bristles that touch the surface.

FIGURE 6-9 On some surfaces, a magnetic-sensitive fingerprint powder is used.

After dusting, the print should be photographed and then lifted from the surface for preservation or later photographic enhancement. One of the more popular fingerprint lifting tapes is the hinge lifter system (FIGURE 6-10). The hinge lifter has a clear adhesive sheet and a backing sheet of either clear or white vinyl; these sheets are attached to each other with a hinge. After separating the two sheets, the investigator places the adhesive side of the clear sheet (which is held on the inside) on the surface holding the print. The print is lifted, and then the two sheets are sealed together to protect the fragile fingerprint and preserve it as evidence.

Traditional dusting powders do not work well on oily or wax-covered surfaces. Instead, small particle reagent (SPR), which is a suspension of finely

FIGURE 6-8 Fingerprint powder is applied with a camel's hair brush.

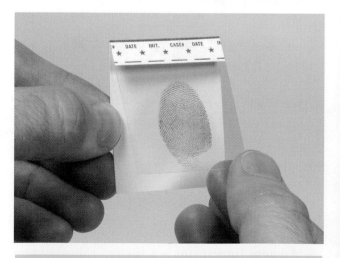

FIGURE 6-10 A hinge lifter is a popular type of fingerprint lifting tape.

ground particles suspended in a detergent solution, is used for these surfaces. These particles adhere to the fatty constituents in the fingerprint to form a visible deposit (FIGURE 6-11). Fingerprints visualized using SPR can be lifted using standard lifting material.

Recovering Fingerprints from Soft and Porous Surfaces

To visualize latent fingerprints on soft and porous surfaces (such as cloth, paper, and cardboard), the crime scene investigator must use some type of chemical treatment. These treatments, which are profiled below, should be used in the following order:

1. Iodine fuming
2. Ninhydrin
3. Silver nitrate
4. Super Glue™ fuming

Iodine Fuming

The oldest chemical method of visualizing fingerprints is **iodine fuming**, which takes advantage of the fact that iodine is prone to sublimation. Sublimation is the process through which a solid, such as crystalline iodine, becomes a vapor when heated without first melting to form a liquid. In the iodine fuming technique, the object containing the latent fingerprint is placed in an enclosed chamber along with iodine crystals (FIGURE 6-12). The temperature in the chamber is then increased until the iodine sublimes; the resulting iodine vapors will react with the body oils in the latent print, making

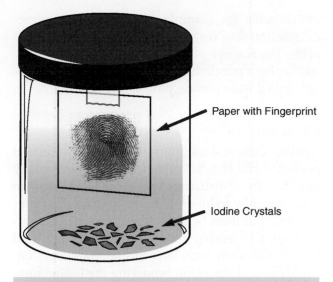

Paper with Fingerprint

Iodine Crystals

FIGURE 6-12 In iodine fuming, the object holding the fingerprint is placed in a chamber where the release of iodine vapors develops the fingerprint.

the print visible. After the heating stops, the print quickly fades. To preserve the image, the developed print must therefore be photographed immediately. Alternatively, it may be sprayed with a 1% solution of starch in water, which will turn the print blue and make it last for several weeks to several months. Because iodine fumes are toxic, care must be taken to protect all laboratory personnel from them.

Ninhydrin

One of the most popular chemical developers is **ninhydrin**. To develop fingerprints on a porous surface, a 0.6% solution of ninhydrin powder that is dissolved in acetone or ethyl alcohol is sprayed onto the surface. Prints usually begin to appear within an hour or two. This reaction can be accelerated if the object with the print is heated at a temperature of 80° to 100°C. This technique is usually not recommended, however, because it can produce a background color owing to the reaction of ninhydrin with the hot substrate (i.e., the surface on which the print resides). If rapid results are needed, specially designed ninhydrin chambers are available that precisely control the temperature and humidity so that development takes only 5 minutes. Ninhydrin has been used to successfully visualize latent prints on paper that were left behind 40 years ago.

All amino acids contain an amino (—NH_2) and an acid (—COOH) group. The eccrine (sweat) glands in human skin secrete a mixture of such

FIGURE 6-11 Small particle reagent (SPR) is used to visualize greasy fingerprints.

amino acids, for example. Ninhydrin is a nonspecific amino acid reagent that reacts with all amino acids. The reaction of ninhydrin with an amino acid, such as from perspiration, forms a purple–blue product, called **Ruhemann's purple** (FIGURE 6-13).

Silver Nitrate

Another chemical used to visualize fingerprints on porous surfaces is silver nitrate. It is particularly effective in visualizing fingerprints that remain undetected by iodine fuming or reaction with ninhydrin—but must be used after those tests are carried out, if necessary. If the silver nitrate test is used first, it will wash away fatty oils and proteins from the surface of the print, rendering the iodine fuming and ninhydrin tests useless.

In this technique, a 3% solution of silver nitrate is brushed onto the suspected object. When the object is then exposed to ultraviolet light, the print turns reddish-brown or black. The color change occurs when sodium chloride (salt), left on the object after evaporation of water from the print, reacts with the silver nitrate to form silver chloride. Silver nitrate produces black stains wherever it

touches the skin, so technical staff must wear protection (e.g., gloves) when applying silver nitrate.

Super Glue Fuming

Super Glue™ fuming has been used as a chemical treatment for fingerprint development since 1982. Super Glue is a polycyanoacrylate ester that releases cyanoacrylate esters when heated; these vapors react with chemicals in the fingerprint to produce a white-appearing latent print. Cyanoacrylate ester fumes are released when Super Glue reacts with sodium hydroxide or when it is heated. Thus, to visualize fingerprints, the Super Glue fumes and the suspect material are allowed to react in an enclosed chamber for up to 6 hours. It is believed that the moisture and salts present in the fingerprint cause the Super Glue to polymerize into a solid.

Super Glue fuming has also been successfully used to visualize fingerprints inside automobiles. In this application, a small hand-held wand is used to heat a small cartridge containing a mixture of the cyanoacrylate and a fluorescent dye (FIGURE 6-14). The wand allows the examiner to direct the fumes vaporizing from the wand to the area of interest. These visualized prints do not fade and are permanently affixed to the surfaces fumed. Because Super Glue fumes are toxic, care must be taken to protect all laboratory personnel from them.

FIGURE 6-13 Ninhydrin reacts with the amino acids in fingerprints to produce a dark purple color, thereby visualizing the fingerprints.

FIGURE 6-14 A hand-held wand heats Super Glue™ to make fumes.

Preservation of Fingerprints

Objects thought to hold latent fingerprints should be carefully collected and submitted to the crime laboratory's latent print section for examination. If it is impossible or impractical to ship the object to the crime lab, then the latent development techniques described in the previous section must be applied at the scene. The developed print must then be carefully photographed, because this photograph may be the only permanent record of the evidentiary fingerprint. Both fingerprint lifts and photographs should be carefully documented to preserve the chain of custody.

Photography of Fingerprints: An Overview

The developing techniques described in the previous section use either powders or chemicals to make the fingerprint visible by taking advantage of a difference in color between the print and the surface on which it sits. While photographing the print, the examiner exploits the selective absorption of light by different colors on the surface.

Visible light is composed of seven colors: red, orange, yellow, green, blue, indigo, and violet. (A handy mnemonic for remembering this sequence is "Roy G. Biv.") The human eye is capable of seeing only visible light, which technically is electromagnetic radiation falling within the wavelength range of approximately 400 to 700 nm (nanometers). A human's perception of color is related to wavelength. For example, radiation at 450 nm is observed as blue, 550 nm as green, and 650 nm as red. When each color (or wavelength segment) of the visible spectrum is present with the same relative intensity, we see white light (FIGURE 6-15). If each color (or wavelength) is not present, or is not present at the same intensity, then we perceive colored light.

If a fingerprint is clearly visible after treatment with a powder or a chemical, a photograph of the fingerprint can be recorded using a white light source such as a tungsten light bulb. In FIGURE 6-16 , a white light source is used to illuminate a black fingerprint on a white surface. The light striking the white surface between the ridges and minutiae is reflected back to the camera as a white light. The ridges and minutiae, which are black, absorb

---

on the CRIME SCENE The Alleged Will of Howard Hughes

In 1976 the death of Howard Hughes led to a famous case involving fingerprint forensics. The eccentric Hughes, whose estate was estimated to amount to between $2 billion and $3 billion, apparently did not file a will with any court. Shortly after his death, speculation began that Hughes had, in fact, drafted a "holographic" will—that is, a handwritten will in his own words that was created without the benefit of an attorney. Not surprisingly, soon after this information became public, an alleged holographic will was recovered that had been left anonymously at the office building of the Church of Jesus Christ of Latter-day Saints (Mormon Church).

The provision that became known as "the Mormon will" stated that $156 million of Hughes's estate should go to Melvin Dummar of Gabbs, Nevada. Dummar and his wife claimed to have no knowledge of how this will came into being or who brought it to the Mormon Church. Furthermore, Dummar claimed that years prior he had picked up a bum in the desert who claimed to be Hughes, gave him a ride into Las Vegas, and donated what spare change he had in his own pocket to the bum.

Dummar's claims set off a media frenzy, but his fame was short-lived. Forensic examinations found fingerprints matching those of Dummar on both the outer envelope containing the alleged will and a copy of the book *Hoax*, which was recovered from the library at the college Dummar attended. This book contained many examples of Hughes's handwriting. A conclusion that the will was a forgery was supported by the fact that Hughes's handwriting had, in fact, undergone significant changes in the two-year period just before the alleged will was written—changes not reflected in the alleged will.

Following a seven-month trial and the expenditure of millions of dollars by the estate of Hughes, the court ruled the Mormon will a forgery. In the end, the billionaire Howard Hughes was found to have died without a legal will.

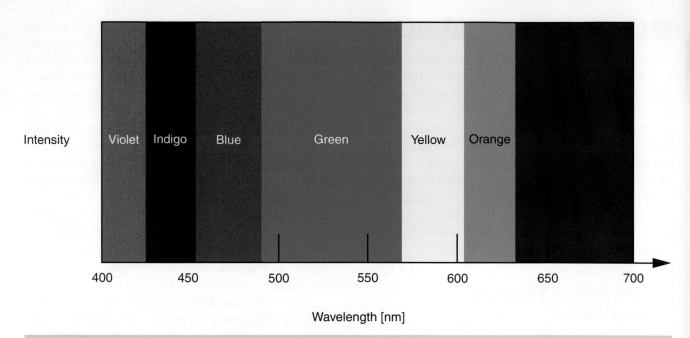

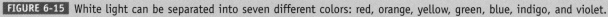

Intensity

Violet Indigo Blue Green Yellow Orange

400   450   500   550   600   650   700

Wavelength [nm]

FIGURE 6-15 White light can be separated into seven different colors: red, orange, yellow, green, blue, indigo, and violet.

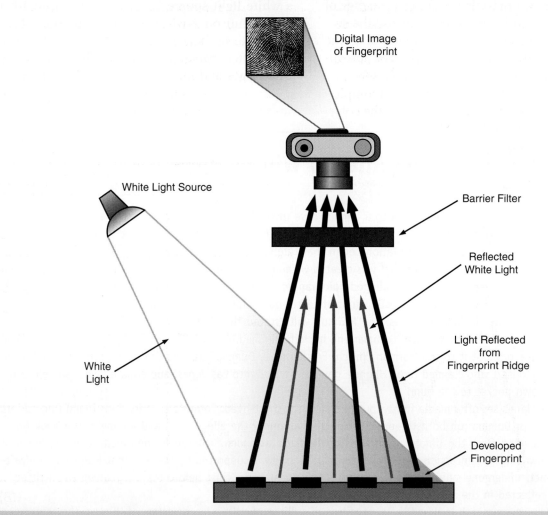

Digital Image of Fingerprint

White Light Source

Barrier Filter

Reflected White Light

Light Reflected from Fingerprint Ridge

White Light

Developed Fingerprint

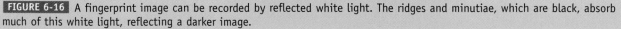

FIGURE 6-16 A fingerprint image can be recorded by reflected white light. The ridges and minutiae, which are black, absorb much of this white light, reflecting a darker image.

much of the white light, so they reflect back a darker image.

A colored fingerprint that appears on a colored surface is much more difficult to see. To make such a print more clearly visible, the examiner should enhance the image so as to maximize the color difference between the fingerprint and the substrate. That is, the fingerprint is darkened by selective absorption of only certain wavelengths of light. The only requirement for enhancement is that the substrate and fingerprint do not absorb light at exactly the same wavelength. Even if the colors of the substrate and fingerprint are similar, a small difference in color will be enough to make this technique work.

To photographically enhance the colored fingerprint on a colored surface (FIGURE 6-17), an excitation filter is placed in front of the white light source (Figure 6-17A). Many different filters are available that transmit only one color of light and filter out all other colors. As can be seen in Figure 6-17B, by directing different colors of light at colored surfaces, the photographer can change the appearance of the background to make the fingerprint more visible. The selection of a filter (or the color of light to be used) to improve contrast is made by referring to a color wheel (FIGURE 6-18). Opposite (complementary) colors on the wheel will darken the color of the fingerprint (increase contrast), whereas adjacent colors will lighten it (reduce contrast). Generally, the filter should be opposite in color to the color of the fingerprint while being as close in color to the background substrate as possible.

This technique can be used only in a dark room where the colored light passing through the filter serves as the sole source of light. Another way to achieve the same result is to move the filter from in front of the white light source to in front of the camera lens.

A ninhydrin-developed fingerprint, when viewed under white light, will appear to be purple. That is, the white light that is reflected from the surface will appear white, and the reflected light from the ridges and minutiae of the fingerprint will be a combination of blue and red light that is reflected and is perceived by the human eye as purple. The purple color results from the absorption of the Ruhemann's purple that forms when the amino acids in the fingerprint react with the ninhydrin.

To obtain the best contrast using colored light in such cases, the photographer needs to have a thorough understanding of the absorption spectra of Ruhemann's purple. FIGURE 6-19 shows the absorption spectrum of Ruhemann's purple. It features two absorption bands: one with a maximum

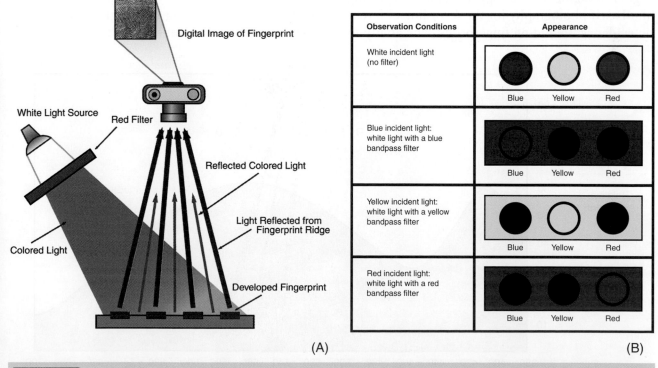

(A)　　　　　　　　　　　　　　　　　　(B)

FIGURE 6-17 (A) The image of some fingerprints on a colored surface can be enhanced with colored light. (B) By illuminating colored surfaces with colored light, their appearance can be altered.

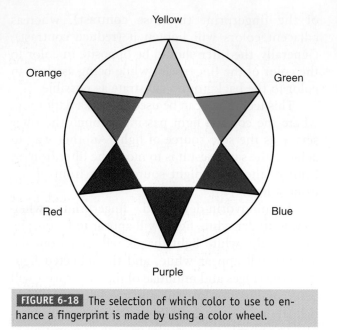

FIGURE 6-18 The selection of which color to use to enhance a fingerprint is made by using a color wheel.

at 410 nm (violet) and another with a maximum at 560 nm (green–yellow). Thus the best contrast will be obtained if the photographer uses a filter that produces either violet light (410 nm) or green–yellow light (560 nm).

Digital Imaging of Fingerprints

Digital imaging technology is rapidly replacing film-based photography for the recording of forensic evidence, including fingerprints. The ability to capture an image quickly and convert it immediately to an electronic file has made digital photography the first choice of examiners who need to search fingerprint files electronically.

In addition, even as the resolution and processing speeds of digital cameras improve, their cost continues to decline. Resolution—the amount of detail that the camera can capture—is measured in pixels. A higher resolution (i.e., more pixels) translates into more detail. Most of today's digital cameras offer a resolution of at least 1216 × 912 pixels—a good choice for an image containing 1.1 mega-pixels (approximately 1,100,000 pixels). Some cameras can handle images that contain up to 10.2 million pixels. When the camera's resolution is too low, however, a picture will appear "grainy" and start to look out of focus when it is enlarged, thereby losing the detail.

Most digital cameras use a charge-coupled device (CCD) as their image sensor. A CCD is a collection of tiny light-sensitive diodes (photosites) that convert photons of light into electrical charges. The brighter the light that hits a particular photosite, the greater the electrical charge that will build up at that site. The camera then reads the amount of accumulated charge in each cell by transporting the charge across the chip and reading it at one corner of the array. CCDs are able to transport charges across the chip without distortion, thereby producing very high-quality images with a minimum amount of background noise.

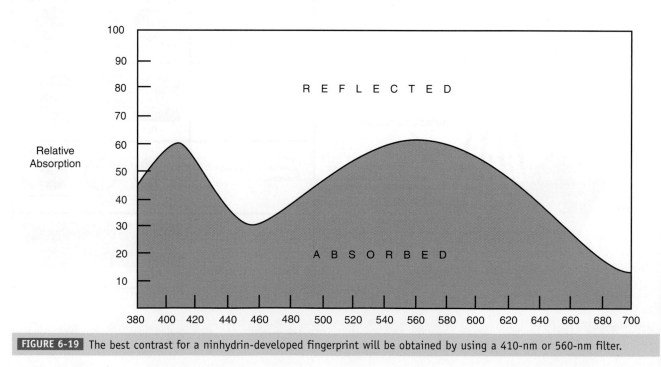

FIGURE 6-19 The best contrast for a ninhydrin-developed fingerprint will be obtained by using a 410-nm or 560-nm filter.

Sometimes valuable forensic information, such as a fingerprint, may be obscured by background information. Sophisticated background suppression techniques can be employed with digital images to remove repetitious patterns on the substrate below the fingerprint. For example, digital images can be modified by addition, subtraction, or multiplication of their underlying data through use of a computer program such as Adobe Photoshop®, thereby greatly reducing or even eliminating background patterns that would otherwise obscure the fingerprint.

A computer program that applies the mathematical operation known as fast Fourier transform (FFT) also may be used to convert the digital image. The computer can then readily recognize and eliminate periodic background noise in the converted image. In addition, the FFT technique is very useful in developing a single print from an object containing multiple overlapping fingerprints. If the ridges from multiple overlapping fingerprints are running in different directions, FFT can enhance those ridges running in one direction. In this way, the ridges of one of the many prints can be enhanced while those of the others are suppressed.

Classification of Fingerprints

After the fingerprint evidence has been collected and recorded, the forensic examiner must classify it. The Henry Classification System of fingerprint identification, which was adopted by Scotland Yard in 1901 and remains in use today, converts forensic ridge patterns from all 10 fingers into two numbers that are arranged as a fraction.

Primary Classification

The Henry system assigns a number to each finger, where the right thumb is given the number 1 and the left little finger has the number 10 (**FIGURE 6-20**). It also assigns a numerical value to fingers that contain a whorl pattern. Fingers 1 and 2 each have a value of 16, and each subsequent set of two fingers has a value that is half of the value of the preceding set. Thus fingers 3 and 4 have a value of 8, and so on, with the final two fingers having a value of 1. Fingers with a non-whorl pattern, such as an arch or loop, are assigned a value of 0. **TABLE 6-1** summarizes how the Henry Classification System assigns finger numbers and finger values.

To determine the fingerprint's primary classification, we calculate a ratio (a fraction) consisting of 1 plus the sum of the values of the whorl-patterned, even-numbered fingers, divided by 1 plus the sum

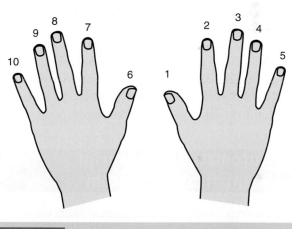

FIGURE 6-20 The Henry system for classifying fingerprints.

SEE YOU IN COURT

In 1998 Byron Mitchell was arrested for robbery. Supporting evidence included the apparent match of his fingerprints with small portions of two latent fingerprints found on the getaway car. In his trial, five full days of testimony were dedicated to the comparison methods used to make this match. During the trial, the public defenders argued that the methods did not meet the criteria established by the U.S. Supreme Court in the *Daubert* decision for inclusion of evidence—specifically that the potential error rate for fingerprint matching is known.

The Mitchell defense team petitioned the court for a Daubert hearing to determine the admissibility of the fingerprint match as scientific evidence. The U.S. District Court for the Eastern District of Pennsylvania denied the defense's motion to exclude fingerprint evidence. However, more and more cases are being appealed on the basis of fingerprint evidence and the *Daubert* decision.

TABLE 6-1

Henry Fingerprint Classification System

Finger	Left little	Left ring	Left middle	Left index	Left thumb	Right thumb	Right index	Right middle	Right ring	Right little
Finger Number	10	9	8	7	6	1	2	3	4	5
Whorl Value	1	1	2	2	4	16	16	8	8	4

of the values of the whorl-patterned, odd-numbered fingers:

Primary classification ratio

$$= \frac{1 + (\text{sum of even-finger values})}{1 + (\text{sum of odd-finger values})}$$

$$= \frac{1 + (\text{right index} + \text{right ring} + \text{left thumb} + \text{left middle} + \text{left little})}{1 + (\text{right thumb} + \text{right middle} + \text{right little} + \text{left index} + \text{left ring})}$$

As simple mathematics reveals, if an individual does not have any whorl-patterned fingerprints, he or she has a primary classification ratio of 1:1. Approximately 25% of the population falls into this category. If all 10 fingers have a whorl pattern, the person's primary classification ratio is 31:31.

If the Henry Classification System were to be used worldwide, the world's population would be divided into 1024 primary groups.

Automated Fingerprint Identification System

The Henry Classification System was an important force driving the development of the Automated Fingerprint Identification System (AFIS). The orig-inal AFIS technology was designed to expedite the manual searching of fingerprint records, with a goal of eventually reducing matching time from months to hours. At that time, most forensic fingerprint cards were sorted according to the Henry system, so the first AFIS solutions attempted to follow this process by comparing prints to Henry-sorted fingerprint cards. Unfortunately, Henry's finger pattern definitions proved extremely difficult to convert into computer descriptions. More important, the Henry-based AFIS suffered from the same limitations as the traditional manual matching system: It required a trained technician and it was not fully automated. Also, the Henry system worked only if all 10 fingerprints were available—a luxury not afforded to most crime scene investigators.

Over time, the AFIS began to classify fingerprints based on the distance between the core and delta, minutiae locations, and pattern type. The pattern type was still based on the Henry system, however. The ongoing limitations led the FBI to overhaul the AFIS in the 1990s. In 1999, the FBI released the most recent version of this system, the Integrated AFIS.

Most modern AFIS systems compare fingerprint minutiae (i.e., identification points). Typically, human and computer investigators concentrate on

---

EXAMPLE

The fingerprint pattern for an individual is recorded as LWAALALWLA, where "L" means loop, "W" means whorl, and "A" means arch. The series begins with finger 1 (the right thumb) and ends with finger 10 (the left little finger). What would be the primary classification ratio for this print?

Solution

Pattern	A	L	W	L	A	L	W	A	A	L
Finger Number	10	9	8	7	6	1	2	3	4	5
Whorl Value	1	1	2	2	4	16	16	8	8	4
Finger Value	0	0	2	0	0	0	16	0	0	0

Primary classification ratio

$$= \frac{1 + (16 + 2) = 19/1}{1 + (0)}$$

This individual belongs to the 19:1 primary group.

points where forensic ridge lines end or where one ridge splits into two (bifurcations). Collectively, these distinctive features are sometimes called **typica**.

The scanner software used to record data about the subject fingerprint uses highly complex algorithms to recognize, analyze, and map the positions of the minutiae. Note that just having the same minutiae is not enough to conclude that two fingerprints match—the individual minutiae must also be located at the same place on each print. Although AFIS points out possible matches to the examiner, the examiner is responsible for performing a visual analysis and confirming that AFIS matched the same bifurcation at each location.

To determine a match, the scanner system must find a sufficient number of minutiae patterns that the two prints share in common. The exact number needed for a positive identification depends on the country in which the crime was committed. In the Netherlands, for example, 12 characteristic minutiae points are required by law for a match. In South Africa, 7 points will make a match. In the United States and the United Kingdom, there is no set number warranting declaration of a match; the expert fingerprint examiner decides how many are needed to make a decision.

Fingerprints for Biometric Identification

The use of fingerprints to solve crimes involves the process of identification—namely, who left the fingerprint behind? In the case of a burglary, was it the homeowner, a neighbor who visited the home earlier in the week, or the thief who ransacked the residence? Fingerprints collected from the scene can be compared with reference fingerprints held in large criminal and civil fingerprint databases in hopes of finding the source of the print, thereby individualizing the evidence.

Biometrics, by contrast, is a way of authenticating (i.e., verifying) the identity of an individual by using fingerprints, palm prints, retinal scans, or other biological signatures. In this case, the goal is simply verification, or confirming that the person is who he says he is—whether an employee who is authorized to enter a sensitive area of the company or the legal owner of a bank account who is trying to access her money at an ATM. This type of biometric matching is easy, because the computer must simply compare the applicant to a relatively small number of possibilities in its database of authorized users. In contrast, identification—matching an unknown fingerprint collected as evidence with thousands or even millions of possibilities in an FBI database, for example—is a much more challenging endeavor that requires considerably more computer processing power.

Finger Scanning

Until a few years ago, fingerprints were typically collected the old-fashioned way, using a technique harkening back to the days of Herschel and Faulds. Each finger was rolled from left to right on an inkpad, then placed carefully on a sheet of paper and rolled again. Many of these paper-based prints have since been scanned electronically and their digital forms stored in electronic databases, where they are available for matching via automated systems.

More recently, the development of inexpensive automated finger scanners has led many police departments to abandon the manual system of recording suspects' fingerprints. Optical finger scanners work much like photo scanners (**FIGURE 6-21**). The heart of an optical scanner is a CCD, the same light sensor system used in digital cameras. The scanner also contains its own light source (typically an array of light-emitting diodes) to illuminate the ridges of the finger. The person whose fingerprint is being recorded places his or her finger on a glass plate, and a CCD camera takes a picture of it. The CCD system actually generates an inverted image of the finger, where darker areas represent more reflected light (the ridges of the finger) and lighter areas represent less reflected light (the valleys between the ridges). If the clarity (how dark or light the

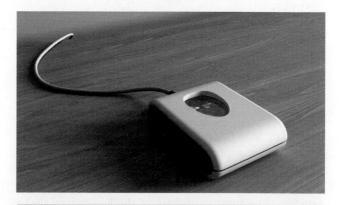

FIGURE 6-21 An optical finger scanner uses a CCD camera to record a digital image of a fingerprint.

Often, the popular *CSI* television shows display electronic devices that only exist on the TV set. Now, "life imitates art." A new hand-held fingerprint identification unit is widely used by law enforcement, military, and government officials to make on-the-spot identifications. Using technology developed for military purposes, this device is a one-person mobile fingerprint or facial scan identifier.

The device provides police officers at the scene of crime the ability to take the fingerprints of a suspect and wirelessly connect to the AFIS database to make a positive identification. The device also is important to identify crime victims without ID who may be in the fingerprint database. With it, police on the street can determine if IDs presented to them are true and accurate or if the person stopped has provided a bogus ID.

The device can make biometric identifications by comparing facial scans of people with their photo IDs to assure a match. It has proven useful in securing borders; the facial scan of an individual can be compared to the passport in hand and to international watchlists both wirelessly and effortlessly. The handheld unit is used in search and rescue missions to help identify rescue workers on the spot, and, by keeping a record of who is available, to help muster volunteers to the site of natural disaster relief.

The device is about the size of a hardback book. It has a small screen upon which a suspect's or victim's fingerprint can be taken and wirelessly sent to compare to available databases. The units have facial scan capabilities, allowing security officers to compare the person in front of them with the photo ID. The unit is becoming a tool in the arsenal of crime solving products in police departments throughout the country. It is widely used in Iraq to help secure military and civilian installations and is being adopted by border and passport controls throughout the world.

image is) and image definition (how sharp the fingerprint scan is) are deemed inadequate, the operator of the optical scanner can adjust the exposure to let in more or less light, alter the positioning of the finger, and retake the picture. If the image is crisp and properly exposed, the computer proceeds to compare the captured fingerprint with fingerprints on file.

Optical finger scanners are starting to be replaced by other scanning methods. For example, a capacitive scanner measures the electrical charge produced when the fingertip comes in contact with an array of tiny capacitors mounted on a silicon microchip. The forensic ridges of the finger will make better contact with these capacitors than the valleys of the fingerprint pattern, because the ridges project outward from the finger. The scanner processor reads the voltage of every cell in the sensor array and determines whether it is characteristic of a ridge or a valley. The computer can then put together a picture of the fingerprint, similar to the image captured by an optical scanner. The resulting image can also be processed in the same way as an image produced by optical scanning. The main advantage of a capacitive scanner is that it requires a real fingerprint-type shape, rather than the pattern of light and dark that makes up the visual impression of a fingerprint. This makes it harder to trick the system with a picture or copy of the known fingerprint. Additionally, because they use semiconductor chips rather than CCD units, capacitive scanners tend to be more compact than optical devices.

Once the fingerprint image has been obtained by scanning, it needs to be compared with previously stored fingerprints to determine whether it is a match. In one approach, a computer algorithm—a program designed to turn raw data into code that can be used more easily by the identification/verification software—identifies minutiae points on the scanned print and locates them relative to other points on the print. It then establishes a mathematical template to serve as a reference. When the person's finger is scanned again at a later time (for example, when the employee wants to enter the company's sensitive area or when the owner of a bank account wants to use an ATM), the computer software compares this template with the newly scanned print. In the case of an ATM card, the template could conceivably be stored on a microchip in the user's card.

Biometrics-Related Issues

Biometric security systems have some disadvantages related to the extent of the damage that can be done when a person's identity information is stolen. If you lose your credit card or someone learns your personal identification number (PIN), you can always get a new card or change your PIN. But if your fingerprints fall into the hands of a criminal, you could have serious problems for the rest of your life. You would be no longer able to use your prints as a form of biometric identification and—although some criminals have tried to do so, as mentioned previously—there's no way to get new fingerprints. Nevertheless, even with this significant drawback, fingerprint scanners and biometric systems are an excellent means of identification.

With the passage of the Enhanced Border Security and Visa Entry Reform Act of 2002, the U.S. Congress mandated the use of biometrics for the issuance of a U.S. visa. This law requires that embassies and consulates abroad must now issue to international visitors "only machine-readable, tamper-resistant visas and other travel and entry documents that use biometric identifiers."

WRAP UP

1. Investigators should focus their efforts first on the area in the bar where the fight took place. Many objects in the bar have hard, nonabsorbent surfaces that might potentially retain fingerprints. Glasses, tableware, dishes, tabletops, and bottles are a few of these nonabsorbent surfaces that should be developed by the application of a powder. If the objects are light in color, then carbon black powder will most likely give satisfactory prints. Prints should be photographed and lifted with tape if possible. Most bars have less soft, porous surfaces. If tablecloths or napkins are found, chemical staining techniques must be used to develop any latent prints left on them.
2. The investigators should begin their investigation in the area where the driver sits. The steering wheel, gear shift knob, outer and inner door handles, and all surfaces on the dashboard and door—all of which are hard and nonporous—should be developed with powder. If more than one bank robber was involved, then other nonporous surfaces of the car (including the trunk and trunk lid) should also be investigated in the same way. If the car was left in the hot sun, there is a high probability that any prints that were originally present will have been lost. The car cannot be stored in a controlled environment to minimize fingerprint loss, so all fingerprint evidence must be processed, photographed, and, if possible, lifted from the surface.

Chapter Spotlight

- A fingerprint is a uniquely individual characteristic; no two fingerprints have yet been found to possess identical characteristics.

- Fingerprints do not change over the course of a person's life.

- The general ridge pattern in a fingerprint allows for effective classification of fingerprints.

- Fingerprints are classified into three major patterns based on the organization of the forensic ridges: loops, whorls, and arches.

- There are three general types of fingerprint impressions: latent (not visible to the naked eye), patent (visible fingerprint), and plastic (an impression left on soft material).

- Prints on hard and nonabsorbent surfaces are usually developed by applying powders with a brush. The powder of choice is the one that shows the best contrast with the surface being examined. Other options include magnetic-sensitive powders and powders that fluoresce when they are viewed under ultraviolet light.

- Latent fingerprints on soft and porous surfaces are made visible by application of chemical developers such as iodine fuming, ninhydrin, silver nitrate, and fumes from Super Glue™.

- If a fingerprint is clearly visible after treatment with a powder or a chemical, a photograph of the fingerprint can be recorded using a tungsten light bulb (white light). A variety of filters can be used to block out other colors.

- The developed latent print must be carefully photographed, usually with a digital camera. Sophisticated background suppression techniques can eliminate periodic background noise in the converted image. Programs such as Adobe Photoshop can perform many of these digital manipulations.

- The Henry system of fingerprint identification was an important force driving the development of AFIS, which enables fingerprint minutiae to be compared via computers.

- Prints can be compared with thousands of fingerprints held in fingerprint databases; the process of finding a match is considered an *identification* technique.
- Biometrics uses fingerprints, palm prints, retinal scans, or other biological signatures to authenticate (verify) the identity of an individual. The process of matching a suspect's fingerprint to a reference fingerprint is considered a *verification* technique.

- The biometric use of fingerprints is expanding thanks to the introduction of inexpensive automated finger scanners. Techniques used to record fingerprints in this way include use of a scanning process that takes a picture of the finger and use of an optical finger scanner that measures the electrical charge produced by the contact of the fingertip.

Key Terms

Anthropometry A method of identification devised by Alphonse Bertillon in the nineteenth century that used a set of body measurements to form a personal profile.

Arch A fingerprint pattern in which ridges enter on one side of the print, form a wave, and flow out the other side.

Biometrics A technology using features of the human body for identification.

Dermis The second layer of skin, which contains blood vessels, nerves, and hair follicles.

Digital imaging The recording of images with a digital camera.

Epidermis The tough outer layer of skin.

Friction ridge characteristics Skin on the soles of the feet, palms of the hands, and fingers in humans that forms ridges and valleys.

Iodine fuming The use of iodine sublimation to develop latent prints on porous and nonporous surfaces. Iodine vapors react with lipids in the oil from the latent print to form a visible image.

Latent fingerprint A friction ridge impression that is not readily visible to the naked eye.

Loop A fingerprint pattern in which ridges enter on one side of the print, form a wave, and flow out the same side.

Minutiae Bifurcations, ending ridges, and dots in the ridge patterns of fingerprints.

Ninhydrin A chemical used to visualize latent fingerprints. Ninhydrin reacts with amino acids in latent fingerprints to form a blue-purple compound called Ruhemann's purple.

Patent fingerprint A fingerprint that is readily visible to the eye.

Plastic fingerprint A fingerprint indentation left by pressing a finger into a soft surface.

Radial loop A loop pattern that flows in the direction of the thumb.

Ruhemann's purple The blue-purple compound formed in latent fingerprints when they are developed with ninhydrin.

Super Glue™ fuming A technique for visualizing latent fingerprints in which fumes from heated Super Glue™ (cyanoacrylate glue) react with the latent print.

Typica Points in a fingerprint where ridge lines end or where one ridge splits into two.

Ulnar loop A loop pattern that flows in the direction of the little finger.

Visible fingerprint A fingerprint that can be seen with the naked eye.

Whorl A fingerprint pattern forming concentric circles in the center of the finger pad.

Putting It All Together

Fill in the Blank

1. The early system of personal identification that was based on measurement of body parts was called _____.

2. Alphonse Bertillon's personal identification system required the measurement of _Bony parts_____, the morphological description of _appearance/shape_ and a description of _marks observed_.

3. The fingerprint classification system that is used in most English-speaking countries was originally developed by _____ (Francis Galton/Edward Henry).

4. The fingerprint classification system that is used in most Spanish-speaking countries was originally developed by_____.

5. Francis Galton estimated the odds of two individual fingerprints being the same as 1 in _____.

6. In 1999, attorneys argued that fingerprints could not be proven to be unique using _____ guidelines. The U.S. District Court _____ the admissibility of fingerprint evidence.

7. The layer of skin nearest the surface is called the _____; the thicker layer below is called the _____.

8. Fingerprints are classified into three general patterns: _____, _____, and _____.

9. Of the three fingerprint patterns, the _____ pattern is the most common, the _____ pattern is the next most common, and the _____ pattern is the least common.

10. A loop pattern on your left hand that enters and exits from the left is called a(n) _____ (radial/ulnar) loop.

11. A loop pattern on your right hand that enters and exits from the left is called a(n) _____ (radial/ulnar) loop.

12. A delta is a location in the fingerprint where two lines run side by side and then diverge to form a(n) _____.

13. Any fingerprint that contains two or more deltas is considered to be a(n) _____ pattern.

14. The center of a loop pattern is called the _____.

15. Whorls can be divided into four classifications: _____, _____, _____, and _____.

16. If a print contains no deltas, then it is a(n) _____ pattern.

17. If a print contains one delta, then it is a(n) _____ pattern.

18. A friction ridge impression that is not visible to the naked eye is called a(n) _____ fingerprint.

19. A friction ridge impression that is left in the surface of a soft material is called a(n) _____ fingerprint.

20. Latent prints on a hard and nonabsorbent surface are usually developed by the application of _____.

21. Fingerprint examiners have found that only two colors of fingerprint powder are necessary: _____ and _____.

22. One of the more popular fingerprint lifting tapes uses a(n) _____ system to protect the lifted print.

23. Latent fingerprints on soft and porous surfaces are made visible by _____ treatment.

24. Fingerprints on waxy or greasy surfaces can be visualized by using _____.

25. Ninhydrin treatment turns latent fingerprints a(n) _____ color.

26. Fumes from Super Glue™ turn latent fingerprints a(n) _____ color.

27. Treatment of latent prints with silver nitrate makes the print visible because of the reaction of _____ in the fingerprint with silver nitrate.

28. Amino acids in fingerprints react with _____ to visualize the print.

29. To photographically enhance a colored fingerprint on a colored surface, the difference in _____ between the two surfaces must be maximized.

30. Once a fingerprint has been made visible, it should be preserved by _____.

31. One fingerprint in a group of multiple overlapping fingerprints can be enhanced by using _____.

32. The Henry system of fingerprint classification converts ridge patterns from all _____ fingers into _____ numbers arranged as a fraction.

33. Most modern AFIS systems compare fingerprint _____ (patterns/minutiae).

34. The use of fingerprints to solve a crime involves the process of _____. When a fingerprint scanner confirms your identity, it is using the process of _____.

35. The term that describes the process of authenticating the identity of an individual by use of fingerprints, palm prints, or retinal scans is _____.

True or False

1. Fingerprints were first used in the United States by the New York City Civil Service Commission to prevent impersonations during examinations.

2. Identification points are more likely to be found at the tip of the finger.

3. Ninhydrin should be used before silver nitrate to visualize a fingerprint.

4. Finger scanners use three different techniques to capture the fingerprint.

5. Because they use semiconductor chips rather than CCD units, capacitive scanners tend to be more compact than optical devices.

Review Problems

1. Latent fingerprints are found on a piece of paper at a crime scene. Describe the techniques you would use to make the fingerprints visible for comparison and identification. In which order should you apply these techniques?

2. Latent fingerprints are found on a water glass at a crime scene. Describe the techniques you would use to make the fingerprints visible for comparison and identification.

3. Refer to FIGURE 6-22 . Classify the ridge patterns in the following fingerprints:

4. Refer to FIGURE 6-23 . Classify the ridge patterns in the following fingerprints:

5. An individual has a fingerprint record with a WWAWWAWAWA pattern series. Calculate the Henry classification ratio for this print.

6. An individual has a fingerprint record with a WAWWLWAWA pattern series. Calculate the Henry classification ratio for this print.

7. An individual has a fingerprint record with a LWAWWAWAWA pattern series. Calculate the Henry classification ratio for this print.

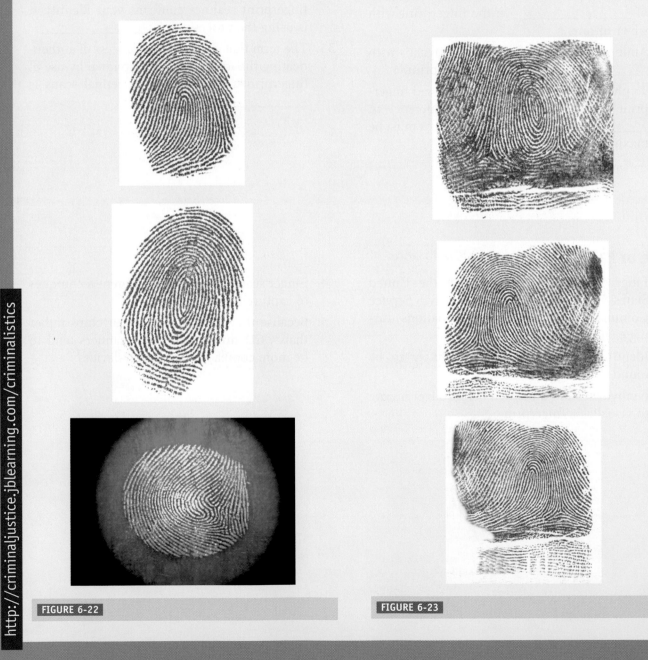

FIGURE 6-22

FIGURE 6-23

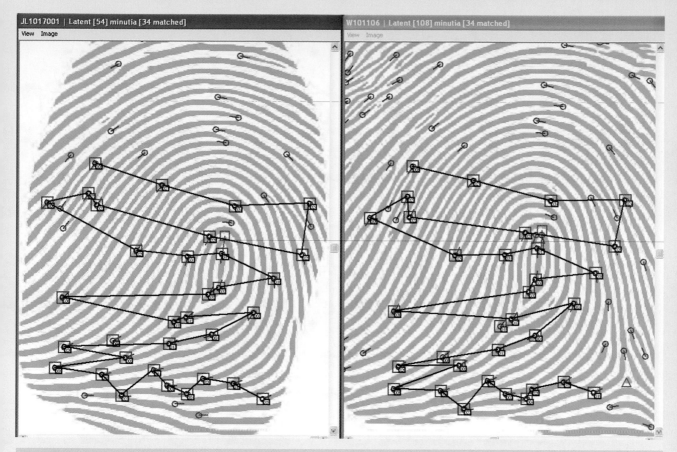

JL1017001 | Latent [54] minutia [34 matched]
View Image

W101106 | Latent [108] minutia [34 matched]
View Image

FIGURE 6-24

8. Refer to **FIGURE 6-24**. Identify the location of the following minutiae on the fingerprint:
   a. Bifurcation
   b. Lake
   c. Crossover
   d. Ridge ending

Further Reading

Budowle, B., Buscaglia, J., Perlman, R. S. Review of the scientific basis for friction ridge comparisons as a means of identification: Committee findings and recommendations. *Forensic Science Communications,* 2006; 8(1). http://www.fbi.gov/hq/lab/fsc/backissu/jan2006/research/2006_01_research02.htm.

Champod, C., Lennard, C., Margot, P., and Stoilovic, M. *Fingerprints and Other Ridge Skin Impressions.* Boca Raton, FL: CRC Press, 2004.

Cole, S. A. *Suspect Identities: A History of Fingerprinting and Criminal Investigation.* Cambridge, MA: Harvard University Press, 2001.

Exline, D. L., Schuler, R. L., Treado, P. J., and ChemImage Corporation. Improved fingerprint visualization using luminescence and visible reflectance chemical imaging. *Forensic Science Communications.* 2003; 5(3). http://www.fbi.gov/hq/lab/fsc/backissu/july2003/exline.htm.

Rhodes, H. T. F. *Alphonse Bertillon, Father of Scientific Detection.* London: George G. Harrap, 1956.

Vince, J. J., and Sherlock, W. E. *Evidence Collection.* Sudbury, MA: Jones and Bartlett, 2005.

http://criminaljustice.jblearning.com/criminalistics

Answers to Review Problems

Interactive Questions

Key Term Explorer

Web Links

http://criminaljustice.jblearning.com/criminalistics

Лекция № 1.

фармакон (греч) - лекарство, logos - учение
1693 году появилось - проф. Dale-ск..
"основы фармакологии" - крен..
фарм - наука, кот. изучает ... и ..
в орг-ме, что происходит при вве..
лекарств. в-в, т.е. изучает действие ..
изучает принципы и возм-ти при..
ма-в в том или иной орг-ме.

Фармакогнозия - наука о ... лек. рас..
Раньше погибало мене много людей ..
инфекции, т.к. не было достаточно лек.
Гален предложил руководство по фар..
в кот. указ. о применении лек..
Далее ... арабы. были откры..
вые аптеки (г. Багдад) Системати..
лек-ва. ... Затем эпоха средне..
Возникла алхимия. Они стреми..
получить золото, и создали ма..
лаборатории. ... полуг. элексир..

Questioned Documents

OBJECTIVES

In this chapter you should gain an understanding of:

- Individualizing characteristics of handwriting and signatures
- Differences between typed and word-processed documents
- Ways to detect erasures, obliterations, and alterations
- Security printing
- The analysis of ink
- The use of biometrics with identity documents

FEATURES

On the Crime Scene

Back at the Crime Lab

See You in Court

WRAP UP

Chapter Spotlight

Key Terms

Putting It All Together

To ensure a complete document examination, a document examiner must scrutinize the documents in question before this evidence is tested in any other way because such testing may potentially destroy or compromise its individualizing characteristics. For example, if the document is tested for latent fingerprints, the reagents used to make any latent prints visible may react with the ink in the document, thereby making the subsequent comparison of this ink with the ink on an exemplar impossible.

Investigators must take great care to preserve questioned documents. They should handle the documents only when wearing gloves to ensure that their own fingerprints are not transferred onto the document. Investigators also should record the location where each document was found, along with the date of its discovery, in a separate log.

Each questioned document should be placed in an envelope that is large enough to hold the document without folding for storage and transportation. If a chain of custody label is affixed to the evidence envelope, all information should be written on this label before the document is inserted into the envelope, so as to prevent any writing indentations from being transferred onto the document. The label should indicate that the contents should be first examined by the document examiner. All documents should also be photographed or scanned—a step that will create another copy but not damage the original in any way. The scanner used should be a model that barely touches the document. Auto-feeding devices should never be used for this purpose because they may crease or mark the document.

Document examination is a highly important area of forensic science that requires years of training to develop true expertise. Because document examiners compare questioned documents to authentic samples of an individual's writing, sometimes the collection and examination of authentic documents for comparison is more difficult than the comparison itself.

1. Several letter-size sheets of paper need to be compared to determine if all of them originated from the same source. The sheets contain printed letters, have been folded, and contain staple marks. Describe the first steps that the document examiner would take when presented with such evidence.
2. A small packet of cocaine that is wrapped in a small piece of scrap paper is recovered from a crime scene. How would an examiner determine whether the paper originated in the suspect's home?

Introduction

The uniqueness of handwriting—that is, the **individual characteristics** that differentiate one person's written word from another's—provides physical evidence whose value is only slightly less useful than the results of a fingerprint comparison. Likewise, typewritten or word-processed documents may prove invaluable when investigating a crime. Before such documents can be presented in court as evidence, however, a document examiner must establish their authenticity. A **questioned document** is an object that contains typewritten or handwritten information but whose authenticity has not yet been definitively proven.

This definition encompasses a wide variety of written and printed documents we commonly use.

For example, it includes the documents we use to manage our financial lives, such as checks, contracts, wills, car and real estate titles, and gift certificates. It also includes the documents we use to establish our identity, such as driver's licenses, passports, voter registration cards, and Social Security cards. Since the terrorist attacks of September 11, 2001, there has been an increased awareness of forged identity documents in the United States. In **forgery**, a person creates a document (or other object) or alters an existing document (or other object) in an attempt to deceive other individuals, whether for profit or for other nefarious purposes, such as entering the country illegally. To counteract this trend, numerous government agencies have stepped up their efforts to establish beyond any doubt the identity of international travelers.

CRIME SCENE   George W. Bush's National Guard Memos

On September 8, 2004, with the presidential election only two months away, CBS News aired a *60 Minutes* report that raised doubts about whether President George W. Bush had fulfilled his obligations to the Texas Air National Guard when he was a young man. This report aired at the time when the Democratic Party had stepped up its criticism of Bush's service with the National Guard between 1968 and 1973. The CBS report revealed four documents that were purported to be written in 1972 and 1973 but had become available only recently. If proven authentic, the documents would have contradicted several long-standing claims by the White House about an episode in Bush's National Guard service in 1972, during which he abruptly moved from Texas to Alabama to take part in a political campaign.

One of these memos was an order for Bush to report for his annual physical exam; another contained advice on how he could "get out of coming to drill." CBS named one of the network's sources as retired Major General Bobby W. Hodges, the immediate superior of the documents' alleged author, Lieutenant Colonel Jerry B. Killian, who died in 1984. The CBS documents asserted that Killian, who was Bush's squadron commander, was unhappy with Bush's performance toward meeting his National Guard commitments and resisted pressure from his superiors to "sugarcoat" Bush's record.

Within hours after the television program aired, the authenticity of the documents was challenged on Internet forums. Several features of the documents suggested that they were generated by a computer or word processor rather than a 1970s-vintage typewriter. Later, experts who were consulted by news organizations pointed out typographical and formatting questions about the four documents.

Forensic document specialists pointed out that the CBS documents raised suspicions because of their use of proportional spacing techniques. Documents generated by the kind of typewriters that were widely used by the military in 1972 spaced letters evenly across the page, such that an "i" took up as much space as an "m." In the CBS documents, each letter used a different amount of space. While IBM had introduced an electric typewriter that used proportional spacing by the early 1970s, it was not widely used in government. The use of the superscripted letters "th" in phrases such as "111th Fighter Interceptor Squadron" (Bush's unit) also presented another anomaly in the documents. Other experts noted that the font used in the CBS documents appeared to be Times Roman, which is widely used by word-processing programs today but was not common on typewriters in the early 1970s.

CBS officials insisted that the network had done due diligence in checking out the authenticity of the documents with independent experts over six weeks prior to the airing of the *60 Minutes* report. A senior CBS official said that the network had talked to four typewriting and handwriting experts "who put our concerns to rest" and confirmed the authenticity of Killian's signature.

Although CBS and Dan Rather (the reporter on the program) defended the authenticity of the documents for a two-week period, scrutiny from independent document examiners and other news organizations eventually led CBS to publicly repudiate the documents and reports on September 20, 2004.

Source: The Washington Post, June 5, 2006.

Handwriting

Most document examiners spend more time examining handwriting than they do printed materials. Document examiners routinely testify under oath that each person's writing is unique to that person, and that no two individuals can produce identical writing. In addition, each writing sample from a given individual is itself unique. The writing of an individual will vary over a natural range, which is a feature of that person's writing. For this reason, handwriting can be used as a means of individual identification, provided that enough samples of the person's writing are available for purposes of comparison.

If the document examiner is to establish a questioned document's authenticity, the known writings, which are called exemplars, should be as similar as possible to the questioned document. The writing implement (e.g., pen, pencil) and type of paper (e.g., heavy, ruled) used to make the exemplar are especially important. A person may grip a pen differently from the way he or she holds a pencil, for example. Likewise, a person may push harder with a ballpoint pen than with a felt-tip pen. Some individuals write more neatly on ruled paper than on blank sheets.

Development of Handwriting

Handwriting is a complex task that requires fine motor skills. Young children begin the development of handwriting skills by copying block letters. The handwriting of all the children in a beginner's class is similar to that of their classmates; thus their handwriting has only **class characteristics** at this point.

As the motor skills of the child increase, the act of writing becomes less demanding. In addition, when the child develops skills in cursive handwriting, certain aspects of writing, such as shape and proportion, become more individualized. These distinctive features, which develop during adolescence, are known as individual characteristics. The two most commonly taught cursive handwriting techniques in the United States are the Palmer method and the Zaner-Blosser method, both of which were introduced into U.S. public schools in the late 1800s.

Once a person reaches adulthood, his or her handwriting usually stays basically the same, with only minor changes occurring. Once a person becomes elderly, however, a decline in motor skills results in a loss of pen control, which significantly alters the individual's handwriting.

The use of some drugs and medications can affect a person's handwriting. Unlike fingerprint analysis, handwriting analysis can be very subjective and must always be carefully considered.

Comparison of Handwriting

Three basic types of handwriting are universally recognized (FIGURE 7-1):

- **Block capitals**: uppercase unjoined letters
- **Cursive**: lowercase letters that are joined together
- **Script**: lowercase unjoined letters

The normal handwriting of most people is a combination of cursive and script writing. For this reason, experts usually use the term "cursive" to denote handwriting in which the letters within a word are mostly joined, and the term "script" to describe handwriting in which the majority of the letters within a word are separated.

(A) BRING THE RANSOM TO

(B) Bring the ransom to

(C) Bring the ransom to

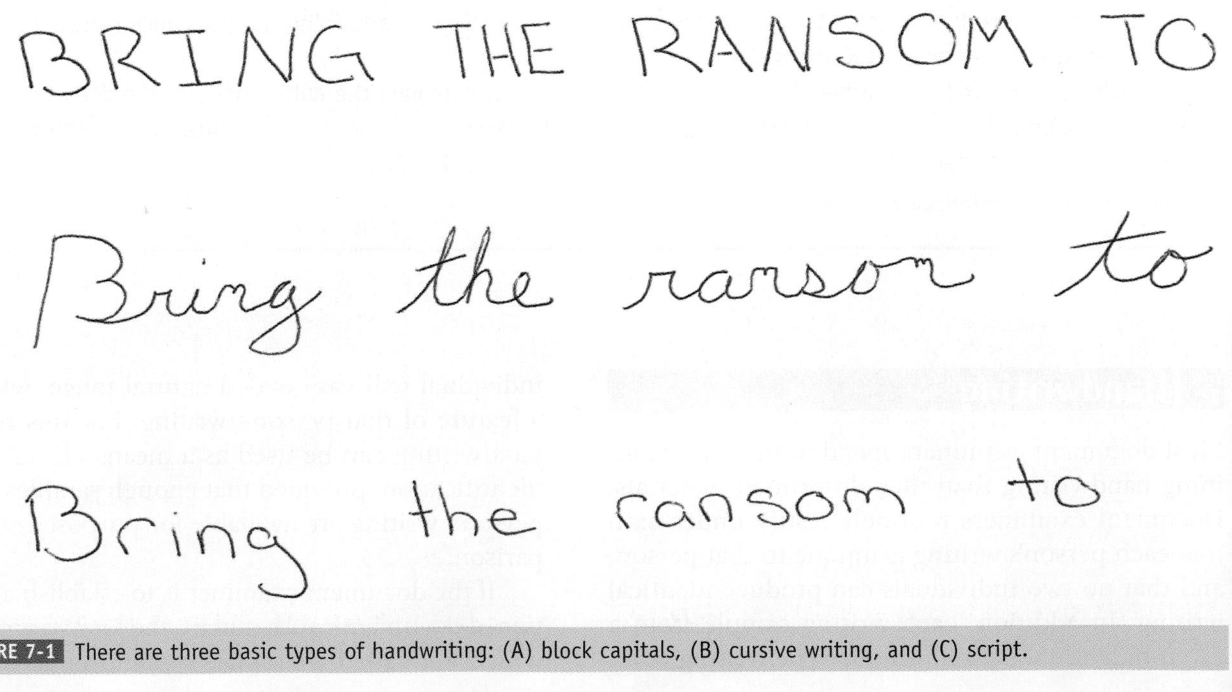

FIGURE 7-1 There are three basic types of handwriting: (A) block capitals, (B) cursive writing, and (C) script.

When analyzing handwritten documents, the document examiner compares all of the characters in the document, using his or her expertise to locate handwriting traits that make the document unique. **TABLE 7-1** lists the important features of handwriting that are analyzed in this kind of examination. **FIGURE 7-2** demonstrates the identification of each of these features in a handwritten sentence.

The proportions and shapes of individual letters are easier to observe when they are placed under magnification. Thus the document examination is typically carried out by using a low-power stereomicroscope or a portable magnifying glass (**FIGURE 7-3**). A magnifying glass used for this purpose may be fitted with a **reticle**, an eyepiece containing a grid of fine lines that is placed on the magnifier to determine the position of the letters being examined (**FIGURE 7-4**). With the reticle in place, the examiner can measure the slant or slope

TABLE 7-1
Important Handwriting Features
Style of writing
Shape of letters
Slope or slant of writing
Spacing of letters
Spacing of words
Initial strokes
Terminal strokes
Connecting strokes

of the letters as well as the spacing between letters and words. Most Americans appear to believe that slanted handwriting looks more pleasing than strictly vertical writing. Handwriting analysis from many individuals has shown that handwriting slope can vary through a range of 50° forward to a slope of 10° backward from vertical.

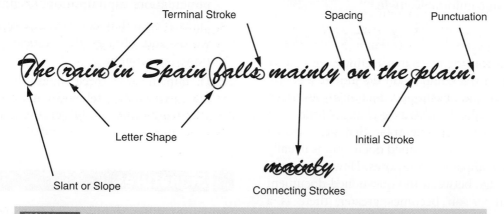

FIGURE 7-2 The important features of this handwritten sentence are identified.

FIGURE 7-3 A portable magnifying glass can be helpful when examining a handwriting sample.

FIGURE 7-4 The reticle can be used to measure slope angles, letter height, writing thickness, and spacing.

In addition, it is important to determine both the direction in which the pen strokes were created and the order in which they were made. The direction of the pen movement can indicate whether the writer is right-handed. For example, circular movement of the pen in a clockwise direction indicates left-handedness.

The document examiner must always remember that the handwriting of an individual shows natural variations. In fact, the same sentence written several times by the same person will never be exactly the same. When making a comparison of a questioned document, the examiner must have enough exemplars available to assess the natural variation in the individual's handwriting. Such variations may arise because of factors such as whether the individual is under the influence of alcohol or drugs, stress, or illness. In some cases, variations may be caused by the writing instrument itself or the writer's creation of the document on a surface that is at an uncomfortable angle.

Collection of Handwriting Exemplars

An important consideration in selecting exemplars is assessing when the comparison document was created relative to the questioned document. As noted earlier, most adults' handwriting changes little from year to year. Thus an exemplar that was created within 2 years of the questioned document is usually adequate for comparison purposes. However, as the difference in age between the questioned document and the comparison becomes greater, there is a greater chance that the comparison document will not be a representative exemplar.

Handwriting specimens fall into two categories:

- Unrequested specimens: Documents created with no idea they would be used for a comparison.
- Requested specimens: Documents produced upon request. The same type of paper and pen can be used, and the suspect may be asked to write out exactly the same text as on the document in question.

In some cases, handwriting specimens may be obtained under court order. Two cases support the constitutionality of taking handwriting specimens:

- *Gilbert v. California* (388 U.S. 263, 1967): The Supreme Court allowed law enforcement authorities to obtain handwriting exemplars from suspects before the suspects obtained legal counsel. The court also ruled that handwriting

samples are identifying physical characteristics that are not protected by the Fifth Amendment.

- *United States v. Mara* (410 U.S. 19, 1973): The Supreme Court ruled that the taking of a handwriting specimen does not constitute an unreasonable search of a person and is not a violation of the individual's Fourth Amendment rights.

Of course, a person may deliberately alter his or her writing when producing a requested handwriting specimen. Trained investigators must therefore know how to minimize the possibility that the suspect will produce a staged writing specimen. The following steps can lower this risk:

1. Furnish the suspect a pen and paper that are similar to those used to create the questioned document.

2. Never show the suspect the questioned document.

3. Do not provide the suspect help with spelling, punctuation, capitalization, or grammar.

4. Dictate text that contains many of the same words and phrases that appear in the questioned document.

5. Dictate the text three times. If the suspect is trying to deliberately alter his or her writing, wide variations will be observed between each specimen.

Signatures

Signatures differ significantly from regular handwriting because we choose to make our signature highly stylized. In fact, some people's signatures are so stylized that the name itself is not readable. Signatures are a type of personal identification that is repeatedly produced in an almost automatic way. Even though we sign our name almost automatically, because of natural variation a person's signature is never identical on two different occasions.

Usually signatures are forged for financial gain. To produce a convincing signature forgery, the forger must possess at least one example of the person's authentic signature. The most common techniques used by forgers are freehand simulation and tracing. A forger who uses the **freehand method** practices making the person's signature until the forger is able to create a reasonably similar signature.

There are two ways to trace a signature: trace-over and use of a light box. In the **trace-over method**, the forger places the document to be forged under a document containing the authentic signature. Next, the writer traces over the signature so that an indentation of the signature is made on the document below. The top sheet is removed, and the indentation on the forged document is then traced with a pen to apply ink to the indentation (FIGURE 7-5). In the light box (or window) method, the document on which the forged signature will be placed is put on top of the document containing the authentic signature. The two sheets are then placed on a light box or held up to the window. The illumination from behind makes the signature in the bottom sheet visible through the document to be forged. The signature is usually traced first by pencil and then inked over later.

Both tracing techniques produce certain characteristics in the signature that allow a document examiner to determine that it is a forgery (TABLE 7-2). In addition, examination of the forged document under infrared light will reveal whether a pencil was used first and ink then applied over the indentation. Examination of the signature under magnification when illuminated with infrared light will reveal parts of the pencil impression that weren't completely inked over. FIGURE 7-6 is an example of a signature produced using a light box. As you can see, the signature was traced with pencil and then retraced with ink.

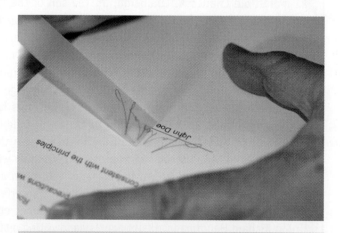

FIGURE 7-6 An example of a forged signature that was made using a light box. The authentic signature was traced first with pencil and then retraced with ink.

Another forgery technique is to lift an authentic signature that was made with a pen containing erasable ink; this technique may work up to an hour after the signature was made. The lift is carried out using Scotch™ frosted tape. The tape with the lifted signature is laid down on the document to be forged, leaving the lifted signature behind (FIGURE 7-7). The signature on the authentic document (which was lifted) remains intact, albeit somewhat faded, and the signature on the forged document looks authentic.

Handwriting comparison software is now available to help the forensic document examiner. Handwriting comparison programs, such as Write-On and Access, have been developed specifically for document examiners who need to scrutinize large amounts of handwriting. These programs assess natural differences in handwriting by finding all possible variations present in questioned and specimen documents. Such a program first creates an index of unique words that serves as the starting point in deciding on which words or character combinations to search. Once the examiner determines

TABLE 7-2
Characteristics of Traced Signatures

Unnatural pen lifts: The forger lifts the pen to check his or her progress.
Retouching: Errors in tracing are covered over.
Shaky handwriting: Tracing slowly produces a less fluid line.
Blunt end strokes: In authentic signatures, pen strokes are tapered at the end.
Two identical signatures: The range of natural variation would not produce two identical signatures.

FIGURE 7-5 An example of a forged signature produced using the trace-over method.

FIGURE 7-7 An example of a signature easily lifted with tape.

the letter combinations that will be investigated, their occurrences are automatically loaded into files that can be compared side by side. The handwriting comparison program also highlights the location of each occurrence of the letter combinations on the actual scanned documents. When the search is completed, the software creates a statistical table that cites the occurrences found within each questioned and specimen document. The statistical correlation of these occurrences helps provide an objective basis for the preparation of forensic conclusions.

Erasures, Obliterations, and Alterations

Important documents can be altered or changed after they are created so that someone other than the intended person may benefit from the forgery. There are many ways to alter a document—some highly sophisticated, others hopelessly amateurish. With practice, however, a scientific investigation can detect even sophisticated forgeries.

One way to alter a document is to erase parts of it using common implements. Rubber erasers, razor blades, and even sandpaper have all been used

to remove written or typed letters. All of these tools scratch or otherwise disturb the paper's surface fibers. When the questioned document is examined under a low-power microscope using a tungsten light bulb to light the surface either directly or at an oblique angle, the abrasion on the surface of the document can be easily seen. Although the magnified image shows an **erasure** has been made, often it doesn't reveal what the original document said. In some cases so much of the surface is removed that the original writing is totally obscured.

Besides physically abrading the document's surface with these mechanical devices, a forger may use chemical treatment to remove writing. For example, chemical oxidizing agents may be carefully applied to ink on a paper surface. An oxidation–reduction reaction occurs between the oxidizing agent and the ink on the surface that produces a colorless product and in essence bleaches out the ink. A careful forger will be able to produce an effect that won't be visible to the naked eye. Examination of the surface under the microscope, however, will almost always reveal discolorations in the paper surface adjacent to the bleached ink.

In addition, less sophisticated forgers intentionally obliterate writing by blacking out an area to conceal the original writing.

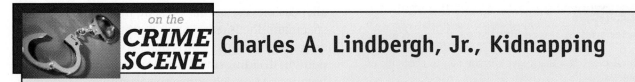

CRIME SCENE on the Charles A. Lindbergh, Jr., Kidnapping

One of the most infamous cases involving ransom notes and the use of secure bank notes was the kidnapping of 21-month-old Charles A. Lindbergh, Jr. The son of the famous aviator Charles A. Lindbergh was kidnapped from the Lindbergh home near Hopewell, New Jersey, on March 1, 1932. The captors left behind a ransom note in the nursery demanding $50,000. A second ransom note, increasing the ransom to $70,000, arrived at the Lindberghs' home on March 6, 1932. The handwriting from the two notes was consistent with both being from the same source. The reason for the increase in the amount demanded remains unknown.

The Lindberghs prepared the $70,000 ransom in gold certificates and recorded all the serial numbers. After the ransom was paid in full, the child's dead body was found four miles from the Lindbergh estate.

The next year, President Franklin D. Roosevelt ordered all gold certificates valued over $100 to be turned in to the FBI by May 1, 1933, specifically to assist with the FBI-led investigation into the Lindbergh kidnapping/murder. This move allowed investigators to determine whether the ransom bills were being spent and where. On September 15, 1934, a gas station manager was given a $100 gold certificate. Having never seen a gold certificate, the gas station manager suspected it was counterfeit and so wrote down the license plate number of the vehicle. Ultimately this bill—whose certificate number matched one from the Lindbergh ransom—was traced to Bruno Richard Hauptmann. Hauptmann's handwriting matched that of the ransom notes. In 1935, Hauptmann was convicted of extortion and first-degree murder; he was put to death in 1936.

Optical Analysis of Ink

Many inks, when they are irradiated with blue–green light, will absorb the blue–green radiation and give off radiation in the infrared spectrum. This phenomenon, which is known as **infrared luminescence**, can be used to find alterations that were made with a different ink than was used to make the original document. Optical analysis, which involves using light in the near-infrared region of the spectrum to illuminate a questioned document, may identify inks that look identical in color but have different chemical compositions, thereby suggesting that one document is a forgery. Document examiners prefer to use optical methods to examine such suspect documents because they are not destructive and hence preserve the evidence.

FIGURE 7-8 illustrates an infrared-sensitive video imaging system used in the optical analysis of documents. Two techniques are commonly used to convert a near-infrared image (which is undetectable by the naked eye) to a visible image, based on near-infrared light and fluorescence, respectively:

- The document is illuminated with near-infrared illumination, and its image is examined sequentially through a series of optical filters. Any difference in the spectrum of the transmitted or reflected light from the ink will become evident when observed through at least one of the filters.
- The document is illuminated with intense visible light that has been filtered to be free of near-infrared radiation; as a consequence, it fluoresces. An examination of the document is conducted as the optical filters are sequentially changed. Any difference in the spectrum of the fluorescence will become evident when viewed through at least one of the filters.

Optical Analysis of Concealed Information

To reveal concealed information on a document, the investigator may examine the document under visible or near-infrared light using a variety of techniques based on the type of inks used (FIGURE 7-9):

- If the masking ink is transparent to near-infrared light and the underlying ink is opaque, the examiner may use infrared light and an infrared filter to illuminate the document and show the concealed information.

FIGURE 7-9 Concealed information becomes visible when it absorbs the infrared fluorescence emitted by the masking ink.

FIGURE 7-8 The Video Spectral Comparator examines documents by using visible light and near-infrared, ultraviolet, and infrared illumination.

- If the masking ink is slightly transparent to visible light and the underlying ink is fluorescent, the examiner may use intense visible light to illuminate the document. Some of this light will penetrate the masking ink, causing the underlying ink to fluoresce. Although it is very weak, this fluorescence signal may be integrated to produce a strong legible image of the concealed information.

- If the masking ink is fluorescent and the underlying ink absorbs the fluorescence, the examiner may study the document under intense visible light to reveal the masked information.

Chemical Analysis of Ink

The most common chemical technique used to analyze ink is **thin-layer chromatography (TLC)**. TLC makes the comparison of different inks very easy. When used in conjunction with the optical tests described previously, the document examiner can reach a more definitive assessment of whether two or more inks were used on the same document.

The ink sample is removed from the questioned document by taking small "punches" from a portion of the written line in question. A blunted hypodermic needle is used to punch out the ink-covered paper plugs. This kind of needle efficiently collects the sample while doing little damage to the document. Ten or more plugs are placed into a small test tube, and the ink is extracted from the paper by adding a drop of the solvent pyridine.

A spot of the ink solvent solution is then placed on a TLC plate, and the solvent is allowed to evaporate. Glass or plastic TLC plates are coated with a thin layer of silica. A small amount of the sample being separated is applied as a spot on the surface of the plate near one end. The TLC plate is then placed in a closed chamber, with its lower edge being immersed in a shallow layer of an alcohol solvent. The solvent climbs up the plate by capillary action. As the solvent ascends the plate, the dissolved sample becomes distributed between the mobile liquid phase and the stationary silica phase. A separation occurs as a result of the interaction taking place between the mobile and stationary phases and the compounds being separated. The more tightly a compound binds to the silica surface, the more slowly it moves up the thin layer plate. The TLC plate is removed from the chamber when the solvent front is about 1 cm from the

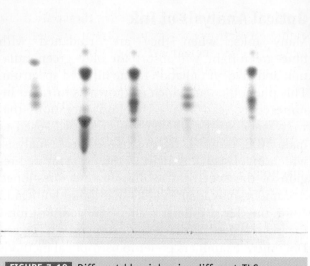

FIGURE 7-10 Different blue inks give different TLC patterns.

top of the plate. The dyes in the ink appear as bands that have moved up the plate (**FIGURE 7-10**).

The U.S. Treasury Department has gathered a complete collection of all commercial pen inks. These inks have been analyzed using TLC and the results catalogued. To aid forensic document examiners, several ink manufacturers have, at the request of the Treasury Department, voluntarily added "tags" to their inks during manufacturing. Document examiners may use these tags to determine both the manufacturer and the year of production of the ink. The tags make it easy for an examiner to date the ink used and determine whether a document was backdated.

Indented Writing

Indented writing (also known as second-page writing) is the impression left by the pen on a second sheet of paper below the page that contained the original writing. It most often is seen in pads of paper and notebooks, when the writer applies pressure while writing on the top sheet. Indented writing produces microscopic damage to the fibers at the surface of a document, which can then be detected and imaged by the document examiner.

To detect indented writing on questioned documents, examiners typically use an **electrostatic detection apparatus (ESDA)**. The ESDA is a nondestructive technique that creates an invisible electrostatic image of indented writing, which is then visualized by the application of charge-sensitive toners. To use this analytical technique, the examiner covers the page bearing the indentations with a cellophane sheet. Both sheets are then pulled into tight contact by a vacuum drawn through the ESDA's porous metal plate. The ESDA then subjects the document and the protective cellophane sheet to a repeated high-voltage static charge by waving an electrically charged wand over the document's surface. This produces a surface with a larger static charge in the impressions than on the surface. Next, toner—similar to that used in photocopy machines—is sprinkled over the cellophane surface. (A soft pad impregnated with toner may also be wiped across the surface of the imaging film to produce visible images of any indented writing; see FIGURE 7-11). The toner is attracted to the static electricity and is caught in the impressions in the cellophane surface. The developed impressions can then be photographed. The indentations can be preserved by placing a clear plastic sheet with an adhesive surface over the cellophane that is held in place by the vacuum.

The ESDA is so sensitive that it can reveal indentations that were made four or more pages below the original writing. Unfortunately, documents that have been previously processed for latent fingerprints with ninhydrin cannot be analyzed using the ESDA.

Typed and Word-Processed Documents

The first typewriter was introduced by the Remington Company in 1873, and a steady improvement in typewriter technology followed throughout the course of the twentieth century. The introduction of computer-based word-processing programs and printers over the past 25 years, however, has greatly reduced the number of typewriters that are in use today. Although the typewriter is not as widely used today as it once was, the investigation of typed documents is still taught to would-be document examiners.

Depressing the key of a manual typewriter causes a typeface to move toward the paper in the machine and to strike an inked ribbon, which then transfers the image of the typeface to the paper beneath. Because it takes physical force to strike the ribbon onto the paper, as manual typewriters are used, their typing mechanism becomes increasingly more worn. This wear, which occurs in a random and irregular way that depends on the habits of the user, imparts to the typewriter individual characteristics. Variations in alignment, both vertical and horizontal, and misalignment of individual letters often are used to identify an individual typewriter as having produced a certain document (FIGURE 7-12). Familiarity with the specific typefaces used by typewriter manufacturers often aids the document examiner in identifying the model and manufacturer of the typewriter used to produce a certain document.

In most homes and offices in the United States, computer-based printers have almost totally replaced typewriters for producing printed documents. In these computer-based systems, the author of a document can easily change the font size, style, and spacing of the typeface by selecting them from a menu that is provided with the word-processing

FIGURE 7-11 The electrostatic detection apparatus (ESDA) creates an invisible electrostatic image of indented writing, which then is visualized by the application of charge-sensitive toners.

Known #1   All questions asked by five watch experts amazed the judge.

Known #2   All questions asked by five watch experts amazed the judge.

Questioned   All questions asked by five watch experts amazed the judge.

FIGURE 7-12 A typewriting comparison can identify the typewriter that created the questioned sample.

software. As a result, an examination of a document may not establish enough individual characteristics to identify the specific computer printer that produced it. Two major types of computer printers are currently used: ink-jet and laser printers.

In ink-jet printers, ink is shot through a nozzle to form the printed character. Two types of ink-jet printers are distinguished:

- A continuous-stream printer delivers a continuous stream of ink that is formed of charged droplets. The droplets needed to create the particular printed character are directed toward the paper, while the other droplets are returned to the ink reservoir.
- A drop-on-demand printer contains a grid of tiny nozzles. When a particular letter is formed, only those nozzles that are needed to create it shoot ink at the paper.

Text created with an ink-jet printer has a blurred edge and is not as sharp as letters formed by a laser printer (**FIGURE 7-13**). Because of this blurring, it is very difficult to find individual characteristics of a document that was printed using an ink-jet printer.

Laser printers produce text that is of a higher quality than that produced by an ink-jet printer. The laser printer uses the same technology as is used in photocopiers. A rotating drum in the laser

printer is electrostatically charged in a grid pattern (**FIGURE 7-14**). The laser in the printer receives instructions from the computer to selectively scan the drum as it rotates and to strike certain spots on the drum. When the laser beam strikes a grid

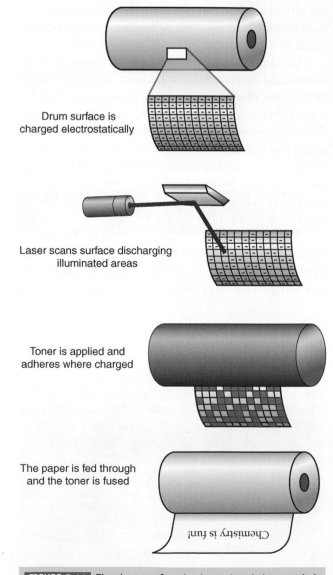

Drum surface is charged electrostatically

Laser scans surface discharging illuminated areas

Toner is applied and adheres where charged

The paper is fed through and the toner is fused

Chemistry is fun!

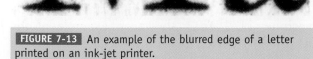

FIGURE 7-13 An example of the blurred edge of a letter printed on an ink-jet printer.

FIGURE 7-14 The drum surface is charged and then, as it is illuminated by the laser beam, discharged.

location on the drum, it causes the site to discharge its electric charge. This process forms a latent charged image on the drum. After the laser image is recorded, the drum rotates farther within the printer, where it encounters the toner (**FIGURE 7-15**). Toner is a waxy carbon material that is attracted to and sticks on the charged area of the rotating drum but not to the uncharged grids. Next, the drum rotates farther within the printer, and the toner is transferred to the paper that is traveling in the opposite direction. The paper then travels through a fuser, a heated strip that melts the waxy coating on the toner and fuses the toner on the paper surface.

An examiner is sometimes able to link a particular document to a suspected laser printer or to show that two documents have been printed on the same laser printer. In cases in which the printer in question is not available, the examiner can analyze the document's class characteristics to identify the make and model of printer that made it. For example, the examiner will usually identify the type of toner, the chemical composition of the toner (both inorganic and organic components), and the type of fusing method used in producing the document. Examination of the toner usually entails a microscopic analysis of the toner's surface morphology—that is, the shape and size of the toner particles. Some printer manufacturers use spherical toner particles, whereas others use irregularly shaped particles.

The results of these tests will allow the examiner to rule out certain printer manufacturers and certain printer models. The examiner can then use

a database to help identify the manufacturer and the model type of the laser printer that made the document. Once a suspect laser printer is identified, the examiner can perform a side-by-side comparison of the questioned and exemplar documents to determine whether they have sufficient identical individual characteristics to conclude that both were printed by the same laser printer.

Photocopied Documents

Photocopiers that use plain paper are used extensively in business offices and are widely available to the general public in copy stores. These devices produce images on paper using technology that is very similar to that employed in laser printers. The page to be copied is illuminated and its image is recorded on the rotating drum. The drum then rotates to pick up the toner, which is later melted on the surface of the paper by heat applied in the fuser of the photocopier.

The document examiner may be able to identify the manufacturer and model of the photocopier by analyzing particular marks that are left on the paper copy. The mechanisms ("grabbers") that move the paper in the photocopier differ between manufacturers. Because the rollers and slides that move the paper through the copies are found in unique locations in specific printer models, characteristic marks will be left on the copies made. In addition, the toner used by each manufacturer has a slightly different chemical composition and morphology. By examining the shape of the toner particles (morphology) under a low-power microscope and the composition of the waxy layer on the toner, the document examiner may be able to identify the manufacturer of the photocopier that produced the questioned document.

It also is possible to connect a particular copy to a specific photocopier. Imperfections on the drum forming the image may produce irregularly shaped characters on the finished copy. In addition, specks or marks in the copy ("trash marks") may result from dirt or scratches on the glass window on which the original document is placed. The location of these trash marks individualizes the photocopier that made the document. Trash marks also can help connect a series of photocopies that were produced by the same photocopy machine. **FIGURE 7-16** shows an example of the trash marks left on two documents made on the same photocopier.

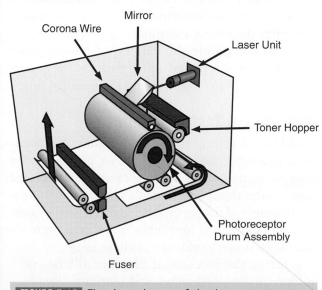

FIGURE 7-15 The charged areas of the drum attract toner, which is then transferred to the paper.

Corona Wire
Mirror
Laser Unit
Toner Hopper
Photoreceptor Drum Assembly
Fuser

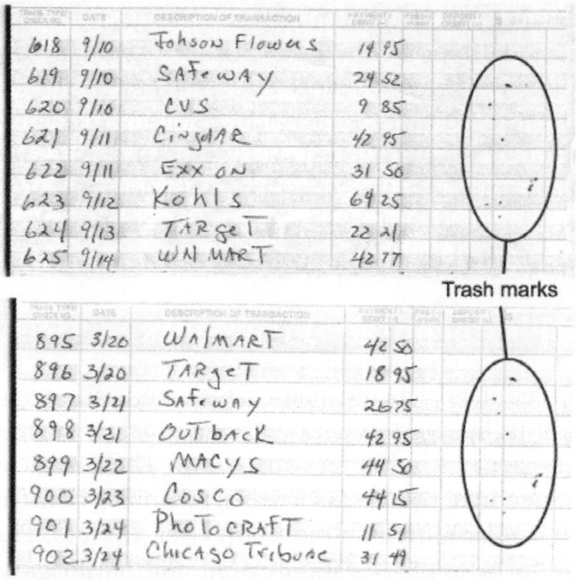

Trash marks

FIGURE 7-16 The trash marks on these two documents match, indicating that the two copies were created on the same photocopier.

Paper

Another area of investigation for the document examiner is the paper on which the document is written. Paper is made from cellulose fibers, either newly processed from freshly cut trees or taken from recycled paper. The most common features of paper that are easily measured are its color, weight, and brightness. A document examiner will first subject the paper to nondestructive tests during his or her analysis before performing any testing that might damage the paper.

Nondestructive Tests for Paper

For home and office use, paper is sold in two standard sizes: letter (8.5 × 11 in.) and legal (8.5 × 14 in.). Even though they are called "standard," however, the papers produced by different paper manufacturers actually vary ever so slightly in terms of their length and width; they are not completely identical. An examiner can determine whether the paper used for two documents was produced by different manufacturers simply by stacking the questioned and standard paper next to each other and observing any differences in dimensions.

Similarly, even "standard"-thickness paper sheets vary considerably from manufacturer to manufacturer. The examiner uses a micrometer to measure the thickness of the sheet to the nearest one-thousandth of an inch (FIGURE 7-17).

FIGURE 7-17 A micrometer is used to determine the thickness of a sheet of paper.

Some expensive paper, when held up to the light, displays a translucent design that is referred to as a **watermark**. The watermark is formed during the paper's manufacturing process. Its design is intended to indicate either the manufacturer of the paper or the company for which the paper was produced. Paper manufacturers keep records of the dates when certain watermarks were used. As a consequence, document examiners can often spot forgeries by comparing the date typed on the paper document to the period during which the watermark on the paper was used by the manufacturer.

Destructive Tests for Paper

If additional testing is warranted, then the document examiner may use a number of destructive tests to complete his or her examination of the paper. Even though these tests require only a very small sample of the paper, some damage to the questioned document is inevitable. Some courts are reluctant to subject evidence to these tests.

During the production of paper, fillers—such as clay, calcium carbonate, and titanium dioxide—are added to make the paper's surface more attractive and bright. In addition, coatings such as rosin are added to make the paper more resistant to ink penetration. Given that the chemicals added by one paper manufacturer differ from the chemicals added by other manufacturers, chemical testing can associate (or dissociate) a questioned document with a known standard. Examination of the surface of the sheet by either scanning electron microscopy with energy-dispersive X-ray or X-ray diffraction will reveal the elements present on the surface.

Security Printing

Every day, we receive in our mailboxes many unimportant advertisements and solicitations that are mass-produced. Some of these solicitations are sent by scam artists who disguise the documents' true origin by making the solicitation appear to be an official document from the government or a legitimate bank. Many other important documents, including currency and bank checks, are also mass-produced.

Because the examination of counterfeit documents is a routine part of a document examiner's work, they must be familiar with common printing methods that are used to mass-produce printed documents. There are four major types of commercial printing:

- **Screen printing**: A nylon screen is stretched across an open box frame, and a stencil is placed over it. Ink is forced through the screen, which transfers the image onto the surface of the object being printed.
- **Letterpress printing** (relief printing): The design to be printed is raised from the surface by cutting away the other parts of the surface. Ink is applied only to the raised area. When this area is brought into contact with the surface of paper, the design is printed. This technique is used to produce newspapers.
- **Lithography**: The design is painted or drawn on an aluminum printing plate with a greasy substance (crayon). A wet roller is used to moisten the area of the plate that has no coating. Next an inked roll is applied; the coated area attracts ink, and the moist area repels ink. The plate is then rolled by a contact rubber cylinder that takes up the inked image. The image on the contact cylinder is then transferred onto paper by rolling the cylinder across the paper.
- **Intaglio** (gravure): The image to be produced is carved into a metal printing plate. The entire printing plate is inked, and then the surface ink

is scraped off by a rubber roller. When a paper sheet is pressed with force against the plate, the ink left in the grooves in the plate transfers the image to the paper. This technique is used for security printing, such as printing of currency, stock certificates, and bond certificates.

Currency

Counterfeiting money is one of the oldest crimes in history. During early U.S. history, banks issued their own currencies. Not surprisingly, the existence of so many types of banknotes proved irresistible to forgers. At the time of the Civil War, an estimated one-third of all currency in circulation in the United States was counterfeit.

After the Civil War, the United States developed a single set of banknotes that was sanctioned by the federal government, ending the proliferation of bank-issued currencies. Since 1877, all U.S. currency has been printed by the Bureau of Engraving and Printing. Most currencies are printed in one of two Washington, D.C., buildings, but the U.S. Treasury also operates a satellite printing plant in Fort Worth, Texas. (Coins are manufactured at several U.S. mints located throughout the country.) Each year, these currency-printing plants produce more than 7 billion notes, worth about $82 million. Ninety-five percent of these bills replace worn-out currency.

U.S. currency is printed on 30 high-speed, sheet-fed rotary presses that are capable of printing more than 8000 sheets of single-denomination bills per hour. This massive operation requires the combined handiwork of highly skilled artists (who create the images that appear on the notes), steel engravers (who carve the design into printing plates), and plate printers (who actually manufacture the notes). During the intaglio (gravure) printing process, printing plates, which are engraved with the image for one side of the banknote, are covered with ink. The surface of each plate is then wiped clean, leaving ink behind in the design and letter grooves on the plates. Each currency sheet is then pressed into the finely recessed lines of the printing plate, so that it picks up the ink in the grooves. The printing impression made in this way is essentially three-dimensional: The surface of the note feels slightly raised, while the reverse side feels slightly indented.

Of course, U.S. currency is not just used in the United States. It also serves as a medium of exchange in many other countries whose own currencies tend to fluctuate in value, such that a black market in U.S. notes thrives around the world. As a consequence, counterfeiting of U.S. currency occurs not just in this country, but in many other countries as well. In 1990, 36% of the dollar value of known counterfeit currency passed in the United States was produced overseas, particularly in Colombia, Italy, Hong Kong, the Philippines, and Bangkok.

The $20 bill, which is the denomination dispensed by most ATMs, is the most commonly counterfeited U.S. banknote. However, counterfeiting of $50 and $100 notes is also on the rise. By contrast, counterfeiting of small bills (such as $1 and $5 bills) is rare because their value is too low to warrant the effort expended.

In 1990, counterfeiters had to engrave plates and print their counterfeit bills. Since the development of the color photocopier, however, the production of counterfeit currency has become much easier for the counterfeiter.

To stop the counterfeiting of currency by photocopy machines, a new series of U.S. currency is being issued. The first of the new bills, the $20 note, entered circulation in late 2003. It was followed by the new $50 note, which was issued in late 2004, and the new $10 note, which was issued in early 2006. The $100 note is also slated to be redesigned, but a timetable for its introduction has not yet been set. Since the life span of a $1 note is only 22 months and that for a $5 note not much longer, the Treasury Department does not plan to introduce new $1 or $5 bills.

The new $20 bill takes advantage of several new technologies as means to thwart counterfeiting (**FIGURE 7-18**). For example, color-shifting ink is used in the number 20 in the lower-right corner on the face of the bill. When the surface of the note is tilted up and down, this ink appears to change color from copper to green. When the bill is held up to the light, a faint image (watermark), similar to the portrait of President Jackson, can be seen from either side of the note.

Since 1990, a polyester thread, which cannot be reproduced by photocopiers, has been woven inside the paper of the $10, $20, $50, and $100 bills. A label that matches the denomination of the note—"USA TEN," "USA TWENTY," and so on—is printed on this polyester security thread. The bills also contain microprinting: "The United States of America" is printed in miniature letters around the border of the presidential portrait. These words appear as a black line to the naked eye and to a photocopier, which cannot resolve the small print and hence prints a solid black line when making a copy of the bill.

Canadian Currency

Beginning in 2001, to improve the security of Canadian bank notes, the Bank of Canada issued the $5, $10, $20, $50, and $100 Canadian Journey series of banknotes. The new notes have enhanced security features intended to help fight counterfeiting as well as a tactile feature to help blind and visually impaired individuals more readily identify

Security Thread: This thread can be seen when the bill is held up to a light. On the thread are the words "TWENTY USA" and a flag. This thread will glow under U/V light.

Microprinting: The words "TWENTY USA" are printed to the right of the portrait in fine print.

Microprinting

Color-Shifting Ink: When closely examined and tilted, the number 20 in the right corner will shift colors from copper to green.

Watermark: A faint image of a portrait can be seen when the bill is held up to a light. This image is embedded in the paper.

FIGURE 7-18 The newly designed $20 bill contains several security features that make it more difficult to counterfeit.

the different denominations (**FIGURE 7-19**). The new notes are the same size and retain the same colors that characterized previous bank notes. Although Queen Elizabeth and Canadian prime ministers are depicted on the same denominations in the new series, the revised bills include newly engraved portraits.

Identity Documents

A wide variety of documents are used to provide identification for individuals: passports, residency papers, and driver's licenses, among others. This section focuses on the passport, because it is the identification document that is most widely accepted around the world and, as such, is a frequent target for forgers.

A passport is a government-issued identity document that identifies its owner as a citizen of a particular country. Passports have a standardized format and include a number assigned by the issuing country. The cover identifies the issuing country. Next, a title page also names the country and

provides information on the bearer, including his or her signature, photograph, date and place of birth, nationality, family name, and given names. On the next page, the government that issued the passport requests permission for the bearer to be permitted to enter and pass through other countries. A number of blank pages follow on which foreign countries may affix visas or stamp the passport on the holder's entrance or exit.

Forged British passports, similar to those in **FIGURE 7-20** , are sold for as little as $250 in Thailand. An Algerian man who was arrested in Bangkok in August 2005 was found to be carrying 180 fake British passports. Between February 2004 and August 2005, Thai police seized 1275 counterfeit passports in separate incidents. A French woman was arrested at a Thai resort in 2005 for selling stolen French passports for $1250 each. Although Thailand is recognized as a center for forgery, false documents are produced throughout the world, including in Albania, Dubai, and Singapore.

Many countries, including the United States, are in the process of developing security features for

① Holographic Stripe
Tilt the note, and brightly colored numerals (20) and maple leaves will "move" within the shiny, metallic stripe. There is a color-split within each maple leaf.

② Watermark Portrait
Hold the note to the light and a small, ghostlike image of the portrait appears to the left of the large numeral (20).

③ Windowed Color-Shifting Thread
Hold the note to the light, and a continuous, solid line appears. From the back of the note, the thread resembles a series of exposed metallic dashes (windows) that shift from gold to green when the note is tilted.

④ See-Through Number
Hold the note to the light and, just like two pieces of a jigsaw puzzle, the irregular marks on the front and back will form a complete and perfectly aligned numeral 20.

© Bank of Canada / Banque du Canada

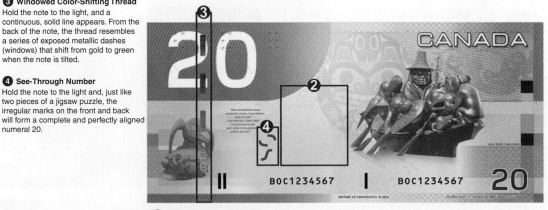

© Bank of Canada / Banque du Canada

FIGURE 7-19 The Canadian $20 bill also contains various anti-counterfeiting features.

FIGURE 7-20 Forged passports confiscated in Thailand.

their passports to further confirm that the person presenting the passport is the legitimate holder. Most new passports can be read by optical scanners and their information checked against a database. Border agents and other law enforcement agents can process such passports quickly, without having to input the information manually into a computer. These machine-readable passports have a standardized presentation in which some information appears as strings of alphanumeric characters (rather than full text); these characters are printed in a manner that is suitable for optical character recognition (a format that is able to be read by a machine, like the characters printed at the bottom of a check).

Eventually, all U.S. passports will include a **radio frequency identification (RFID) chip** containing a duplicate of the information printed on the passport's physical pages, which the U.S. government hopes will strengthen national security. The RFID passport is activated when an electronic reader sends it a signal on a designated frequency. The chip channels that radio energy and responds by sending back the passport holder's name, address, date and place of birth, and digital photograph.

Use of Biometrics for Identity Authentication

In the future, authentication of identity will likely use biometrics in addition to identity documents to verify that a person is who he says he is. As discussed in Chapter 6, **biometrics** uses methods to recognize a person that are based on one or more of the individual's intrinsic physical or behavioral traits. A variety of biometric technologies that measure and analyze physical characteristics are being developed, including fingerprint, eye (retina and iris), facial pattern, and hand geometry scanners. Technologies that measure mostly behavioral characteristics, such as signature dynamics and walking gait, are considered to be less accurate. Voice recognition is also considered less accurate than other biometrics-based approaches, because a person's speech is considered to be a mix of both physical and behavioral characteristics.

US-VISIT

The U.S. Visitor and Immigrant Status Indicator Technology (US-VISIT) program was introduced in July 2003. This program, which is under the U.S. Department of Homeland Security, is currently being used at numerous air, sea, and land ports with international arrivals and at border crossings to verify the identity of all visitors who seek to enter the United States, regardless of their country of origin or whether they are traveling on a visa (**FIGURE 7-21**). The visitor's identity is confirmed by his or her passport and by US-VISIT's biometric procedures—digital, inkless finger scans of the index fingers, and a digital facial photograph—upon entry into the United States. Biometric identifiers should also protect visitors to the United States because they should make it virtually impossible for anyone else to claim their identity if their travel documents are stolen or duplicated.

The US-VISIT program uses a computer network known as IDENT, which requires travelers to have prints of both index fingers taken at U.S. consulates and embassies overseas. When the passenger arrives in the United States, IDENT scans the passenger's index fingers and matches his or her finger scans against a database that contains the finger

FIGURE 7-21 The US-VISIT program uses fingerprint scans and facial recognition biometrics to confirm the identity of visitors entering the United States.

scans recorded earlier at the embassy abroad. If the two finger scans match, the visitor is allowed to enter the country. In addition, US-VISIT uses a biometric digital photograph of the individual's face that is matched by a federal agent at the border crossing.

The index finger scanning system included in US-VISIT has been criticized because it is not linked to the Federal Bureau of Investigation (FBI) fingerprint files or the government terrorist watch list file. Both of these databases rely on state-of-the-art, 10-fingerprint systems, rather than just fingerprints taken from the two index fingers. In addition, the discrepancy in the number of fingerprints found in the IDENT and FBI databases are striking: The IDENT database contains fingerprints for 15,000 suspected terrorists and their alleged associates and about 1 million known criminals or deportees overall; the FBI fingerprint database contains data for 47 million people. According to the U.S. Justice Department, Homeland Security officials expected to check about 800 people out of the roughly 118,000 visitors a day who should be screened against the FBI database in 2006. The failure to check visitors' information against the much more extensive FBI database suggests that the US-VISIT program might inadvertently allow terrorists to slip into the United States.

Testimony of Document Examiners

Testimony from document examiners has been accepted by the courts for almost 100 years. In 1989, however, three law school professors—D. Michael Risinger, Mark P. Denbeaux, and Michael J. Saks—wrote an article challenging the use of expert handwriting testimony. These authors argued that documentation examination had an unacceptably high error rate and that no evidence had ever been presented proving that trained document examiners would make fewer errors when examining documents than would nonexperts. This article had an immediate effect, prompting courts to subject expert document testimony to more scrutiny and defense attorneys to more aggressively challenge expert document testimony. In a 1995 case, *United States v. Starzecpyzel*, the court ruled that the document examiner provides technical—rather than scientific—testimony. This decision marked a major turning point in how expert document testimony was viewed under the Federal Rules of Evidence.

A subsequent study of the proficiency of document examiners was published in 1997 (Kam, Fielding, & Conn, 1997). It revealed that when trained document examiners were compared with nonexperts in terms of their proficiency in handwriting identification, the experts erred only 6.5% of the time, whereas the error rate for nonexperts was 38.3%. The nonexperts mistakenly matched handwriting samples at a rate more than 5 times higher than the experts did. In 2001, another study by the same authors found that when examining false signatures, trained document examiners had an error rate of less than 1%, whereas nonexperts had an error rate greater than 6% (Kam, Fielding, & Conn, 2001). Thus, both studies demonstrated that trained document examiners possess skills that exceed those of nonexperts.

In 1999, the U.S. Court of Appeals for the 11th Circuit upheld a ruling in *United States v. Paul* that had admitted expert handwriting testimony. In addition, it upheld the lower court's decision to exclude the testimony of Mark Denbeaux, who had attempted to be qualified as a document expert. The court ruled that Denbeaux's education and training as a lawyer did not qualify him to testify about the reliability of handwriting examination because he had no specific training or education in handwriting examination.

BACK AT THE CRIME LAB

Questioned document examination often involves studying obliterated writing. Examiners must take care with this type of evidence to ensure that the sample is not further altered or destroyed.

For example, organic solvents used to clean surfaces may destroy trace evidence that was invisible to the naked eye but might be readily viewed under ultra-violet or infrared light. Infrared photographs of questioned documents are commonly taken, even when writing is visible, to see whether the document has been altered. Looking at questioned documents under various types of light can help examiners identify writing imprints, altered writing, and writing that may have been covered by paint.

YOU ARE THE FORENSIC SCIENTIST SUMMARY

1. When comparing printed documents for a common origin, the forensic document examiner would compare the questioned document with specimens that are known to be authentic. In this case, the examiner would compare the paper, ink, and printing on the documents. In addition, the placement of folds and staple marks may provide similarities between the questioned page and the exemplars. Careful examination will also show if there are erasures, obliterations, or indented writing that is common to the set of documents. Care must be taken when handling the evidence contained in documents, since it can be easily compromised by sloppy investigative work.

2. The document examiner would begin by carefully examining the scrap paper. The type of paper, its color, writing or printing on the paper, and type of ink used would all be recorded. Next, investigators would obtain a warrant to search the suspect's home. Once that warrant was issued, a search of the suspect's home would be made to determine if paper matching the scrap could be found there.

Chapter Spotlight

- Highly trained document examiners can differentiate individuals' handwriting, a finding that may provide physical evidence that is almost as useful as fingerprint matches.

- As a child grows and develops fine motor skills, his or her handwriting evolves from having class characteristics (common to all children) to having discriminating individual characteristics.

- The three basic types of handwriting are block capital, cursive, and script. Most individuals' handwriting is a combination of the latter two styles.

- Document examiners, who typically are the first to examine questioned documents, should wear gloves to avoid compromising the evidence for further analysis.

- Reference writings known as exemplars are intentionally generated to be similar to the questioned document.

- Requested exemplars may require the examiner to take steps to minimize intentional changes in the suspect's handwriting.

- Software may help examiners distinguish forged signatures that are generated using a trace-over or freehand method.

- Microscopic analysis can often detect whether a document has been erased or chemically treated to remove ink.

- Optical and chemical analyses are sensitive methods used by document examiners to effectively differentiate various inks.

- Impressions left behind when the original document is missing, called indented writing, can be examined by a nondestructive technique known as ESDA.

- While typed documents often produce hallmark "signatures," computer printers produce documents with fewer identifiable features.

- Roller marks characteristic of a given photocopier can be used as a discriminating characteristic of photocopies.

- Paper type and watermarks are other useful identifiers of a document.

- Mass-produced printed documents are manufactured using one of four major methods: screen printing, letterpress, lithography, or intaglio.

- While readily available technologies have resulted in an increase in document counterfeiting, new printing processes, advances in RFID methods, and biometrics are being developed as means to counteract these activities.

Biometrics A technology using features of the human body for identification.

Block capitals Uppercase, unjoined letters.

Class characteristic A feature that is common to a group of items.

Connecting stroke A line joining two adjacent letters.

Cursive A type of writing in which the letters are joined and the pen is not lifted after writing each letter.

Electrostatic detection apparatus (ESDA) An instrument used to visualize writing impressions.

Erasure The removal of writing or printing from a document.

Forgery When a person creates a document (or other object) or alters an existing document (or other object) in an attempt to deceive other individuals, whether for profit or for other nefarious purposes, such as entering the country illegally.

Freehand method A forged signature that was created by simulation rather than by tracing.

Individual characteristic A feature that is unique to a specific item.

Infrared luminescence Light given off in the infrared region when ink is irradiated with visible light.

Initial stroke The first stroke leading into a letter or signature.

Intaglio A printing method in which a metal printing plate is engraved, producing a raised image on the document. Intaglio is used for security printing, such as for money, passports, and identity documents.

Letterpress printing A printing method in which detail is raised off the printing plate, which results in a corresponding indentation of the characters of the printed document.

Lithography A printing method in which an image is transferred from a printing plate to an offset sheet and then onto the document. The resulting image is not raised or indented.

Questioned document An object that contains typewritten or handwritten information but whose authenticity has not yet been definitively proven.

Radio frequency identification (RFID) chip A computer chip that uses communication via a radio frequency to uniquely identify an object, such as a visa.

Reticle A network of fine lines in the eyepiece of a microscope that allows the examiner to measure the distance between magnified objects.

Screen printing A printing method in which ink is passed through a screen.

Script A writing style characterized by lowercase unjoined letters.

Terminal stroke The final stroke trailing away from a letter, word, or signature

Thin-layer chromatography (TLC) A technique for separating components in a mixture. TLC is used to separate the components of different inks.

Trace-over method The copying of a genuine signature by tracing over it.

Watermark A translucent design impressed into more expensive paper during its manufacture.

Putting It All Together

Fill in the Blank

1. A(n) _____ is a document (or other object) that contains typewritten or handwritten information whose authenticity is in question.

2. _____ is the process of making or adapting objects or documents with the intention to deceive.

3. Handwriting samples that are used as a means of individual identification are called _____.

4. The handwriting of young children who write block letters has _____ characteristics.

5. Cursive handwriting exhibits _____ characteristics.

6. There are three basic types of handwriting: _____, _____, and _____.

7. A(n) _____ is an eyepiece that contains a grid of fine lines that is placed on the magnifier to determine the position of the letters being examined.

8. It is important to determine both the _____ in which the pen strokes were created and the order in which they were made.

9. Handwriting analysis that indicates circular movement of the pen in a clockwise direction indicates _____ (left/right)-handedness.

10. The trace-over method places the document to be forged _____ (under/over) the paper containing the authentic signature.

11. Examination of a forged document under _____ light will reveal if a pencil was used first and ink applied over it.

12. It is possible to lift an authentic signature that was made with a pen containing _____ ink up to an hour after the signature was made.

13. The removal of ink from a document by chemical bleaching can be determined by examination of the discolorations in the paper surface _____ to the bleached ink surface under the microscope.

14. Many inks, when they are irradiated with blue–green light, will absorb the blue–green radiation and then give off radiation in the _____ .

15. A document examiner uses optical methods to examine inks first because these techniques are _____.

16. Inks that look identical in color but have different chemical compositions may provide evidence of an alteration when examined using illumination in the _____ region of the spectrum.

17. When examining ink, the document may be illuminated with near-infrared illumination and its image then examined sequentially through a series of _____.

18. If the masking ink used to conceal underlying writing is transparent to near-infrared light and the underlying ink is opaque, the examiner may reveal the underlying information by using _____ light and a suitable infrared filter.

19. If the masking ink used is slightly transparent to visible light and the underlying ink is fluorescent, the examiner may reveal the masked information by illuminating the document with intense _____ light, some of which will penetrate the masking ink and induce the underlying ink to fluoresce.

20. The most common chemical technique used to analyze ink is _____.

21. A(n) _____ is the leading technology for detecting indented writing on questioned documents.

22. Letters formed by a typewriter produce _____ (individual/class) characteristics.

23. Examination of the laser printer toner usually involves a microscopic analysis of the surface _____, followed by identification of the inorganic and organic components present.

24. Specks or marks on photocopies called _____ are caused by dirt or scratches on the glass window on which the original document is placed.

25. The most common features of paper that are easily measured are the _____, _____, and _____.

26. An examiner can determine whether the paper used for two documents was produced by different manufacturers by stacking the questioned and standard paper next to each other and observing any differences in _____.

27. The thickness of paper sheets can be measured by using a(n) _____ .

28. Some expensive paper, when held up to the light, displays a translucent design called a(n) _____.

29. U.S. currency is printed by the Bureau of Printing and Engraving using _____ printing.

30. In the future, U.S. passports will contain a(n) _____ chip, which the government hopes will make passports harder to counterfeit.

True or False

1. An adult's handwriting usually stays basically the same over the course of his or her lifetime, with only minor changes occurring.

2. The same sentence, when written several times by the same person, will always be exactly the same.

3. The techniques most commonly used by forgers to duplicate signatures are tracing and freehand simulation.

4. It is very easy to find individual characteristics of a document that was printed using an ink-jet printer.

5. It is possible to link a particular document to a suspected laser printer or to show that two documents may have been printed on the same laser printer.

Review Problems

1. For the following written sentence, list important features that may individualize this writing sample. Refer to Table 7-1 for a list of the important handwriting features and to Figure 7-2 for examples.

Parting is such sweet sorrow; shall I say goodnight, 'til it be morrow?

2. For the following written sentence, list important features that may individualize this writing sample. Refer to Table 7-1 for a list of the important handwriting features and to Figure 7-2 for examples.

Birds of a feather, flock together, and so will pigs, and swine.

3. For the following written sentence, list important features that may individualize this writing sample. Refer to Table 7-1 for a list of the important handwriting features and to Figure 7-2 for examples.

And may we, like the clock, keep a face clean and bright, with hands ever ready, to do what is right.

4. Specimens of handwriting (exemplars) that are used for comparison with a questioned document may be requested or unrequested specimens. List the steps that should be taken to minimize the possibility of staged writing for each type of specimen.

5. List the major types of commercial printing. Describe the commercial printing technique used for printing currency.

6. List the security features in the new $20 bill. How do these features combat photocopier-based forgeries?

7. A document examiner is asked to analyze the ink used on a bank check. How would the examiner determine if more than one ink was used to make the check? List one nondestructive test and one destructive test that could be used.

8. A document examiner is asked to analyze a signature on a credit application for an erasure. How would the examiner determine if the document contains an erasure?

9. Two ransom demands are received. Both were printed on a laser printer. List how a document examiner would determine if the two notes were printed by the same laser printer.

10. A notepad recovered from a hotel room last occupied by a murder victim may contain the phone number of a suspect. Describe how the document examiner might reveal the indented writing on the notepad.

11. A set of documents consists of letter-size white paper containing a legal agreement that was printed using a word processor and a laser printer. One of the parties to the agreement claims that page 4 was switched for the original page 4 after the agreement was signed but before it was photocopied. Explain how a document examiner would establish whether page 4 was the original or a switched page.

12. List four biometric devices that detect a person's physical characteristics.

13. List two biometric devices that measure a person's behavioral characteristics.

14. Using resources you can find on the Web, describe biometric devices that are designed to scan the iris of the eye. From your investigation, do you think a scan of the iris should be used with a passport for entry into the United States? What are the pros and cons?

15. Do voice recognition devices measure physical characteristics, behavioral characteristics, or both?

Further Reading

Brunelle, R. L., and Crawford, K. R. *Advanced in the Forensic Analysis and Dating of Writing Ink*. Springfield, IL: Charles C. Thomas, 2003.

Harrison, W. R. *Suspect Documents: Their Scientific Examination*. London: Sweet & Maxwell, 1966.

Hilton, O. *Scientific Examination of Questioned Documents*. Boca Raton, FL: CRC Press, 1993.

Kam, M., Fielding, G., and Conn, R. Writer identification by professional document examiners. *Journal of Forensic Science*. 1997; 42:778.

Kam, M., Fielding, G., and Conn, R. Signature authentication by forensic document examiners. *Journal of Forensic Science*. 2001; 46:844.

Kelley, J. S., and Lindblom, B. S. *Scientific Examination of Questioned Documents*. Boca Raton, FL: CRC Press and Taylor & Francis, 2006.

Morris, R. *Forensic Handwriting Identification: Fundamental Concepts and Principles*. New York: Academic Press, 2000.

Morkrzycki, G. M. Advances in document examination: The video spectral comparator 2000. *Forensic Science Communications*. 1999; 1(3). http://www.fbi.gov/hq/lab/fsc/backissu/oct1999/mokrzyck.htm.

Risinger, D. M., Denbeaux, M. P., and Saks, M. J. Exorcism of ignorance as a proxy for rational knowledge: The lessons of handwriting identification expertise. *University of Pennsylvania Law Review*. 1989; 137:731.

Scientific Working Group for Forensic Document Examination (SWGDOC). Guidelines for forensic document examination. *Forensic Science Communications*. 2000; 2(2). http://fbi.gov/hq/lab/fsc/backissu/april2000/swgdoc1.htm#Introduction.

http://criminaljustice.jblearning.com/criminalistics

Answers to Review Problems

Interactive Questions

Key Term Explorer

Web Links

http://criminaljustice.jblearning.com/criminalistics

Firearms

OBJECTIVES

In this chapter you should gain an understanding of:

- The use of rifling marks on a bullet to identify the manufacturer of the gun that fired it
- The differences between rifles and shotguns
- The differences between revolvers and semi-automatic pistols
- The components of ammunition (cartridges)
- The Integrated Ballistic Identification System (IBIS)
- Ways to restore obliterated firearm serial numbers

FEATURES

On the Crime Scene

Back at the Crime Lab

See You in Court

WRAP UP

Chapter Spotlight

Key Terms

Putting It All Together

YOU ARE THE FORENSIC SCIENTIST

Whenever fatalities occur involving firearms, the investigation must first establish whether death resulted from a homicide, a suicide, or an accident. The investigating team usually includes a firearms expert, a medical examiner, a fingerprint examiner, and a pathologist. It is their job to identify, if possible, what happened.

The firearm will always be found at the scene in cases of suicide and accidental self-inflicted shootings. If the weapon is not found at the scene, then investigators can logically infer that the shooting was caused by another individual, either by accident or intentionally. However, simply finding the gun at the scene does not always imply that the wound was self-inflicted or that the particular weapon discovered there caused the fatal injury. Thus, the first thing that the investigator must establish is that the weapon found was, in fact, the one that caused the death.

In self-inflicted injury cases, trace evidence such as the fingerprints or blood of the victim should be found on the weapon; gunshot residue (GSR) will be left on the body. Bullets found at the scene, either in the body or nearby, should be taken into custody and compared with the firearm and any unspent bullets found at the scene. Finding these elements of evidence can help the investigators determine how the individual died.

If these key elements are not present, then the investigator can conclude that the victim's wounds were self-inflicted. At this point, the investigator must establish whether the shooting was homicidal or accidental. Any bullets or cartridges found near or in the body will help the firearms expert determine what kind and what caliber of weapon was used.

1. Describe how the investigating team should proceed at a crime scene where there has been a death caused by a firearm.

2. Suppose that a suspect admits that he held the gun that caused the death, but says that the gun discharged accidentally. What evidence (if any) would substantiate his claim?

Introduction

In crimes involving the use of a firearm, the recovery of the firearm or related items, such as bullets and cartridge cases, is an extremely important part of the investigation. This evidence can be used to establish that a crime was committed, and it can frequently be used to link the victim with the weapon or to link the weapon with a suspect. In addition, such evidence may play a critical part in any later attempt at crime scene reconstruction.

Although forensic examiners will encounter a wide range of weapons during their careers, most of these weapons fall into one of four major categories: handguns, shotguns, rifles, and submachine guns. This chapter describes these firearms, their ammunition, rifling marks on bullets, the Integrated Ballistic Identification System (IBIS), serial number restoration, and the collection and preservation of firearm evidence. The analysis of gunshot residue (GSR) is covered in Chapter 9.

To begin our exploration of firearms, we consider the technical aspects of their operation. The development of firearm technology has focused on four major problems: delivering the projectile accurately, increasing the force delivered by the projectile, increasing the rate of firing, and improving the reliability of the firing mechanism.

Firearm Accuracy

The accuracy of a firearm can be greatly improved by lengthening its barrel. A gun barrel is manufactured by boring through a solid rod of steel in a lengthwise fashion. The drill inevitably leaves small marks on the barrel's inner surface; these marks make each barrel unique.

As the barrel of a gun becomes longer, the weapon becomes more awkward to handle, and two hands may be needed to control it. Generally, rifles require two hands and are usually placed against the shoulder when fired.

In the eighteenth century, gunsmiths also discovered that putting spiral **grooves** in the **bore** would cause the bullet to spin, which improved the

accuracy of the firearm. Gunsmiths had originally cut grooves inside the walls of the barrel to reduce the buildup of residue from unburned gunpowder. When spiral grooves, which are called **rifling**, were shown both to reduce residue buildup and to improve accuracy, they became a standard feature of all firearms except shotguns. The rifling is slightly different for each weapon, because different gun manufacturers use different spiral grooves. Given that the rifling imprints different scratch patterns on the bullet, the forensic scientist can use these patterns to identify the manufacturer of the weapon used in a crime.

The interior surface of a gun barrel contains both grooves and **lands** (the area between the grooves), as shown in FIGURE 8-1 . The internal diameter of the gun barrel, which is measured from a land on the left to a land on the right, is known as the **caliber** of the weapon. Caliber is measured in hundredths of an inch, such as 0.38 in., but is described as simply ".38 caliber" by U.S. manufacturers. Thus a bullet that is 22 hundredths of an inch (0.22 in.) in diameter is called a .22-caliber bullet. Measurement of European guns is made in the same way but the results are stated in millimeters. Many law enforcement officers use 9-mm handguns. A

.38-caliber weapon takes a 9.65-mm bullet, so a .38-caliber handgun and a 9-mm handgun have barrel diameters that are close to the same size.

Before 1940, the grooves were cut in the barrel of a gun one at a time, by drawing a scraper repeatedly down the inside of the barrel as it was rotated either left or right. Today, mass production of weapons requires a more rapid process. The use of rifling **broach cutters**, carbide tools that are harder

FIGURE 8-1 The interior surface of a gun barrel contains both grooves and lands; lands are the area between the grooves.

than steel (**FIGURE 8-2**) and have teeth that are positioned at fixed distances around the circumference of the device, has improved the process. The diameter of each successive ring is larger than that of the preceding ring. As the broach cutter is pushed through the gun barrel, it simultaneously cuts all the grooves at a fixed depth. The broach cutter is rotated as it is pushed through the barrel, producing the desired rifling twist.

The number of lands and grooves and their direction and rate of twist in a gun are characteristic of a particular product from a specific manufacturer. **FIGURE 8-3** shows four different rifling variations and the marks each leaves on a bullet. Such land and groove configurations are class characteristics and will allow the forensic examiner to determine the manufacturer of the weapon.

During the gun's manufacture, as the broach cutter is pushed through the barrel to cut the grooves, it liberates shards of steel from the side of the barrel. Depending on how worn it is, the broach cutter may also produce fine lines (called **striations**) that run the length of the barrel in the lands of the barrel. Striations are formed either from imperfections on the surface of the broach cutter or from steel shards that become wedged between the broach cutter and the barrel wall. The random distribution of striations

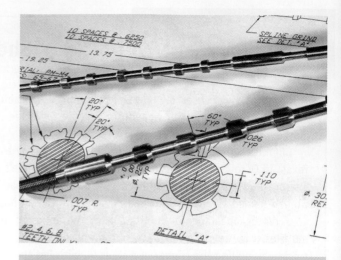

FIGURE 8-2 Rifling broach cutters for two different caliber weapons.

in the wall of the barrel is impossible to duplicate in any two barrels. Even barrels that are made in succession will not have identical striations. Because no two gun barrels will have identical striations, the individual marks left on bullets by the gun's striations will produce individual characteristics that will allow a forensic examiner to match a bullet to the gun barrel it was fired from.

The set of spiraling lands and grooves in the gun barrel gives a bullet from the gun a rotational

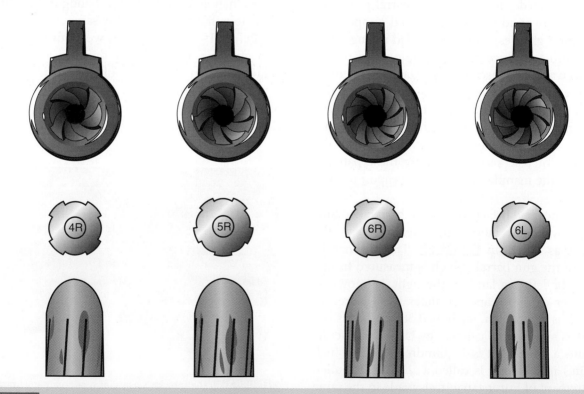

FIGURE 8-3 Four different rifling variations: 4 right, 5 right, 6 right, and 6 left. A forensic examiner can use these class characteristics to determine the manufacturer of the weapon.

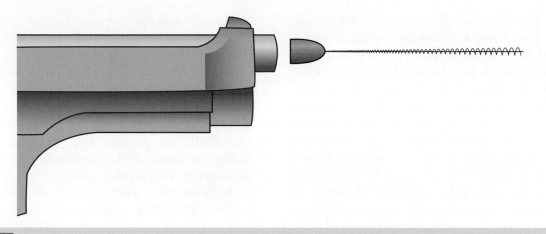

FIGURE 8-4 The bullet is given a spin by the lands and grooves in the gun barrel.

motion that helps the bullet fly straighter and farther. Bullets do not typically fly in a straight line to the target. An important rotational effect, called precession, is shown in **FIGURE 8-4**.

Increasing the Force of the Bullet

The force delivered by a bullet is proportional to the kinetic energy (E_k) imparted to it, given by the following formula:

$$E_k. = 1/2\ mv^2$$

where m is mass and v is velocity. As the caliber of the weapon increases, so does the kinetic energy of the bullet fired from the gun. The larger the caliber, the wider the bore of the barrel, and the larger the bullet needed to fill that bore. The larger the caliber, the greater the mass of the projectile.

Another way that modern weapons increase the kinetic energy of the projectile is by increasing its velocity. Modern gunpowder, which produces much more force than black gunpowder, also increases the velocity of the projectile. Given that kinetic energy increases as a square of velocity, increases in velocity generally produce much larger increases in kinetic energy.

Improving the Rate of Firing and Firing Reliability

The flintlock rifle, which was first developed in the early 1600s, used a spark to ignite the gunpowder that was loaded in the barrel of the gun (**FIGURE 8-5**).

To create this spark, the flintlock used flint (a hard rock) and steel. When flint strikes iron (which is present in steel), the force of the blow creates tiny particles of red-hot iron. If these hot sparks come near gunpowder, they will ignite.

The flintlock was a single-shot rifle—that is, it needed to be reloaded after each shot. To load the gun, first the hammer was half-cocked and a measure of gunpowder was poured down the barrel. Next a lead ball (the bullet) and a small piece of cloth or paper were rammed down the barrel on top of the gunpowder so that the bullet–cloth plug would fit tightly against the lands and grooves of the rifle barrel. A small amount of gunpowder was then placed in the flintlock's pan, which was located next to a hole in the barrel that leads to the gunpowder inside. The hammer, the jaws of which held the flint, was cocked. When the trigger was

FIGURE 8-5 Flintlock rifles suffered from two major disadvantages: They only could fire a single shot, and they were unreliable, especially in the rain.

pulled, the flint struck the iron and created sparks. The pan's gunpowder ignited, and it flashed through a small hole in the side of the barrel to ignite the gunpowder inside the barrel, causing the gun to fire.

The flintlock rifle had two serious shortcomings. First, its firing mechanism was temperamental and difficult to fire in the rain (no spark was possible with damp powder). Second, its rate of firing was slow, because it took even an experienced marksman precious time to reload. Advances in firearms technology eventually addressed both of these deficiencies—namely, the development of the percussion lock, prepackaged ammunition (**cartridges**), and multiple-shot firearms.

Handguns

The handgun is a compact weapon that can be fired by using one hand. It was originally designed for the cavalry soldier, who had one hand occupied controlling the horse. These guns were designed to solve some of the problems with flintlocks. First, the handgun delivered multiple shots before it needed to be reloaded, solving the firing rate problem; this ability was crucial for members of the cavalry, who rarely had a perfect single shot or faced a single attacker. Second, the handgun delivered projectiles with enough kinetic energy to stop an attacker, yet were not so powerful that the recoil generated knocked the gun from the shooter's hand.

Because handguns have shorter barrels than rifles, and they can be more easily concealed, they are controlled in most states. The two most commonly encountered types of defensive handguns are revolvers and semiautomatic pistols.

Revolvers

Sam Colt first conceived the handgun that is now called the revolver. He was issued a U.S. patent in 1836 for a firearm with a revolving cylinder containing six bullets. Prior to Colt's invention, only one- and two-barrel flintlock pistols were available. According to a Colt advertisement in 1870s, "Abe Lincoln may have freed all men, but Sam Colt made them equal."

Compared with semiautomatic handguns, revolvers have both advantages and disadvantages. On the positive side, revolvers are less expensive, simpler, more reliable, easier to use, and more accurate than semiautomatic handguns. On the neg-ative side, most revolvers are limited to six shots, take more time to reload, and require a more forceful trigger pull than semiautomatics.

The first Colt revolvers were single-action models. These guns required the hammer to be cocked before the trigger would release the hammer. As the hammer was pulled back, the mechanism rotated the cylinder to put the live cartridge in place. Pulling the trigger on the single-action Colt released the hammer, which subsequently struck the primer on the cartridge. By contrast, in a Colt double-action revolver (FIGURE 8-6), the trigger does all the work. As the trigger is pulled, it raises the hammer, rotates the cylinder, and then releases the hammer to strike the primer in the cartridge.

In a revolver, barrel length is made (1) smaller than in other guns to make the revolver easier to conceal and (2) relatively longer to improve the revolver's accuracy. A revolver may weigh less than 1 lb but usually weighs no more than 4 lb. The cylinder contains a number of holes (most commonly six) for the cartridges and can be swung out for easy reloading. Cartridge cases (the back part of the bullet) remain in the cylinder after firing. When reloading, the empty cartridge cases from the expended bullets must be removed before new cartridges can be inserted. Thus, at the scene of a crime committed with a revolver, empty cartridge cases will be found only if the assailant stopped to reload.

A gap between the revolver's cylinder (containing the bullets) and its barrel allows the cylinder to turn freely. This movement also allows gases to escape laterally. Thus, when the gun is fired at close range, the gases may explode from the gun and deposit GSR on both the shooter and surrounding structures. GSR, as we will learn in Chapter 9, may help the forensic examiners to reconstruct the scene.

Semiautomatic Pistols

Semiautomatic pistols were first introduced in the nineteenth century, largely through the efforts of John Browning. Today, almost every semiautomatic handgun on the market is similar to the Colt 45 semiautomatic pistol (FIGURE 8-7). Colt is a major supplier to law enforcement officials, delivering approximately 2.5 million Colt 45 pistols to the U.S. government alone.

Semiautomatic pistols make use of the recoil generated by the fired cartridge to operate the ac-

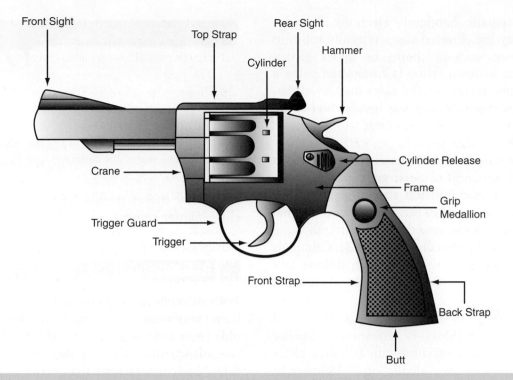

Front Sight
Top Strap
Rear Sight
Cylinder
Hammer
Cylinder Release
Frame
Grip Medallion
Crane
Trigger Guard
Trigger
Front Strap
Back Strap
Butt

FIGURE 8-6 The Colt double-action revolver was the standard handgun used by law enforcement for more than 50 years.

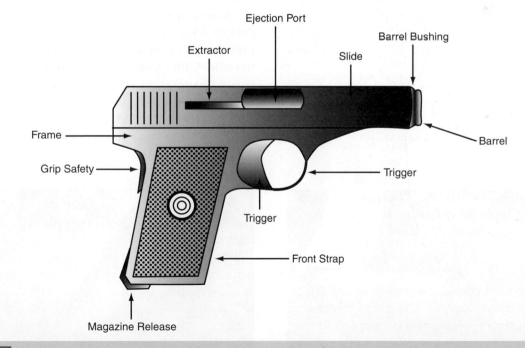

Ejection Port
Barrel Bushing
Extractor
Slide
Frame
Barrel
Grip Safety
Trigger
Trigger
Front Strap
Magazine Release

FIGURE 8-7 The Colt 45 semiautomatic is a popular weapon among law enforcement personnel today.

tion, which will eject the empty cartridge case, load the next cartridge, and cock the hammer. Making this series of steps automatic lessens the time needed to fire multiple shots, so many semiautomatics are designed to carry 15 to 19 rounds.

Semiautomatic handguns have several disadvantages relative to revolvers, however. For instance, they have more complicated mechanisms than revolvers and are more prone to jam. Semiautomatics also require more practice to use effectively. The cartridge cases (back of the bullet) must be kept short if they are to work well; thus semiautomatic cartridges are smaller and less powerful than revolver cartridges.

Semiautomatic handguns eject the cartridge cases as they fire. Ejected cases typically roll into hiding places, such as drains or under garden plants. Thus, when a crime is committed using a semiautomatic pistol, ejected cases may be left behind at the scene—but they may be difficult to find.

Silencers

A silencer is a device that is attached to a handgun to reduce the amount of noise generated by firing the weapon. This cylindrical-shaped metallic tube is fitted onto the barrel of the firearm with internal baffles to reduce the sound of firing by allowing the gas to expand internally (**FIGURE 8-8**). Criminals have been known to shoot through pillows and empty plastic soda bottles as make-do silencers.

Submachine Guns

A submachine gun (SMG) is a firearm that can fire automatically like a machine gun but uses pistol ammunition and is small and lightweight (**FIGURE 8-9**). Such weapons are issued to special forces of the military, security details guarding important government officials, and police units. Two of the most well-known submachine guns are the Israel Military Industries Uzi (Figure 8-9A) and the Ingram MAC-10 (Figure 8-9B).

Security details typically prefer the SMGs as their primary weapons because they are small and easy to conceal. Although these firearms are not very accurate at moderate distances, their rapid dis-

FIGURE 8-8 A silencer is attached to the end of the barrel to muffle the sound of the gunshot.

charge rate makes them a very effective weapon in close-quarter combat.

Rifles

Rifles differ from handguns in that they have a long barrel and a butt stock, much like the flintlocks of old. Their larger size makes them hard to conceal, so consequently rifles are more loosely regulated than handguns. Despite the greater effort required to carry them, these firearms are much more accurate than handguns.

Rifles may come in both single-shot and multiple-shot models. The bolt-action version is particularly popular for use as a large-caliber hunting rifle. Military rifles may be either semiautomatic or automatic; these models typically have a detachable magazine that holds 5 to 50 rounds. Pump-action and lever-action rifles, which usually have a smaller caliber, have magazines below the barrel.

(A)

(B)

FIGURE 8-9 (A) The Israel Military Industries Uzi and (B) the Ingram MAC-10 are two semiautomatic machine guns often used by government security details.

Shotguns

Shotguns were designed for bird hunters. Since birds are small and may be 50 to 100 yards from the hunter, a hunter would have to be a *very* good shot to hit a bird by firing a rifle. A shotgun, by contrast, shoots a large number of small pellets that disperse into a circular pattern; this area becomes increasingly larger as the pellets travel farther away from the end of the barrel. Hunters use a device called a **choke**, which screws into the muzzle end of the barrel on some shotguns, to change the area of pellet dispersion (shot spread). Given that the pellets may disperse over a 20-yd-wide area, the hunter's probability of hitting a bird in flight increases greatly when he or she uses a shotgun. A small bird is easily brought down when hit by a single pellet.

Like rifles, shotguns have a butt stock and long barrel; unlike rifles, however, they lack rifling inside the barrel. As a consequence, projectiles fired from a shotgun will not be marked with any characteristic scratches like those produced by a rifle.

The diameter of the shotgun barrel is expressed as a **gauge**. The higher the gauge, the smaller the gun's barrel and the less kinetic energy it delivers. A 12-gauge shotgun has a diameter of 0.73 in.; a 16-gauge shotgun has a diameter of 0.670 in. Shotguns are available in single-shot (break action), double-barrel, bolt-action, pump-action, and semi-automatic models (FIGURE 8-10).

Criminals often modify shotguns by cutting down (shortening) the barrel and the stock (the wooden part that usually rests against the shooter's shoulder). These modifications make the sawed-off shotgun easier to conceal under a jacket or in a boot. A sawed-off shotgun delivers more damage per shot at close range but is less effective when fired over longer distances. In the United States, it is illegal to possess a sawed-off shotgun (a barrel length less than 18 in. or 46 cm) without a tax-stamped permit from the U.S. Treasury Department, which requires an extensive background check before it will issue such a permit. FIGURE 8-11 shows how a firearms examiner determines the barrel length of a modified shotgun.

Ammunition

To reduce air resistance as it travels, the ideal bullet should be long, sharp, and heavy. Such a projectile, however, would pass through the target without delivering much of its energy and without causing much damage. Projectiles that are shaped as spheres, by comparison, deliver more energy and cause more damage. Of course, since they offer more wind resistance, they do not travel as far as narrower bullets and might not reach the target. A compromise between these two extremes is an aerodynamic shape that has a parabolic front and a wind-splitting shape.

The ammunition used in modern rifles and handguns is called cartridges. Cartridges typically have a brass case that contains a primer, gunpowder, and a bullet (projectile). By contrast, a shotgun shell may contain a bore-size projectile (called a slug), pellets of large shot, or many tiny pellets (BBs).

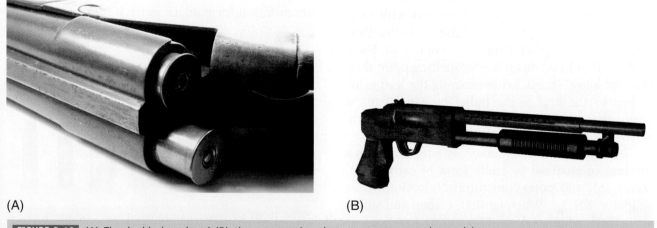

(A) (B)

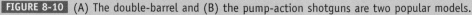

FIGURE 8-10 (A) The double-barrel and (B) the pump-action shotguns are two popular models.

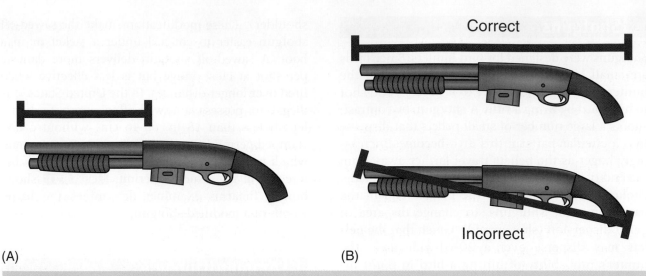

Correct

Incorrect

(A)   (B)

FIGURE 8-11 It is illegal to possess any shotgun with a barrel length of less than 18 in. or on overall length less than 26 in. (A) The barrel length is measured by placing a ruler next to the barrel. (B) The overall length is measured by placing a ruler on the table parallel to the gun and taking the measurement from the butt of the firearm to the end of the barrel.

Cartridges

Lead (Pb), a high-density yet inexpensive material, has been used to make projectiles (musket shot, bullets) since muzzle-loading rifles were first introduced. Because of its high density, a small lead ball delivers a lot of mass. At the same time, lead has disadvantages: In particular, it has a low melting point of 327.5°C (621.43°F), and it softens at velocities greater than 1000 feet per second (fps), causing it to smear the inner surface of the gun barrel and decreasing the accuracy of the shot. At velocities greater than 2000 fps, lead melts completely.

To produce a metal with a higher melting point, gun manufacturers make alloys of lead by adding a small amount of antimony (Sb) to molten lead. Even so, the melting point of the lead alloy is not high enough for modern weapons with high muzzle velocities. For these weapons, the lead/antimony alloy is placed inside a copper or brass "jacket" that has a much higher melting point than the lead alloy. The jacket containing the lead is attached to the end of the bullet by a mechanical crimp that holds the bullet together but is easily separated when fired (**FIGURE 8-12**).

Black powder, also known as gunpowder, is a mixture of charcoal (a crude form of carbon, C), sulfur (S), and potassium nitrate (also known as saltpeter, KNO_3). When ignited, carbon and sulfur act as fuel, whereas potassium nitrate functions as an oxidizer. No oxygen (or air) is required for ignition of gunpowder, because the potassium nitrate provides the oxygen that drives the reaction:

$$2\ KNO_3 + C + S \rightarrow K_2SO_4 + CO_2 + N_2$$

In this reaction, the solid carbon is oxidized (gains oxygen) and becomes gaseous carbon dioxide. The potassium nitrate is reduced (loses oxygen) and is converted to gaseous nitrogen. The conversion of the solid reactants into gaseous products ($CO_2 + N_2$) provides the pressure necessary for the expulsion of the bullet from the barrel of the gun. The optimal proportion of these components for gunpowder is 75% potassium nitrate, 10% sulfur, and 15% charcoal (by weight).

Although black powder is no longer used in ammunition for modern guns, it remains the most

FIGURE 8-12 Copper-jacketed bullets are available in different calibers.

common explosive used in fireworks. The composition of black powder manufactured for pyrotechnic use today is 75% potassium nitrate, 15% softwood charcoal, and 10% sulfur.

Black powder has a number of problems when used as ammunition. First, it produces lots of smoke. In Civil War battles, it was easy to identify the shooter who was using black powder—you just looked for the cloud of smoke.

Second, several different chemical reactions can take place during the detonation of black powder. Charcoal can be produced from a variety of sources, including corn, ash, and oak. In addition, different manufacturers use different processes to turn this charcoal into a fine, rapidly reacting powder. Every brand of charcoal has its own composition and produces different reaction products such as CO, CS_2, K_2O, nitrogen oxides (NO_x), and sulfur oxides (SO_x). The ratio in which these products are produced depends on the reaction conditions (the temperature and pressure inside the barrel) and the type of charcoal. Not surprisingly, the presence of so many variables made it difficult to standardize the performance of black powder for use as ammunition.

Third, some of the reaction products are solids. Indeed, for black powder only 40% of the reaction products are gases. Potassium sulfate (K_2SO_4), for instance, is a solid product that is deposited within the barrel of the gun. Solid products leave residue that must be cleaned out of the gun. Even worse, for every solid product produced, one less gaseous product is made. Given that gas pressure is what pushes the bullet out of the barrel, black powder does not produce enough gas for high-performance weapons.

Smokeless powder is the name given to any propellants used in firearms that produce little smoke when fired, unlike the older black powder (which smokeless powder replaced). Smokeless powder is made from two components—cordite and ballistite. Cordite and ballistite are made by mixing together two high explosives, **nitrocellulose** and **nitroglycerin**. Nitrocellulose, also known as guncotton, is a highly flammable compound formed by nitrating cellulose with nitric acid. Nitroglycerin is an explosive liquid obtained by nitrating glycerol. In its pure form, it is shock-sensitive (physical shock can cause nitroglycerin to explode), and it can degrade over time to even more unstable forms. This makes nitroglycerin highly dangerous to use in its pure form. By adding 10% to 30% of ethanol, acetone, or dinitrotoluene, however, one can chemically "desensitize" nitroglycerin to a point where it can be considered safe for use in modern ammunition.

The explosive power of nitroglycerin is derived from its ability to detonate. This chemical reaction sends a shock wave through the smokeless powder at a supersonic speed, which generates a self-sustained cascade of pressure-induced combustion that grows upon itself exponentially. An explosion is essentially a very fast combustion, and ongoing combustion requires both fuel and an oxidant. Nitroglycerin essentially contains both of these components (**FIGURE 8-13**).

If it is detonated under pressure, nitroglycerin explodes to form thousands of times its original volume in hot gas. One of the gases that results from this reaction is nitrogen gas (N_2). Nitrogen is very stable, so its production is highly exothermic (energy releasing); this characteristic explains why nitrogen is a main constituent of most explosives.

Modern ammunition uses a **primer** to set off the smokeless powder (**FIGURE 8-14**). A primer is a small copper or brass cup that holds a precise amount of stable, but shock-sensitive explosive mixture, such as lead azide, $Pb(N_3)_2$, or potassium perchlorate, $KClO_4$. The most common primers today contain a mixture of lead styphnate ($PbC_6H_1N_3O_8$), barium nitrate $Ba(NO_3)_2$, and antimony sulfide (Sb_2S_3). The primer was a crucial invention needed to make a gun that could fire in any weather.

In a handgun, the hammer (firing pin) of the gun strikes the primer, which dents and crushes the shock-sensitive chemical inside the cup against the anvil. This material explodes, igniting a secondary charge of gunpowder or other explosive.

If lead azide is used as the primer, the explosive reaction releases lead as very fine soot. Many indoor firing ranges are moving to ban lead primers because of the potential health risks they pose. Lead-free primers have been on the market for the past 10 years. Lead-free primers were originally less sensitive (and thus less reliable) and had a much greater moisture sensitivity and correspondingly shorter shelf life than lead primers. Since their introduction,

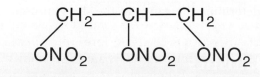

FIGURE 8-13 When detonated, a nitroglycerin molecule converts to CO_2 and N_2.

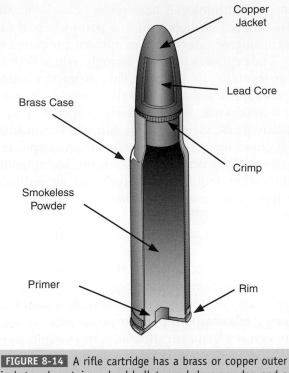

FIGURE 8-14 A rifle cartridge has a brass or copper outer jacket and contains a lead bullet, smokeless powder, and a primer. A revolver cartridge contains the same components, but it is shorter than a rifle cartridge.

however, lead-free primers have improved to the point that they are nearly equal in performance to lead-based primers.

Shotgun Ammunition

Shotgun cartridges consist of a metal base with a primer cup embedded in the middle and sides that are made of paper or plastic, which is crimp sealed at the top (**FIGURE 8-15**). These cartridges differ from regular gun cartridges in that they contain one or more disk-shaped wads placed between the smokeless powder and the shot. The **wad** can be made of felt, cardboard, or plastic. The wad is designed to make an almost gas-tight seal with the barrel and to cushion the shot while it is accelerated out of the gun barrel during firing. In some cartridges, a plastic cup holds the shot and acts as a wad to keep the shot away from the barrel when fired. At crime scenes where a shotgun is thought to be the weapon used, recovery of the wad will help forensic examiners determine the gauge of the shotgun used.

Ballistics

Ballistics is the study of bullet motion. Three aspects of ballistics are important to crime investigators: in-

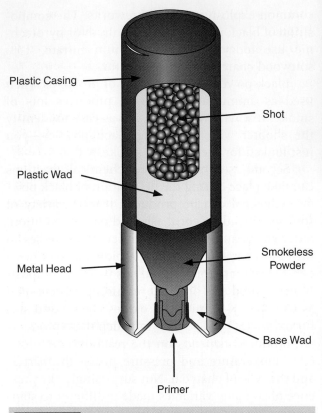

FIGURE 8-15 A shotgun shell has a plastic casing and a metal head. It contains a primer, smokeless powder, a plastic wad, and shot.

ternal ballistics, external ballistics, and terminal ballistics.

The events that transpire within the firearm are known as **internal ballistics**. The striking of the firing pin, the scratching of the sides of the bullet by the rifling, and any other imperfections on the inner walls of the gun's barrel will usually provide individualized characteristics that may later help identify the weapon that fired the bullet.

External ballistics describes the events that occur after the bullet leaves the barrel of the gun but before it strikes its target. An understanding of external ballistics will help investigators understand where the perpetrator was when he or she fired the gun. The bullet may have traveled in a straight path to its target or it may have ricocheted as a result of striking another object. If it can be proved that the victim died from a ricochet rather than a shot aimed at the victim, this evidence may convince a jury that the perpetrator is not a murderer.

Terminal ballistics refers to what happens when the bullet strikes its target. When a bullet hits a person, it enters the body, forming an entry wound. If the caliber of the weapon used is large enough, the bullet will have so much kinetic energy that it will not be totally stopped upon entry. Instead, it will pass through the victim, leaving an exit wound. A careful examination of the wounds by a medical examiner will establish whether the bullet passed through the victim from front to back, or vice versa.

The damage caused by a firearm being fired is influenced by all three aspects of ballistics. Luckily for the firearm examiner, if the same gun containing the same ammunition is fired multiple times, many of these aspects will remain constant. If the suspected firearm is located, the firearm examiner may conduct a series of test firings, usually into a water tank or a block of gel. These test firings produce fired specimen bullets that may then be compared with bullets recovered from the crime scene (**FIGURE 8-16**). The large amount of water contained in the tank used for test firings slows the bullet, but does not damage or distort the bullet or alter the impressions it carries, thereby generating an ideal known bullet that can be later compared with the bullet from the crime scene.

Collection and Preservation of Firearm Evidence

Collection and Preservation of Firearms

The first step a responding officer should take when recovering a firearm is to ensure that it is rendered safe. The officer must make sure the weapon is safely unloaded, the cylinder open or magazine is removed, and no round is in the chamber. Although it is desirable to keep the firearm, its ammunition, and its magazine preserved for possible fingerprint recovery, safety must take precedence. The firearm

FIGURE 8-16 To produce a sample bullet and/or cartridge, a forensic examiner can fire a gun into blocks of gel.

should be examined, unloaded, and cleared (for removal from the crime scene) by a qualified officer at the scene who is also familiar with the weapon.

The officer recovering the gun should carefully examine the weapon and prepare an evidence tag at the scene that contains a full description of the firearm, including the manufacturer's name, the caliber, the model, and the serial number. The tag is usually attached to the trigger guard. If the firearm is not a common model, additional marking may be found on the frame or barrel of the weapon. These letters and symbols may prove helpful in identifying the weapon's origin.

If the firearm was loaded, the number and type of rounds should be noted. When a revolver is recovered, each chamber should be examined and marked. A diagram should be made with each chamber being designated with a number; as each cartridge or spent casing is removed, its position should be marked on the diagram. Information about the positions of spent cartridge cases in the

BACK AT THE CRIME LAB

Parts of a gun, ballistic material, silencers, and wadding may all be used as firearm evidence. Wadding is the material placed between the propellant and the projectile; it is used in shotgun shells and muzzle-loader firearms to distribute force on the projectile. Today, the wadding and shot cup are usually constructed as a single unit. Shotgun wounds can be characterized by massive tissue destruction and embedded wadding if the shot was fired within 10 ft. By examining the wadding, the investigator can determine the type of shot, gauge of gun, and other possible evidence to identify the gun. Because wadding and shotgun shells are placed together in a single unit, the wadding shape and composition can be used to identify the source of the projectile.

chamber may be useful in establishing a sequence of events later. Each round removed should be placed in a separate paper envelope with a label that references the chamber number from which it came.

Once the firearm has been safely unloaded and properly tagged, it is ready for transportation to the crime laboratory or the evidence custody location. When custody of the firearm is passed to another location, the seizing officer should get a receipt for the evidence to establish a proper chain of custody.

Weapons that are found underwater should be transported to the lab in a container filled with water from the same source. This step minimizes rusting of the gun.

Serial Number Restoration

Every gun that is manufactured has a serial number stamped into the metal of the gun's frame. Some weapons have this serial number stamped in a number of locations. Federally licensed gun dealers are required by law to maintain records of the serial numbers of all of their firearms as well as the names of individuals who purchased guns from them. Police may use these serial numbers to track the ownership of weapons that are used in committing a crime.

All firearms taken into custody should immediately be checked to see whether they have been stolen. The Recovered Gun File of the National Crime Information Center allows an investigator to search for the status of a gun or to check whether a suspect's gun is being sought by other law enforcement agencies. Unfortunately, firearms cannot be traced without a serial number, although sometimes they may be traced with a partial serial number.

Criminals often try to destroy a firearm's serial number to prevent police from tracing the weapon. Studies have shown that as many as 15% of all firearms recovered in a year by major police departments have had their serial numbers removed or partially removed. Not only is the removal of the serial number a crime, but it is also a clear indication that the criminal is attempting to conceal the origin and history of the firearm. Often forensic labs are asked to restore firearm serial numbers that perpetrators have attempted to make unreadable by grinding or sanding of the weapon (**FIGURE 8-17**).

To stamp serial numbers and other information on the gun, manufacturers use hard metal dies that strike the metal body with enough force to leave a deep imprint of the manufacturer's name and the

FIGURE 8-17 The serial number has been ground off this semiautomatic Arcus 94 9-mm handgun.

gun's make, model, caliber, and serial number (**FIGURE 8-18**). The metal around and below the impressed serial number is placed under strain by this stamping process. Consequently, the strain makes this area of the metal more chemically reactive.

Several techniques have been used successfully to restore obliterated serial numbers on recovered firearms. The two most commonly used methods are the **magnaflux method** and the **acid etching method**, which are described next. If the criminal grinds deep enough into the metal, however, he or she may obliterate all of the strain recorded in the metal during the manufacturer's stamping process. In these cases the serial number cannot be recovered by using any of the methods described here. Research is ongoing to develop other means of restoring serial numbers with the use of acids.

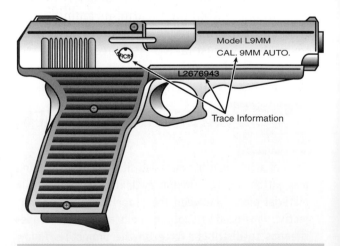

FIGURE 8-18 The arrows on this Lorcin semiautomatic pistol indicate the positions of some of the information required to complete a firearm trace: make, model, caliber, and serial number.

Magnaflux Method

The magnaflux method is designed to restore serial numbers that were imprinted on iron or steel. This method begins by using pretreatment with a small grinding tool to smooth the obliterated surface. A magnet is then attached to the opposite side of the firearm, directly behind the obliterated area. A mixture of fine iron filings and light oil is then applied to the obliterated area. The metal filings arrange themselves in the oil to provide a shadow of each number, allowing visualization of the serial number. The major advantage of the magnaflux method is that it is a nondestructive technique. If the serial number cannot be restored with this method, then the acid etching method can subsequently be used.

Acid Etching Method

The acid etching method is used on firearms that are made of both ferrous (iron) and nonferrous materials (most likely aluminum). The obliterated area is pretreated with a small grinding tool to smooth the obliterated area. Fry's reagent is commonly used for the restoration of serial numbers on iron and steel components. Fry's reagent—a mixture of hydrochloric acid, copper(II) chloride, ethyl alcohol, and water—is carefully applied to the area of interest. It then slowly dissolves away the extraneous scratches and markings, revealing some of the numbers imprinted in the metal (FIGURE 8-19).

Vinella's reagent is used to restore serial numbers on aluminum alloys; it is a mixture of glycerol, hydrofluoric acid, and nitric acid. If the aluminum alloy contains silicon, Hume-Rothery solution can be used. Hume-Rothery solution is a mixture of copper(II) chloride, hydrochloric acid, and water. Often forensic examiners alternate successive treatments of Vinella's reagent followed by treatment with Hume-Rothery solution. As with Fry's reagent, these solutions are carefully applied, and then slowly dissolve away the extraneous scratches and markings to reveal the numbers pressed into the metal.

Because many applications of these chemicals are necessary, the acid etching method is painstakingly slow. Also, once the number is visible, the acid may continue to react with the metal, which will eventually destroy the number. Extreme care should be made to record the procedures and results (photographically), because etching cannot be repeated.

Collection and Preservation of Ammunition

Any markings on fired bullets and cartridge cases must be preserved for later comparison. Each fired bullet or cartridge case found at a crime scene should be placed in a separate paper envelope so that it is protected from additional marking or damage. This envelope should be properly labeled with the date, case number, exhibit number, description of item in the envelope, and examiner's initials, and all chain of custody procedures should be carefully followed when handling it.

Fired bullets often are difficult to find at the crime scene. They may be lodged in walls, floors, or ceilings. Once found, only rubber or plastic objects should touch the bullet. It is good practice for investigators to cut out the section of the wall (or floor or ceiling) that holds the bullet, being careful not to touch the bullet. This wall section should then be properly packaged, labeled, and shipped to the crime lab, where the forensic firearms examiner can remove the bullet. Many police departments teach investigators to place a bullet in a sealed plastic box, which is then marked for identification.

Laboratory Examination of Firearm Evidence

Ballistic comparisons of fired bullets and cartridge cases have provided evidence that has been useful to law enforcement since the mid-1920s. With the advent of the comparison microscope (described in Chapter 4), forensic examiners have been able to make comparison and identification

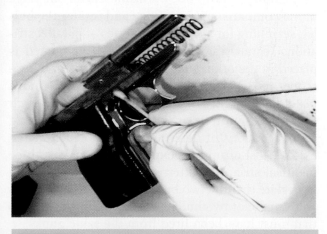

FIGURE 8-19 Forensic examiners often are asked to restore serial numbers from handguns. Here, the forensic examiner uses the acid etching method.

For firearm testimony to be accepted in court, there must be an absolute match between the marks being compared. Marks may include firing-pin marks on the bullets, muzzle marks on the bullets, and marks produced via in-laboratory caliber tests. Repeated use of a weapon can degrade the metal of the muzzle and cause slight variations in the marks left on either the shell casing or the actual bullet. In many cases, the recovered bullet is so damaged that only a partial match can be made.

Marks on shells cannot be matched easily, if at all, to smooth-bore weapons such as shotguns or muskets because the muzzles of these firearms do not leave identifying marks on the shell. Even without marks, however, it is often possible to get a ballistic match from a shotgun. A metallurgical comparison between a shot recovered from a body or other target and a shot recovered in association with the suspect's property may be acceptable, for example.

matches with fired bullets and discharged cartridge cases from crime scenes using the Integrated Ballistics Identification System (IBIS) (FIGURE 8-20). More recently, automated ballistics comparisons have become possible.

To provide a sample bullet or cartridge case for comparison with the evidence from the crime scene, the forensic firearms expert fires the suspect gun into a long water tank or a dense gel to retrieve an undamaged reference bullet. This test firing must be done carefully to preserve the markings made by the firearm's barrel. If the suspect gun is a semiautomatic pistol, the discharged cartridge case will also be marked by the gun's mechanism.

FIGURE 8-20 The Integrated Ballistics Identification System (IBIS) allows for automated ballistics comparisons of either bullets or cartridge cases; an organized workstation creates a smooth process.

Laboratory Examination of Fired Bullets

The goal of the initial examination of a fired bullet is to determine the general rifling characteristics of the firearm that discharged it. By determining the caliber, number of lands and grooves, direction of twist, and width of the lands and grooves, the search for the weapon will be narrowed to only certain models and manufacturers. If the fired bullet is not deformed, the caliber is determined by measuring its diameter with a micrometer. If it is deformed, the caliber can sometimes be determined by weighing the bullet. If the bullet has fragmented, determination of caliber becomes much more difficult. When an intact, fired bullet is examined under a compound microscope, the number of lands and grooves, as well as their direction of twist, can be determined.

During the visual analysis of bullet evidence, the examiner places one of the bullets on one stage of the comparison microscope and rotates the bullet, searching for a distinctive series of parallel striations (scratches). The bullet to be compared is placed in the other stage of the comparison microscope and then rotated in an attempt to find a matching pattern. The examiner continues to search the surfaces until he or she either finds a matching set of striations or determines that no such match exists. Most examiners expect to find at least three identical striation patterns before they are willing to declare that the two bullets match. FIGURE 8-21 is an example of two bullets with matching striations that must have been fired from the same gun.

Unfortunately, these comparisons require a thoroughly qualified forensic examiner who expends untold hours painstakingly comparing a

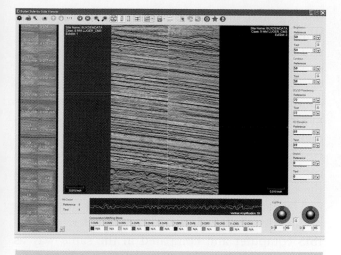

FIGURE 8-21 A "match" of striations on two bullets shows that these were fired from the same gun.

FIGURE 8-22 An IBIS match.

bullet from a crime scene to one fired from the questioned gun. Fifty years ago, when shootings in major cities were rare, this simple technology was adequate for processing evidence. Since the 1980s, the number of gun-related crimes has skyrocketed, so the manual method of comparison is no longer fast enough to handle the sheer volume of evidence being generated.

Automated Ballistics Comparisons

In recent times, the IBIS has emerged as the standard for examining projectiles and cartridge cases (**FIGURE 8-22**). In the United States, this technology is being offered to local police departments through the National Integrated Ballistic Information Network. IBIS is a single-platform computer identification system that correlates and matches both fired bullets and discharged cartridge case evidence. The comparison procedure involves placing the crime scene–fired bullet on the left-hand stage of the microscope and the test-fired bullet from the suspect gun on the right-hand stage (**FIGURE 8-23**). While observing the bullets through the eyepiece, the firearms examiner brings the stage into focus and rotates the exhibits. The IBIS also has an auto-focusing feature.

To perform a comparison with the IBIS, a pair of high-speed computers is used: the Data Acquisition Station, which the forensic firearms examiner uses to

(A)

(B)

FIGURE 8-23 To compare bullets, the forensic firearms examiner uses a comparison microscope. Here, the fired bullet appears on the left-hand side of the microscope and the test-fired bullet appears on the right-hand side. (A) The striations are a match, and (B) the striations are a non-match.

enter his or her data on the suspect bullets and cartridges, and the Signature Analysis Station, which automatically correlates the examiner-provided images with existing images found in the IBIS database. The heart of the system is digital image recognition software (aptly named Bulletproof) that converts the markings on both ammunition samples into a set of binary numbers. That is, the IBIS stores bullet (and cartridge) images as strings of binary numbers and looks within its existing binary number sets to find a bullet (or cartridge) that matches the mathematical description of the one just entered by the local firearms examiner as evidence. It ranks its findings according to the likelihood of a match. Once the IBIS has completed its analysis, the firearms examiner can then use his or her specialized skills to optimize the microscopic comparisons and focus on promising areas in the bullet or cartridge image where matches may be found.

Laboratory Examination of Expended Cartridges

Expended cartridges found at a crime scene usually indicate that the perpetrator used a semiautomatic pistol, semiautomatic rifle, or submachine gun. Most cartridges are identified with numbers and letters that are impressed onto the head by the manufacturer. When the gun is fired, the cartridge case is forced back against the gun's **breechblock**. This case becomes marked by its contact with the weapon's metal surfaces (**FIGURE 8-24**). Thus, inspection of cartridges may provide valuable class evidence, such as the caliber of the weapon, the location of its firing pin, the size and shape of the firing pin, and the sizes of **extractors** and **ejectors**. With this information, the search for the weapon will be narrowed to only certain calibers, models, and manufacturers.

As mentioned earlier, the IBIS can be used to compare cartridge cases. A separate microscope mount is used for this purpose. The cartridge case is digitally photographed by an auto-focusing camera (using the Brasscatcher software program), which captures the markings made by contact between the cartridge and the firing pin/ejector mechanism and by contact between the cartridge and the breech of the gun. The resulting data are transferred to the IBIS computer and compared in the same way as bullet-related data (i.e., the images are translated into sets of binary numbers and compared with the IBIS database of cartridge images).

The individualization of cartridge cases, via comparison of the sample cartridge and the existing images in the IBIS database, is possible because the cartridge case is imprinted when the gun is fired. First, the firing pin will leave a mark in the soft metal of the primer on the cartridge. The firing pin impression may reveal imperfections in the firing pin that will be captured by the mark left on the cartridge. As the bullet fires, the cartridge is thrust backward, where it hits the breechblock. The breechblock will inevitably have individual imperfections from its manufacture that will be imprinted on the bottom of the cartridge. If a semiautomatic or automatic firearm was used, there also will be marks on the sides of the cartridge from the ejector and extractor mechanism or chambering marks from the cartridge (**FIGURE 8-25**).

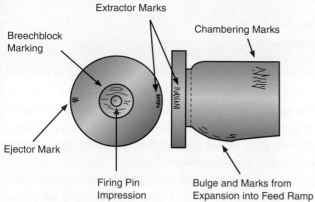

Breechblock
Marking

Extractor Marks

Chambering Marks

Ejector Mark

Firing Pin
Impression

Bulge and Marks from
Expansion into Feed Ramp

FIGURE 8-24 A semiautomatic pistol leaves marks on the cartridge casing as it contacts the internal mechanism of the gun. These marks are class information.

FIGURE 8-25 A fired cartridge case showing extractor marks.

WRAP UP

 YOU ARE THE FORENSIC SCIENTIST SUMMARY

1. The investigators should search carefully for the weapon. When they find it, the investigators should have a firearms expert inspect and disarm the weapon. The location of the firearm should be recorded by both notebook sketch and photography. Examination of the weapon should include identification of the type of firearm, cartridges present, manufacturer, serial number, caliber, and any modifications that may have been made to it.

2. Examination of the firearm will determine the amount of pressure it takes to discharge. This examination will determine whether the weapon had a "hair trigger." Next, the suspect's hand and clothes should be carefully examined for GSR. The location of GSR on his clothes may substantiate his claim.

Chapter Spotlight

- Structural irregularities in bullets result from scratches, nicks, breaks, and wear in a gun's barrel and can be used to match a bullet to a gun.

- The manufacturing of a gun barrel leaves characteristic grooves inside it. No two barrels will be identical, even if the guns are manufactured in succession.

- The comparison microscope is the most important tool for examining firearm evidence. It allows two bullets or cartridge cases to be observed side by side.

- Like bullets, cartridge cases are uniquely marked by the source gun when the firing pin, breech-block, and ejector and extractor leave their markings on the cartridge.

- GSR particles and other discharge residues around a bullet hole can be used to assess the distance from which the gun was fired.

- Restoration of a serial number that has been ground away is possible through chemical etching.

- The IBIS has emerged as the standard for examining projectiles and cartridge cases. It allows the local forensic firearm examiner to send an image of the bullet or cartridge case to the IBIS computer, which compares the suspect image with images in the IBIS database and ranks any matches it finds.

Acid etching method A method in which strong acid is applied to a firearm to reveal the serial number.

Black powder The oldest gunpowder, which was composed of a mixture of potassium nitrate (saltpeter), charcoal, and sulfur.

Bore The interior of a gun barrel.

Breechblock The metal block at the back end of a gun barrel.

Broach cutter A tool that is pushed through the gun barrel to form the rifling.

Caliber The diameter of the bore of a firearm (other than a shotgun). Caliber is usually expressed in hundredths of an inch (.38 caliber) or in millimeters (9 mm).

Cartridge Ammunition enclosed in a cylindrical casing containing an explosive charge and a bullet, which is fired from a rifle or handgun.

Choke A device placed on the end of a shotgun barrel to change the dispersion of pellets.

Ejector The mechanism in a semiautomatic weapon that ejects the spent cartridge from the gun after firing.

External ballistics A description of the events that occur after the bullet leaves the barrel of the gun but before it strikes its target.

Extractor The device that extracts the spent cartridge from the gun's chamber.

Gauge The unit used to designate size of a shotgun barrel. The interior diameter of a shotgun barrel is determined by the number of lead balls that fit exactly in the barrel and are equivalent to 1 lb. For example, a 16-gauge shotgun would have a bore diameter of a lead ball that is 1/16 lb.

Grooves The cutout section of a rifled barrel.

Internal ballistics A description of the events that transpire within the firearm.

Lands The raised section of a rifled barrel.

Magnaflux method A method of restoring a gun's serial number.

Nitrocellulose A cotton-like material produced when cellulose is treated with sulfuric acid and nitric acid; also known as "guncotton." Nitrocellulose is used in the manufacture of explosives.

Nitroglycerin An explosive chemical compound obtained by reacting glycerol with nitric acid. Nitroglycerin is used in the manufacture of gunpowder and dynamite.

Primer An igniter that is used to initiate the burning of gunpowder.

Rifling The spiral grooves on the inside surface of a gun barrel that make the bullet spin.

Semiautomatic pistol A firearm that fires and reloads itself.

Smokeless powder An explosive charge composed of nitrocellulose or nitrocellulose and nitroglycerin (double-base powders).

Striations Fine scratches left on bullets, formed from contact of the bullet with imperfections inside the gun barrel.

Terminal ballistics A description of what happens when the bullet strikes its target.

Wad A plastic, cardboard, or fiber disk that is placed between the powder and the shot in a shotgun shell.

http://criminaljustice.jblearning.com/criminalistics

Putting It All Together

Fill in the Blank

1. The accuracy of a firearm can be greatly improved by making the barrel _____ (longer/shorter).

2. The spiral grooves in a gun barrel are referred to as _____.

3. The rifling on the interior of a gun barrel has spiral grooves and _____.

4. The diameter of a gun barrel, as measured from a land on the left to one on the right, is known as the _____ of a weapon.

5. A broach cutter is used to cut the _____ in the interior of the gun barrel.

6. The lands and grooves in a gun barrel are _____ (class/individual) characteristics.

7. Besides grooves, the broach cutter produces fine lines called _____.

8. As bullets emerge from the gun barrel, they rotate, a phenomenon that is called _____.

9. The kinetic energy of a bullet _____ (increases/decreases) as the caliber of the weapon increases.

10. The diameter of a shotgun barrel is expressed as a(n) _____.

11. The higher the shotgun gauge, the _____ (larger/smaller) the diameter of the barrel.

12. A sawed-off shotgun is one that has a barrel length less than _____ inches.

13. In the past, lead was used to make bullets because it has a high ___ and it is _____.

14. Lead's low _____ makes it soften at higher velocities.

15. The element _____ is added to lead to raise its melting point.

16. The jacket of a cartridge is made from _____ or _____.

17. Black powder is a mixture of _____, _____, and _____.

18. Smokeless powder is made from the chemicals _____ and _____.

19. Nitroglycerine is desensitized by adding _____, _____, or _____.

20. The primer of a cartridge holds a(n) _____ explosive mixture.

21. The _____ was a crucial invention needed to make a gun that would fire in any weather.

22. Many firing ranges have banned _____ primers because of their potential health risks.

23. In a shotgun shell, the _____ is designed to make an almost gas-tight seal with the barrel.

24. The matching of fired bullets is carried out by using a(n) _____ microscope.

25. IBIS is an acronym for _____ _____ _____ _____.

26. Serial numbers on handguns can be restored by two methods: the _____ method and the _____ method.

True or False

1. No two gun barrels will have identical striations.

2. The handgun was originally designed for infantry soldiers.

3. Rifles are more accurate than handguns.

4. One disadvantage of black powder is that its explosion releases a cloud of smoke.

5. For black powder, only 40% of the reaction products are gases.

Review Problems

1. You find a handgun at the scene of the crime. Describe the procedure you would follow to collect and preserve the firearm evidence.

2. You find a bullet lodged in the wall at the scene of a crime. Upon examination, you determine that it is a .32-caliber bullet with a left 6 twist. Which company manufactured this weapon?

3. You find a bullet lodged in the wall at the scene of a crime. Upon examination, you determine that it is a .32-caliber bullet with a right 5 twist. Which company manufactured this weapon?

4. To increase the force delivered by a bullet, would it be better to double the mass of the projectile or to double the velocity of the projectile? Justify your answer.

5. What would be the increase in kinetic energy expected if you doubled the mass and velocity of a projectile?

6. Evidence at a crime scene indicates a shotgun was fired. After a careful search, you locate a wad from one of the expended shells. Its diameter measures slightly less than 0.75 in. What gauge shotgun was probably used to commit the crime?

Further Reading

Boorman, D. *The History of the Colt Firearms.* Guilford, CT: Lyons Press, 2001.

Boorman, D. *The History of Smith & Wesson Firearms.* Guilford, CT: Lyons Press, 2002.

Boorman, D. *The History of Winchester Firearms.* Guilford, CT: Lyons Press, 2003. An Introduction to Forensic Firearm Identification. http://www.firearmsid.com.

Rowe, W. F. Firearms identification. In R. Saferstein (Ed.), *Forensic Science Handbook, vol. 2.* Upper Saddle River, NJ: Prentice-Hall, 2005.

Schehl, S. A. Firearms and toolmarks in the FBI laboratory. *Forensic Science Communications.* 12(2); 2000. http://www.fbi.gov/hq/lab/fsc/backissu/april2000/schehl1.htm.

Vince, J. J., and Sherlock, W. E. *Evidence Collection.* Sudbury, MA: Jones and Bartlett, 2005.

http://criminaljustice.jblearning.com/criminalistics

Answers to Review Problems

Interactive Questions

Key Term Explorer

Web Links

http://criminaljustice.jblearning.com/criminalistics

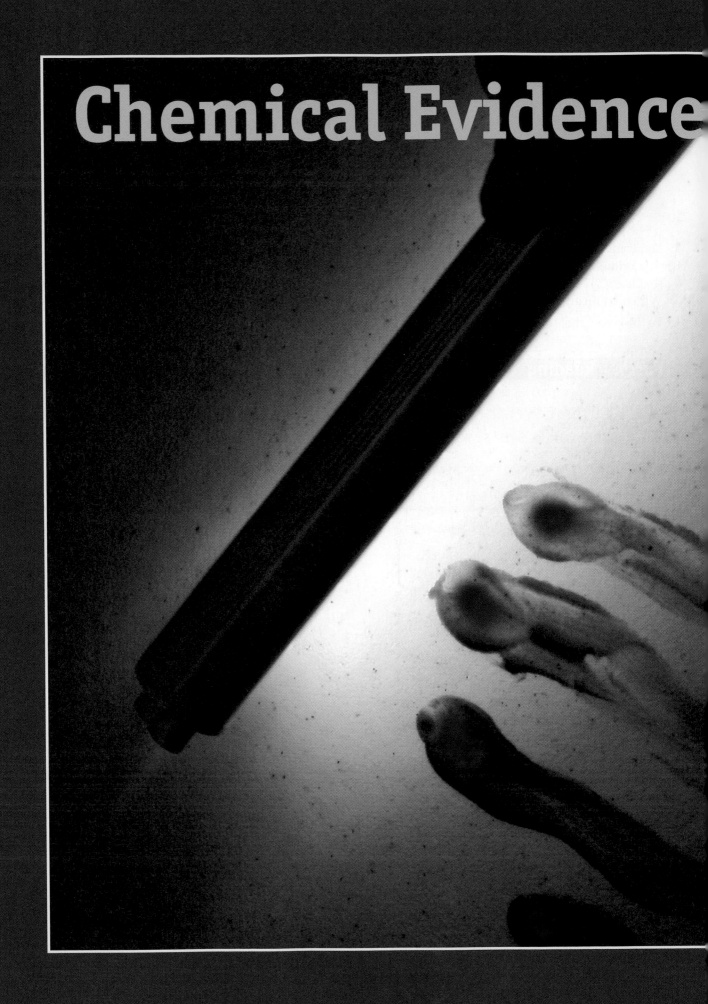

Chemical Evidence

SECTION

4

CHAPTER 9

Inorganic Analysis: Forensic Determination of Metals and Gunshot Residue

CHAPTER 10

Arson

CHAPTER 11

Drugs of Abuse

Inorganic Analysis: Forensic Determination of Metals and Gunshot Residue

OBJECTIVES

In this chapter you should gain an understanding of:

- The names and symbols that chemists use for elements and compounds
- Properties of electrons, protons, and neutrons—the fundamental particles of all matter
- The organization of the periodic table based on the arrangement of electrons in the elements
- The use of inductively coupled plasma optical emission spectroscopy (ICP-OES) to elucidate the composition of metals
- The use of ICP-OES and X-ray fluorescence (XRF) spectrometry to identify the presence of trace metals
- The detection of gunshot residue (GSR) by atomic absorption spectrophotometry (AAS) and by scanning electron microscopy (SEM) with energy-dispersive X-rays (EDX)

FEATURES

On the Crime Scene

Back at the Crime Lab

See You in Court

WRAP UP

Chapter Spotlight

Key Terms

Putting It All Together

Investigators will undoubtedly encounter inorganic materials at crime scenes. Inorganic metals, such as copper, iron, aluminum, nickel, and lead, are strong and easily forged into useful shapes—a characteristic that leads to their use in tools, guns, knives, pipes, locks, and automobiles.

Consider the average burglary. Often the examination of the suspect's clothes will yield metal shards (dust or scrapings) that may have been deposited there during the course of the break-in. An examiner should vacuum this clothing using a special vacuum with a small nozzle attached to a filter to collect all trace evidence. Note that all metal shards must be collected from the crime scene or suspect before dusting for fingerprints.

The examiner may then attempt to physically match these shards to tools or metallic items damaged (window frames, door locks, and hinges) during the break-in. Given that it is very difficult to compare the shapes of these small fragments to the much larger metallic object that made them, other forensic methods must be used to prove the link between the two pieces of evidence. In this case a chemical analysis of the shards will determine their composition (which elements are present and in what ratio). The results of this analysis can then be compared with the compositions of the objects that may have produced the shards during the crime. The analysis will often identify useful class characteristics, such as that the shards found on the suspect match the metallic object at the crime scene.

Brass is an alloy that is commonly used in the manufacturing of locks, keys, and hinges. Alloys are mixtures of metals, and brass is a mixture primarily made of copper and zinc. As the ratio of copper and zinc varies, and as other metals such as lead, aluminum, manganese, iron, and tin are added to the mix, the weight and strength of the alloy change. For instance, forensic scientists have determined that different parts of a Yale® lock, such as the latch, cylinder, and retaining ring, are manufactured from brass with different copper-to-zinc ratios (Rendle, 1981). The analysis of brass dust found on a suspect may, therefore, link that dust with a particular type of lock.

1. Burglars entered a bank at night and broke into the safe by drilling through the iron front plate of the safe. Four suspects are apprehended, and all claim they know nothing of the robbery. Is there a way to quickly assess whether any of the suspects is covered with metal powder?

2. Suspect B's shirt is found to be covered in dust. He claims that it isn't dust, but rather really bad dandruff. What does the evidence tell you?

Introduction

Chemists categorize all chemicals into two classes—organic and inorganic. The instruments and methods used to measure the properties of inorganic materials are distinctly different from those used to measure the properties of organic materials. For that reason, this chapter will focus on inorganic substances and the techniques used to measure them. Chapter 10 covers organic substances.

Before we can describe the techniques of inorganic analysis, however, we need to learn more about the composition of matter and the chemical properties of various substances. This chapter describes the composition of inorganic substances and the techniques and instrumental methods that forensic scientists use to examine inorganic evidence, such as GSR and the inorganic elements in glass.

Elements and Compounds

All substances can be classified as either elements or compounds. **Elements** are the building blocks from which all matter is constructed. Elements cannot be broken into simpler substances by ordinary chemical means. **Compounds**, by contrast, are substances that are composed of two or more elements combined chemically in fixed proportions.

Compounds can be decomposed by chemical means to yield the elements from which they are constructed. The properties of a particular compound are always the same under a given set of

While overseeing a chemistry lab renovation at Wilkes University in August 1991, contractor Robert Curley, 32, began to grow ill. He entered the hospital for tests to identify the cause of his burning skin, numbness, weakness, repeated vomiting, and rapid hair loss. Robert was tested for heavy metal exposure because he also became more agitated and aggressive. The tests showed his body contained elevated levels of thallium, a chemical that was once used in rat poisons but banned in 1984. Robert died shortly thereafter, in September 1991.

The initial investigation into his death started at the chemistry lab in which Robert worked, with the belief that he was accidentally exposed to thallium at this worksite. A search of the chemistry building turned up sealed bottles of thallium salts in the chemistry stockroom, but no other worker in the building showed any symptoms of thallium poisoning. The amount of thallium in Robert's body at autopsy was found to be extremely high. It was clear to investigators that Robert had been deliberately poisoned, so his death was ruled a homicide.

The police naturally began their investigation at the Robert home. Tests showed that Robert's wife and step-daughter also had elevated levels of thallium, albeit not lethal amounts. A search of the home uncovered a Thermos bottle that contained traces of thallium, but there the trail turned cold. The case languished for three years without any new leads until authorities requested permission to exhume Robert's body to remove hair shafts, toenails, fingernails, and skin samples for testing.

Forensic scientists used the hair samples to conduct a segmental analysis. Hair samples can provide a clear history of heavy metal poisoning. Heavy metals accumulate in hair roots; since hair grows at a constant rate, the accumulated poisons remain in the root and can be analyzed to reveal a timeline for the poisoning. The hair strand samples taken from Robert were sufficiently long that the forensic laboratory was able to reconstruct a nine-month history of his poisoning. Some of the segments contained very little thallium, whereas others showed a very high concentration, suggesting a lengthy, systematic poisoning that began months before Robert started to work on the renovation job at Wilkes University.

Forensic scientists also found that the segment of hair grown just before Robert's death contained a massive concentration of thallium. Remembering that Robert was hospitalized prior to his death, investigators noted that hospital records indicated that his wife, Joann Curley, visited him often and brought in meals for Robert, making her the only person with the opportunity to poison him. Curley was consequently charged with her husband's murder.

Curley later confessed to the crime to avoid a trial and a possible death sentence. She admitted using old rat poison to kill her husband. In her statement, Curley confessed to administering small doses of the rat poison to herself and her daughter to shift suspicion away from her. She said that she was motivated to kill her husband for his $300,000 life insurance policy. In a twist of fate, Curley received an insurance payment of $1.2 million just two days before Robert's death—a settlement from the death of her first husband.

conditions but are different from the properties of the elements from which they are constructed. Examples of simple compounds (together with the elements from which they are composed) are water (hydrogen and oxygen), carbon dioxide (carbon and oxygen), salt (sodium and chlorine), and sulfuric acid (hydrogen, sulfur, and oxygen). In all of these compounds, two or more elements are chemically combined in fixed proportions.

There are 92 naturally occurring elements; 17 other elements have been created by scientists in nuclear reactions. Every compound that exists is made up of some combination of these 109 elements. **FIGURE 9-1** shows the elements that are the most abundant in the Earth's crust and the human body. Just 5 of the 92 naturally occurring elements account for more than 90% of all the matter in the Earth's crust. Just 3 elements—oxygen, carbon, and hydrogen—make up more than 90% of the human body.

Industrial manufacturers obtain most of their raw materials from the Earth's crust. These materials are extracted from mines in the form of rock, with the rock varying in composition depending on

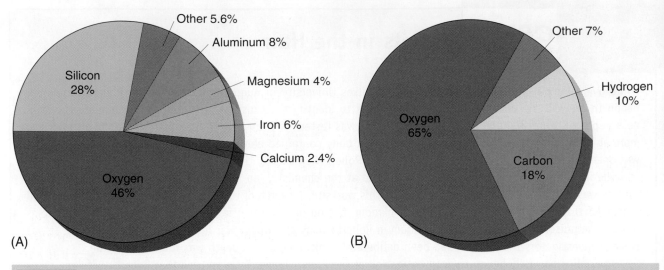

FIGURE 9-1 The relative abundance (in terms of percent by mass) of the major elements in (A) the Earth's crust, including the oceans and the atmosphere, and (B) the human body.

Adapted from: Chemistry: The Central Science, 5th ed., by Theodore L. Brown, Eugene LeMay, and Bruce Bursten. Englewood Cliffs, NJ: Prentice Hall, 1991.

the geographic area in which the mine is located. The rock is then processed to extract the elements present, although the material processed inevitably contains a number of other elements as impurities. For many industrial items, it is not economically feasible to remove all the impurities because they will not affect the performance of the manufactured item. The presence of these trace elements—that is, those impurities present at less than 1% concentration—provides the forensic scientist with a basis for comparing materials.

Elements and Molecules

Chemists use one- and two-letter symbols to represent the different elements. The first letter of the symbol is always capitalized; the second letter (if present) is always lowercase. Thus, Co is the symbol for the element cobalt, whereas CO denotes the compound carbon monoxide (carbon + oxygen). **TABLE 9-1** lists some common elements with their symbols.

The first (or, in some cases, only) letter of a symbol is usually the first letter of the element's name (O for oxygen; Mg for magnesium). The second letter is often the second letter in the element's name (Br is bromine; Si is silicon). Where elements share the same second letter, another letter from the name of the element is chosen as the second letter (Ca is calcium; Cd is cadmium). Some elements—

mainly those that have been known for hundreds of years—have symbols that are derived from their Latin names (**TABLE 9-2**). Thus the symbol for lead, Pb, comes from the Latin name "plumbum"; the symbol for copper, Cu, comes from "cuprium."

TABLE 9-1

Common Elements and Their Symbols

Element	Symbol	Element	Symbol
Aluminum	Al	Iron	Fe
Antimony	Sb	Lead	Pb
Argon	Ar	Lithium	Li
Arsenic	As	Magnesium	Mg
Barium	Ba	Mercury	Hg
Beryllium	Be	Neon	Ne
Bismuth	Bi	Nickel	Ni
Boron	B	Nitrogen	N
Bromine	Br	Oxygen	O
Cadmium	Cd	Phosphorus	P
Calcium	Ca	Platinum	Pt
Carbon	C	Potassium	K
Chlorine	Cl	Radium	Ra
Chromium	Cr	Selenium	Se
Cobalt	Co	Silicon	Si
Copper	Cu	Silver	Ag
Fluorine	F	Sodium	Na
Gold	Au	Sulfur	S
Helium	He	Tin	Sn
Hydrogen	H	Uranium	U
Iodine	I	Zinc	Zn

TABLE 9-2

Elements that Have Symbols Derived from Their Latin Names

Element	Latin Name	Symbol
Antimony	Stibium	Sb
Copper	Cuprum	Cu
Gold	Aurum	Au
Iron	Ferrum	Fe
Lead	Plumbum	Pb
Mercury	Hydrargyrum	Hg
Potassium	Kalium	K
Silver	Argentum	Ag
Sodium	Natrium	Na
Tin	Stannum	Sn

EXAMPLE

Which of the following are elements and which are compounds: Be, NO, CO, Cr?

Solution
Be, beryllium, and Cr, chromium, are elements. NO is a compound of N (nitrogen) and O (oxygen). CO is a compound of C (carbon) and O (oxygen).

Practice Exercise
Which of the following are elements and which are compounds: Si, KI, He, HI?

Solution
Si and He are elements, KI and HI are compounds.

Very few elements exist in nature in their elemental form—that is, as single atoms. Instead, most matter is composed of **molecules**. A molecule is a combination of two or more atoms. It can be defined as the smallest unit of a pure substance (element or compound) that can exist and still retain its physical and chemical properties. Molecules are the particles that undergo chemical changes in a reaction. They are composed of atoms held together by forces known as chemical bonds.

Molecules may be composed of two or more identical atoms or two or more different kinds of atoms. For example, two atoms of oxygen bond together to form a molecule of ordinary oxygen gas, O_2. A molecule of water consists of two atoms of hydrogen bonded to a single atom of oxygen, H_2O. A molecule of potassium perchlorate, $KClO_4$ (a compound that is used to make pipe bombs), has one atom of potassium, K; one atom of chlorine, Cl; and four atoms of oxygen, O.

Physical Properties of Inorganic Substances

The physical properties of a substance are those properties that can be observed without changing the substance into another substance. Color, odor, taste, hardness, density, solubility, melting point (the temperature at which a substance melts), and boiling point (the temperature at which a substance boils or vaporizes) are all physical properties.

As an example, consider the physical properties of pure copper. Copper is a bright, shiny metal. It is malleable (it can be beaten into thin sheets) and ductile (it can be drawn into fine wire); it is also a good conductor of electricity. Copper melts at 1083°C (1981°F) and boils at 2567°C (4653°F); its density is 8.92 grams per milliliter (g/mL). No matter what its source, pure copper always has these properties.

To determine the melting point of a substance, we must change it from a solid to a liquid. When we do so, however, we do not change the composition of the substance. For example, to determine the melting point of ice, we must change the ice into water. In this process, the appearance of the H_2O changes, but there is no change in its molecular composition.

By contrast, in a chemical reaction, a substance changes its molecular composition. The new molecule formed in this way has different physical properties than the original molecule. The chemical properties of a substance are the properties that can be observed when the substance undergoes this kind of change in its molecular composition. For example, gasoline has several notable chemical properties: It can be ignited, it reacts with oxygen in air, and it can be used as an accelerant by an arsonist. The fact that a substance does not react with another substance is also a chemical property. For example, one of gold's chemical properties is that, unlike copper, it does not tarnish or change when it is exposed to the atmosphere.

Matter can be classified as being in one of the three familiar states of matter: solid, liquid, or gas. Whether a substance exists as a solid, a liquid, or a gas at any particular time depends on its temperature, the surrounding pressure (the force per unit area exerted on it), and the strength of the forces holding the substance's internal components together. Water—the most common liquid in our environment—is a very unusual substance in that it exists naturally on Earth in all three states: ice, liquid water, and water vapor. Many substances, however, can be changed from one state to another only by the use of extremes of temperature or pressure, or both. Iron, for example, is a solid at normal temperatures and must be heated to 1535°C (2795°F) to change it to a liquid and heated to 3000°C (5432°F) to change it to a vapor.

Elements and Subatomic Particles

It is difficult to believe that everything found on Earth—whether living, nonliving, natural, or synthesized by humans—is made up from fewer than 92 naturally occurring elements. Now we have to accept the even more startling fact: Everything on Earth is constructed from just three basic particles—electrons, protons, and neutrons (**TABLE 9-3**).

An **electron** (e⁻) is a negatively charged (−1) particle. It is the smallest of the subatomic particles.

Compared with a proton or a neutron, the mass of an electron is negligible and is taken to be zero for most purposes.

A **proton** (p) is a positively charged (+1) particle. The charge on a proton is opposite, but exactly equal to, the charge on an electron. A proton, although an extremely small particle, is approximately 1800 times heavier than an electron.

A **neutron** (n) is a neutral particle; that is, it has no electrical charge. The mass of a neutron is very slightly greater than that of a proton, but for almost all purposes, it can be taken to be equal to that of a proton.

An atom has two distinct regions: the center, called the **nucleus**, and an area surrounding the nucleus that contains the electrons. The very small, extraordinarily dense nucleus is made up of protons and neutrons closely packed together. Because it contains protons, the nucleus is positively charged. The nucleus is so dense that 1 cubic centimeter (cm³) of nuclear material would have a mass equal to 90 million metric tons. Almost all (99.9%) of the mass of an atom is concentrated in the nucleus.

TABLE 9-3

Characteristics of the Three Basic Subatomic Particles

Particle Name	Symbol	Location in Atom	Relative Electrical Charge	Mass (g)	Approximate Relative Mass (amu)
Electron	e⁻	Surrounding the nucleus	−1	9.1×10^{-28}	0
Proton	p	Nucleus	+1	1.7×10^{-24}	1
Neutron	n	Nucleus	0	1.7×10^{-24}	1

 *on the* **CRIME SCENE** Radioactive Leftovers

In January 2000, MIT graduate student Yuanyuan Xiao used a Geiger counter to determine whether any contamination had occurred during her experimentation with and analysis of radioactive materials. Her results, which showed high levels of radioactivity, prompted a thorough study of her laboratory space.

When that search did not provide any reason for the increased radioactivity, the search extended to Yuanyuan's home. Some leftovers within her refrigerator were found to be laced with I-125, a radioactive isotope of iodine with a reasonably short half-life and weak gamma and X-ray emissions. The leftovers had been prepared by Yuanyuan's ex-boyfriend, Cheng Gu, who also worked in the MIT laboratory. When questioned, Gu admitted trying to poison Xiao.

Investigators eventually discovered that Gu had stolen the I-125 sample from a laboratory at Brown University. Brown University was able to trace the loss and connect it with Gu, which, combined with his delivery of the food to Yuanyuan's apartment, was sufficient for Gu's prosecution for poisoning, theft, and domestic violence. Luckily, the dose ingested by Yuanyuan was not sufficient to cause significant damage.

Moving around the nucleus are the electrons. The region occupied by the electrons, often termed the electron cloud, is mostly empty space. Compared with the size of the nucleus, this region is very large.

All atoms are made up of electrons, protons, and neutrons. These three basic units come together in various ways to form the atoms of the different elements.

Atomic Number

All atoms of a particular element have the same number of protons in their nuclei. This number, which determines the identity of an element, is the **atomic number** of that particular element.

Because atoms are electrically neutral, it follows that the number of protons in the nucleus of an atom will be balanced by an equal number of electrons in the space surrounding the nucleus:

Number of protons = number of electrons
= atomic number

The sum of the number of protons and the number of neutrons in the nucleus of an atom is termed the **atomic mass number** of the element:

Mass number = number of protons +
number of neutrons

The mass of a proton is taken to be 1. Given that the mass of a neutron is for all practical purposes identical to that of a proton, the mass of a neutron also is 1.

The arrangement of protons, neutrons, and electrons in the atom of the element carbon (C) is illustrated in FIGURE 9-2.

EXAMPLE

An atom of iron (Fe) has 26 electrons. Its mass number is 56. What is the atomic number of Fe? How many protons and neutrons does the nucleus contain?

Solution

1. Number of protons = number of electrons = atomic number
   Number of protons = 26 = atomic number
   Both the atomic number and the number of protons are 26.
2. Mass number = number of protons + number of neutrons
   Number of neutrons = mass number − number of protons
   Number of neutrons = 56 − 26 = 30
   The number of neutrons is 30.

Practice Exercise
An atom of copper (Cu) has a mass number of 63 and an atomic number of 29. How many electrons, protons, and neutrons are contained in this atom?

Solution

$$e^- = 29;\ p = 29;\ n = 34$$

Periodic Table

On the **periodic table**, the elements are arranged in order of atomic number (i.e., number of protons) instead of atomic mass. The atomic number gives each element its own unique atomic number.

Each element in the periodic table is represented by its symbol; included with the symbol are the element's atomic number and atomic mass (FIGURE 9-3). The first 92 elements—hydrogen through uranium—occur naturally; the remainder have been made in the laboratory.

Periods and Groups

The periodic table is divided into 7 horizontal rows, called **periods**, and 18 vertical columns, called **groups** (FIGURE 9-4). Although in many ways it is simple to use the new international system of numbering groups 1 through 18, most textbooks still use the older system in which Roman numerals identify the groups, and the letters A and B distinguish families within a group.

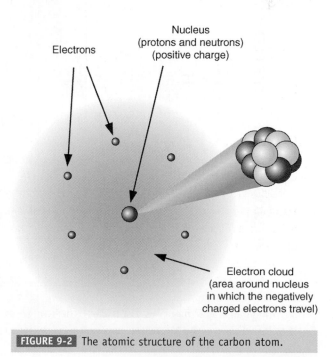

FIGURE 9-2 The atomic structure of the carbon atom.

PERIODIC TABLE OF THE ELEMENTS

Element	hydrogen
Atomic Number	1
Symbol	**H**
*Atomic Mass	1.01

1 IA	2 IIA		3 IIIB	4 IVB	5 VB	6 VIB	7 VIIB	8 VIII	9 VIII	10 VIII	11 IB	12 IIB	13 IIIA	14 IVA	15 VA	16 VIA	17 VIIA	18 VIIIA
																		helium 2 **He** 4.00
lithium 3 **Li** 6.94	beryllium 4 **Be** 9.01												boron 5 **B** 10.81	carbon 6 **C** 12.01	nitrogen 7 **N** 14.01	oxygen 8 **O** 16.00	fluorine 9 **F** 19.00	neon 10 **Ne** 20.18
sodium 11 **Na** 22.99	magnesium 12 **Mg** 24.31												aluminum 13 **Al** 26.98	silicon 14 **Si** 28.09	phosphorus 15 **P** 30.97	sulfur 16 **S** 32.07	chlorine 17 **Cl** 35.45	argon 18 **Ar** 39.95
potassium 19 **K** 39.10	calcium 20 **Ca** 40.08		scandium 21 **Sc** 44.96	titanium 22 **Ti** 47.88	vanadium 23 **V** 50.94	chromium 24 **Cr** 52.00	manganese 25 **Mn** 54.94	iron 26 **Fe** 55.85	cobalt 27 **Co** 58.93	nickel 28 **Ni** 58.69	copper 29 **Cu** 63.55	zinc 30 **Zn** 65.39	gallium 31 **Ga** 69.72	germanium 32 **Ge** 72.61	arsenic 33 **As** 74.92	selenium 34 **Se** 78.96	bromine 35 **Br** 79.90	krypton 36 **Kr** 83.80
rubidium 37 **Rb** 85.47	strontium 38 **Sr** 87.62		yttrium 39 **Y** 88.91	zirconium 40 **Zr** 91.22	niobium 41 **Nb** 92.91	molybdenum 42 **Mo** 95.94	technetium 43 **Tc** (99)	ruthenium 44 **Ru** 101.07	rhodium 45 **Rh** 102.91	palladium 46 **Pd** 106.42	silver 47 **Ag** 107.87	cadmium 48 **Cd** 112.41	indium 49 **In** 114.82	tin 50 **Sn** 118.71	antimony 51 **Sb** 121.75	tellurium 52 **Te** 127.60	iodine 53 **I** 126.90	xenon 54 **Xe** 131.29
cesium 55 **Cs** 132.91	barium 56 **Ba** 137.33		lanthanum 57 **La** 138.91	hafnium 72 **Hf** 178.49	tantalum 73 **Ta** 180.95	tungsten 74 **W** 183.85	rhenium 75 **Re** 186.21	osmium 76 **Os** 190.2	iridium 77 **Ir** 192.22	platinum 78 **Pt** 195.08	gold 79 **Au** 196.97	mercury 80 **Hg** 200.59	thallium 81 **Tl** 204.38	lead 82 **Pb** 207.2	bismuth 83 **Bi** 208.98	polonium 84 **Po** (209)	astatine 85 **At** (210)	radon 86 **Rn** (222)
francium 87 **Fr** (223)	radium 88 **Ra** (226)		actinium 89 **Ac** (227)	rutherfordium 104 **Rf** (261)	dubnium 105 **Db** (262)	seaborgium 106 **Sg** (263)	bohrium 107 **Bh** (262)	hassium 108 **Hs** (265)	meitnerium 109 **Mt** (266)	ununnilium 110 **Uun** (269)	unununium 111 **Uuu** (272)	ununbium 112 **Uub** (277)						

Lanthanide Series

cerium 58 **Ce** 140.12	praseodymium 59 **Pr** 140.91	neodymium 60 **Nd** 144.24	promethium 61 **Pm** (147)	samarium 62 **Sm** 150.36	europium 63 **Eu** 151.97	gadolinium 64 **Gd** 157.25	terbium 65 **Tb** 158.93	dysprosium 66 **Dy** 162.50	holmium 67 **Ho** 164.93	erbium 68 **Er** 167.26	thulium 69 **Tm** 168.93	ytterbium 70 **Yb** 173.04	lutetium 71 **Lu** 174.97

Actinide Series

thorium 90 **Th** 232.04	protactinium 91 **Pa** (231)	uranium 92 **U** 238.03	neptunium 93 **Np** (237)	plutonium 94 **Pu** (244)	americium 95 **Am** (243)	curium 96 **Cm** (247)	berkelium 97 **Bk** (247)	californium 98 **Cf** (251)	einsteinium 99 **Es** (252)	fermium 100 **Fm** (257)	mendelevium 101 **Md** (258)	nobelium 102 **No** (259)	lawrencium 103 **Lr** (260)

*Note: For radioactive elements, the mass number of an important isotope is shown in parenthesis; for thorium and uranium, the atomic mass of the naturally occurring radioisotopes is given.

FIGURE 9-3 In the periodic table, each box contains the element's name, symbol, atomic number, and atomic mass.

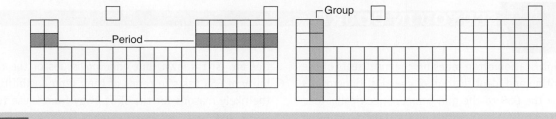

The periodic table is divided into periods (horizontal rows) and groups (vertical columns).

For historical reasons, some groups of representative elements are often referred to by common names. For example, Group IA elements are known as the alkali metals; Group IIA elements are called the alkaline earth metals; Group VIIA elements are known as the halogens; and Group VIIIA elements are called the noble gases.

Elements within a group, particularly if they are representative elements, have similar physical and chemical properties. For example, the alkali metals—lithium (Li), sodium (Na), potassium (K), rubidium (Rb), and cesium (Cs)—are all soft, shiny, silvery metals with low densities and low melting points. These elements are very reactive and combine readily with many other elements, including oxygen, sulfur, phosphorus, and the halogens. Because they react vigorously with water, the alkali metals must be stored under oil. The halogens—fluorine (F), chlorine (Cl), bromine (Br), and iodine—are all so reactive that they do not occur free in nature, but they combine readily with many other elements to form a very large number of compounds that are widely distributed.

EXAMPLE

Refer to the periodic table to determine which two elements in the following list would be most alike in their physical and chemical properties: sodium (Na), carbon (C), chlorine (Cl), sulfur (S), bromine (Br).

Solution

Chlorine (Cl) and bromine (Br) would be most alike because they are in the same group (VIIA) of the periodic table.

Practice Exercise

Which of the following elements would you expect to show similar physical and chemical properties: magnesium (Mg), potassium (K), aluminum (Al), fluorine (F), calcium (Ca)?

Solution

Magnesium (Mg) and calcium (Ca), which are both in Group IIA, have similar physical and chemical properties.

Metals and Nonmetals

Elements are of two main types: metals and nonmetals. **Metals** have many distinctive properties. They are good conductors of heat and electricity, and most metals have a characteristic lustrous (shiny) appearance. Metals are both ductile (they can be drawn out into a fine wire) and malleable (they can be rolled out into thin sheets). All metals are solids at room temperature, with the exception of mercury (Hg), which is a liquid at room temperature. Familiar metals include sodium (Na), aluminum (Al), calcium (Ca), chromium (Cr), iron (Fe), copper (Cu), silver (Ag), tin (Sn), platinum (Pt), and gold (Au).

Nonmetals usually do not conduct heat or electricity to any significant extent. They have little or no luster and are neither ductile nor malleable. At room temperature, many nonmetals—including hydrogen (H), nitrogen (N), oxygen (O), chlorine (Cl), and the noble gases exist as gases. One nonmetal, bromine (Br), is a liquid at room temperature. Solid nonmetals include carbon (C), phosphorus (P), sulfur (S), and iodine (I).

The periodic table shows metals as purple and nonmetals as green. The change from metal to nonmetal properties is not abrupt but gradual, and the elements shown in yellow in the table have properties that lie between those of metals and nonmetals. These elements, which are called **semimetals** or **metalloids**, include boron (B), which conducts electricity well only at a high temperatures, and the semiconductors—silicon (Si), germanium (Ge), and arsenic (As)—which conduct electricity better than nonmetals but not as well as metals like copper and silver. The special conducting ability of the semiconductors, particularly silicon, accounts for their use in computer chips and electronic calculators.

Techniques for the Analysis of Inorganic Materials

To identify an inorganic material, the forensic scientist chooses among a variety of analytical techniques that can probe the chemical differences between

GSR identification often relies on chemical tests of the composition and identification of chemical fingerprints that link the GSR to the gun source. Specifically, lead (Pb), barium (Ba), and antimony (Sb) are characteristic components of GSR particles. These elements are also abundant in brake pads and can be found on the hands of legal gun owners. In 2006, the Federal Bureau of Investigation (FBI) stopped GSR testing altogether, citing a shift in agency priorities. What the new priorities are isn't known, but ongoing issues of the reliability of GSR tests in terms of their reproducibility are the likely reasons for the shift away from these tests. State and local forensic laboratories soon followed suit. Due to the issues associated with the accuracy of bullet lead matching and the time and expense of the tests, courts are now seeing an increase in appeals of convictions that hinged on GSR evidence.

different elements. Organic molecules always contain the element carbon. Organic materials of forensic importance are described in the next two chapters.

The criminalist must also know if a quantitative analysis or qualitative analysis is required. A qualitative analysis simply confirms the presence of an element or molecule of interest. For instance, a quantitative analysis may determine whether white powder found at a crime scene involving a poisoning contains arsenic or is powdered sugar, as the suspect claims. A quantitative analysis, by contrast, determines the compounds present and their concentrations. For example, as we will see later, the quantitative analysis of the inorganic metals in an automotive paint chip found at a hit-and-run can provide evidence of which model of car may have been involved.

Many forensic measurements are made using spectrophotometry, a technique for identifying or measuring a substance based on its absorption or emission of different wavelengths of light. Spectrophotometry is used first to identify the presence of an element or molecule in a questioned sample and then to measure the amount present, if necessary, by determining how much light is absorbed at different wavelengths. Forensic scientists use the selective absorption of ultraviolet, visible, or infrared radiation as a basic measurement tool.

We now look at two important analytical techniques that are used in forensic laboratories to determine the chemical composition of physical evidence—atomic absorption and emission spectrophotometry. We will begin by describing how different wavelengths of light are separated from one another.

Electromagnetic Radiation and Spectra

Light is a form of **electromagnetic radiation (EMR)**. EMR is a general term that is used to describe energy that we encounter daily. In FIGURE 9-5, the EMR spectrum starts on the left with high-energy cosmic rays and ends on the right with low-energy radio waves. Familiar terms such as "ultraviolet," "visible," and "infrared" are used to describe regions within this energy spectrum.

EMR travels in waves that can be compared to the waves that ripple outward on the surface of a pond when a stone is dropped into the water. Electromagnetic waves differ from water ripples, however, in that they travel outward in all directions. All electromagnetic waves, whether they are light waves, microwaves, radio waves, or any other type of waves, travel at the same rate. This rate is at a maximum (300,000 kilometers per second or 186,000 miles per second) when the waves travel in a vacuum; electromagnetic waves are slowed very slightly when they travel through air or any medium where they encounter atoms or molecules.

For convenience, the electromagnetic spectrum is divided at arbitrary intervals into specific types of radiation. In reality, the spectrum is continuous, with the **wavelength** of the radiation gradually increasing from cosmic rays to radio waves. The wavelength (λ) is the distance between adjacent wave crests, and it is usually measured in nanometers (nm) (FIGURE 9-6). The number of crests that pass a fixed point in 1 second is termed the **frequency** (υ). A frequency of one cycle (passage of one complete wave) per second is equal to l hertz (Hz).

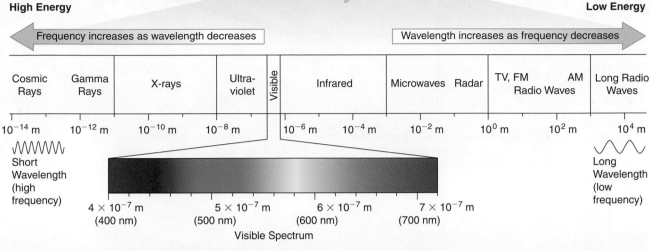

Sun

High Energy   Low Energy

Frequency increases as wavelength decreases   Wavelength increases as frequency decreases

| Cosmic Rays | Gamma Rays | X-rays | Ultra-violet | Visible | Infrared | Microwaves | Radar | TV, FM Radio Waves | AM | Long Radio Waves |

10^{-14} m 10^{-12} m 10^{-10} m 10^{-8} m 10^{-6} m 10^{-4} m 10^{-2} m 10^{0} m 10^{2} m 10^{4} m

Short Wavelength (high frequency)

Long Wavelength (low frequency)

4×10^{-7} m (400 nm) 5×10^{-7} m (500 nm) 6×10^{-7} m (600 nm) 7×10^{-7} m (700 nm)

Visible Spectrum

FIGURE 9-5 The electromagnetic spectrum. As the wavelength of electromagnetic radiation increases from cosmic rays to radio waves, the energy and frequency of the waves decrease. Light that is visible makes up only a very small part of the electromagnetic spectrum. (One nanometer is equal to 10^{9} meters.)

Wavelength and frequency are related as shown in the following equation:

$$\upsilon = \frac{c}{\lambda} \text{ (where } c = \text{ the speed of light in a vacuum)}$$

Although the various forms of radiation seem very different, they are actually all manifestations of the same phenomenon. They differ from one another only in terms of their energy, wavelength, and frequency. The shorter the wavelength (and the higher the frequency) of the radiation, the greater the energy it transmits. High-energy radiation is very damaging to living tissue.

Visible light—the kind we see—makes up only a small part of the entire electromagnetic spectrum. Its wavelength ranges from approximately 400 nm (violet) to approximately 700 nm (red). Infrared radiation (which has a wavelength greater than 700 nm) is invisible to the eye and of lower energy than visible light. Ultraviolet (UV) radiation is higher in energy than visible light.

When white light from an incandescent light bulb is passed through a glass prism, the light is separated into a **continuous spectrum** containing all of the visible colors—violet, blue, green, yellow, orange, and red. These colors merge smoothly into one another in an unbroken band (**FIGURE 9-7**).

If the light produced from putting an element into a flame is passed through a prism and focused onto a photographic film, a series of lines separated by black spaces is seen (**FIGURE 9-8**). This type of spectrum is called a discontinuous spectrum or a **line spectrum**. The pattern of lines produced in this way is unique for each element and, therefore, can be used to identify that element. In such cases, the

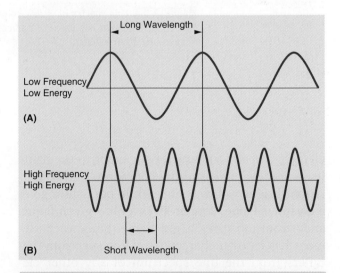

FIGURE 9-6 The wavelength and frequency of electromagnetic radiation. The frequency of the high-frequency (high-energy) radiation shown in (B) is three times that of the low-frequency (low-energy) radiation shown in (A).

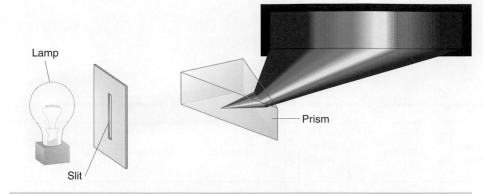

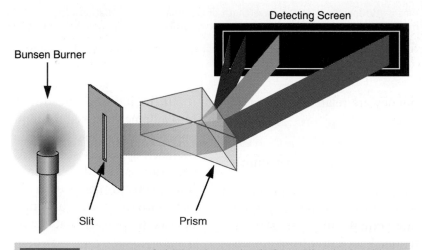

FIGURE 9-7 A continuous spectrum forms when a narrow beam of light (from either sunlight or an incandescent bulb) is passed through a glass prism, which separates the light into its component colors. Adjacent colors merge into one another in an unbroken band.

FIGURE 9-8 Line spectra of visible colors emitted when sodium is placed in a flame. Each element has its own characteristic line spectrum, which differs from that of any other element.

element imparts a characteristic color to the flame itself. For example, sodium compounds give a persistent bright yellow flame, while potassium gives a lavender flame, copper gives a blue–green flame, and strontium gives a red flame. Since each element has its own characteristic line spectrum that differs from the line spectrum of any other element, forensic scientists can use this characteristic to identify elements in specimens of evidence.

Our current model of the atom is able to explain the line spectra produced by the different el-

ements. The electrons in an atom orbit around the nucleus in much the same way that the planets in our solar system revolve around the sun. According to this model, each orbit is associated with a definite **energy level**.

The analogy of a bookcase can be used to explain the concept of an energy level (FIGURE 9-9). A book can be placed on any shelf in a bookcase, but it cannot be placed between the shelves. In a similar way, an electron can reside in one energy level or another around the nucleus, but it cannot exist

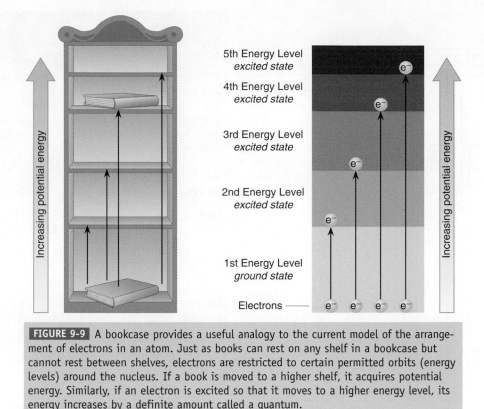

FIGURE 9-9 A bookcase provides a useful analogy to the current model of the arrangement of electrons in an atom. Just as books can rest on any shelf in a bookcase but cannot rest between shelves, electrons are restricted to certain permitted orbits (energy levels) around the nucleus. If a book is moved to a higher shelf, it acquires potential energy. Similarly, if an electron is excited so that it moves to a higher energy level, its energy increases by a definite amount called a quantum.

Adapted from: Modern Analytical Chemistry, by J. N. Miller. Englewood Cliffs, NJ: Prentice Hall, 1992.

between energy levels. Furthermore, a book can be moved from a lower shelf to a higher one, but a certain amount of energy must be expended to do this. Similarly, if sufficient energy is supplied to an electron, the electron can move from a lower energy level to a higher one. That definite fixed amount of energy, called a **quantum** of energy, is needed to excite an electron so that it will jump from one energy level to a higher one. In going to a higher level, the electron—like the book—acquires potential energy (energy of position). If the electron drops back to a lower level, it emits energy in the form of light—an experimental fact that could be recorded as a line spectrum.

The line spectrum of sodium can be explained as follows. When a sample containing sodium is put into the flame, the sodium atoms become energized and the electrons in individual atoms jump from their original energy level—the one closest to the nucleus—to a higher energy level called the **excited state**. Depending on the degree of excitation, electrons jump to different energy levels. Some reach the second energy level; other, more excited ones reach the fifth or sixth energy level (**FIGURE 9-10**). The characteristic line spectrum of sodium results when the excited electrons drop back down to lower energy levels in a series of jumps, eventually returning to the lowest level, the **ground state**. During each transition—from level 5 to level 1, from level 3 to level 2, from level 6 to level 3, and so on—a definite amount of light energy is released. Because light energy is related to the wavelength of the light, which in the visible range is related to the color of the light, those transitions that produce energy in the form of visible light are seen as lines in the spectrum.

Although electrons are confined to definite energy levels, the **principal energy levels** also contain sublevels called **orbitals**. These orbitals are identified by the letters s, p, d, and f. Each successive principal energy level going outward from the nucleus has one more orbital than the one before. Each orbital is described by a number, referring to the energy level, and a letter, referring to the orbital. Superscripts are used to indicate the number of electrons in each orbital. Thus hydrogen (H), which

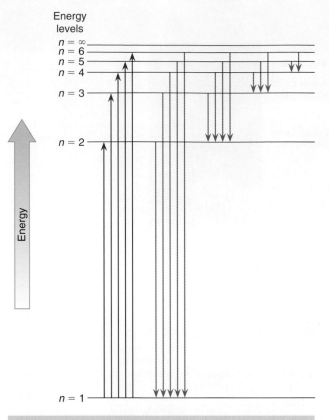

Energy levels

$n = \infty$
$n = 6$
$n = 5$
$n = 4$

$n = 3$

$n = 2$

$n = 1$

Energy

FIGURE 9-10 The upward-pointing arrows show transitions of ground-state electrons from allowed energy levels to higher energy levels as they absorb energy from either the flame or the electric arc. When the electrons return to their original energy levels (downward-pointing arrows), they release their energy as light and produce line spectra. Some of the lines are in the visible part of the spectrum; others are in the ultraviolet and infrared regions.

has one electron, is designated 1s¹; helium (He), which has two electrons, is designated as 1s²; lithium (Li), which has three electrons, is designated as 1s²2s¹. The electron configurations of the first 20 elements are shown in the last column of **TABLE 9-4**. With argon (Ar, atomic number 18), the 3p orbital is filled. The 3d orbital fills as electrons are added for the elements in the first row of transition elements—scandium (Sc, atomic number 21) through zinc (Zn, atomic number 30).

EXAMPLE

Write the electron configuration of sulfur, S (Group VIA), atomic number 16.

Solution
1. Each S atom has 16 electrons.
2. Refer to Figure 9-11 to determine the order in which the orbitals are filled.
3. Refer to Table 9-4 to determine how many electrons can be placed in each orbital.

Order of filling orbitals	1s	2s	2p	3s	3p	4s	3d
Maximum number of electrons per orbital	2	2	6	2	6	2	10

Total number of electrons placed when orbitals 1s through 3s have been filled: $2 + 2 + 6 + 2 = 12$.

Four electrons remain to be placed. The next orbital to be filled, 3p, can hold a maximum of six electrons. The four remaining electrons are therefore placed in the 3p orbital. The electron configuration of sulfur (atomic number 16) is

$$S = 1s^2 2s^2 2p^6 3s^2 3p^4$$

Note that the sum of the superscripts equals 16.

Practice Exercise
Write the electron configurations for the following elements:

1. Sodium, Na, (Group IA)
2. Argon, Ar (Group VIIIA)
3. Zinc, Zn (Group IIB)

Solution

1. $1s^2 2s^2 2p^6 3s^1$
2. $1s^2 2s^2 2p^6 3s^2 3p^6$
3. $1s^2 2s^2 2p^6 3s^2 3p^6 4s^2 3d^{10}$

Electron Configuration and the Periodic Table

Once the electron configurations in the atoms of the different elements were understood, it became evident that electron configuration was an appropriate basis for the arrangement of the elements in the periodic table (refer to Table 9-4). When the elements are arranged in order of increasing atomic number, elements with similar properties recur at periodic intervals because elements with similar electron configurations recur at periodic intervals.

FIGURE 9-11 shows the correlation between the periodic table and the filling of electron orbitals. The alkali metals and alkaline earth metals (Groups IA and IIA, respectively) are formed as one and two electrons are added to the s orbitals. The other representative elements (Groups IIIA through VIIIA) are formed as the p orbitals are filled. The main block of transition elements is formed as the d orbitals are filled, and the two rows of inner transition elements (the lanthanides and actinides) are formed as the f orbitals are filled.

Now that we understand the arrangement of electrons in atoms, we can use this information to identify which elements are present in a questioned sample. As we saw earlier, when we heat an element it gives off EMR (light), a process referred to as emission. Thanks to their knowledge of the

TABLE 9-4

Electron Arrangements of the First 20 Elements in the Periodic Table

Element	Atomic Number	Principal Energy Levels	Electron Configuration
Hydrogen (H)	1	1e⁻	$1s^1$
Helium (He)	2	2e⁻	$1s^2$
Lithium (Li)	3	2e⁻ 1e⁻	$1s^2 2s^1$
Beryllium (Be)	4	2e⁻ 2e⁻	$1s^2 2s^2$
Boron (B)	5	2e⁻ 3e⁻	$1s^2 2s^2 2p^1$
Carbon (C)	6	2e⁻ 4e⁻	$1s^2 2s^2 2p^2$
Nitrogen (N)	7	2e⁻ 5e⁻	$1s^2 2s^2 2p^3$
Oxygen (O)	8	2e⁻ 6e⁻	$1s^2 2s^2 2p^4$
Fluorine (F)	9	2e⁻ 7e⁻	$1s^2 2s^2 2p^5$
Neon (Ne)	10	2e⁻ 8e⁻	$1s^2 2s^2 2p^6$
Sodium (Na)	11	2e⁻ 8e⁻ 1e⁻	$1s^2 2s^2 2p^6 3s^1$
Magnesium (Mg)	12	2e⁻ 8e⁻ 2e⁻	$1s^2 2s^2 2p^6 3s^2$
Aluminum (Al)	13	2e⁻ 8e⁻ 3e⁻	$1s^2 2s^2 2p^6 3s^2 3p^1$
Silicon (Si)	14	2e⁻ 8e⁻ 4e⁻	$1s^2 2s^2 2p^6 3s^2 3p^2$
Phosphorus (P)	15	2e⁻ 8e⁻ 5e⁻	$1s^2 2s^2 2p^6 3s^2 3p^3$
Sulfur (S)	16	2e⁻ 8e⁻ 6e⁻	$1s^2 2s^2 2p^6 3s^2 3p^4$
Chlorine (Cl)	17	2e⁻ 8e⁻ 7e⁻	$1s^2 2s^2 2p^6 3s^2 3p^5$
Argon (Ar)	18	2e⁻ 8e⁻ 8e⁻	$1s^2 2s^2 2p^6 3s^2 3p^6$
Potassium (K)	19	2e⁻ 8e⁻ 8e⁻ 1e⁻	$1s^2 2s^2 2p^6 3s^2 3p^6 4s^1$
Calcium (Ca)	20	2e⁻ 8e⁻ 8e⁻ 2e⁻	$1s^2 2s^2 2p^6 3s^2 3p^6 4s^2$

Inorganic analytical techniques can identify and quantify a particular substance in a sample. The following techniques are often used by forensic examiners:

ICP-OES

- Measures elements by emission of UV or visible light
- Measures the concentrations of as many as 70 elements simultaneously
- Provides total composition of sample
- Uses a liquid medium

XRF

- Measures element concentrations through emission of X-rays
- Determines range of elements simultaneously
- Provides composition of sample surface only
- Uses a solid medium

AAS

- Measures elements by absorption of UV or visible light
- Determines one element at a time
- Determines total composition of sample
- Uses a liquid medium

SEM

- Detects elements by the emission of backscattered electrons
- Determines concentrations of elements simultaneously
- Determines composition of sample surface only
- Uses a solid medium

EDX spectroscopy

- Is a specific detector attached to SEM
- Determines concentrations of elements by the emission of X-rays
- Measures elemental values simultaneously
- Determines composition of sample surface only
- Uses a solid medium

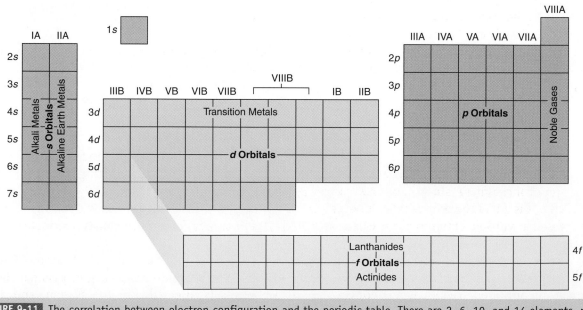

FIGURE 9-11 The correlation between electron configuration and the periodic table. There are 2, 6, 10, and 14 elements, respectively, in each period in the s, p, d, and f areas of the periodic table. These numbers indicate the maximum numbers of electrons that each type of orbital can hold.

arrangement of electrons in elements, chemists have been able to correlate the energy of this emitted light with the element present.

Forensic Determination of Metals

Iron, aluminum, and copper are the three metals most commonly used for fabricating everyday objects. The wide range of applications for these metals increases the probability that they will be found as part of weapons, tools, and metal shards at crime scenes.

Many tools and common household items, such as hinges and locks, are made of alloys. As noted earlier, alloys are mixtures of metals designed to have properties that are more desirable than those of their components. For instance, steel is stronger than iron, one of its main elements, and brass is stronger than copper, its main component.

Alloys are made by mixing metals in different ratios. Brass, for instance, is a mixture of copper and zinc. Museums will often carefully analyze the composition of antiquities to authenticate their age. A case reported in England involved old brass navigational instruments that were alleged to be eighteenth-century antiques. When a small scraping from the instruments was removed and analyzed, it showed the brass to be composed of 65% copper and 35% zinc. Given that brass with zinc content greater than 30% was not commercially produced until the nineteenth century, the instruments were obviously made of more modern brass and could not be as old as claimed. In addition, the analysis failed to find trace elements that would have normally been present in eighteenth-century brass.

Tools are often used in robberies to cut or pry metal. The investigation of such crimes often reveals that metal shards are clinging to the clothing of a suspect. If forensic scientists can determine the composition of these metal shards and match it to the metal that was cut at the crime scene, it provides a strong associative link between the suspect and the crime.

Other metals, such as arsenic, mercury, lead, and cadmium (all of which are poisonous), are sometimes encountered in industrial workplaces and occasionally as poisons. Fans of Sherlock Holmes will recall that he solved several cases in which the cause of death was arsenic poisoning.

Arsenic salts were commonly used to control rodents and other pests at that time in England, so arsenic-containing pesticides could be found in most homes. Today, arsenic is no longer used to control pests, so it is difficult to obtain and is rarely used as a poison.

In addition to these obvious uses of metal, the forensic scientist may encounter lower concentrations of metals used as pigments in paints and dyes, as components of improvised explosives, in gunpowder, and in common everyday materials such as glass.

Atomic Spectroscopy

In atomic spectroscopy, metal ions in solution are transformed into gaseous atoms by a flame, furnace, or plasma that operates at temperatures between 2000°C and 11,000°C. A **plasma** is a mixture of gases that conducts electricity because it contains significant concentrations of cations (positively charged ions) and electrons. A flame, furnace, and plasma all have enough energy to cause electrons to be elevated from the ground state to the excited state. When an excited-state electron returns to the ground state, it emits a photon of light along with energy that is used to identify the element present. The concentration of each metal is measured by the absorption or emission of UV or visible light from these gaseous atoms.

Inductively Coupled Plasma Optical Emission Spectroscopy

Decades ago, forensic analysis of trace metals often consisted of **neutron activation analysis (NAA)**. NAA had two major drawbacks: It was extremely expensive, and it required sending the sample to a nuclear reactor (only a handful of crime laboratories had access to a nuclear reactor).

The NAA technique has since been replaced in crime labs by ICP-OES, which can determine the concentrations of as many as 70 elements simultaneously in some samples and has been used in the FBI laboratory for forensic analysis for the past 15 years.

FIGURE 9-12 is a schematic of a typical ICP torch. The ICP technique uses an argon plasma; that is, argon gas is injected into the system, and argon ions and electrons conduct the electricity. Argon ions, once formed in the plasma, are capable of absorbing sufficient power from the induction coil

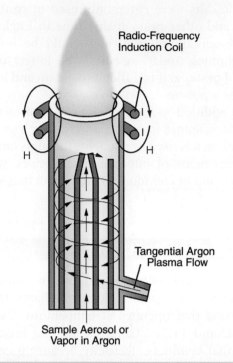

Radio-Frequency
Induction Coil

Tangential Argon
Plasma Flow

Sample Aerosol or
Vapor in Argon

FIGURE 9-12 Using an ICP torch in this GSR detection technique, samples containing metal atoms absorb energy, move to an excited state, and then fall back to the ground state, releasing a photon of energy.

that sends EMR into the torch in the form of a 2-kW radiofrequency (RF) signal at 27 MHz. The argon ions are accelerated by the oscillating RF and form a closed annular "torch" that reaches temperatures as high as 10,000°C and has the shape of a doughnut.

Samples are carried into the torch by argon flowing through the central quartz tube. Once in the torch, the solvent is stripped from the metal ions, electrons are captured by the ions, and the extreme heat of the torch thermally excites the metal atom into an excited state. As the excited metal atom leaves the torch, it cools down and relaxes to the ground state. In so doing, the atom releases a photon of light—a process called optical emission.

As an example, strontium (Sr) in solution absorbs energy in the torch, forming an exited-state strontium (Sr*) atom. The exited-state strontium then cools and emits a photon of light (with wavelength of 392 nm) as it is leaving the argon plasma.

$$Sr + heat \rightarrow Sr*$$
$$Sr* \rightarrow Sr + photon \ (\lambda = 392 \ nm)$$

The major gas in the plasma is the inert gas argon. The excited-state atom Sr* therefore has a long

half-life because no other reactive gases (such as oxygen) are present that might react with it. Different metals present in the sample will undergo a similar transition to the excited state, but since each has a different arrangement of electrons, the wavelength of the photon released will be different for each element. The forensic scientist can determine which elements are present by measuring the different wavelengths given off.

The most commonly used device for sample injection is a nebulizer (**FIGURE 9-13**). In the nebulizer, the sample solution is sprayed into a stream of argon. This technique produces small droplets that are carried by the argon stream into the plasma.

One type of ICP-OES spectrometer is a simultaneous multichannel instrument (**FIGURE 9-14**). In this type of instrument, many photo detectors can be positioned along the curved focal plane of a concave grating monochromator. The grating monochromator isolates the light emitted by each element in the sample and disperses each individual wavelength (λ) to the appropriate photo detector. The emission from each metal ion in solution

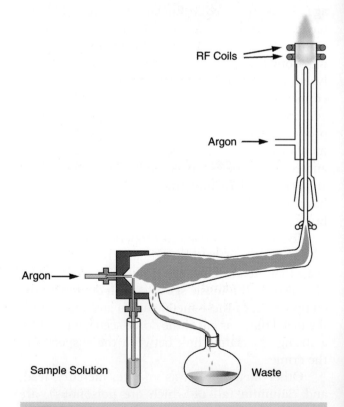

RF Coils

Argon

Argon

Sample Solution

Waste

FIGURE 9-13 When using a typical nebulizer for sample injection into a plasma source, the sample solution is sprayed into a stream of argon, which produces small droplets that are carried by the argon stream into the plasma.

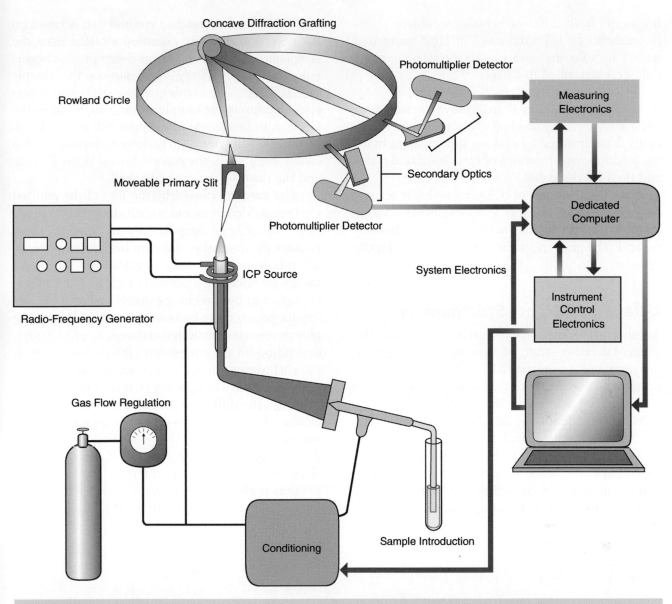

Concave Diffraction Grafting

Photomultiplier Detector

Rowland Circle

Measuring
Electronics

Moveable Primary Slit

Secondary Optics

Photomultiplier Detector

Dedicated
Computer

ICP Source

System Electronics

Radio-Frequency Generator

Instrument
Control
Electronics

Gas Flow Regulation

Conditioning

Sample Introduction

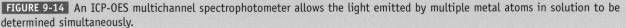

FIGURE 9-14 An ICP-OES multichannel spectrophotometer allows the light emitted by multiple metal atoms in solution to be determined simultaneously.

can be determined simultaneously. Each photo detector has to be placed in exactly the precise location on the Rowland circle to measure the specific emission from the element of interest.

Determining the Elemental Composition of Glass by ICP-OES

Originally, the ICP-OES methods developed by forensic scientists were used primarily to classify glass samples. Although glass is composed mostly of silicon and oxygen, it is the trace elements that are of forensic interest.

The ICP-OES method most widely used today measures the concentration of 10 elements (Al, Ba, Ca, Fe, Mg, Mn, Na, Ti, Sr, Zr) in glass fragments weighing only 20 to 80 mg. This method permits the classification of glass fragments into one of two categories: sheet glass or container glass. ICP-OES has also been able to link food containers to the plants in which they were manufactured. The probability that two glass fragments from different sources will have the same concentrations of all 10 elements is extremely small. A more recent study of 10 elements in automobile side-window glass, for example, found that the probability that two glass fragments from different vehicles would be indistinguishable by ICP-OES was 1 in 1080, compared with 1 in 5 for the refractive index measurements that were described in Chapter 5.

Prior to its analysis, each glass fragment must be scrupulously cleaned to remove surface contamination and any residual contamination. This is easily accomplished by soaking the fragment in nitric acid followed by rinses of first de-ionized water and then ethyl alcohol.

One disadvantage of ICP-OES is that it accepts only liquid samples. Thus, glass fragments must be dissolved in high-purity acids, such as hydrofluoric acid. The glass fragment can take up to 3 days to dissolve.

X-Ray Fluorescence Spectrometry

Another technique that is used to measure the elemental composition of materials is **X-ray fluorescence spectrometry (XRF)**. Because XRF is a nondestructive method that can identify a large number of elements simultaneously, it is an excellent tool for identifying all kinds of materials inspected in forensic analysis. Because XRF can measure small, solid samples, these instruments are a popular choice in the forensic laboratory. Many materials can be identified and matched by analyzing their "elemental fingerprints."

XRF is a spectrometric method that is based on the detection of X-ray radiation emitted from the sample being analyzed. This two-step process begins when a focused X-ray beam strikes the sample (FIGURE 9-15). The incident X-ray strikes an inner-shell electron in the sample atoms, which causes the electron to be ejected like a pool ball being struck by the cue ball (step 1). The lowest electron shell is called the K shell, the second lowest is the L shell, and the third lowest is the M shell.

The vacancy caused by the loss of the emitted electron is filled by an outer-shell electron that drops to the lower level (step 2; refer to Figure 9-15A). Because the outer-shell electron has a higher energy, when it drops to the lower level it loses the excess energy by releasing a photon of EMR. The fluorescent photon has an energy that is equal to the difference between the two electron energy levels. The photon energies are designated as K, L, or M X-rays, depending on the energy level filled. For example, a K-shell vacancy filled by an L-level electron results in the emission of a K_α X-ray (refer to Figure 9-15B).

Since the difference in energy between the two electron levels is always the same, an element in a sample can be identified by measuring the energy of the emitted photon (or photons, if more than one electron is emitted). The intensity of the emitted photons is also directly proportional to the concentration of the element emitting the photon in the

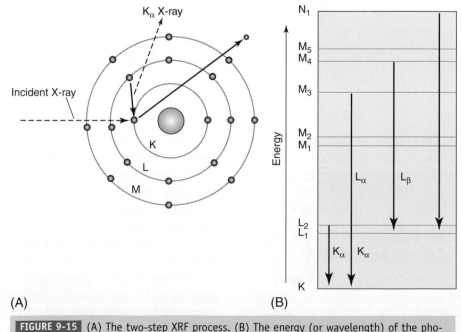

(A)   (B)

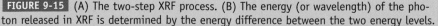

FIGURE 9-15 (A) The two-step XRF process. (B) The energy (or wavelength) of the photon released in XRF is determined by the energy difference between the two energy levels.

sample. Thus the XRF instrument measures the photon energy to identify which element is present and measures the intensity of that photon to quantify the amount of the element in the sample.

Given that the XRF technique measures energy differences from inner-shell electrons, it is insensitive to how the element (being measured) is bonded. The bonding-shell electrons are not involved in the XRF process. XRF will not detect every element; the elemental range is limited to elements larger than beryllium, and the detection of elements with a low atomic number ($Z < 11$, Na) is difficult.

FIGURE 9-16 provides a schematic diagram of the XRF instrument. As shown in this figure, an X-ray tube produces X-rays that are directed at the surface of the sample. The incident X-rays cause the sample to release photons. The emitted photons released from the sample are observed at 90° to the incident X-ray beam. An **energy-dispersive X-ray** fluorescence detector collects all the photons simultaneously. This technique is very sensitive and can identify metals even when they are present at only a parts per million level.

XRF is considered to be a bulk analysis technique: The X-ray source does not focus well into a narrow beam, so it bombards the entire surface of the sample with X-rays. XRF is also a surface technique: It accurately reports the elements present on the surface of the sample. If an analysis of the total composition of the particle is needed, then the sam-

ple must be dissolved in acid and the metals present analyzed by ICP-OES.

Gunpowder Residue

Composition of Gunshot Residue

As described in Chapter 8, when a gun is fired, the primer undergoes a chemical reaction that leads to the detonation of the smokeless powder in the cartridge. This reaction does not always consume all of the primer and powder, however. Thus, any remaining materials, as well as all of the products of the combustion reaction, can be used to detect a fired cartridge. This material is described as **gunshot residue (GSR)**.

The reaction of the primer typically releases three elements: lead, barium, and antimony. A few cartridges include primers that do not use antimony and barium; they are mostly .22 caliber and are marked "lead free" and "clean fire." Less commonly encountered elements in primer include aluminum, sulfur, tin, calcium, potassium, chlorine, and silicon. Mercury (from mercury fulminate) is often released when the shooter uses ammunition manufactured in Eastern Europe.

GSR may be found in many places—for example, on the skin or clothing of the person who fired the gun or on the entrance wound in the shooting victim. Unfortunately, the pattern of GSR dispersal does not always definitively point out the shooter, because the discharge of a firearm—particularly a revolver—can spread the GSR on all nearby objects (including people). Although the amount of residue deposited tends to decrease with increasing range of fire, the actual deposits close to the weapon can vary widely. For example, while the majority of GSR is typically found at the location where the gun was fired, primer residues may adhere to fired bullets and be found at a considerable distance from the point of firing (up to 200 m).

The presence of lead residues at a crime scene may prove challenging to the forensic examiner, because these residues mimic GSR. Lead residues may travel as far 30 ft from the muzzle and are always present on the opposite side of a penetrated target.

Detection of Gunshot Residue

Even though GSR tests are usually performed on individuals who are strongly suspected of having fired a gun, they more often give a negative result

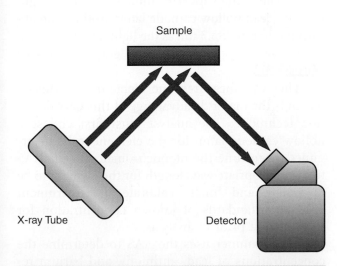

FIGURE 9-16 The highly sensitive technique of using an XRF instrument can identify metals even when they are present at a parts per million level.

than a positive one (i.e., a result confirming that the individual was in close proximity to a fired gun). This tendency reflects the fact that GSR is a perishable type of evidence: Delay in obtaining residues, movement, or washing of the body prior to autopsy may diminish or destroy GSR. Primer residues on the surface of the hands, for example, are easily lost. Simply washing the hands will remove the GSR.

For the most popular GSR detection techniques (described later in this section), the GSR samples to be analyzed must be obtained from the exposed skin of a victim at the scene. Thus time is of the essence when collecting GSR evidence. One study has shown that a GSR sample taken from an individual 2 hours after he or she fired a gun will give negative results. This study has prompted some crime laboratories to reject GSR samples that were taken from a living suspect more than 8 hours after the time when he or she was suspected of firing a gun. **FIGURE 9-17** includes instructions issued to forensic examiners by the Illinois State Police. In cases of suicide, a higher rate of positive results for the presence of GSR can be obtained when the sample is taken before the body is moved or if the victim's hands are placed in paper bags.

GSR tests are most commonly performed by state laboratories, so these laboratories tend to dictate which type of collection kit investigators in the state should use. Four methods are typically used to detect primer residues: neutron activation analysis (NAA), **atomic absorption spectrophotometry (AAS)**, **inductively coupled plasma (ICP)**, and **scanning electron microscopy with energy-dispersive analysis (SEM-EDX)**. While atomic absorption is the most commonly used GSR test at this time, the SEM gives more conclusive results. Somewhere between 20% and 30% of crime labs now use the SEM method. The high cost of neutron activation analysis has limited its application to only a few crime labs, so we will not discuss it further here. ICP was discussed earlier in this chapter.

Atomic Absorption Spectroscopy Analysis of Gunshot Residue

Sample Collection GSR is collected for AAS analysis by using a special kit that contains vials, cotton-tipped swabs, and dilute nitric acid. The forensic examiner begins by treating the swabs with four drops each of the dilute nitric acid (kits may vary), followed by sequentially rubbing the swabs on skin at different locations on the body of the individual being tested. The GSR remaining on the skin will be dissolved in the nitric acid on the swab.

Specifically, the forensic examiner rubs the right back of the hand with one set of swabs and places the two swabs in a labeled vial. He or she then rubs the right palm with another set of swabs and places them in a labeled vial. The examiner follows the same procedure for the left hand. One additional swab, which also was treated with four drops of nitric acid, is taken out, but it is not allowed to contact any surface. This swab is placed in its own container and is labeled as a control. The swabs are then sealed in their labeled vials and packaged in a plastic box to be properly marked with the suspect's name, the date and exact time of the test, and an approximation of the time that passed between the firing of the gun and the application of the swabs.

Sample Analysis The kit containing the swabs is sent to a crime lab, where an analyst removes each swab from its vial and carefully treats it with a dilute nitric acid solution to remove the dissolved GSR. The dilute nitric acid solution from the swab is carefully collected and water is added to give a known total volume. The diluted GSR solution is now ready to be injected into the flameless atomic absorption spectrophotometer.

Under normal circumstances, the dissolved GSR solution will not contain much lead, antimony, or barium. Thus, to determine the elements present in the sample, the forensic examiner uses an atomic absorption spectrometer with an acetylene-air flame. The flame vaporizes the metal atoms. Light from the lead hollow cathode lamp produces an absorption line at 283.3 nm as the light travels through a window at each end of the graphite furnace (**FIGURE 9-18**).

The AAS measures one element at a time—which is the major disadvantage of this GSR detection technique. The analyst must first choose a hollow cathode lamp for the element to be determined, then tune the monochromator of the AAS to the appropriate wavelength for the element to be measured, and finally calibrate the instrument based on standards of known concentration for that element prior to analysis.

The examiner uses the AAS to determine the concentrations of lead, antimony, and barium released from the primer when the gun was fired and to determine whether gunshot residue is, in fact, present. He or she reports the specific amount of

(B)

ILLINOIS STATE POLICE FORENSIC SCIENCE LABORATORY
GUNSHOT RESIDUE ANALYSIS INFORMATION FORM

(Fill out all information requested then return form to kit envelope)

Collecting Officer's Name: _____ Badge No.: _____

Collecting Agency's Name: _____ Agency Case No.: _____

☐ Homicide ☐ Suicide ☐ Assault ☐ Drive By ☐ Other: _____ (Describe)

SUBJECT INFORMATION

Subject's full name: _____ DOB: _____

Subject is: ☐ Living ☐ Dead

Subject is: ☐ Right-handed ☐ Left-handed ☐ Unknown

Any debris and/or blood on subject's hands? ☐ Yes ☐ No

If yes, describe: _____

Has subject washed his/her hands since shooting? ☐ Yes ☐ No ☐ Unknown

Was Control Collection Device used? ☐ Yes ☐ No

Were additional collection devices used? ☐ Yes ☐ No

If yes, why and what area was sampled?: _____

_____

Subject's occupation: _____

Subject's hobbies: _____

Write a brief description of subject's activity between the time of the shooting and the time of the GSR collection: _____

SHOOTING INFORMATION

Date and time shooting occurred: Date: _____ Time: _____ am pm

Place (example—in kitchen, parking lot, indoors, outdoors): _____

Type of firearm used: _____ Caliber: _____

Caliber of ammunition used: _____ Manufacturer of ammunition:* _____

Number of shots fired: _____

*Note: If cartridge manufacturer is unknown, draw head stamp here:

Base of Cartridge

Collecting Officer's Signature: _____

Date: _____ and Time: _____ am pm of GSR collection.

GSRSEMISP: ANA.2 4/01

(A)

ILLINOIS STATE POLICE

GUNSHOT RESIDUE
EVIDENCE COLLECTION KIT

FROM

SUBJECT'S NAME: _____

OFFICER'S NAME: _____

OFFICER'S BADGE NUMBER: _____

AGENCY'S NAME: _____

AGENCY CASE NUMBER: _____

CHAIN OF POSSESSION

RECEIVED FROM: _____

DATE: _____ TIME: _____ am pm

RECEIVED BY: _____

DATE: _____ TIME: _____ am pm

RECEIVED FROM: _____

DATE: _____ TIME: _____ am pm

RECEIVED BY: _____

DATE: _____ TIME: _____ am pm

DELIVER SEALED KIT TO CRIME LABORATORY AS SOON AS POSSIBLE

FOR CRIME LABORATORY PERSONNEL ONLY

CRIME LABORATORY CASE NUMBER: _____

GSRSEMISP: ENV.1 4/96

(C)

ILLINOIS STATE POLICE FORENSIC SCIENCE LABORATORY
INSTRUCTIONS FOR COLLECTING GUNSHOT RESIDUE (GSR)

(READ ENTIRE INSTRUCTION SHEET BEFORE USING KIT)

NOTE

(A) If you should have any questions concerning the use of this kit, do not hesitate to call the GSR Analysis Laboratory at 312/433-8000 Ext. 2029.

(B) In control test firings, it has been shown that the concentration of gunshot residue significantly declines on living subjects after approximately 8 hours. In view of these findings, if more than 8 hours have passed since the shooting, it is recommended that you check with your crime laboratory before submitting samples for analysis.

(C) When the cap is removed from the clear plastic vials containing the SEM stubs, the adhesive collecting surface is exposed and care must be taken to not drop the stub or contaminate the collection surface by allowing the surface to come in contact with an object other than the area that is to be sampled. (See Figure 1.)

(D) Heavily soiled or bloody areas should be avoided if possible.

(E) When pressing the stubs on the questioned areas, use enough pressure to cause a mild indentation on the surface of the subject's hand.

STEP 1 Fill out all information requested on the enclosed Gunshot Residue Analysis Information Form.

STEP 2 Put on the disposable plastic gloves provided in this kit. *Do not substitute with other gloves!*

STEP 3 **CONTROL:**

NOTE: The CONTROL stub is used for collecting any GSR that is in the air and *should not be used for any other purpose.*

(A) Carefully remove the cap from the vial labeled CONTROL.

(B) Place cap (collecting surface facing up) on a flat surface in the area you will be collecting GSR from the subject's hands.

STEP 4 **RIGHT BACK:**

NOTE: If there is blood on the subject's hands or clothing, the investigating officer should put on latex or other approved barrier gloves to protect him/her from bloodborne pathogens, then put on the plastic gloves provided in this kit.

(A) Carefully remove the cap from the vial labeled RIGHT BACK.

(B) While holding the vial cap, press the collecting surface of the stub onto the back of the subject's right hand until the area shown below in Figure 2 has been covered.

(C) After sampling the back of the subject's right hand, return the cap, with metal stub, to the RIGHT BACK vial.

Fig. 1 VIAL CAP METAL STUB COLLECTING SURFACE

STEP 5 **LEFT BACK:**

For collection from the left hand, repeat Step 4 using the vial labeled LEFT BACK.

STEP 6 Carefully pick up the CONTROL cap and return cap, with metal stub, to the CONTROL vial.

Fig. 2

NOTE: (A) If the situation requires additional GSR collection from "other" specific areas, open another kit and use those collection devices.

(B) **Write the area the additional GSR samples were collected on the vial label and state the reason and describe the area sampled on the GSR Information Form where requested.**

FINAL INSTRUCTIONS

(A) Fill out all information requested on the front of the kit envelope.

(B) Return completed GSR Analysis Information Form and the three (or more) capped vials to the kit envelope.

(C) Moisten kit envelope flap, then seal envelope. Affix Police Evidence Seal where indicated, then initial seal.

(D) Mail or hand deliver sealed kit to the Illinois State Police Forensic Science Laboratory for analysis. (If mailed, package kit in a cardboard box to prevent damage in transit.)

GSRSEMISP: INS.2 4/01

FIGURE 9-17 (A) Illinois State Police GSR evidence collection kit. (B) Illinois State Police analysis information form. (C) Illinois State Police Science laboratory instructions for collecting GSR.

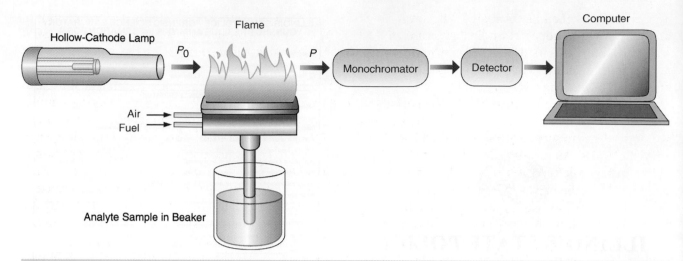

FIGURE 9-18 A flameless atomic absorption spectrometer (AAS) is used in conjunction with a flameless graphite furnace atomizer, which vaporizes the metal atoms. Their spectra then can be measured to identify the metal.

each element present in the nitric acid wash in terms of parts per million (ppm) or micrograms per milliliter (μg/mL).

Scanning Electron Microscopy Analysis of Gunshot Residue

As noted earlier, scanning electron microscopy with energy-dispersive X-ray (SEM-EDX) has emerged as the preferred method for the detection of GSR. Its collection method for GSR is very simple and easy to carry out in the field; the investigator just applies an adhesive tape directly to the suspect's hands (or other surface) to lift the GSR.

Sample Collection The GSR sampling kits used in conjunction with SEM-EDX contain several tubes. In the cap of each tube is an aluminum "stub" consisting of an electrically conductive adhesive layer protected with a cover (FIGURE 9-19). After removing the cover, the investigator dabs the stub over the suspect's hand; the gummed surface of the adhesive removes the GSR from the skin. The investigator then replaces the protective cap and labels the tube. In the laboratory, the forensic examiner removes the stub from the cap and places it directly into a scanning electron microscope for analysis.

An alternative collection method uses polyvinyl alcohol as the basis for removing the GSR. This approach allows the investigator to preserve the topical distribution of GSR and to measure other trace elements if the skin is partially covered in blood.

Sample Analysis A scanning electron microscope operates on the same basic principle as the light microscope but uses electrons rather than light. Electrons, which are produced in the electron source, travel through a specimen in a way that is similar to a beam of light passing through a sample in a light microscope. Instead of glass lenses directing light wavelengths through a specimen, however, the electron microscope's electromagnetic lenses direct electrons through the specimen. Because the wavelength of electrons is much smaller than the wavelength of light, the resolution achieved by the SEM is many times greater than that of the light microscope. Thus, the SEM can reveal the finest details of structure.

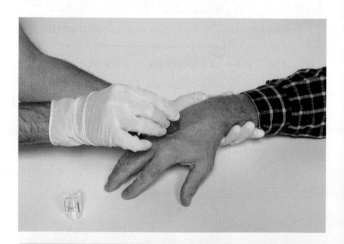

FIGURE 9-19 The adhesive "stub" for collecting GSR, a SEM-EDX method, is very simple and easy to carry out in the field.

These data may then be compared with known examples of GSR in hopes of finding a match. In this way, the large particles of partially burned powder and primer residues can be distinguished from contaminant materials (FIGURE 9-20).

Some of the electrons that strike the surface of the GSR sample are immediately reflected back toward the electron source; these are called backscattered electrons. An electron detector can be placed above the sample to scan the surface and produce an image from the backscattered electrons. Some of the electrons striking the surface penetrate the sample, and they can be measured by an electron detector that is placed beneath the sample. In addition, some of the electrons striking the surface cause X-rays to be emitted. By measuring the energy of the emitted X-rays, the elemental composition of the surface can be determined by EDX.

Energy-Dispersive X-Ray Spectroscopy If the electrons bombarding the atoms on the surface of the sample are able to strike an inner-shell electron in a sample atom, the electron will be ejected from the atom like a pool ball being struck by the cue ball. This is the same effect we saw earlier with XRF. The only difference is that XRF uses a beam of X-rays to dislodge the inner electron, whereas SEM uses an electron beam.

5kV X7000 16mm | 1um

FIGURE 9-20 Use of SEM-EDX allows a large GSR particle to be differentiated from contaminant material.

An EDX detector that measures the energy of the X-ray photons can be attached to the SEM (FIGURE 9-21). The spectrum that is obtained in this way may be used to identify which element is present, and the intensity of that photon may used to measure the amount of the element in the sample. Given that this technique measures energy differences from inner-shell electrons (not bonding-shell electrons), it is insensitive to what kinds of bonds are present in the element being measured. EDX will not detect every element; its elemental range is limited to elements larger than beryllium, and the detection of elements with a low atomic number ($Z < 11$, Na) is difficult.

FIGURE 9-22 shows an SEM image of a GSR particle that was collected on an adhesive stub and an EDX spectrum of the X-rays ejected from the surface of the particle. The x-axis of the spectrum shows the energy of the X-ray; the y-axis shows the intensity of the X-ray signal. The spherical shape of the particle is typical of GSR particles that arise from the high-temperature environment of a gun barrel. The X-ray spectrum reveals the presence of lead, barium, potassium, and strontium presenting the residue. As mentioned earlier, ammunition manufacturers use different ratios of chemicals in their primers, so it may be possible to distinguish between manufacturers based on the elemental composition of the gunshot residue.

The SEM method can be used to search the surface of the tape for individual particles of residue. These particles are usually 3 to 10 microns in size but are often 1 micron or smaller. Sometimes, a complete search of the tape yields only a few GSR particles. It is not likely that the GSR in the sample shown in Figure 9-22 would have been detected by the AAS method, because AAS can only quantify the amount of metal present; it cannot measure the shapes and sizes of particles. Thus, SEM is both more sensitive and definitely more specific than AAS; it also permits the finest details of the sample to be photographed.

Gunpowder Residue: Greiss Test

In crimes involving guns, it is often necessary to determine the distance between the firearm and its target. Often the individual suspect in the shooting claims self-defense as the reason the gun was discharged. Determining the distance between the firearm and its target can support the self-defense claim or show that it is untrue. In a suspected gun-related suicide, the distance between the

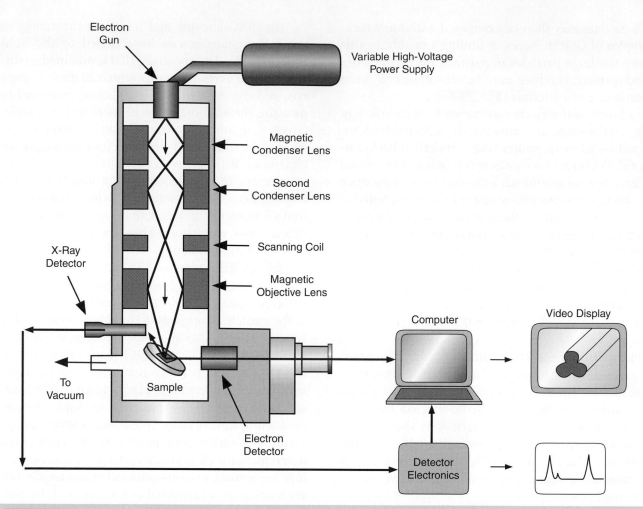

FIGURE 9-21 A schematic representation of the components of the SEM-EDX.

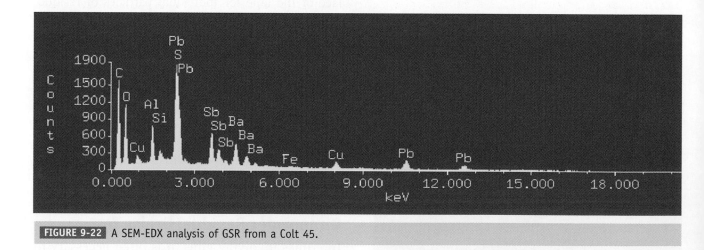

FIGURE 9-22 A SEM-EDX analysis of GSR from a Colt 45.

weapon and the victim will confirm if the shot was self-inflicted or indicate the possibility of a homicide.

When a gun is fired, not all of the smokeless powder in the cartridge is consumed in the reaction. The incomplete combustion of nitrocellulose in the gunpowder releases nitrites (in the form of sodium nitrite, $NaNO_2$). These discharged, inorganic nitrite particles are expelled from the muzzle of a firearm and can become embedded in or deposited on the surface of the target or anyone in close proximity.

One chemical test for locating such powder residues is the **Modified Greiss Test**. The Modified Greiss Test is the first test performed on any evidence because it will not interfere with later tests for lead residues. It is the primary test used by firearms examiners to determine the muzzle-to-garment distance.

To perform the Modified Greiss Test, the forensic examiner uses desensitized photographic paper. Photographic paper is desensitized by being treated with a hypo solution, which makes the photographic paper no longer light sensitive. The paper is then treated with a solution of sulfanilic acid in distilled water and alpha-naphthol in methanol that will make the paper reactive to nitrite residues.

In the Modified Greiss Test, the object being processed (e.g., clothing) is placed face down against a piece of treated photographic paper with the bullet hole centered on the paper. The back of the exhibit being examined is then heated with a steam iron that has been filled with a dilute acetic acid solution. The acetic acid vapors penetrate the object, and a reaction takes place between any nitrite residues on the exhibit and the chemicals contained in the photographic paper. Areas containing nitrites will appear as orange specks on the photographic paper (**FIGURE 9-23**).

If the investigator is able to recover the gun in question and the type of ammunition used in the shooting, the precise distance from which the gun was fired can be determined by a careful comparison of the powder residue pattern on the victim's clothing or skin, because the spray of gunshot residue varies widely among different weapons and types of ammunition. A piece of cloth comparable to the victim's clothing is used as the target. Test firings of the confiscated weapon are made at various distances from the cloth, and the density of residue (space between spots) is recorded. The distribution of gun powder particles around the bullet hole increases as the distance between the target and the gun increases. By comparing the test firings and the evidence, the investigator may find a similar shape and density pattern that will allow him or her to precisely determine the distance from which the shot was fired.

Without the gun in question, the investigator must focus on more general characteristics near the bullet hole. These are approximations with some important characteristics: If the gun is fired at 1 in. or closer to the target, a ring of lead and smoke surrounds the entrance hole of the bullet. Firing 12 to 18 in. from the target will produce a ring of deposited smoke and lead as a halo around the bullet hole. As the distance increases to about 2 ft., the presence of scattered specks of unburned powder are seen without the deposition of unburned powder. A gun that has been fired more than 3 ft. from its target does not usually deposit any powder residues on the target surface (**FIGURE 9-24**).

The Modified Greiss Test detects both organic nitrites and nitrates. Given that nitrates are found in many household chemicals, forensic scientists must be aware of any other chemicals a suspect might have handled.

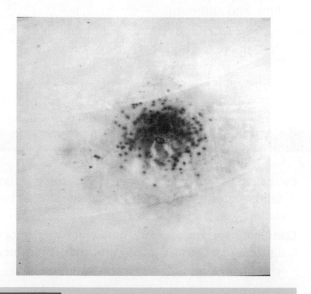

FIGURE 9-23 Applying heat and dilute acetic acid with an iron to develop the Modified Greiss Test.

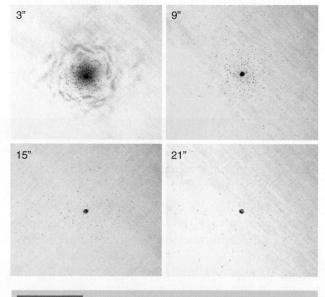

FIGURE 9-24 Powder patterns made at different distances.

YOU ARE THE FORENSIC SCIENTIST SUMMARY

1. Because the front plate of the safe is iron (ferrous), rubbing a magnet over the suspect's clothes and hair will quickly remove any iron filings or dust.
2. Dandruff is dried scalp skin. It is organic and nonmetallic. A magnet wouldn't lift it from the shirt. The material that is lifted by the magnet is definitely a metal.

Chapter Spotlight

- Elements are building blocks from which all matter is built. Each element consists of three types of subatomic particles: protons, neutrons, and electrons. Elements cannot be broken into their subatomic particles by ordinary chemical means.
- In the periodic table, each element is represented by its symbol, atomic number, and atomic mass. Elements are arranged in the periodic table based on their electron configuration and are listed in increasing atomic number.
- The periodic table is divided into periods and groups. Elements in a group have similar physical and chemical properties.
- Compounds are composed of two or more elements in fixed proportions. A molecule is the smallest unit of a substance that can exist and retain its physical and chemical properties.
- Physical properties of inorganic substances include color, odor, taste, hardness, density, solubility, melting point, and boiling point.
- A qualitative analysis confirms the presence of an element or compound of interest.
- A quantitative analysis determines which compounds are present and measures their concentrations.

- Light is a form of electromagnetic radiation that travels in waves in all directions. It can be separated into either a continuous spectrum or a line spectrum.
- The line spectra are caused by changes in energy levels of the electrons and can be used to identify elements.
- ICP-OES identifies and quantifies elements based on their emission of UV or visible light. This technique can measure the concentrations of as many as 70 elements simultaneously.
- XRF measures the photon energy to identify which element is present on the surface of a sample; it measures the intensity of the photon to quantify the amount of the element in the sample.
- GSR may be left on the person who fired the gun, on the victim, or elsewhere at the crime scene. The major elements found in GSR are lead, barium, and antimony.
- AAS, SEM, SEM-EDX, and the Modified Greiss Test all are techniques used to detect GSR.

Key Terms

Atomic absorption spectrophotometry (AAS) A quantitative analysis technique that measures absorption of light by vaporized elements in a sample.

Atomic mass number The sum of the number of protons and the number of neutrons in the nucleus of an atom.

Atomic number The number of protons in the nucleus of an atom of an element.

Compound A substance formed by the chemical combination of two or more elements in fixed proportions.

Continuous spectrum A pattern in which one color of light merges into the next color that is produced when white light is passed through a prism.

Electromagnetic radiation (EMR) A general term that is used to describe energy that is encountered daily.

Electron A subatomic particle with a negative charge of one (−1) and negligible mass.

Element A fundamental building block of matter that cannot be decomposed into a simpler substance by ordinary chemical means.

Energy-dispersive X-ray A technique that measures the energy of X-rays emitted in the scanning electron microscope.

Energy level Any of several regions surrounding the nucleus of an atom in which the electrons move.

Excited state The state of an atom in which an electron has acquired sufficient energy to move to a higher energy level.

Frequency The number of crests that pass a fixed point in 1 second. A frequency of one cycle (passage of one complete wave) per second is equal to 1 hertz (Hz).

Ground state The state of an atom in which all of its electrons are in their lowest possible energy levels.

Group One of the vertical columns of elements in the periodic table.

Gunshot residue (GSR) Material discharged from a firearm other than the bullet.

Inductively coupled plasma (ICP) A high-energy argon plasma that uses radio-frequency energy.

Line spectrum A spectrum produced by an element that appears as a series of bright lines at specific wavelengths, separated by dark bands.

Metals Malleable elements with a lustrous appearance that are good conductors of heat and electricity and that tend to lose electrons to form positively charged ions.

Modified Greiss Test A test for the presence of nitrites in gunshot residue.

Molecule The smallest unit of a compound that retains the characteristics of that compound.

Neutron An electrically neutral subatomic particle found in the nuclei of atoms.

Neutron activation analysis (NAA) An analysis method that bombards the sample with neutrons and then measures the isotopes produced.

Nonmetals Elements that lack the properties of metals.

Nucleus The small, dense, positively charged central core of an atom, composed of protons and neutrons.

Orbital The sublevels of principal energy levels, identified by the letters *s*, *p*, *d*, and *f*.

Period One of the horizontal rows of elements in the periodic table.

Periodic table The elements arranged in order of atomic number (i.e., number of protons). Each element has a unique atomic number.

Plasma A high temperature gas that has a portion of its atoms ionized which makes the gas capable of conducting electricity.

Principal energy level One of the energy levels to which electrons are limited.

Proton A subatomic particle with a relative positive charge of one (+1) and a mass of 1; a hydrogen ion.

Quantum The smallest increment of radiant energy that may be absorbed or emitted.

Scanning electron microscopy with energy-dispersive X-ray (SEM-EDX) The preferred method for the detection of GSR. The investigator applies an adhesive tape directly to the suspect's hands (or other surface) to lift the GSR.

Semimetal (metalloid) Elements with properties that lie between those of metals and nonmetals.

Wavelength The distance between adjacent wave crests, usually measured in nanometers (nm).

X-ray fluorescence spectrometry (XRF) A technique that measures the emission of an X-ray when a sample is exposed to higher-energy X-rays. The energy of the emitted X-ray indicates which element released it.

Putting It All Together

Fill in the Blank

1. The building blocks from which all matter is constructed are called _____ .

2. Substances that are composed of two or more elements are called _____ .

3. The number of naturally occurring elements is _____.

4. A molecule is a combination of _____ or more elements.

5. A proton has a(n) _____ charge.

6. The nucleus of an atom is composed of _____ and _____.

7. The atomic number indicates the number of _____ in the nucleus.

8. In a neutral atom, the number of protons is equal to the number of _____.

9. The periodic table is divided into 18 vertical columns called _____.

10. Elements within a group on the periodic table have similar _____ properties.

11. The elements that are good conductors of electricity and heat and have a shiny appearance are called _____.

12. Elements that do not conduct heat or electricity are called _____.

13. White light from an incandescent light bulb, when passed through a prism, produces a(n) _____ spectrum.

14. An element that is placed in a flame emits light that, when passed through a prism, produces a(n) _____ spectrum.

15. A line spectrum is produced by electrons returning to the ground state from the _____ state.

16. The alkali metals (Group IA) all have _____ electron(s) in their outermost energy levels.

17. The halogens (Group VIIA) all have _____ electron(s) in their outermost energy levels.

18. The three metals most commonly used for fabricating everyday objects are _____, _____, and _____.

19. _____ are mixtures of metals that are created to have properties that are more desirable than those of their components.

20. Glass is composed mostly of the elements _____ and _____.

21. A(n) _____ is a mixture of gases that conduct electricity.

22. The X-ray beam of the XRF strikes a(n) _____-shell electron, which causes it to be ejected.

23. The XRF measures the composition of the _____ of the sample.

24. Collecting GSR for AAS analysis involves the use of swabs containing _____ that are rubbed on the skin of the hand.

25. The atomic absorption spectrometer uses a(n) _____ furnace to vaporize the collected GSR.

26. The collection of gunshot residue for the scanning electron microscope analysis uses a(n) _____ _____ to pick the GSR from the skin.

27. A scanning electron microscope operates on the same principle as the light microscope but uses _____ rather than light to illuminate the sample.

28. In SEM, some of the electrons striking the surface cause _____ to be emitted.

29. By measuring the energy of the X-rays emitted in SEM, the identity and amount of each _____ on the GSR can be determined.

30. The incomplete combustion of gunpowder releases a nonmetal _____.

True or False

1. Glass fragments can be individualized by analyzing 10 trace elements.

2. The ICP-OES method of GSR detection destroys the sample.

3. X-ray fluorescence spectrometry is a nondestructive technique for detecting GSR.

4. The XRF technique can be used to focus on one specific spot on a piece of evidence.

5. The AAS technique is more sensitive than the SEM technique for detecting GSR.

Review Problems

1. Consult the periodic table and complete the table below:

Atom	Atomic Mass Number	Mass Number	Number of Protons	Number of Neutrons	Number of Electrons
Zn	30	64	_____	_____	_____
Pb	_____	207	_____	_____	_____
Sb	_____	121	_____	_____	_____
Au	_____	197	_____	_____	_____

2. A small safe was removed from a company office. The robbers took the safe to a deserted warehouse, where they cut through its door using a torch and a saw with a metal cutting blade. After taking the safe's contents, they left the torch, saw, and safe behind. Three suspects were later apprehended. Describe in detail what evidence should be collected from the company office, the warehouse, and the suspects. How would you analyze the metal shards and dust found on the suspects' clothes?

3. At one time, the mumps, measles, and rubella (MMR) vaccine contained the preservative thiomersal, a mercury compound. A mother who has a sick child suspects that the clinic administered an older dose of the MMR to her child. Since that older dose contained mercury, she contends that her child has mercury poisoning. Describe how you would determine whether the child was poisoned with mercury.

4. A homicide investigation in a home reveals that although the victim died of a gunshot wound, fragments of brown glass somehow became lodged in his skull. A broken beer bottle is found on the floor. When dusted, the broken bottle provides no fingerprints. The victim's wife and her brother are both suspects in the murder. What evidence should investigators seek from the two suspects? What evidence might they find at the crime scene? What should be done with the evidence next?

5. Police responding to a report of a gunshot fired find a dead woman in her home. Her husband is present, and a handgun is on the kitchen table in plain view. What evidence should investigators seek from the suspect? How should the evidence be analyzed, and what results might be expected from this analysis?

Further Reading

Alexander, J. Effect of hair on the deposition of gunshot residue. *Forensic Science Communications.* 2004; 6(2). http://www.fbi.gov/hq/lab/fsc/backissu/april2004/research/2004_02_research02.htm.

Hill, John W. *Chemistry for Changing Times* (ed 12). Upper Saddle Brook, NJ: Prentice-Hall, 2009.

Koons, R. D., and Buscaglia, J. The forensic significance of glass composition and refractive index measurements. *Journal of Forensic Sciences.* 1999; 44: 496–503.

Koons, R. D., Fiedler, C., and Rawalt, R. C. Classification and discrimination of sheet and container glasses by inductively coupled plasma–atomic emission spectrometry and pattern recognition. *Journal of Forensic Sciences.* 1988; 33: 49–67.

Rendle, D. F. Analysis of brass by x-ray powder diffraction. *Journal of Forensic Sciences.* 1981; 343–351.

Schwoeble, A. J., and Exline, D. L. *Forensic Gunshot Residue Analysis.* Boca Raton, FL: CRC Press, 2000.

Scientific Working Group for Materials Analysis (SWGMAT). Elemental analysis of glass. *Forensic Science Communications.* 2005; 7(1). http://www.fbi.gov/hq/lab/fsc/backissu/jan2005/standards/2005standards10.htm.

http://criminaljustice.jblearning.com/criminalistics

Answers to Review Problems

Interactive Questions

Key Term Explorer

Web Links

http://criminaljustice.jblearning.com/criminalistics

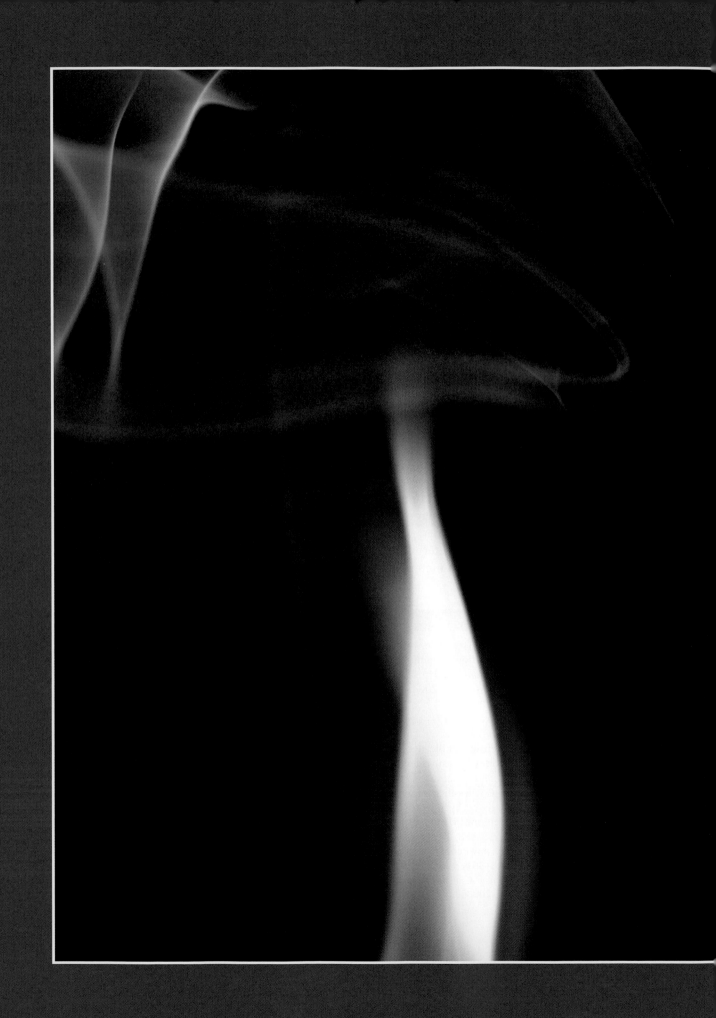

Arson

CHAPTER

10

Fires are common occurrences, and more than 80% of all fires are accidental. In many states, legal statutes require fire marshals to presume that all fires are accidental until proven otherwise. Even though investigators may assume that a fire started accidentally, the remains of the fire should be treated in the same way that evidence is treated at a crime scene.

Arson investigators begin their examination of a fire scene as soon as the fire has been extinguished. Their first task is to determine the location where the fire originated (point of origin). When the suspected point of origin is identified, fire debris should be collected from this location and stored in airtight containers. Next, investigators should establish the path that heat and smoke traveled during the event.

Based on data from previous, similar situations, the investigators should be able to estimate how long the fire burned prior to being extinguished. They must also determine which materials were present in the structure. Based on this evaluation, investigators can answer a critical question: Did this fire burn as expected or was some accelerant used?

As part of their work, investigators should confirm the information obtained during their examination of the fire scene with the information obtained from witnesses and people associated with the scene (e.g., owners, renters). In particular, once the point of origin is determined, its location may confirm or contradict the statements of witnesses. After sifting through all of the evidence, investigators then determine the cause of the fire through a process of elimination.

1. Investigation of a burned home reveals that most of the damage occurred in the kitchen. Further investigation of the path that heat and smoke traveled indicates the most intense heat was encountered under the gas stove. Does this fire appear to be intentionally set? What evidence should investigators search for?

2. A fire in a retail store at midnight causes a total loss of the store's contents. Inquiries by police reveal that the owner of the store has severe financial problems, but the contents of the store are insured for $10 million. Does this fire appear to be intentionally set? What evidence should investigators search for?

Introduction

Arson crimes are hard to solve—in 2003, only 17% of such cases ended in arrests. Vandalism is the leading cause of arson. An estimated 25% of all arson fires are drug-related; sadly, nearly half of these fires in the United States were set by children. The National Fire Protection Association reports that in 2003 (the most recent year for which statistics are available), there were 37,500 intentionally set structure fires in this country. These fires caused $692 million in property damage and were responsible for the deaths of 305 people. In addition to these structure fires, 30,500 intentionally set vehicle fires occurred in 2003. These fires caused an estimated $132 million in property damage.

Oxidation: The Heart of Fire

Fire is the result of an **oxidation** reaction. Oxidation, in its strictest sense, is a chemical reaction between a substance and oxygen. The oxygen source may be air (approximately 20% O_2) or it may be a substance that contains oxygen, such as sodium hypochlorite, potassium chlorate, or ammonium nitrate, which are oxygen-containing chemicals used in making pipe bombs.

A chemical reaction can be described by a chemical equation, a type of shorthand that places the reacting chemicals on the left side of the equation and the chemical products on the right side of the equation. The arrow ($\rightarrow$) means "reacts to produce."

$$\text{reactants} \rightarrow \text{products}$$

A chemical equation includes the chemical formulas of all reactants and products as well. For example, this chemical equation describes the burning of methane, the major component of natural gas:

$$CH_4 + 2\,O_2 \rightarrow CO_2 + 2\,H_2O$$
$$\text{methane} + \text{oxygen} \rightarrow \text{carbon dioxide} + \text{water}$$

Notice that the total number of C, O, and H atoms on the left side of the arrow is the same as the number of those atoms on the right side. There are one C, four O, and four H atoms on each side of the equation. Because equal numbers of atoms appear on each side, this is considered to be a balanced equation.

Oxidation reactions are differentiated according to their rate. "Slow" (controlled) oxidation reactions include processes such as rusting or bleaching. "Fast" oxidation reactions are characterized by the rapid release of heat and are termed combustion or fire. "Extremely fast" (uncontrolled) oxidations are called explosions.

Chemical Reactions: What Makes Them Happen?

Spontaneous Reactions

As we all know, water flows downhill until it reaches level ground. This is a spontaneous process; once started, it proceeds without any outside intervention. Conversely, we cannot get the water back to the top of the hill without some human (or outside) intervention and expenditure of **energy**. That is, carrying the water to the top of the hill is a nonspontaneous process.

As it flows downhill, water gives up potential energy and achieves greater stability. In this and other spontaneous mechanical processes, we recognize that the driving force is the tendency of the system to move to a lower energy state. The same principle applies to most—but not all—chemical reactions. As the reaction proceeds, energy in the form of heat is released, and the energy of the products formed is less than the energy of the reactants.

In a chemical substance, potential energy is stored in the chemical bonds that hold the constituent atoms together. In any chemical reaction, bonds between atoms in the reactants break and the atoms then recombine in a different way to form the products. We will use the following equation to represent a spontaneous chemical reaction:

$$AB + CD \rightarrow AC + BD$$

If the principle that explains spontaneous mechanical processes also applies to spontaneous chemical reactions, we expect the potential energy in the bonds in the products (AC and BD) to be lower than the potential energy in the bonds in the reactants (AB and CD). Thus, in accordance with the first law of thermodynamics, which states that energy can be neither created nor destroyed, energy should be released in the reaction. In the great majority of spontaneous chemical changes, energy is, indeed, released.

In a natural-gas fire, for example, large amounts of heat energy are produced:

$$CH_4 + 2\ O_2 \rightarrow CO_2 + 2\ H_2O + \text{heat energy}$$

Complete combustion of hydrocarbons or oxygenated hydrocarbons produces carbon dioxide, water, and heat (as shown in the chemical equation). Incomplete combustion produces carbon monoxide, soot (carbon), or unburned hydrocarbons. Incomplete combustion is the result of an inadequate supply of oxygen. High-temperature combustion reactions for which air is the source of oxygen also produce oxides of nitrogen $(NO)_x$. Chlorine-substituted plastics or hydrocarbons containing chlorine will produce toxic products when burned, such as hydrochloric acid and phosgene.

Another, albeit less obvious, example of a spontaneous reaction is the rusting of iron. In this process, heat is produced, but the reaction proceeds so slowly that the heat can be detected only with very sensitive instruments. Reactions in which heat is produced are called **exothermic reactions**.

Although the majority of chemical reactions are exothermic, **endothermic reactions**—reactions in which heat is absorbed—also are possible. Obviously, the tendency to achieve a lower energy state is not the sole driving force in a spontaneous chemical reaction.

Getting a Reaction Started

Before any reaction between two or more substances can occur, the reactant particles (atoms, molecules, or ions) must make contact or collide with one another. The more frequently collisions occur, the more rapidly a product is formed. However, not every collision necessarily leads to the formation of a product. To be successful, collisions must have sufficient energy to break the bonds that hold atoms together in the reactants.

The minimum amount of energy that reactant molecules must possess to react is called the **activation energy**. It can be likened to the energy required to lift or push a boulder over a barrier or hill before it will roll down the other side by its own accord (FIGURE 10-1). Lighting a match provides another example. If used as intended, a match will not burst into flame until it is drawn quickly over

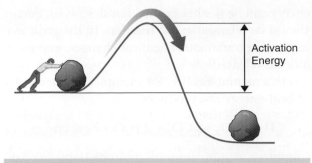

Activation
Energy

FIGURE 10-1 Activation energy is analogous to the energy required to push a boulder over a rise. Once over the top, the boulder will pick up speed as it descends the hill.

Adapted from: Chemistry: The Central Science, 5th ed., by Theodore L. Brown, Eugene LeMay, and Bruce Bursten. Englewood Cliffs, NJ: Prentice Hall, 1991.

the rough striking surface. It needs the heat generated by friction to supply the activation energy. Once lit, the match keeps on burning by obtaining energy from the burning wood in the match. Similarly, if methane from natural gas and oxygen from air are mixed together at room temperature, they do not react automatically. If a spark is introduced into the mixture, however, an explosive reaction occurs immediately.

Factors That Influence the Intensity of a Fire

As noted earlier, fire is a chemical reaction. As the rate of this reaction increases, so does the intensity of the fire. There are three ways to increase or decrease the rate of a chemical reaction: by manipulating the temperature, by changing the concentration of reactants, or by adding a catalyst. When discussing arson, therefore, we need to focus on the temperature, the state of the fuel, the concentrations of the reactants (fuel and oxygen), and the use of accelerants—the means by which the arsonist starts the fire (chemical reaction) and quickly increases its temperature.

The Effect of Temperature

A fuel will produce a flame only when a sufficient number of fuel molecules exist in the gaseous state. That's because only when the fuel molecules are in the gas phase are they free to move and react with oxygen molecules and support a flame. For liquid fuels, the temperature must be high enough to vaporize a significant portion of the fuel. The vapor that is released mixes with oxygen in the atmosphere and can support a flame.

Chemists have measured the **flash points** of a wide variety of liquids. The flash point is the minimum temperature at which a liquid fuel will produce enough vapor to burn. **Flammable liquids**, such as gasoline, acetone, and ethyl alcohol, have flash points of less than 100°F (38°C). **Combustible liquids**, such as kerosene and fuel oil, have flash points of 100°F (38°C) or more.

Once a liquid reaches its flash point, the fuel can be ignited by some outside source of heat. The ignition temperature is always much higher than the flash point (**TABLE 10-1**). For example, gasoline has a flash point of −45°F (−43°C) and an ignition temperature of 536°F (280°C); thus at any temperature at or above its flash point, gasoline will produce a sufficient number of molecules in the gaseous state (i.e., vaporize) such that these molecules can be ignited if exposed to an open flame, spark, or any heat source that is at or above its ignition temperature.

An **accelerant** is a material that is used to start or sustain a fire. Most liquid accelerants tend to form flammable or explosive vapors at room temperature. That explains why arsonists often use gasoline as an accelerant: A significant fraction of gasoline molecules are in the vapor state at the temperatures normally encountered in buildings. The temperature at which these vapors will ignite is called the ignition temperature (or auto ignition temperature). If the source of the heat is an open flame or spark, it is referred to as piloted ignition.

TABLE 10-1

Flash Points and Ignition Temperatures of Common Accelerants

Accelerant	Ignition Temperature	Flash Point
Acetone	869°F (465°C)	−4°F (−20°C)
Carbon disulfide	212°F (100°C)	−22 °F (−30°C)
Ethyl alcohol	689°F (365°C)	55 °F (13°C)
Ethyl ether	356°F (180°C)	−49°F (−45°C)
Kerosene (fuel oil #1)	410°F (210°C)	110°F (43°C)
Heating oil (fuel oil #2)	494°F (257°C)	126°F (52°C)
Gasoline	536°F (280°C)	−45°F (−43°C)
Isopropyl alcohol (IPA)	750°F (399°C)	54°F (12°C)
Methyl alcohol	867°F (464°C)	54°F (12°C)
Methyl ethyl ketone (MEK)	759°F (404°C)	16°F (−9°C)
Paint thinner	473°F (245°C)	104°F (40°C)
Turpentine	488°F (253°C)	90°F (32°C)

Source: inter-Fire Online. Fire and arson investigations. http://www.interfire.org.

Liquid accelerants do not ignite spontaneously. Thus corn oil, which has a flash point of 490°F, could not be used as an accelerant. Conversely, someone using corn oil to cook a turkey in a deep fryer on his or her deck could easily set the deck on fire if not careful to keep the flame of the fryer away from the hot oil. In this case, the flame acts as a spark that ignites the accelerant.

Burning solids is a more complicated process. Wood or plastic materials will burn when heated to a temperature that is hot enough to decompose the solid and produce combustible gaseous products. The breakdown of solids by heat is called **pyrolysis**. When a solid is pyrolyzed, a large variety of gaseous products are released that combine with oxygen to produce a fire. In this process, fire can be considered the result of a chain reaction: A source of flame initiates the pyrolysis of the solid fuel; the gaseous products then combine with oxygen in the air in an oxidation reaction that releases heat and light; the heat released is then absorbed by the solid, which releases more volatile gases; and so on.

Yellow flames usually indicate incomplete combustion. Smoldering is combustion at a surface that occurs without a visible flame. It is characteristic of solids with a relatively large surface area, such as charcoal and heavy fabrics. In the smoldering reaction, pyrolysis occurs at a rate just barely sufficient to supply the fuel necessary to support combustion. Smoldering tends to produce relatively large quantities of carbon monoxide. When a log burns in the fireplace, the wood burns with a flame until all the gaseous products have been expended, but the carbon remaining continues to smolder long after the flame has been extinguished.

The Effect of Concentration

In the same way that increasing the number of cars on the highway increases the chance of a collision occurring, an increase in the concentration of reactant molecules increases the frequency of collisions and thus the rate of the reaction. We know that fuel and air must mix in the gaseous state to initiate a fire, but the ratio of the two is also critically important in driving this reaction. Put simply, if the ratio of gaseous fuel to air is too low (lean) or too high (rich), the mixture will not burn.

The concentration of gaseous fuel (i.e., the percentage of fuel vapors present in the air) that will support combustion is called the **flammable range**. The lowest concentration that will burn is called the

lower explosive limit (LEL); the highest concentration that will burn is called the **upper explosive limit (UEL)**. For example, the LEL for propane is 2.15 and the UEL is 9.6. This means that any mixture of propane and air in which propane accounts for between 2.15% and 9.6% of the total amount of gases present will ignite if exposed to an open flame, spark, or other heat source equal to or greater than its ignition temperature, which is between 920°F (493°C) and 1120°F (604°C). For gasoline, the LEL is 1.3 and the UEL is 6.0.

Typically, the rate of a chemical reaction increases as the temperature rises. The magnitude of this increase depends on the specific reaction mechanism. As a general rule, the rate of a chemical oxidation doubles for every 10°C rise in temperature. This fact helps explain why fires spread so rapidly. As the fire burns, it raises the temperature of the fuel–air mixture, which in turn increases the rate of the reaction. As the rate of the reaction increases, more heat is released, which increases the rate again and makes the fire spread. Firefighters spray water on fires in an attempt to lower the temperature and gain control of the spread of the fire.

This brings us to another important point: In most structural fires, oxygen from air is a necessary reactant to keep the fire burning. Firefighters are well aware of this fact, so they try to limit the amount of oxygen reaching the fire. They are taught to "smother" fires by cutting off the source of oxygen. A fire is halted only when the fuel or oxygen is totally consumed.

The Sequence of Events During a Fire

A fire in a room will usually advance through three stages: the incipient (growth) stage, the free-burning (development) stage, and the smoldering (decay) stage. To determine the origin and cause of a fire, the investigator must understand the normal sequence of events that take place during a fire.

Incipient Stage

The **incipient stage** begins with the ignition of the fire. Initially, the fire remains confined to a limited area because its growth is controlled by the amount of fuel immediately available to it. As the combustion reaction produces gases, however, these lighter-than-air vapors begin to rise in the room. Rushing in to replace them is denser (i.e., heavier) oxygen, which dives to the bottom of the flames. As the fire continues to burn, it produces a characteristic "V" pattern on vertical surfaces (**FIGURE 10-2**).

Free-Burning Stage

In the free-burning stage, the fire consumes more fuel and intensifies. Flames spread upward and outward from the point of ignition, and a dense layer of smoke and fire gases (which include poisonous carbon monoxide) accumulate near the ceiling. As the temperature of the fire builds, this collection of smoke and gases dips lower and heats up until one (or more) of the fuels reaches its ignition temperature.

At a temperature of approximately 1100°F (593°C), sufficient heat is generated to cause all fuels in the room to simultaneously ignite, a process called **flashover**. Temperatures in the superheated space may range from 2000°F (1093°C) at the ceiling to more 1000°F (538°C) at the floor. Not surprisingly, a person cannot survive in such an environment for more than a few seconds.

The length of time necessary for a fire to go from the incipient stage to flashover depends on the fuel available and the ventilation. That is, prior to flashover, the fire's growth is controlled by the amount of fuel available. If the fire stays confined to the area where it originated, its growth then depends on the amount of oxygen available. As a gen-eral rule, however, once flashover occurs, the entire structure quickly becomes engulfed in flames. For a typical residential accidental fire, the leap from incipient stage to flashover may be as short as 2 to 3 minutes.

Smoldering Stage

In the smoldering stage, all of the fuel has been consumed and the fire's open flames have disappeared. The chemical reactions driving the fire have produced a vast amount of soot and combustible gases, and the temperatures of any parts of the structure that have survived to this point may be greater than the ignition temperatures of the accumulated gases. If oxygen suddenly enters the area (from a window breaking or structural collapse, for example), the soot and fire gases can ignite with explosive force as the rush of oxygen regenerates the chemical reactions and produces a **backdraft**. The pressures generated by a backdraft can cause significant structural damage and endanger the lives of firefighters.

Hydrocarbon Accelerants

Table 10-1 lists the ignitable liquids that are most commonly used as accelerants. More than half of these liquids are hydrocarbons that are derived from petroleum. Other accelerants include common alcohols, ethers, and ketones. We will now investigate the chemical structures of the organic liquids that are used as accelerants.

Hydrocarbons

Hydrocarbons are composed of the two elements carbon and hydrogen. Here we focus on the class of hydrocarbons known as **alkanes**, which contain —C—C— single bonds. Some of the most commonly used accelerants, including gasoline, kerosene, fuel oil, turpentine, and paint thinner, are mixtures of hydrocarbons.

The simplest hydrocarbon is methane, CH_4, the major component of natural gas. The structure of this molecule can be represented in several different ways (**FIGURE 10-3**). The expanded structural formula includes the four covalent bonds; the ball-and-stick model and the space-filling model show the spatial arrangement of the atoms. As the ball-and-stick model indicates, the methane molecule is shaped like a tetrahedron. The space-filling model gives the most accurate representation of the actual

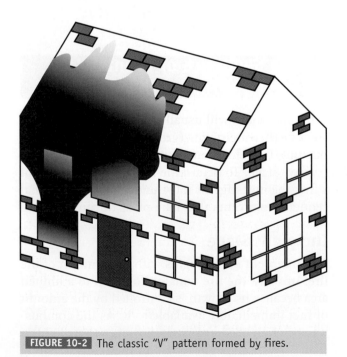

FIGURE 10-2 The classic "V" pattern formed by fires.

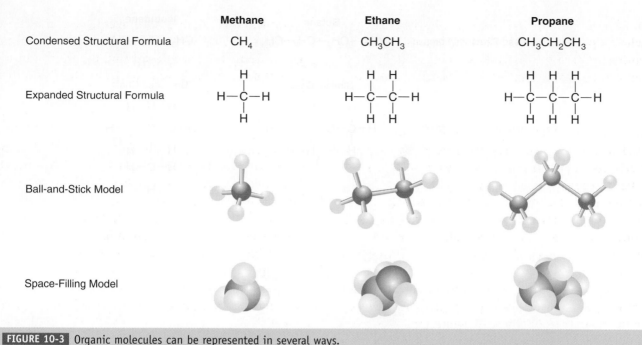

	Methane	**Ethane**	**Propane**
Condensed Structural Formula	CH_4	CH_3CH_3	$CH_3CH_2CH_3$
Expanded Structural Formula			
Ball-and-Stick Model			
Space-Filling Model			

FIGURE 10-3 Organic molecules can be represented in several ways.

shape of the molecule. For simplicity, the condensed structural formula, which does not show the bonds or the bond angles, is usually used to represent the molecule.

The next, more complex hydrocarbons include ethane (C_2H_6) and propane (C_3H_8) (Figure 10-3). Propane is a major component of bottled gas. The next member of the hydrocarbon series is butane (C_4H_{10}). Two structures are possible for butane (FIGURE 10-4): The four carbon atoms can be joined in a straight line (n-butane) or the fourth carbon atom can be added to the middle carbon atom in the —C—C—C— chain to form a branch (isobutane).

Methane, ethane, propane, and butane are collectively named alkanes. The general formula for the alkane series is C_nH_{2n+2}, where n is the number of carbon atoms in a member of the series. Thus, each alkane differs from the one preceding it by the addition of a —CH_2 group. TABLE 10-2 shows the first 10 straight-chain saturated alkanes, where n stands for "normal" and signifies a straight chain; the first four alkanes are gases, and the remainder of those are liquids.

As the carbon chains in an alkane grow longer, the properties of the members change systematically with increasing molecular weight. As Table 10-2 shows, the boiling points of the straight-chain alkanes increase quite regularly as the number of carbon atoms increases.

EXAMPLE

Write the structural formula for the straight-chain alkane C_5H_{12}.

Solution
Write five carbon atoms linked together to form a chain.

$$C—C—C—C—C$$

Attach hydrogen atoms to the carbon atoms so that each carbon atom forms four covalent bonds.

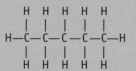

Practice Exercise
Write the structural formula for the straight-chain hydrocarbon represented by C_nH_{2n+2}, where n = 7.

We have described nonbranched chains as straight chains. Because carbon atoms form tetrahedral-shaped bonds, the carbon atoms are not actually in a straight line (as shown in Table 10-2) but rather are staggered in the ball-and-stick models for propane and n-butane (refer to Figure 10-3 and Figure 10-4). Compounds such as butane and isobutane that have the same molecular formula (C_4H_{10}) but different structures are called **structural isomers**. Because their structures are different, the properties of the isomers—for example, their boiling points and melting points—are also different.

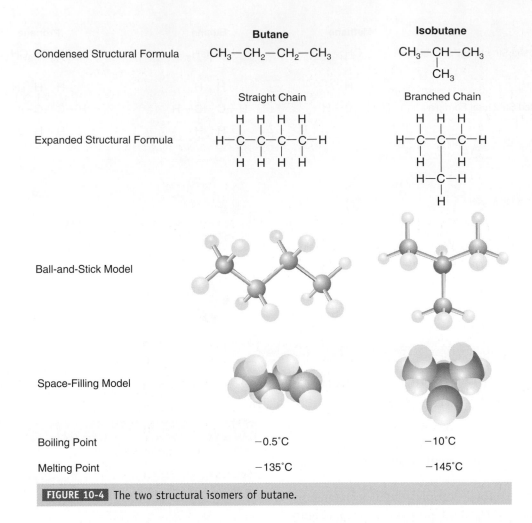

	Butane	Isobutane
Condensed Structural Formula	CH_3—CH_2—CH_2—CH_3	CH_3—CH—CH_3 (with CH_3 below)
	Straight Chain	Branched Chain
Expanded Structural Formula		
Ball-and-Stick Model		
Space-Filling Model		
Boiling Point	−0.5°C	−10°C
Melting Point	−135°C	−145°C

FIGURE 10-4 The two structural isomers of butane.

All alkanes containing four or more carbon atoms form structural isomers. The predicted number of possible isomers increases rapidly as the number of carbon atoms in the molecule increases. For example, butane, C_4H_{10}, forms 2 isomers; decane, $C_{10}H_{22}$, theoretically forms 75 isomers; and $C_{30}H_{62}$ theoretically forms more than 400 million isomers. Of course, most of these isomers do not exist naturally and have not been synthesized. Nevertheless, the large number of possibilities helps explain the abundance of carbon compounds. In isomers with large numbers of carbon atoms, crowding of atoms renders most of the structural isomers too unstable to exist.

EXAMPLE

Give the structural and condensed formulas of the isomers of pentane.

Solution

Pentane has five carbon atoms. The carbon skeletons of the three possible isomers are:

C—C—C—C—C   C—C—C—C   C—C—C
                          | |
                          C   C

Attach hydrogen atoms so that each carbon atom forms four bonds.

Each isomer has the same condensed formula, C_5H_{12}.

Practice Exercise

Draw the structural formulas of the possible isomers of hexane, C_6H_{14}.

The Composition of Petroleum

Petroleum, which is a complex mixture of hydrocarbons, is the main source of most organic compounds used by industry. As noted earlier, crude petroleum is a complex mixture of thousands of organic compounds, most of which are hydrocarbons. Prior to being processed, petroleum also contains small amounts of sulfur-, nitrogen-, and oxygen-containing compounds.

TABLE 10-2

The First 10 Straight-Chain Alkanes

Name	Formula	Boiling Point (8C)	Structural Formula
Methane	CH_4	−162	
Ethane	C_2H_6	−88.5	
Propane	C_3H_8	−42	
η-Butane	C_4H_{10}	0	
η-Pentane	C_5H_{12}	36	
η-Hexane	C_6H_{14}	69	
η-Heptane	C_7H_{16}	98	
η-Octane	C_8H_{18}	126	
η-Nonane	C_9H_{20}	151	
η-Decane	$C_{10}H_{22}$	174	

Refining Petroleum

Crude petroleum is sent to a refinery and separated into fractions by **fractional distillation**, a process that separates the hydrocarbons according to their boiling points.

FIGURE 10-5 shows the essential parts of a fractional distillation tower. Such a tower may be as tall as 30 m (100 ft.). When crude oil at the bottom of the tower is heated to about 930°F (499°C), the majority of the hydrocarbons vaporize. The mixture of hot vapors rises in the fractionating tower; components in the mixture then condense at various levels in the tower and are collected. The temperature in the tower decreases with increasing height,

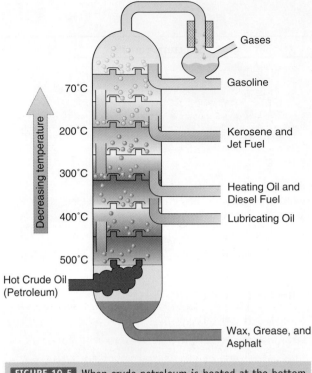

FIGURE 10-5 When crude petroleum is heated at the bottom of a distillation tower, most of the hydrocarbon constituents vaporize. Inside the tower, the temperature decreases with increasing height. Components of the petroleum (called fractions) with higher boiling points condense in the lower part of the tower; components (including the gasoline fraction) with lower boiling points condense near the top. Fractions are removed at various heights as they condense.

so that the more volatile fractions (i.e., those with lower boiling points) condense near the top of the tower. Less volatile fractions (i.e., those with higher boiling points) condense near the bottom. Any uncondensed gases are drawn off at the top of the tower. Residual hydrocarbons that do not vaporize accumulate at the bottom of the tower and are transferred to a vacuum distillation tower, where they are vaporized under reduced pressure to yield

further fractions. In this way, the full range of hydrocarbons in crude oil becomes available for useful purposes.

TABLE 10-3 lists the properties and applications of typical petroleum fractions. Each of these fractions is a complex mixture of hydrocarbons that have roughly the same boiling point. Gasoline, for instance, is a mixture of straight-chain and branched C_6 to C_{10} hydrocarbons, whereas **kerosene** is a mixture of straight-chain and branched C_{11} to C_{12} hydrocarbons. Given that gasoline contains smaller hydrocarbons, it is easier to ignite and more likely to explode compared with kerosene. As we will see later, because each fraction contains a different mix of hydrocarbons, the presence of certain hydrocarbons in fire debris may indicate that a certain accelerant was used to set a fire.

Determining the Origin and Cause of a Fire

In a fire investigation, the place where a fire starts—either the precise spot or the general area—is referred to as the point of origin. In almost all cases, the point of origin must be located if the investigator hopes to identify the cause of the fire. In general, fires tend to burn longer at or near the point of origin, so this area is typically the site of the greatest amount of damage.

In determining the origin and cause of a fire, investigators typically use a "backwards theory." In this technique, investigators systematically work their way from the exterior of the structure to its interior, and from the least damaged areas to the most heavily damaged areas. They examine the entire area surrounding the fire scene, including sites where no fire damage occurred. This search may turn up en-

TABLE 10-3

Typical Petroleum Fractions

Fraction	Boiling Point (°C)	Composition	Uses
Gas	Up to 20	Alkanes from CH_4 to C_4H_{10}	Synthesis of other carbon compounds; fuel
Petroleum ether	20–70	C_5H_{12}, C_6H_{14}	Solvent; gasoline additive for cold weather
Gasoline	70–180	Alkanes from C_6H_{14} to $C_{410}H_{22}$	Fuel for gasoline engines
Kerosene	180–230	$C_{11}H_{24}$, $C_{12}H_{26}$	Fuel for jet engines
Heating oil (fuel oil #2)	230–305	$C_{13}H_{28}$ to $C_{17}H_{36}$	Fuel for furnaces and diesel engines
Heavy gas oil and light lubricating distillate	305–405	$C_{18}H_{38}$ to $C_{25}H_{452}$	Fuel for generating stations; lubricating oil
Lubricants	405–515	Higher alkanes	Thick oils, greases, and waxy solids; lubricating grease; petroleum jelly
Solid residue			Pitch or asphalt for roofing and road material

A Serial Arsonist Terrorizes Washington, D.C.

Beginning in March 2003 and lasting for two years, a serial arsonist terrorized sections of Washington, D.C., by setting fires in more than 46 occupied houses and apartments. Unlike most arsonists, who carefully plan to set fires when buildings are unoccupied, this criminal torched houses while the family was at home asleep. An 86-year-old resident died in one of the fires.

The perpetrator's methods became so familiar to investigators that they rarely needed forensic tests to determine whether it was set by this arsonist. The fires were set using a plastic jug filled with gasoline and with a gasoline-soaked sock stuck in the opening as a wick. The accelerant was usually planted during the hours between midnight and 6 A.M. outside an occupied dwelling. Authorities from the Bureau of Alcohol, Tobacco, Firearms, and Explosives (ATF) believe it took up to 25 minutes before the fire bombs exploded, enabling the arsonist to escape unnoticed.

ATF investigators recovered the fire-setting device and a single human hair—their first piece of forensic evidence—at the site of an attempted arson in northeast Washington, D.C., in September 2003. They also recovered DNA from a sock and the piece of fabric from pants found in two other containers left at fire scenes. But their biggest break came when Arlington County firefighters recovered Marine Corps dress pants and hat near a small deck fire. ATF officials learned about the Arlington fire two days after it occurred. Although they were not initially convinced this evidence was connected to the serial arsonist, they nevertheless decided to submit the pants to the agency's crime lab for testing. DNA from the pants matched DNA recovered from the two earlier fires and attempted arsons in Maryland and Washington, D.C.

Investigators approached the Naval Criminal Investigative Service in hopes of identifying the owner of the clothing. Naval authorities told task force members that they had investigated several suspicious car fires set near the Marine barracks at Eighth and I streets SE in Washington, D.C., about 30 months before the serial arsons began. While the Navy investigators questioned Thomas Sweatt as a suspect in the car blazes, they did not have enough evidence to charge him. The DNA from the clothing found at the site of the deck fires, however, matched Sweatt's DNA.

Agents kept Sweatt under surveillance for several days, learning his routine. Declaring him to be a "person of interest," authorities questioned Sweatt and took a voluntary sample of his DNA. The Montgomery County Police Crime Laboratory reported that Sweatt's DNA matched the samples recovered from the three fires and one attempted arson.

During further questioning, Sweatt admitted to authorities that he was addicted to setting fires. He eventually confessed to igniting 37 residential blazes in the area, including the one that had killed the elderly woman. Sweatt also told investigators that he often targeted places that he knew would be occupied and that he kept setting fires even though he knew authorities were engaged in a massive search for the arsonist. Sweatt was convinced that "at some point he would be caught" after investigators came up with a composite sketch of a suspect. Even so, he couldn't stop setting fires.

In September 2005, Sweatt was sentenced to life in prison. Denise Giles, whose house in District Heights was struck in June 2003, said that her family still cannot put the trauma behind them, saying, "We are no longer heavy sleepers."

Source: The Washington Post.

lightening bits of fire debris and yield valuable information such as the direction of heat flow, the lowest point of burning, overhead damage, and fire patterns—all of which may suggest something about the origin, cause, and path of the fire.

Burn Patterns

Fire patterns are the physical marks and char that remain after a fire. Lines or darkened areas are often found on surfaces after a fire, and they record the boundaries between different levels of heat and smoke produced at the fire scene. Markings left on surfaces depend on the composition of the material.

The direction in which the fire traveled can be determined by examining any burn holes in walls or ceilings. Generally, the fire will have originated from the wider side of the burn hole. For instance, if the hole in a wall is wider at the left side and smaller on the right side, it indicates that the fire came from the left. Also, damage to surfaces will be more severe on the side of that surface from which the fire was set. If the fire moved upward, the damage to the underside of the surface will be more severe. If the fire moved downward, then the top of the surface will typically show more damage.

The remains of damaged material at the fire scene may also provide clues about the source of the fire. For example, inspection of these remains may reveal lines of demarcation from the fire pattern. **FIGURE 10-6** shows the top of fire-damaged wall studs. The studs closest to the fire are shorter than the ones farthest away. The studs in the middle display the fire's progression through the wall. Likewise, the extent of damage on either side of a door can be used to determine the direction of fire travel.

Wood undergoes a chemical decomposition when it is exposed to elevated temperatures. This reaction produces gases, water vapor, and a variety of pyrolysis products, such as smoke. **FIGURE 10-7** shows the remaining solid residue after wood undergoes extended or intense heat exposure; most of these remains are carbon, often referred to as char. By measuring the depth and extent of charring, the investigator may be able to determine which part of a wooden structure was exposed longest to a heat source, which can be used to estimate the duration of the fire (**FIGURE 10-8**). To take these measurements, a blunt probe (e.g., a tire-tread depth gauge or small metal ruler) is placed in the char; to ensure consistency in the resulting measurements, the investigator should use the same measuring tool throughout the entire fire site.

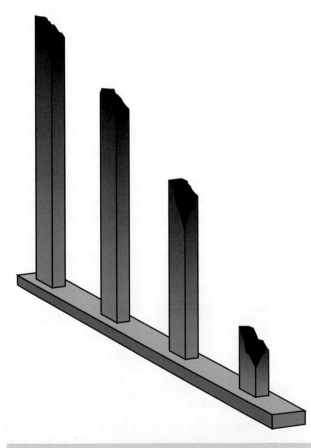

FIGURE 10-6 This burn pattern indicates that the fire spread from right to left.

FIGURE 10-7 There is more charring on the top of these boards.

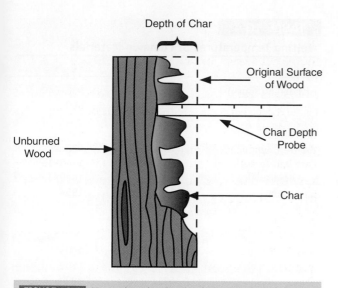

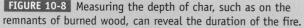

FIGURE 10-8 Measuring the depth of char, such as on the remnants of burned wood, can reveal the duration of the fire.

Burn Pattern Geometry

Distinctive geometry or shapes that are left on the walls or floor are an indication of how the arsonists handled the accelerant (**FIGURE 10-9**). If the accelerant burns in a container, it will leave the "V" pattern that is caused by the upward and outward movement of flames and hot fire gases (refer to Figure 10-9A). As a general rule, the wider the angle of the V, the longer the material has been burning.

If an hourglass pattern is found on a vertical surface, it was most likely formed by a pool of burning liquid (refer to Figure 10-9B). The soot pattern is wider at the floor and ceiling. A burn found in the middle of the room is called a pour pattern; it often indicates that an accelerant was poured onto the floor at that location (refer to Figure 10-9C). A burn

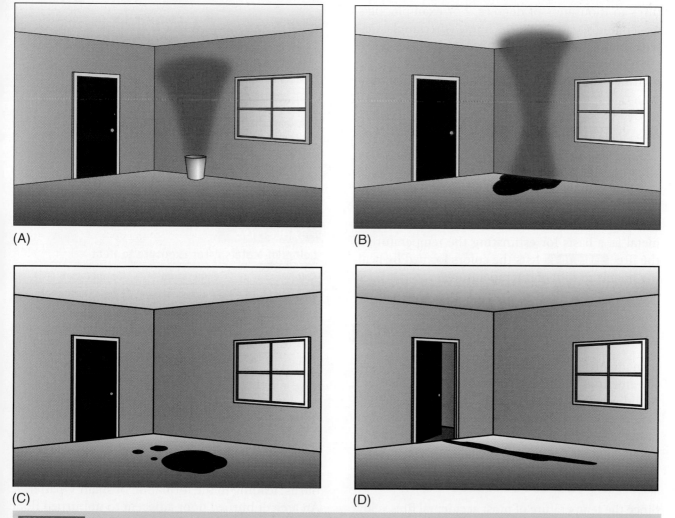

(A)

(B)

(C)

(D)

FIGURE 10-9 (A) The "V" pattern indicates that a container of accelerant was used. (B) The hourglass pattern indicates that a pool of accelerant was here. (C) A pour pattern indicates that accelerant was poured at this location. (D) The trailer pattern indicates that the accelerant was spread from one location to another.

pattern on the floor that resembles a flowing stream, called a trailer pattern, indicates where an incendiary liquid was spread on the floor from one location to another (refer to Figure 10-9D).

Melting of Materials

Investigators can estimate the temperature of the fire by carefully searching for melted objects. Melting—the transformation of a solid into a liquid—is a change in the physical state of a material brought about by its exposure to heat. Knowing the melting points of various materials can help establish the temperatures that were reached during the fire. This information assists investigators in determining the intensity and duration of the heating, the extent of heat movement, and the relative rate of heat release from fuels.

Melting temperatures of materials may range from slightly over normal ambient room temperatures to thousands of degrees. For example, plastics commonly found in the home (thermoplastics) have melting temperatures that range from 200°F (93°C) to near 750°F (399°C). **TABLE 10-4** lists the melting temperatures of a variety of common materials.

Discolored Metals

The presence of melted metals in fire debris usually indicates that the fire was extremely hot. Metal surfaces that are discolored by extreme heat retain that color after the fire ends and they cool down. Investigators can, therefore, use the color of the metal as a basis for estimating the temperature of the fire. **TABLE 10-5** lists the colors formed by heating metals to different temperatures.

Indicators of Arson

Some frequently encountered indicators may assist the investigator in determining the presence of a liquid accelerant. Specifically, certain charring and smoke patterns, odors, the presence of liquid containers, and chemical residues may indicate that an accelerant was used.

Charring of Floor Surfaces

Since the temperature of most accidental fires is low, very little floor charring is observed. Liquid accelerants, by contrast, produce very hot fires and can flow into cracks in the floor. There, the accelerant

TABLE 10-4
Melting Temperatures of Common Materials

	°F	°C
Aluminum (alloys)	1050–1200	566–649
Aluminum	1220	660
Brass (yellow)	1710	932
Brass (red)	1825	996
Bronze (aluminum)	1800	982
Cast iron (gray)	2460–2550	1349–1399
Cast iron (white)	1920–2010	1049–1099
Chromium	3550	1954
Copper	1981	1083
Fire brick	2980–3000	1638–1649
Glass	1100–2600	593–1427
Gold	1945	1063
Iron	2802	1539
Lead	621	327
Magnesium (AZ31B alloy)	1160	627
Nickel	2651	1455
Paraffin	129	54
Platinum	3224	1773
Porcelain	2820	1549
Pot metal	562–752	294–400
Quartz	3060–3090	1682–1699
Silver	1760	960
Solder (tin)	275–350	135–177
Steel (stainless)	2600	1427
Steel (carbon)	2760	1516
Tin	449	232
Wax (paraffin)	120–167	49–75
White pot metal	562–752	294–400
Zinc	707	375

Source: National Fire Protection Association. *NFPA 921, Guide for Fire and Explosion Investigations.* Quincy, MA: Author, 2001.

TABLE 10-5
Colors of Metals After Exposure to Heat

Color	Temperature (°F)	Temperature (°C)
White	2200	1204
Lemon	1800	982
Pink	1600	871
Dark red	1400	760
Red	1100	593
Pale red	900	482
Blue	600	316
Brown	550	288
Yellow	450	232

Source: Adapted from Factory Mutual Engineering Corporation. *A Pocket Guide to Arson Investigation,* 2nd ed. Norwood, MA: Author, 1979.

burns, leading to the formation of small V patterns. An area of burned floor adjacent to unburned floor is a reliable sign that a liquid accelerant was used.

Because a liquid accelerant will flow to the lowest point in a floor, it can often be detected in the

If catching an arsonist is difficult, catching a wildfire arsonist is even more challenging. Fewer than 10% of the wildfire arson cases reported annually are ever solved.

One case in which investigators beat the odds was the Hayman fire, which started in Colorado in June 2002. The massive blaze destroyed 133 homes and caused $29 million in damage.

A forest service worker, Terry Barton, initially reported the wildfire as an illegal campfire that had begun to burn out of control. Later she confessed to burning a letter from her estranged husband in a campfire ring, leaving the site "after the fire was extinguished" and returning to find the fire had spread. Arson investigators were not satisfied, however, because they were unable to find any trace of paper ash at the point of origin. They did find two incriminating pieces of evidence. The first and most important was a triangle of matches, stuck head first into the ground, placed about an inch apart. The second piece of evidence, based on the lack of paper ash and the recovered matches, was a separate, staged campfire Barton had built to corroborate her story. These findings ultimately led to the conviction of Barton for intentionally starting the Hayman fire.

corners of a room or along the base of a wall. Unburned accelerant sometimes flows under the baseboard of the wall, where it becomes trapped. Investigators may find it by removing the baseboard.

Containers

The presence of liquid containers may indicate that a flammable accelerant was used. Liquid containers found at the scene should be retained as evidence for laboratory analysis comparison samples. If the container is not damaged, the arsonist may have left latent fingerprints on the container. If plastic containers were used, they may be melted or totally consumed in the fire. Even if melted, however, they may still contain accelerant residue. For this reason, the container should not be cut open at the scene but rather taken back to the crime lab intact.

Odors

Firefighters may have noticed liquid accelerant odors while they fought the fire. If so, evidence should be collected as soon as possible: Most accelerants evaporate rapidly, so any delay may result in their loss. Areas of carpet or fabric contaminated with liquid accelerants may be detected by ultraviolet light, which may suggest areas where samples should be collected.

A hydrocarbon detector is a very sensitive instrument that measures accelerant vapors and can assist the investigator in pinpointing the presence of an accelerant (FIGURE 10-10). This device sucks in air, which it then passes over a heated filament. If a flammable vapor is present in the air sample, the vapor is oxidized immediately; this oxidation increases the temperature of the filament in the detector. The rise in temperature is recorded by the detector and displayed on the screen. Although very sensitive, the vapor detector is not a selective instrument, so it cannot identify precisely which accelerant is present.

Results from the hydrocarbon detector should never be presented as proving an accelerant was used "to a reasonable degree of scientific certainty" in a court of law. This instrument is merely a screening tool used to identify a location for future testing. Specially trained dogs are also used to detect accelerant residue.

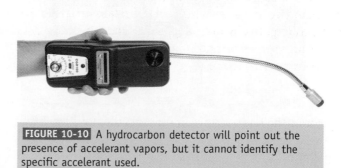

FIGURE 10-10 A hydrocarbon detector will point out the presence of accelerant vapors, but it cannot identify the specific accelerant used.

Collection and Preservation of Arson Evidence

Scenes that potentially involve arson should be investigated much like any other crime scene. In particular, you should search for, identify, record, and photograph all evidence and note its location in a rough fire scene sketch. Also photograph and document the condition of doors, windows, and locks in the fire area. Look for fire doors that were left open, blocked open, or tied open. Check the overall condition and degree of housekeeping in undamaged areas close to the fire. Pay special attention to seemingly out-of-place electrical appliances.

Many commonly used hydrocarbon liquid accelerant vapors are heavier than air, so these vapors may flow downward into stairwells, drains, and cracks. Although heavier than air, they are lighter than water; thus they appear as a "rainbow" when floating on water.

In a fire set by using an accelerant, the highest temperatures will be observed above the center of the burning liquid pool and the coolest temperatures at its edge. As a consequence, residues are more likely to be found at the edges of the burn pattern.

Debris suspected of containing accelerants should be collected and placed in clean metal paint cans, if possible, or sealable glass jars (**FIGURE 10-11**). As mentioned earlier, containers suspected of containing accelerants should be transported intact to the laboratory.

At the laboratory, the forensic examiner begins the analysis of the evidence by briefly opening the evidence container and inspecting its contents. This initial examination must be brief, however, so as to minimize loss of a volatile accelerant that may be present only at trace levels. If the forensic examiner detects a very strong odor suggesting the presence of a particular accelerant, such as gasoline, he or she may choose to perform a particular chemical test based on this clue. Likewise, the nature of the debris (e.g., wood, carpet, soil) may dictate that the examiner carry out a certain type of chemical test on the evidence. The major problem encountered when analyzing fire debris analysis is that of "finding a needle in a haystack": Only a small quantity of accelerant may be hiding in a large amount of solids or liquids.

Analysis of Flammable Residue

Once the debris samples have been delivered to the analytical lab, they must be prepared for analysis. The accelerant sample must be removed from the debris before it can be tested. This step is usually accomplished by either **headspace** sampling or passive headspace diffusion (charcoal sampling).

Heated Headspace Sampling

Any accelerants present in the debris will vaporize if heated in a sealed container; these vapors can then be tested and identified. To increase the concentration of vapors in the air cavity (headspace) of the container, the examiner can heat the sample container.

When collecting evidence at the fire scene, investigators usually put each piece of fire debris in a special metal container that has a small rubber septum (rubber seal) in the lid. Once in the laboratory, this container is heated, and then a syringe with a needle is used to puncture the rubber septum and remove a small (0.1 mL) sample of the headspace vapor (**FIGURE 10-12**). This vapor is then injected directly into the gas chromatograph (GC). Because the GC can accept only a small-volume sample, just a small portion of the headspace gases can be sampled, which limits the sensitivity of this analytical method. To increase the sensitivity, some crime labs use a vapor concentration technique that relies on a charcoal strip to concentrate the sample.

Passive Headspace Diffusion (Charcoal Sampling)

Passive headspace diffusion is another technique that captures the vapor in the headspace of a sealed

FIGURE 10-11 Investigators should use clean paint cans to store residue suspected to contain accelerants.

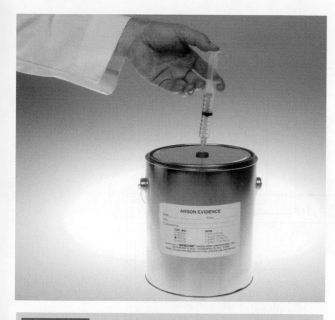

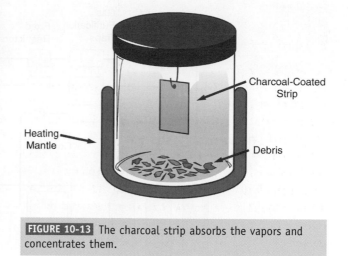

container. In this case, a small strip of carbon fiber mat is suspended in the container with the debris, and the container is then heated to about 140°F (60°C) to turn the liquid accelerant into a gaseous form (**FIGURE 10-13**). The charcoal strip absorbs the accelerant vapor released into the headspace, and after 30 minutes it will concentrate a significant amount of the accelerant vapor. The carbon strip is then removed from the container and the accelerant extracted from it with a small volume of solvent, such as carbon disulfide or diethyl ether. This extract is then injected into the gas chromatograph with a syringe.

Solid-Phase Microextraction

A variation on the charcoal sampling technique, called solid-phase microextraction, replaces the carbon fiber mat with a solid-phase absorbent bonded to a fiber. This material is inserted into the container holding the fire debris via a hollow needle. As in the other techniques for analyzing flammable residues, the container is then heated, causing the accelerants to vaporize and, in this case, be absorbed by the solid-phase absorbent. After it has been exposed to the heated contents of the can for 5 to 15 minutes, the absorbent material is withdrawn from the can and inserted directly into the injection port of the gas chromatograph. There the accelerants are heated until they undergo yet another phase change, turning back into vapors that can be analyzed via the GC. The technique offers two major advantages: It is simple (because it does not require any manipulation of the sample or extract), and it is inexpensive (because it does not require the use of any solvents or any modifications of a conventional GC).

Gas Chromatography

Gas chromatography (GC) is a method used to separate and detect complex mixtures of volatile organic compounds. It separates the gaseous components of a mixture by distributing them between an inert-gas mobile phase and a solid stationary phase. If one component of the gaseous mixture is more strongly absorbed by the stationary phase, then it will move through the system more slowly than components that are not so strongly absorbed by that phase.

As shown in **FIGURE 10-14**, the major components of a GC include the carrier gas, the injection port, the separation column, the detector, and the data acquisition system. The column is held inside an oven whose temperature can be raised to 662°F (350°C). The injector and detector both have separate heaters. The sample is usually injected with a syringe through a septum (rubber seal) into a heated injection port, where it is rapidly vaporized. The gaseous sample is swept out of the injector and into the separation column by helium (a carrier gas that flows at a constant rate). After the components of the sample are stratified in the separation column, they are isolated individually as they pass through the detector. The individual components produce symmetrically shaped peaks, and the area under each peak is proportional to the concentration of that component.

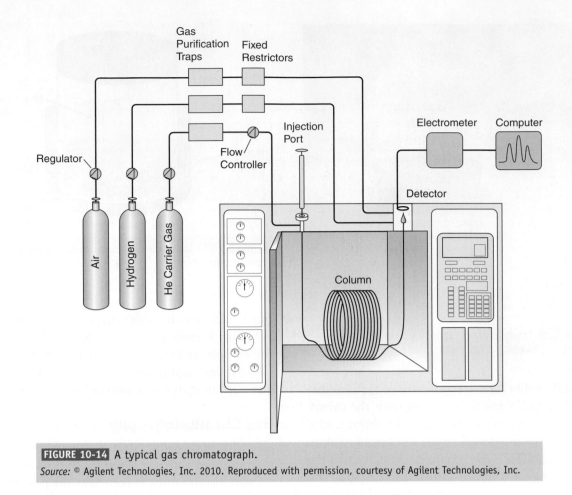

FIGURE 10-14 A typical gas chromatograph.

Source: © Agilent Technologies, Inc. 2010. Reproduced with permission, courtesy of Agilent Technologies, Inc.

The separation column is a long capillary tube (at least 30 m in length, with a 0.25-mm internal diameter) made of fused silica (SiO_2). The inner wall of the capillary is coated with a 1.0-μm-thick film of stationary liquid phase. Silicon stationary phases are commonly used for this purpose, because they remain stable at the elevated temperatures found in the GC oven. Generally, the component of the sample with the lowest boiling point travels through the GC the most rapidly, while the component with the highest boiling point comes out last.

Once the forensic examiner has set the flow rate and temperature of the oven for the desired separation, he or she uses a syringe to inject a reference sample into the GC; the GC then determines the composition of the sample (**FIGURE 10-15**). When a complex mixture, such as gasoline, is analyzed, a complex pattern of peaks is observed. Figure 10-15A

BACK AT THE CRIME LAB

Most arson cases yield only circumstantial evidence. From a forensic scientist's perspective, numerous problems may be encountered in the recovery of accelerant materials from submitted specimens. These problems limit the potential identification of accelerants and usually stem from a lack of care taken during collection and handling of the evidence. For instance, many samples submitted for testing are inappropriate: Recall that arson evidence usually requires a cost- and labor-intensive chemical test to identify the accelerant.

Difficulties also may be encountered because of the absence of an accelerant or simply because of the material itself. For example, in a heavily charred wood, all flammable accelerant may have been lost. Also, if the fire department has sprayed water throughout the region of the fire, the rags that are now damp may have been inadvertently cleaned of accelerant. Critical to the job in the lab is the knowledge of how, when, and where the samples were collected so that the appropriate tests can be performed.

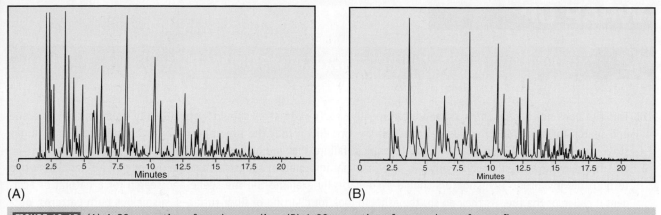

FIGURE 10-15 (A) A GC separation of regular gasoline. (B) A GC separation of an accelerant from a fire scene.
Source: © Agilent Technologies, Inc. 2010. Reproduced with permission, courtesy of Agilent Technologies, Inc.

shows the GC results for regular gasoline (i.e., gasoline as it is obtained at the pump). Gasoline contains hundreds of different hydrocarbons. Each of the peaks in this figure represents a different hydrocarbon component, and the area under each peak is proportional to its concentration.

Figure 10-15B shows the GC results obtained from the headspace analysis of debris taken from the site of a suspicious fire. Notice how similar these results are to the results for gasoline. The peaks that come out of the GC first (the components with the lowest boiling points) in the gasoline sample are missing in the debris sample, but the latter results contain peaks that are almost identical to those seen for gasoline. Thus the debris GC indicates that gasoline was used as an accelerant in the fire. As the fire burned, the components of gasoline with the lowest boiling points were naturally the first to be consumed by the flames, which explains why they are missing from Figure 10-15B.

Usually a forensic analyst will compare the pattern generated by the sample in question to existing reference chromatograms of commonly used accelerants. The GC patterns generated by different grades of gasoline, kerosene, and paint thinner can be stored in a computer database, and the GC's computer can search for a match between the sample of interest and known reference patterns.

Some arsonists may use a mixture of accelerants, which makes comparison of the fire debris results to the existing patterns in the GC database more difficult. Additionally, as the fire burns plastic objects contained in the burning structure, organic decomposition products may be formed that will inevitably be extracted from fire debris along with the accelerant in the laboratory analysis. In these cases the GC pattern obtained from the sample cannot be matched with one of the reference patterns in the database of known accelerants.

To overcome these problems, the GC is often attached to a mass spectrometer; this technique is called—not surprisingly—**gas chromatography–mass spectrometry (GC-MS)**. As each component of the sample is isolated, it enters the mass spectrometer, where its mass is measured. The forensic scientist running the GC-MS will be able to tell from the mass spectrum which components of the evidence sample are contained in a certain accelerant and which came from other sources. The GC-MS instrument is described in more detail in Chapter 11.

WRAP UP

1. This fire does not appear to have been intentionally set. Investigators should systematically examine the fire scene and search for a natural or accidental cause of the fire. Given that the fire appears to have originated under a gas stove, investigators should search for a mechanical failure that might have occurred in the gas line or valves.

2. This fire may have been intentionally set. The financial information suggests that the owner had a financial motive to burn his inventory. Investigators should systematically examine the fire scene and search for a natural or accidental cause of the fire. As they do so, they should look for charring of floor surfaces, suspicious burn patterns, out-of-place objects, and containers that might have held flammable liquids (accelerants). Any of these findings may indicate this fire was arson.

Chapter Spotlight

- Fire is caused by an oxidation reaction. Oxidation is the process central to all combustion reactions, whether they are rapid (fire) or uncontrolled (explosions).

- Fire is a chemical reaction. As the rate of this reaction increases, so does the intensity of the fire.

- Temperature, the state of the fuel, and the concentration of the reactants govern the intensity of the fire.

- Accelerants start or sustain fires. Two types exist: flammable liquids (such as gasoline, acetone, and ethyl alcohol) and combustible liquids (such as kerosene and fuel oil).

- Pyrolysis is the process of breaking down a solid using heat.

- The flammable range is the range of concentrations of a gaseous fuel that will support combustion.

- A fire goes through several stages of development. The incipient (growth) stage begins with the fire's ignition.

- In the free-burning (development) stage of a fire, more fuel is consumed and the fire intensifies. Flashover occurs when the upper layer of the fire reaches approximately 593.3°C and all fuels in the room are ignited.

- In the final stage of a fire, smoldering (decay), all of the fuel has been consumed and the fire's open flames have disappeared. At this point, the fire may nevertheless be reignited if it is revived by an influx of oxygen.

- Hydrocarbons—that is, compounds containing hydrogen and carbon—are the most common accelerants and include many petroleum-derived liquids as well as alcohols.

- The origin and cause of fires can be examined by analyzing the extent of damage to different types of materials, the fire and burn pattern geometry, the melting of materials, and the discoloration of metals found at the scene.

- Indicators of arson include charring of floor surfaces, suspicious containers left at the scene, and odors associated with accelerants.

- Techniques for analyzing arson evidence include headspace sampling, where the sample is heated under controlled conditions, and the vapor emitted by the sample is collected and analyzed by gas chromatography, flame ionization detection, or mass spectrometric detection. The chemical fingerprint of the sample is compared with the fingerprint of known compounds and identified based on the time of its elution from the GC column or by the mass spectrum.

Key Terms

Accelerant Any material that is used to start a fire, but usually an ignitable liquid.

Activation energy The amount of energy that must be applied to reactants to overcome the energy barrier to their reaction.

Alkane A hydrocarbon containing only carbon–carbon single bonds; the general formula is C^nH_{2n+2}.

Backdraft An event in which an oxygen-starved fire suddenly receives oxygen. The sudden rush of oxygen causes the smoldering materials present to reignite at the same time, which causes an explosion.

Combustible liquids Liquids with flash points of 100°F (38°C) or higher, such as kerosene and fuel oil.

Endothermic reaction A chemical reaction in which heat is absorbed.

Energy The ability to do work. Energy has many forms—for example, heat, chemical, electrical, or mechanical.

Exothermic reaction A chemical process that releases heat to the surroundings.

Flammable liquids Liquids with flash points of 100°F (38°C) or lower.

Flammable range The range of vapor concentrations in air that is capable of burning.

Flashover The temperature at which a fire begins unrestrained growth and can cause complete destruction.

Flash point The minimum temperature at which a liquid fuel will produce enough vapor to burn.

Fractional distillation The separation of the components of petroleum by boiling, followed by condensation into fractions with similar boiling ranges. Small molecules with low boiling points emerge first, followed by larger molecules with higher boiling points.

Gas chromatography (GC) A technology used to separate complex mixtures of hydrocarbons, alcohols, ethers, and ketones.

Gas chromatography–mass spectrometry (GC-MS) Use of a gas chromatograph, which separates components of a mixture, in conjunction with a mass spectrometer, which identifies each component by measuring its mass.

Headspace The space above fire debris that has been stored in a sealed container.

Hydrocarbon A chemical that contains only two elements, carbon and hydrogen.

Incipient stage The growth phase of a fire, which begins at ignition.

Kerosene A petroleum fraction that boils at temperatures between 300°F (149°C) and 550°F (288°C). Kerosene is used in space heaters, cook stoves, and wick lamps.

Lower explosive limit (LEL) The lowest concentration of vapor in air that will burn.

Oxidation A chemical reaction in which oxygen is combined with other substances.

Pyrolysis The decomposition of a substance by the application of heat.

Structural isomers Organic compounds that have the same formula but different molecular structures.

Upper explosive limit (UEL) The highest concentration of vapor in air that will burn.

Putting It All Together

Fill in the Blank

1. _____ is the leading cause of arson.

2. Oxidation is a chemical reaction between a substance and _____.

3. _____ reactions produce heat.

4. _____ reactions absorb heat.

5. The rate of a chemical reaction is the _____ at which reactants are converted into products.

6. The minimum amount of energy that a reactant molecule must possess to react is called the _____.

7. The three factors that affect the rate of a chemical reaction are _____, _____, and _____.

8. A fuel will produce a flame only when a sufficient number of fuel molecules are in the _____ state.

9. The _____ is the minimum temperature at which a liquid fuel will produce enough vapor to burn.

10. The temperature at which vapors ignite is known as the _____ temperature.

11. Liquids with flash points of 100°F (38°C) or lower are known as _____ liquids.

12. Liquids with flash points of 100°F (38°C) or higher are known as _____ liquids.

13. The breakdown of solids by heat is called _____.

14. Yellow flames indicate _____ combustion.

15. _____ is a term that describes combustion at a surface that occurs without flames.

16. The concentration of gaseous fuel that will support combustion is called the _____.

17. The lowest concentration of gaseous vapor that will burn is called the _____ _____ _____.

18. A vapor density that is _____ (greater/less) than 1 means the vapor is lighter than air.

19. The rate of chemical oxidation (burning) _____ for every 212°F (100°C) rise in temperature.

20. The first stage of a fire is called the _____ stage.

21. The second stage of a fire is called the _____ stage.

22. _____ occurs when there is spontaneous combustion of all fuels in a room.

23. If oxygen suddenly comes in contact with smoldering material, an explosive reaction can take place that is commonly called _____.

24. Heating wood causes it to decompose by releasing gases and leaving a solid residue called _____.

25. A common fire pattern has a distinctive _____ shape.

26. In determining the origin of a fire, the _____ theory helps identify the source of the fire.

27. Hydrocarbons that contain only single bonds are called _____.

28. Crude petroleum is separated by _____ _____ in the refinery.

29. Accelerant residue in fire debris is removed by one of two techniques: _____ or _____.

30. In gas chromatography, the sample is usually introduced into the GC instrument by a(n) _____.

31. Generally, the component of the sample with the _____ boiling point travels through the GC the most rapidly and the component with the _____ boiling point comes out last.

32. Individual components of a sample produce symmetrically shaped GC peaks, and the _____ is proportional to the concentration of that component.

33. Each component of a sample has a unique GC _____ _____.

True or False

1. The highest temperature in accelerant fires occurs at the edge of the pool of burning accelerant.

2. A line of demarcation between burned and unburned flooring is a reliable sign of an accidental fire.

3. Charring under a coffee table is a sign that an accelerant was used.

4. The hydrocarbon detector is a definitive test for an accelerant, and its results can be used in a court of law.

5. Debris suspected of containing accelerants should be collected and placed in zip-lock bags.

Review Problems

1. Firefighters fighting a natural gas (CH_4) fire try to control the following reaction:

   $$CH_4 + 2\,O_2 \rightarrow CO_2 + 2\,H_2O + heat$$

   Explain how each of the following will slow the spread of the fire.

   a. Slowing the release of methane (CH_4)

   b. "Smothering" the fire

   c. Applying water

   d. Using a carbon dioxide fire extinguisher

2. As can be seen in Table 10-3, the gasoline fraction of petroleum is composed of C_6 to C_{10} hydrocarbons. Figure 10-15 shows a gas chromatography separation of the many hundreds of hydrocarbons in gasoline.

   a. Why are the earlier peaks missing from the GC of the fire debris in Figure 10-15B?

   b. Describe the pattern of peaks that the GC result would show for kerosene.

   c. Describe the pattern of peaks that the GC result would show for fire debris from an arson site at which kerosene was used as an accelerant.

3. Fire investigators need to establish whether a fire was started accidentally or deliberately. Which characteristics of a fire scene would indicate that it was set deliberately? How would these characteristics differ from those at an accidental fire?

4. Using Table 10-1 and Table 10-3, describe why heating oil #2 is not commonly used as an accelerant.

5. An unsuccessful businessman who has a failing business decides to torch the business and then submit a claim for reimbursement to his insurance company. Late one night, he disables the fire alarm, pours gasoline over the carpets and on furniture, and lights the gasoline. His business burns to the ground. You are the fire investigator. Describe how you will proceed with the investigation.

Further Reading

Allen, Stephen P., et al. The National Center for Forensic Science Ignitable Liquids Reference Collection and Database. Forensic *Science Communications*. 2006; 8(2). http://www.fbi.gov/hq/lab/fsc/backissu/april2006/standards/2006_04_standards01.htm

DeHaan, J. D. *Kirk's Fire Investigation* (ed 4). Upper Saddle River, NJ: Prentice-Hall, 1997.

Faith, N. *Blaze: The Forensics of Fire*. New York: St. Martin's Press, 2000.

Fire Laboratory Standards and Protocols Committee Scientific/Technical Working Group for Fire and Explosions. Quality Assurance Guide for the Forensic Analysis of Ignitable Liquids. *Forensic Science Communications*. 2006; 8(2).

Lentine, J. J. *Scientific Protocols for Fire Investigation*. Boca Raton, FL: Taylor & Francis, 2006.

Midkiff, C. R. Arson and explosive investigation. In R. Saferstein (Ed), *Forensic Science Handbook, vol. 2*. Upper Saddle River, NJ: Prentice-Hall, 2002.

National Fire Protection Association. NFPA 921, *Guide for Fire and Explosion Investigations*. Quincy, MA: NFPA, 2001. http://www.fbi.gov/hq/lab/fsc/backissu/april2006/standards/2006_04_standards02.htm

http://criminaljustice.jblearning.com/criminalistics

Answers to Review Problems

Interactive Questions

Key Term Explorer

Web Links

http://criminaljustice.jblearning.com/criminalistics

Drugs of Abuse

OBJECTIVES

In this chapter you should gain an understanding of:

- The classification of drugs under the Controlled Substances Act
- The drugs that are commonly abused
- Screening tests that are used to detect illicit drugs
- Confirmatory tests for illicit drugs

FEATURES

On the Crime Scene

Back at the Crime Lab

See You in Court

WRAP UP

Chapter Spotlight

Key Terms

Putting It All Together

The penalties for drug possession vary depending on the type of drug, the amount of drug, and the location where the drug possession took place. Drug possession penalties are harsher for possession of drugs with a higher potential for abuse. Likewise, larger quantities of drugs often result in longer sentences because these amounts are typically intended for sale. Drug trafficking charges may, therefore, accompany drug possession charges in some cases. Individuals who are charged with drug possession and trafficking or intent to distribute receive even longer sentences. Those who sell to minors often receive twice as harsh a punishment in drug possession and distribution cases. Drug possession or distribution that occurs within a particular proximity to a school, university, or day-care center may also result in stiffer drug possession penalties.

Forensic scientists subject material seized by police to qualitative tests that are designed to identify which, if any, controlled substances are present in the seized samples. Testing begins with presumptive tests, such as an examination of the substance's color, and microcrystalline tests that will quickly indicate the possibility that the sample contains a specific drug. After obtaining a positive presumptive test, the forensic scientist begins confirmatory testing with infrared or gas chromatography–mass spectrometry (GC-MS).

Once the identity of the drug is confirmed, the purity of the sample is determined by a quantitative test. Given that most street drugs are "cut" with an adulterant before being distributed, the concentration of drugs in the sample could be as low as 10%. Once the forensic scientist weighs the seized material and determines its purity, the quantity of drug seized is easily calculated. This information helps the court establish whether the suspect is guilty, how the suspect should be charged, and which penalties should be applied.

1. Using the U.S. Drug Enforcement Administration (DEA) website, determine the penalties for a person convicted for the first time for trafficking these drugs:
   a. Heroin (500 g)
   b. Cocaine (500 g)
   c. PCP (10 g)
2. Using Catholic University's website, determine the penalties for a person convicted for the first time for "manufacturing, distributing, or dispensing or possessing with the intent to manufacture, distribute, or dispense" 500 g of cocaine in the following jurisdictions:
   a. Maryland
   b. Virginia
   c. District of Columbia

Introduction

Humans have used naturally occurring drugs for thousands of years. Alcohol has flowed in wine from before the time of the early Egyptians. Narcotics derived from poppies have been used since 4000 B.C., and marijuana was used as a medicinal product in China as early as 2700 B.C.

In the nineteenth century, the active compounds in poppies and coca plants were extracted from the plant and purified. The resulting substances—morphine and cocaine—were completely unregulated and were prescribed by physicians for a wide variety of ailments. They were available in patent medicines and sold in traveling shows, in drugstores, or through the mail. During the U.S. Civil War, morphine was commonly used as a painkiller, and wounded veterans were sent home with morphine and hypodermic needles. Opium dens flourished. By the early 1900s, there were an estimated 250,000 drug addicts in the United States.

History of Drug Regulation

Drug control legislation in the United States has been an outgrowth of national and local law enforcement customs, as well as the result of moral and political philosophies. The problems of addiction were acknowledged only slowly, however.

Amphetamines Drugs that have a stimulant effect on the central nervous system and can be both physically and psychologically addictive when overused. The street term "speed" refers to stimulant drugs such as amphetamines.

Anabolic steroids Molecules that promote the storage of protein and the growth of tissue; sometimes used by athletes to increase muscle size and strength.

Analgesic A medicine used to relieve pain.

Barbiturates A group of barbituric acid derivatives that act as central nervous system depressants and are used as sedatives.

Bulk samples Samples of drugs that are large enough to be weighed.

Central nervous system (CNS) The vertebrate nervous system, consisting of the brain and the spinal cord.

Cocaine A white crystalline alkaloid that is extracted from coca leaves and is widely used as an illicit drug for its euphoric and stimulating effects.

Codeine An alkaloid narcotic that is derived from opium and is used as a cough suppressant and analgesic.

Confirmation test A test that identifies the presence of an illegal drug. Its results may be used as confirmatory evidence in court proceedings.

Depressant A drug that decreases the rate of vital physiological activities.

Hallucinogen A drug that induces hallucinations or altered sensory experiences.

Heroin A white, odorless, bitter crystalline compound that is derived from morphine and is a highly addictive narcotic.

Immunoassay A test that makes use of the binding between an antigen and its antibody to identify and quantify the specific antigen or antibody in a sample.

Infrared spectrophotometry A technique that measures the absorption of infrared radiation by a drug sample at different wavelengths.

Lysergic acid diethylamide (LSD) A crystalline compound derived from lysergic acid and used as a powerful hallucinogenic drug.

Marijuana A hallucinogenic drug made from the dried flower clusters and leaves of the cannabis plant. It is usually smoked or eaten to induce euphoria.

Mass spectrometry An instrument used to identify chemicals in a substance by their mass and charge.

MDMA (3,4-methylenedioxymethamphetamine) A drug that is chemically related to amphetamine and mescaline and is used illicitly for its euphoric and hallucinogenic effects. MDMA reduces inhibitions and was formerly used in psychotherapy but is now banned in the United States.

Mescaline An alkaloid drug, obtained from mescal buttons, that produces hallucinations. Also called peyote.

Methadone A synthetic narcotic drug that is less addictive than morphine or heroin and is used as a substitute for morphine and heroin in addiction treatment programs.

Microcrystalline tests Tests used to identify drugs based on the formation of crystals during a chemical reaction. When certain illegal drugs are mixed with testing reagents, they produce colored crystals with specific geometry that can then be used to identify the drug.

Morphine A bitter crystalline alkaloid, extracted from opium, that is used in medicine as an analgesic, a light anesthetic, or a sedative.

Narcotic An addictive drug that reduces pain, alters mood and behavior, and usually induces sleep or stupor.

National Institute on Drug Abuse (NIDA) A U.S. government agency that implements the drug testing program that has become a condition of federal employment.

OxyContin A prescription painkiller that has become a popular and dangerous recreational drug. Known as "oxy" on the street, this time-released morphine-like narcotic is prescribed by physicians to relieve chronic pain.

Phencyclidine (PCP) An anesthetic used by veterinarians; an illicit hallucinogen known as "angel dust."

Physical dependence The situation in which regular use of a drug causes a physiological need for it. When the drug is stopped, withdrawal sickness begins.

Psychological dependence A pattern of compulsive drug use caused by continual craving.

Screening test A preliminary test performed to detect illegal drugs.

Stimulant A drug that temporarily increases alertness and quickens some vital processes.

Tetrahydrocannabinol (THC) The psychoactive substance present in marijuana.

Thin-layer chromatography (TLC) A technique for separating components in a mixture. TLC is used to separate the components of different inks.

Putting It All Together

Fill in the Blank

1. Under the Controlled Substances Act (CSA), drugs that have no recognized medical use are placed in Schedule _____.

2. Under the CSA, the least dangerous controlled substances are placed in Schedule _____.

3. Under the CSA, a designer drug that is similar to a controlled substance triggers the same penalties as if it were a controlled substance in Schedule _____.

4. The two most abundant alkaloid drugs that are extracted from the poppy are _____ and _____.

5. Drugs that produce dependence act on the _____ _____ _____.

6. A drug that numbs the senses and places the user in a stupor is a(n) _____.

7. Amines are derived from _____, NH_3.

8. A secondary amine has _____ hydrogen on the nitrogen.

9. Heroin is made by reacting _____ with acetic anhydride.

10. Drug-sniffing dogs are trained to sniff for _____, which indicates heroin may be present.

11. The material most commonly used to dilute heroin is _____.

12. The synthetic narcotic that is widely used to treat heroin addiction is _____.

13. A(n) _____ is a substance that induces changes in mood, thought, or perception.

14. _____ is the most commonly used illicit drug in the United States.

15. The resin found on the flower clusters of marijuana is used to make _____.

16. The hallucinogen that is found in the peyote cactus is _____.

17. Substances that diminish a person's functional activity are called _____.

18. The proof of an alcoholic beverage is equal to _____ the percentage of alcohol in the drink.

19. Amphetamines are _____ (stimulants/depressants).

20. The drug that was introduced as an appetite suppressant for weight loss is ____.

21. The drug that is easily made from pseudoephedrine, which is in Sudafed, is _____.

22. The naturally occurring alkaloid that is extracted from the leaves of the coca plant is _____.

23. If the hydrochloride is removed from cocaine hydrochloride, the free base is produced, which is known on the street as _____.

24. Anabolic steroids are designed to mimic the body-building traits of _____.

25. A seized quantity of drugs that is large enough to be weighed is called a(n) _____ sample.

26. The test cited by a forensic chemist in court to confirm the identification of a drug sample is known as the _____ test.

27. In thin-layer chromatography (TLC), compounds are separated by their _____, _____, and _____ in a solvent.

28. In TLC, the more tightly a compound binds to the stationary silica plate, the more _____ it moves up the thin-layer plate.

29. Drugs can be identified by taking an IR spectrum that produces a "fingerprint" spectra that is a _____ (screening/confirmation) test.

30. It is necessary to remove all _____ and _____ from a suspected drug sample before measuring an IR spectrum.

31. The mass spectrum of a suspected drug sample is considered a _____ (screening/confirmation) test.

32. If the NIDA screening test indicates a drug of abuse is present in the individual's urine, a(n) _____ confirmation test is then performed.

True or False

1. Unlike heroin, methadone is not addictive.
2. OxyContin is a time-release drug designed to be taken intravenously.
3. Abusers of PCP often become violent and are very dangerous.
4. Withdrawal from barbiturates can be more difficult than withdrawal from heroin.
5. Methaqualone (Quaalude) is a barbiturate.
6. The white crystalline form of cocaine is cocaine hydrochloride.
7. A screening drug test conclusively proves that the substance tested is a controlled substance.
8. Thin-layer chromatography (TLC) is a confirmation test.

Review Problems

1. A crime scene technician discovers a bag of white powder at a crime scene. When she performs color tests, she observes a purple color when she adds the Marquis reagent to the powder. Is the powder cocaine? Explain.
2. Explain why drug-sniffing dogs are taught to sniff for vinegar and solvents.
3. From which common drug is methamphetamine made?
4. You perform a TLC separation on a sample collected from a bag of white powder that was seized at a crime scene. To do so, you dissolve the powder in solvent and spot it on the TLC plate. The solvent moves 60 mm on the plate, and your spot moves 38 mm. The R_f of heroin is 0.63 and that for cocaine is 0.85. Is the unknown sample an illicit drug? Which drug and which schedule applies?
5. You perform a TLC separation on a sample collected from a bag of white powder that was seized at a crime scene. To do so, you dissolve the powder in solvent and spot it on the TLC plate. The solvent moves 60 mm on the plate, and your spot moves 51 mm. The R_f of heroin is 0.63 and that for cocaine is 0.85. Is the unknown sample an illicit drug? Which drug and which schedule apply?
6. During a screening test of a bulk sample for illicit drugs, the forensic analyst performs two color tests. When he adds the Marquis reagent to the sample, it doesn't change color. When he adds the Scott reagent, it turns solution A blue. Which drug is indicated by the results of these tests?
7. You are applying for a government job and must take a drug test. The technician performing the immunoassay screening test tells you that the test was negative for all drugs. Do you have to wait for a GC-MS test to confirm the results or have you passed the drug test? Explain.

Further Reading

Hanson, G. R., Venturelli, P. J., and Fleckenstein, A. E. *Drugs and Society* (ed 10). Sudbury, MA: Jones and Bartlett, 2009.

Scientific Working Group for the Analysis of Seized Drugs (SWGDRUG). Methods of analysis and drug identification. *Forensic Science Communications*. 2005; 7(1). http://www.fbi.gov/hq/lab/fsc/backissu/jan2005/standards/2005standards11.htm.

Scientific Working Group for the Analysis of Seized Drugs (SWGDRUG). Code of professional conduct for drug analysts. *Forensic Science Communications*. 2005; 7(1). http://www.fbi.gov/hq/lab/fsc/backissu/jan2005/standards/2005standard13.htm.

Siegel, J. A. Forensic identification of illicit drugs. In R. Saferstein (Ed.), *Forensic Science Handbook,* vol. 2, pp. 111–174. Upper Saddle River, NJ: Prentice-Hall, 2005.

Smith, F., and J. A. Siegel (Eds.). *Handbook of Forensic Drug Analysis*. Amsterdam: Academic Press, 2004.

http://criminaljustice.jblearning.com/criminalistics

Answers to Review Problems

Interactive Questions

Key Term Explorer

Web Links

http://criminaljustice.jblearning.com/criminalistics

Biological Evidence

SECTION

5

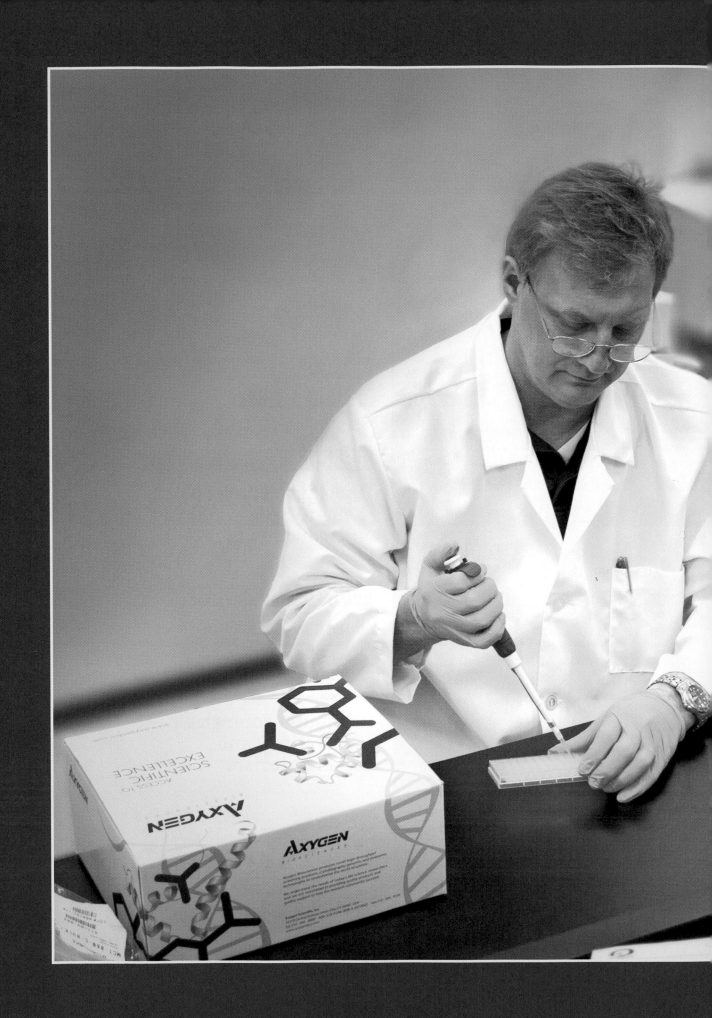

Forensic Toxicology

OBJECTIVES

In this chapter you should gain an understanding of:

- The postmortem analysis performed by a toxicologist
- The many factors that make postmortem toxicological information difficult to interpret
- Alcohol's effects on human performance
- The way in which alcohol is absorbed and eliminated from the body
- The measurement of blood alcohol concentration (BAC)
- Alcohol-related vehicle laws

FEATURES

On the Crime Scene

Back at the Crime Lab

See You in Court

WRAP UP

Chapter Spotlight

Key Terms

Putting It All Together

YOU ARE THE FORENSIC SCIENTIST

When a corpse is discovered, the first measurements made are to determine when the person died. Living humans maintain a body temperature of 98.6°F (37°C). To determine the postmortem interval, a rectal temperature is measured. Although many environmental factors can affect the rate of cooling after death (algor mortis), the simplest estimate is that body temperature declines at a rate of 1.5°C per hour after death. Much later, when decomposition starts, the internal body temperature begins to rise again.

Another indicator of the postmortem interval is rigor mortis—that is, the stiffening of the body that occurs about 4 to 6 hours after death. Rigor mortis begins in the face and jaw, and then extends toward the arms, trunk, and limbs. Within 12 to 18 hours, the entire body is rigid. Rigor mortis disappears 24 to 36 hours after death in the same order as it appeared, leaving the body limp.

1. A corpse discovered at a crime scene is found to have a temperature of 88°F. How long ago did this person die?
2. The corpse in question 1 was found to have rigor mortis of only the face and jaw. Does this observation suggest a postmortem interval that is consistent with the estimate made in problem 1?

Introduction

Forensic toxicology can be divided into two categories: postmortem toxicology and human performance toxicology. In performing a postmortem analysis, a toxicologist investigates the presence of drugs, gases, metals, and other toxic chemicals in human fluids and organs and determines their role, if any, as contributing factors in the individual's death. Human performance toxicology, by contrast, measures the amount of alcohol or drugs in a living person's blood or breath and estimates their role in modifying human performance or behavior.

This chapter begins with a description of how postmortem toxicological investigations are carried out. We then consider the role of alcohol in human performance toxicology, the analytical methods used to measure BAC in the breath and blood, and alcohol-related motor vehicle laws.

Postmortem Toxicology

Today we are surrounded by thousands of chemicals that would be lethal if we inhaled or ingested them. Because we routinely encounter so many potential poisons, a postmortem toxicological investigation always begins with a case history of the deceased. This case history will include the person's age, sex, weight, medical history, any medica-

tion that might have been administered before death, autopsy findings, drugs that were available to the decedent, and the interval between the onset of symptoms and death. A thorough investigation of the area where the body was found may also lead to a tentative identification of which poisons or drugs were involved. Typically, a postmortem toxicology laboratory will perform analyses for poisons as diverse as prescription drugs, drugs of abuse, chemical products (e.g., antifreeze, insecticides, weed killers), and gases (e.g., carbon monoxide).

Collection of Postmortem Specimens

During the autopsy, a pathologist collects postmortem specimens for analysis. Owing to their different physical and chemical properties, drugs and poisons accumulate in different parts of the body. Given that detection of a poison is more likely in an organ in which it accumulates, all body fluids and organs in which chemicals tend to concentrate must be collected during the autopsy. TABLE 12-1 lists the organs and biological fluids that hold special interest for the toxicologist.

During the autopsy, the pathologist places each specimen in a labeled container. On the label, the pathologist records the date and time of the autopsy, the name of the decedent, the type of specimen, and a case identification number; the pathologist also signs his or her name.

on the CRIME SCENE The Fentanyl Patch

Drug abusers are always seeking ways to get "high." Over the past 30 years, fentanyl—a powerful painkiller—has found its way into the underground drug market. Sometimes, however, its use remains hidden from the view of investigators.

Fentanyl was first synthesized in Belgium in the 1950s as an analgesic that was 80 times more effective than morphine. Hospitals and hospice care centers use fentanyl patches for patients with chronic pain from diseases or injuries. Each patch can deliver medicine over a 3-day period. An overdoses of fentanyl, however, can cause the individual to fall into a coma and stop breathing. Fentanyl gained international notoriety in 2002 when the drug was used in gas form to end a hostage crisis in a Moscow theater. Although the crisis ended, 129 hostages and 41 terrorists died from breathing the gas.

During the 1970s fentanyl was used to develop so-called designer drugs in California. During the manufacture of these compounds, the components are altered slightly so that they are not restricted by the Drug Enforcement Administration (DEA). The resulting drugs are extremely powerful—so much so that fentanyl derivatives often are used as substitutes for heroin.

A label required by the Food and Drug Administration (FDA) on fentanyl's packaging warns patients to use caution when taking this drug because anything that can raise body temperature—such as alcohol or exposure to heat from hot tubs, saunas, heating pads or blankets, or heated water beds—can cause an overdose reaction. This information, in combination with the use of forensic analysis and postmortem toxicology, allowed investigators to unravel the mystery surrounding the death of a 32-year-old man on a fishing trip. The man suddenly became dizzy and nauseated, and he collapsed into a coma. Eventually, he suffered cardiac arrest and died. An autopsy found no signs of a traumatic injury or disease, but a toxicological analysis found the presence of fentanyl; no fentanyl patches were found on the man's body.

Investigators began to look for the source of the fentanyl that had caused the man's death. When they examined his employment records, they discovered that the man worked at a funeral home. They also learned that he had a history of drug abuse. An analysis of the funeral home's records revealed that the decedent had recently been in contact with a female client of the mortuary who had died of cancer. Two fentanyl patches were missing from the dead woman's body. Investigators presumed that the drug abuser stole the patches from the woman when her body was brought to the funeral home. He then removed the liquid in the patch with a syringe and injected it to achieve a high.

Sources: Fenton, J. J. *A Case-Oriented Approach*. Boca Raton, FL: CRC Press, 2002.Lilleng, P. K., Mehlum, L. I., Bachs, L., and Morild, I. Deaths after intravenous misuse of transdermal fentanyl. *Journal of Forensic Science*. 2004; 49(b): 1364–1366.

TABLE 12-1

Specimens Collected at Autopsy

Specimen	Quantity
Brain	50 g
Liver	50 g
Kidney	50 g
Heart blood	25 mL
Peripheral blood	10 mL
Vitreous humor (eye)	All available
Bile	All available
Urine	All available
Gastric contents	All available

Source: Forensic Toxicology Guidelines, 2006 Version, Society of Forensic Toxicologists and the American Academy of Forensic Sciences.

Specimens should be collected before the body is embalmed, because the embalming process may destroy poisons still remaining in the corpse. In addition, the alcohols used as a constituent of embalming fluid may give an incorrect indication of the deceased's sobriety.

Occasionally, toxicological analysis is needed for exhumed or burned bodies. Because of the advanced decay of the body, it is sometimes necessary to focus on unusual parts of the corpse when collecting specimens. For instance, many drugs have been successfully identified in bone marrow from skeletal remains even after being buried for years. Additionally, **vitreous humor** (fluid in the eye) is more isolated from the natural mechanisms of decay than other body fluids; as a result, chemicals in the vitreous humor decompose more slowly. Poisons such as drugs, alcohols, and antifreeze have all been detected in the vitreous humor of corpses in advanced stages of decay. Antibiotics, antipsychotic drugs, and drugs of abuse have also been found in the hair of a corpse and have been accepted in court as indicators of drug use.

In cases involving a badly decomposed corpse, the lack of blood and the scarceness of tissue have prompted toxicologists to collect and analyze maggots feeding on the body. Toxicologists have successfully argued in court that when drugs or poisons are detected in the maggots, these substances could have originated only from the victim's tissue on which the maggots were feeding.

Analysis of Toxicology Specimens

The case history may help the toxicologist choose a strategy for an analysis that will yield the maximum amount of information. For instance, in cases in which the poison may have been taken orally, the gastrointestinal (GI) contents should be analyzed first. If the person died before all the poison left the stomach, the GI contents may still hold unabsorbed poisons or drugs.

Urine should be analyzed next. The kidney is the major pathway for the **excretion** of most poisons, and the urine may contain either the intact poison or one of its **metabolites**. The FDA requires that pharmaceutical companies thoroughly study the metabolic breakdown products (metabolites) of all prescription drugs, so toxicology laboratories know which metabolites are produced by which pharmaceutical drugs. Additionally, the Federal Bureau of Investigation (FBI) and the Drug Enforcement Administration (DEA) laboratories have identified the metabolites of all drugs of abuse.

After the poison is absorbed from the GI tract, it is carried to the liver. Because the liver has the ability to detoxify (render harmless) many chemicals, people can tolerate small amounts of some poisons. The liver detoxifies a chemical in one of two ways: (1) by either an **oxidation** reaction or a reduction reaction or (2) by binding the poison to a sugar that is naturally present in the liver. If a drug is suspected to have caused the death, the toxicologist focuses first on the tissues of the organs where the concentration of the drug may be the greatest. For example, after its ingestion, cocaine accumulates in such a way that its highest concentrations are found in the kidney and spleen (**FIGURE 12-1**).

The analysis of organ tissue and biological fluids is made more difficult by the chemical reactions that occur during the decomposition of the corpse. For this reason, the autopsy and analysis of specimens should be completed as soon after death as possible. Although the natural process of decomposition will lower the concentration of a poison that was present at death, many poisons—such as mercury, arsenic, and barbiturates—are very stable and are often detectable in the corpse many years after death.

Forensic toxicology laboratories begin testing with presumptive tests that are not specific for a single poison but rather are designed to detect the presence or absence of one member of a group of drugs. Immunoassay tests are used to test the specimens for opiates, barbiturates, and other classes of prescription drugs.

A positive immunoassay test leads to a confirmation test that is usually performed by gas chromatography–mass spectrometry (GC–MS). The gas chromatograph uses the difference in chemical properties between different components of a mixture to separate the molecules. The molecules take different amounts of time (**retention time**) to emerge from the gas chromatograph, and each compound separated in this way then enters the mass spectrometer. The mass spectrometer breaks each molecule into charged fragments and sepa-

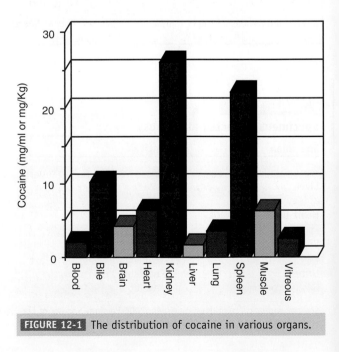

FIGURE 12-1 The distribution of cocaine in various organs.

rates the charged fragments according to their mass. Each compound gives a characteristic fragment spectrum (mass spectrum) that is recorded by the computer attached to the GC–MS. The GC–MS computer also contains a reference library of the mass spectra of more than 100,000 different compounds. To identify which compound is present in the specimen sample, the GC–MS compares the spectra it generated during the analysis with the spectra stored in its database until it finds a match for the specimen's mass spectrum, thereby identifying the unknown compound in the specimen.

If the case study suggests the poison is an inorganic metal, such as mercury, arsenic, or thallium, then the toxicologist tests for inorganic substances (using the techniques described in Chapter 9). Because inductively coupled plasma optical emission spectroscopy is able to routinely determine the presence and concentration of as many as 20 metals simultaneously, it is the perfect tool to screen specimen samples for the presence of metals.

Interpretation of Toxicological Information

Once the specimens are analyzed, the forensic toxicologist must assemble the results, study them, and determine the cause of death. Interpreting the analytical data is often the most difficult task faced by the toxicologist. He or she must determine how the poison entered the body (i.e., the route of administration), how much poison was administered (i.e., the dose), and whether enough poison was ingested to cause death directly or indirectly.

To determine how the poison entered the body, the toxicologist usually compares the concentration of poison found in each of the specimens collected. Generally, the highest concentration of poison is found where it entered the body. If the highest concentration of a poison is found in the GI tract or liver, for example, it indicates that the poison was taken orally. Smoking is a popular way to ingest drugs of abuse (cocaine, phencyclidine, and heroin). A higher concentration of these chemicals in the lungs than in other organs would suggest they were inhaled. For drugs that may have been administered intravenously, the highest concentration will be found near the site of injection. The toxicologist should be aware of any emergency medical treatment administered before the person's death, such as blood and plasma transfusions or diuretics, because these measures may have diluted or flushed out poisons.

Finding a poison in the GI tract does not necessarily establish that it was the cause of death. To verify its lethal effect, analysis of other specimens must confirm that absorption of the poison from the GI tract occurred and that the vascular system (i.e., the blood) then transported the poison to other organs. An exception to this rule is poisoning by extremely strong acids, such as sulfuric acid, which destroy tissue on contact and cause severe bleeding and shock.

In the past, pathologists assumed that the concentration of drugs in postmortem blood was uniform throughout the body, so blood samples were collected from any convenient location. More recently, investigators have recognized that heart blood contains a much higher concentration of

on the CRIME SCENE DNA Testing

Was Napoleon Bonaparte poisoned? In 2001 French scientists revived the theory that Napoleon's death in 1821 was caused by arsenic poisoning. Although the official cause of his death was stomach cancer, testing of several hairs belonging to Napoleon revealed high levels of arsenic. Arsenic in normal hair is at a concentration of about 1 nanogram per milligram of hair. Napoleon's hair contained 38 ng/mg hair. Such a concentration implies sinister actions.

But here is the real question: Are the hairs tested truly those of Napoleon Bonaparte? DNA tests require more hairs than those used to test for arsenic. There is a quest to exhume Napoleon to get enough hairs to test, but the French government has been reluctant to grant permission for the exhumation of his body. Until DNA tests show conclusively that the hairs tested for arsenic were, in fact, those of Napoleon, the jury, as they say, remains out.

residual drugs than does blood collected from other parts of the body. Additionally, the concentration of drugs in heart blood has been found to increase as the interval between death and autopsy lengthens.

The analysis of tissue specimens allows a toxicologist to estimate the "minimal administered dose" of a drug or poison. To make this calculation, the toxicologist analyzes the concentration of the drug or poison in as many different tissue samples as is practical. Next, the concentration of each drug or poison in each separate specimen is multiplied by the total weight or volume of that organ. The amounts of the drug found in all of the organs are then added together to give the total amount of drug or poison in the body. This total amount is an estimate of the "minimal administered dose" of the drug or poison.

In cases of intentional poisoning, toxicological analysis may be able to provide an estimate of when the poison was dispensed. In cases involving arsenic poisoning, for example, the concentration of arsenic in the root and first few millimeters of a person's hair increases within a few hours of ingesting the poison. Given that hair grows at a constant rate of 12.5 mm per month, it will record the concentration of arsenic in the body during the past year. Analysis of hair segments in cases of suspected arsenic poisonings has been used to show that increases in arsenic concentration in specific areas of the hair coincide with times when the suspect was with the victim and had an opportunity to administer the arsenic poison.

Human Performance Testing

Forensic toxicologists also measure the concentration of ethyl alcohol and other drugs in blood and breath to evaluate their role in modifying human performance and behavior. The most commonly used human performance tests are those employed to determine whether an individual is driving a car under the influence of alcohol (**DUI**) or drugs (**DUID**). While most adult Americans drive cars, few of us truly appreciate the complex mental and physical functioning that driving requires. The driver coordinates visual stimuli from the front, side, and rear of the vehicle while simultaneously judging distance, speed, and the intentions of close-by drivers. Alcohol and drugs diminish our ability to perform these complex driving skills.

The amount of alcohol in a person's body is expressed in terms of the **blood alcohol concentration (BAC)**. Many studies have shown a direct relationship between an increased BAC and an increased risk of accidents. In the United States, an estimated 17,419 people died in alcohol-related traffic crashes in 2002, or an average of 1 person every 30 minutes. These deaths accounted for 41% of the 42,815 U.S. traffic fatalities for that year. These deaths, as well as the millions of dollars of damage to property caused by alcohol-related accidents, are evidence of the tragic consequences of alcohol abuse.

Field Sobriety Testing

The field sobriety test is performed to assess the degree of physical impairment of a driver who is behaving erratically, perhaps because of DUI or DUID. When a police officer suspects that a driver has consumed alcohol or drugs before taking the wheel, he or she conducts several presumptive tests before moving on to a confirmation test (i.e., a blood or breath test). The following steps can be elements of a field sobriety test:

1. Interview with officer
2. Preliminary examination (structured questions)
3. Eye examination for nystagmus (irregular tracking) or inability of the eyes to converge
4. Divided attention psychophysical tests
   a. Walk and turn
   b. One-leg stand
   c. Romberg balance
   d. Finger to nose
5. Vital signs exam
   a. Pulse
   b. Blood pressure
   c. Body temperature
6. Dark room exam (pupils of eyes)
7. Muscle rigidity exam
8. Examination for injection sites
9. Suspect's statements
10. Opinions of the evaluator
11. Toxicology exam (breath, blood, and/or urine)

When the officer determines that the suspect has failed the field sobriety test, then a BAC test is administered.

TABLE 12-2

The Effects of Various Blood Alcohol Concentrations on the Human Body

Effect	Blood Alcohol Concentration (BAC)
Sober	0.00–0.06 % BAC
Euphoria	0.03–0.12 % BAC
Excitement	0.09–0.25 % BAC
Confusion	0.18–0.30 % BAC
Stupor	0.27–0.40 % BAC
Coma	0.35–0.50 % BAC
Death	0.45% BAC*

*By respiratory paralysis.

BAC Levels

As discussed in Chapter 11, alcohol is a depressant that acts on the central nervous system and, in particular, on the brain. It affects cerebral function first at a low BAC. At a high BAC, medullar function is affected as well. TABLE 12-2 shows the effects of alcohol on the body at various BACs. Up to a BAC of 0.06%, the individual is considered sober; at 0.18%, the individual is easily confused; by 0.27%, the person is in a stupor; and by 0.35%, he or she is in a coma.

In the United States, laws that regulate blood alcohol levels almost exclusively use the unit "percent weight per volume" (% w/v). A BAC of 0.08% (w/v) is equivalent to 0.08 g of alcohol per 100 mL of blood, or 80 mg of alcohol per 100 mL of blood.

Alcohol and the Law

Laws against "drinking and driving" actually predate the automobile. In fact, the first such sanctions were imposed by the railroad industry in the 1800s. Not until the end of Prohibition (1933–1934) did the relationship between drinking alcohol and driving automobiles become a source of public concern—after all, it was not until then that both alcohol and automobiles became widely available to the average person. In 1939, Indiana became the first state to enact a "drunk driving" law in which the BAC was used as the basis determining a person's sobriety.

With the growing awareness of this issue and the need to justify laws dealing with this social problem came more research into how alcohol affects the skills needed to navigate a car (such as clear vision and good hand–eye coordination). In

1943 researchers discovered that abstainers were more sensitive to alcohol and, when they were given alcohol, they suffered from impairment much sooner than moderate or heavy drinkers. Nevertheless, "experience" was not found to protect drinkers from alcohol's effects: Even heavy drinkers showed deterioration in some abilities, such as vision, at BACs as low as 0.04%.

Although research continued over the next two decades, the watershed moment occurred with the publication of the so-called Grand Rapids Study in 1964. This study had three primary goals: (1) to determine how a driver's BAC affected his or her chances of being in a crash; (2) to identify the percentage of drivers who routinely or sporadically consumed alcohol before taking the wheel; and (3) to characterize the type of person who drives after drinking. To gather data, the researchers interviewed drivers at roadblocks and measured their BACs with Breathalyzers to create a control sample of more than 17,000 subjects. The data for these drivers was then compared with the data for another sample—namely, more than 3300 drivers who had been in crashes.

One of the major accomplishments of the Grand Rapids Study was that it established the scientific foundation for "per se" (meaning "by itself") legislation, which triggers sanctions for *all* drivers who are found to have a certain BAC. Per se laws make the mere fact that a person has a BAC at or above a prescribed level acceptable as conclusive evidence of intoxication, regardless of whether the individual appears to be or acts intoxicated.

The Grand Rapids Study also had another important outcome: It validated the Breathalyzer's effectiveness and accuracy. This device measured the BAC by analyzing the suspect's breath (as discussed in detail later in the chapter). For the first time, it was possible to measure a person's BAC instantaneously without taking a blood sample.

Finally, this study was important for what it *didn't* accomplish because of what the lead author, Robert F. Borkenstein (inventor of the Breathalyzer), described as the "formidable gap between law and science." The probability of a driver being in a crash increases geometrically as his or her BAC rises (FIGURE 12-2). Indeed, the Grand Rapids Study showed that a BAC as low as 0.04% was enough to significantly increase a driver's chance of being involved in a collision. With a BAC of 0.04%, a person's risk increases by 4 times. By contrast, with a BAC of 0.15%, a person has a 25 times greater risk

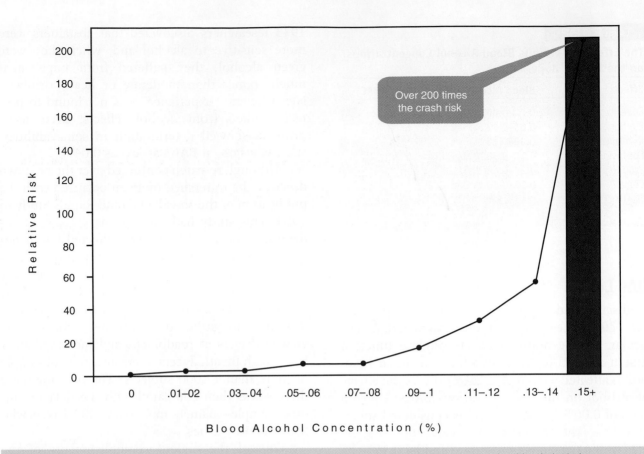

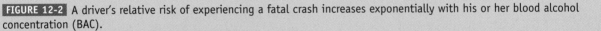

Blood Alcohol Concentration (%)

FIGURE 12-2 A driver's relative risk of experiencing a fatal crash increases exponentially with his or her blood alcohol concentration (BAC).

Adapted from: H. M. Simpson and D. R. Mayhew, *The Hard Core Drinking Driver*. Ottawa, Ontario, Canada: Traffic Injury Research Foundation, 1991, pp. 23–24.

of being in a crash. Yet politics and public opinion intervened when the time came to actually formulate legislation, such that the legal BAC limit for the first per se laws inspired by this study was 0.15%.

Over time, individual state legislatures passed their own per se laws. Eventually, they also strengthened the penalties for DUI by adopting "implied consent" laws, which require motorists suspected of driving while intoxicated to undergo a Breathalyzer test or face sanctions. By 1973 all 50 states had adopted implied consent laws that state that a driver of a motor vehicle on a public highway has the choice of either submitting to a test for alcohol intoxication if requested or being subject to the loss of his or her driver's license for a designated period—usually 6 months to 1 year.

Despite the laws against DUI, drivers continued to consume alcohol and wreak havoc on the roads. The ongoing carnage prompted the spontaneous formation of citizen-based advocacy groups over the next few decades. The best-known and largest of these groups is Mothers Against Drunk Driving (MADD), which was founded in 1980 by

Candace Lightner in Fair Oaks, California, after her daughter was killed in a drunk-driving collision. MADD's emergence came at a fortuitous time, coinciding with the Reagan administration's emphasis on a morality-based conservatism. Its agenda included laws imposing more jail time, vehicle forfeiture, and other stiff measures for DUI offenders. These measures had significant positive effects: DUI-related fatal crashes fell from 60% of all fatal accidents in 1980 to 43% in 1993. Simultaneously, stepped-up police patrols and the implementation of sobriety checkpoints ensured that arrests for DUIs increased by more than 50% to nearly 2 million per year in 1982.

Over time, other issues—such as increased homicide rates in major cities—claimed the spotlight from the activist movement. As public attention to this problem dwindled, law enforcement personnel turned to other, more pressing concerns. By 1992 DUI arrests had dropped to 1.6 million per year. In the same year, the U.S. Department of Transportation recommended that all states adopt a 0.08% BAC as the legal definition of drunk driv-

FIGURE 12-3 Illegal BAC levels vary by state. Minors, under the age of 21, are subject to "zero tolerance" statutes for BAC.

ing. This recommendation became law in 2000, and it gave states until 2003 to enact new legislation enforcing this limit. Since 2003, states that have not adopted the 0.08% per se level have begun to lose federal highway construction funds. The limit is even lower for some professional drivers: The federal government has set a BAC of 0.04% for commercial drivers (e.g., long-haul drivers of "18-wheelers"). **FIGURE 12-3** shows the zero tolerance BAC level for each state.

Many countries, such as Canada, Japan, Italy, Switzerland, and the United Kingdom, have set 0.08% as the BAC level above which it is an offense to drive a motor vehicle. Australia has established 0.05% as the maximum allowable BAC, and Sweden has set its BAC limit at 0.02%.

Alcohol Metabolism

Absorption of Alcohol into the Body

Like any beverage that is consumed orally, ethyl alcohol passes from the mouth, to the esophagus, to the stomach, to the small intestine. As it travels in this way through the GI tract, ethyl alcohol is absorbed into the bloodstream, largely from the stomach (approximately 20%) and the small intestine (80%). This absorption produces the BAC. Once in the bloodstream, the alcohol is carried to the brain; thus the concentration of alcohol in the brain is proportional to the BAC.

A number of factors influence the rate at which ethyl alcohol enters the bloodstream: the rate of gastric emptying, the presence of food in the stomach, the concentration of the ethyl alcohol taken in, the type of alcohol-containing beverage, and the rate at which the alcohol is consumed (i.e., more drinks in a shorter time leads to a higher BAC). In general, the faster the stomach empties, the more rapidly alcohol is absorbed into the bloodstream and the higher the BAC. It is well-known that consuming alcohol with food makes a person feel "less drunk." That's because the presence of any type of food delays the emptying of the stomach, thereby slowing the absorption of ethyl alcohol into the bloodstream. It takes the stomach 1.5 to 2 hours to digest a meal. Other factors that delay gastric emptying, such as physical exercise

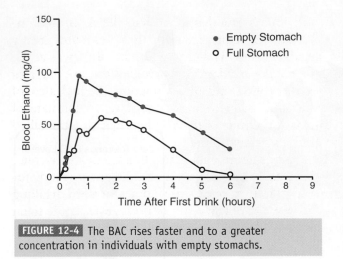

FIGURE 12-4 The BAC rises faster and to a greater concentration in individuals with empty stomachs.

and use of drugs that affect the GI tract (e.g., nicotine, marijuana, ginseng), likewise lead to slower absorption of alcohol into the bloodstream.

How much of an effect does delaying gastric emptying have on the rate at which ethyl alcohol enters the bloodstream? FIGURE 12-4 shows the results of a study in which healthy men drank 0.8 g of alcohol per kilogram of body weight immediately upon awakening; they repeated the experiment immediately after breakfast (Forney & Hughes, 1963). The researchers found that the rate at which the BAC increased was much faster when alcohol is consumed on an empty stomach and that the peak BAC reached was higher on an empty stomach. They also found that the men with full stomachs had lower peak BACs and that the alcohol was cleared from their systems faster.

In another study, two groups of male subjects (weighing 150 lb) were given alcohol (Jones & Jonsson, 1994). The first group consumed 1 oz of 100-proof whiskey at the rate of one drink per hour. The second group consumed 2 oz of 100-proof whiskey at the rate of one drink per hour. Blood samples were taken every 30 minutes for BAC determination. The results of this study revealed that the BAC of the group consuming only one drink per hour never exceeded 0.05%, even after 8 hours of drinking (FIGURE 12-5). By comparison, the group consuming two drinks per hour had a much more rapid increase in BAC, which rendered them intoxicated with a BAC of 0.18% in 5 hours. The rate of elimination for the slower drinkers was sufficient to keep the BAC from rising to a point where the subjects were legally intoxicated.

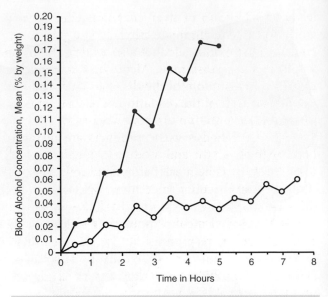

FIGURE 12-5 The rise in BAC depends on the rate at which alcohol is consumed. Members of the first group (open circles) consumed 1 oz of 100-proof whiskey at the rate of one drink per hour. Members of the second group (solid circle) consumed 2 oz of 100-proof whiskey at the rate of one drink per hour.

Although ethyl alcohol gets into the body by absorption, this is a slow process, and distributing the alcohol throughout the body solely via absorption would take a long time. As it travels in the bloodstream, however, ethyl alcohol crosses into cells' internal membranes along with water (from the body's lymph system), which speeds up the distribution of alcohol to the body's tissues. More specifically, the actual concentration of ethyl alcohol in a particular tissue in the body depends on the water content of that tissue: The more water the organ contains, the higher the concentration of alcohol expected. Given that ethyl alcohol mixes freely with water, it would be expected that even within the blood, alcohol distribution would parallel the distribution of water in the blood. Furthermore, ethyl alcohol concentrations in the tissues rapidly reach equilibrium with the BAC.

Elimination of Alcohol by Oxidation–Reduction Reactions

After its absorption by the body's blood and tissues, ethyl alcohol—like any toxin—is eliminated. During the elimination process, it is transported to the liver, where enzymes catalyze the oxidation of the alcohol—first into acetaldehyde, then into acetic acid, and finally into carbon dioxide and water. Chronic alcohol consumption, however, leads to a buildup of liver enzymes, which must work overtime to eliminate ethyl alcohol. The same enzymes that oxidize alcohol also oxidize the male sex hormone testosterone. Alcoholic impotence, a well-known symptom of the disease of alcoholism, is a direct result of the oxidation of testosterone by the high concentration of liver enzymes.

When the body's waste products are metabolized by the liver, the end product is not always less toxic than the chemical being oxidized. For instance, methyl alcohol (methanol, CH_3OH) is oxidized to the more toxic chemical formaldehyde (HCHO). Methyl alcohol poisoning is known to cause blindness, respiratory failure, and death. In this case, it is not the methyl alcohol itself that causes the problems but rather its oxidation product, formaldehyde. Ethyl alcohol is often administered intravenously as an antidote for methyl alcohol poisoning and for ethylene glycol (CH_2OHCH_2OH, antifreeze) poisoning. Because ethyl alcohol competes with these other alcohols to undergo oxidation reactions in the liver, it can be used to slow their conversion to more harmful oxidation products, giving the body a chance to excrete these toxins before damage is done.

More than 90% of the ethyl alcohol that enters the body is completely oxidized to acetic acid, primarily in the liver. The remaining alcohol that is not metabolized is excreted in either the sweat, urine, or breath. The latter route of excretion is the basis of the Breathalyzer test used in law enforcement. Police officers often smell alcohol on the breath of someone who has been drinking recently and then administer a field sobriety test.

Measurement of Blood Alcohol Concentration

Estimating the BAC from Absorption–Elimination Data

The most obvious way to measure intoxication is to record the amount and type of beverage consumed. Unfortunately, most arrests are made after an auto crash, so information about the driver's past drinking is not readily available. Often BAC analyses are performed hours after an accident, and the results of those tests simply report the driver's BAC at the time the blood sample was taken. That is, the body's normal metabolic processes have reduced the concentration of alcohol in the

individual's blood by the time the tests are conducted. Of course, what the prosecutor really needs to know is what the driver's BAC was at the time of the accident. Toxicologists can use the rates of absorption and of elimination of ethyl alcohol from the body to make an accurate estimate of BAC at the time of the accident by a process called back extrapolation.

The rate at which the body removes alcohol from the blood (β) is constant for any one individual. Therefore, once a person has stopped drinking and the alcohol absorption process has come to an end, the BAC at the time of the accident (C_a) can be calculated from the BAC that is measured later (C_t)—provided both the rate at which the body removes alcohol from the blood and the time interval (t) are known. The time interval is the time that has elapsed from the accident to the taking of the blood sample. The following equation will give the alcohol concentration at the time of the accident:

$$C_a = C_t + (\beta)\, t$$

where:

C_a and C_t are measured in mg alcohol/100 mL blood.

β is measured in mg alcohol/100 mLh blood.

t is measured in hours.

A study by Jachau, Sauer, Krause, and Wittig (2004) found that the rate at which the body removes alcohol from the blood varies significantly from one person to the next. It is possible, however, to establish a range for β: The lowest probable rate is 12.5 mg/100 mLh; the highest rate is 25 mg/100 mLh; and the average rate is 18.7 mg/100 mLh.

Example

Consider a traffic accident that occurs at 3:00 A.M. Reliable eyewitnesses place the driver in a bar and state that he consumed his last drink 2 hours before the accident, at 1:00 A.M. After the crash, the driver is taken to a local hospital, where a blood sample is taken at 5:00 A.M. for BAC determination by gas chromatography (GC). The GC analysis reports a BAC of 0.06% (i.e., 60 mg/100 mL). A BAC of 0.06% at 5:00 A.M. is less than the 0.08% limit, but what was the BAC at 3:00 A.M.?

Solution

Because the driver stopped drinking 2 hours before the accident, we can assume that the alcohol absorption process from the intestine to the blood was completed. To determine the concentration at the time of the accident, we would use the previously given equation and first determine the alcohol concentration at 3:00 A.M., using the lowest elimination rate and a time of 2 hours:

$$C_a = C_t + (\beta)\ t$$
$$C_a = 60\ \text{mg/100 mL} + (12.5\ \text{mg/100 mLh})(2\ \text{h})$$
$$C_a = 85\ \text{mg/100 mL} = 0.085\%$$

Using the same equation and the highest rate of elimination, the BAC would be 0.11% (110 mg/100 mL). Using the average rate of elimination, the BAC would be 0.097% (97 mg/100 mL). These results indicate that the driver's BAC was most likely between 0.085% and 0.11% at the time of the accident—above the legal limit.

Estimating the Amount of Alcohol in the Circulatory System

The quantity of alcohol present in a person's blood can be determined either by direct chemical analysis of a blood sample to measure its alcohol content or by measuring the concentration of alcohol in the person's breath. Although the direct chemical analysis of blood gives a more reliable BAC, blood must be drawn under medically accepted conditions by a qualified individual. Usually a blood sample is not taken at the scene of an accident, but rather the person involved is transferred to a hospital, where the blood sample is drawn— something that may not happen until hours after the accident. A breath sample, by comparison, is easy to collect and process by use of a testing instrument at the scene of the accident. This kind of test can be administered by the responding officer close to the time of the accident. Certain assumptions are made when converting the concentration of alcohol in expired breath to BAC. Although both of these tests measure the movement of alcohol through the circulatory system, the blood alcohol test is considered to be more reliable.

To understand how these tests work, we must first understand how blood moves around the body. In the human circulatory system, blood vessels called arteries carry blood away from the heart, and blood vessels called veins carry blood back to heart (FIGURE 12-6). As noted earlier, when a person drinks an alcoholic beverage, the alcohol first passes through the mouth, then moves down the esophagus, and finally enters the stomach. There, approximately 20% of the alcohol is absorbed through the stomach walls into the portal vein of the circulatory system. The remaining alcohol stays in the stomach until the pyloric valve opens and allows the contents of the stomach to pass into the small intestine, where most of the remaining alcohol is absorbed into the bloodstream. Once in the blood, the alcohol is carried to the liver, where oxidation reactions begin to remove it from the bloodstream (the elimination process discussed earlier). From the liver, the blood moves toward the right side (right atrium or auricle) of the heart and is forced into the lower right chamber (right ventricle). At this point the blood contains large amounts of carbon dioxide but very little oxygen. The pumping of the heart sends the blood through the pulmonary artery to the lungs.

In the lungs, the pulmonary artery branches into numerous small capillaries. These tiny blood vessels pass close by myriad (about 250 million!) pear-shaped sacs called alveolus bronchioles (commonly called **alveoli**) that are located at the ends of the bronchial tubes (FIGURE 12-7). The bronchial tubes are connected to the trachea, which in turn

BACK AT THE CRIME LAB

In the lab, collection of samples for "alcohol testing" often extends beyond venous blood samples. Postmortem blood alcohol measurements are tricky because ethanol may have been ingested by the victim prior to death, but ethanol also is a product of tissue composition. Putrefactive blisters can be used as samples for determining BAC. In fact, research on putrefactive blister ethanol concentrations from samples collected from putrefied corpses (3 to 23 days old) shows a correlation between ethanol values of femoral blood/muscle and putrefactive blister fluid. Although ethanol concentrations from putrefactive blisters must be judged with caution, the measured ethanol concentration corresponds to blood levels to a degree sufficient to allow the use of these values in criminal cases.

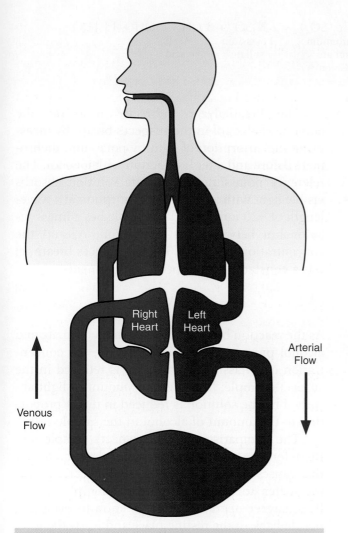

FIGURE 12-6 In the human circulatory system, venous blood flows to the lung, and arterial blood flows away from the lung.

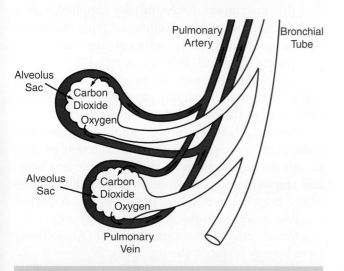

FIGURE 12-7 In the human respiratory system, the trachea connects the nose and mouth to the bronchial tubes. The bronchial tubes are divided into branches that terminate in the alveolus sac.

leads to the mouth and nose. In the alveoli, blood flowing through the capillaries comes in contact with oxygen in the air. A rapid exchange occurs in which air enters the alveoli and carbon dioxide exits the blood. If any volatile chemical is present in the blood (such as alcohol), it will pass into the alveoli during this exchange process as well. Thus, as the person who has consumed alcohol breathes, the carbon dioxide and alcohol are expelled through the nose and mouth, the alveoli are replenished with fresh air, and the blood becomes enriched with oxygen thanks to this new supply.

The distribution of alcohol between the blood and the air in the alveoli can be determined by applying Henry's law: When a volatile chemical (alcohol) is dissolved in a liquid (blood) that is in contact with air (alveoli air), the volatile chemical is always present in a fixed concentration in the air at a given temperature. The normal temperature of expired air is 93°F (34°C). At this temperature, the ratio of alcohol in the blood to alcohol in alveoli air is approximately 2100 to 1. In other words, 1 mL of blood will contain the same amount of alcohol as 2100 mL of expired breath. Thus a toxicologist can use this ratio to determine a person's BAC based on the concentration of alcohol in his or her expired breath.

Breath Tests for Alcohol

In 1954 Dr. Robert Borkenstein of the Indiana State Police invented the Breathalyzer, which was widely used by law enforcement agencies to estimate individuals' BAC until it was replaced in the 1990s by other devices (**FIGURE 12-8**). To begin our discussion of breath test instruments, we describe the principles underlying the Breathalyzer and its limitations.

Breathalyzer

The Breathalyzer device includes two parts: the instrument that calculates the BAC, which includes two vials containing chemicals and a photocell indicator of the test results, and an attached mouthpiece that samples the breath of the suspect. In this test, the suspect blows into the mouthpiece. This breath sample passes to the measurement instrument, where it is bubbled in one vial through a mixture of sulfuric acid, potassium dichromate, silver nitrate, and water. The following chemical reaction takes place:

$$2 \text{ K}_2\text{Cr}_2\text{O}_7 + 3 \text{ CH}_3\text{CH}_2\text{OH} + 8 \text{ H}_2\text{SO}_4 \rightarrow 2 \text{ Cr}_2(\text{SO}_4)_3 + 2 \text{ K}_2\text{SO}_4 + 3 \text{ CH}_3\text{COOH} + 11 \text{ H}_2\text{O}$$

| Potassium dichromate Reddish orange | Ethyl alcohol | Sulfuric acid | Chromium sulfate Green | Potassium sulfate | Acetic acid |

In this reaction, sulfuric acid adsorbs the alcohol from the expired breath and transfers it to a liquid solution. The alcohol then reacts with potassium dichromate to produce chromium sulfate and acetic acid. Silver nitrate acts as a **catalyst**, a substance that makes the reaction go faster without itself being consumed (it is not shown in the chemical equation because it does not actually take part in it). The preceding chemical equation tells us that a set relationship exists between the number of potassium dichromate molecules that react with the expired ethyl alcohol molecules: Two molecules of potassium dichromate always react with three molecules of ethyl alcohol. If we determine how much potassium dichromate has been converted to chromium sulfate, then we can indirectly compute how much ethyl alcohol must have been present in the reaction chamber.

During this reaction, when the reddish-orange dichromate ion reacts with ethyl alcohol, it produces green chromium ion. The degree to which this color change takes place is directly related to the concentration of alcohol in the suspect's breath. To determine the precise amount of alcohol present in that expelled air, the reacted mixture is compared with a vial of unreacted potassium dichromate mixture in the photocell system.

The Breathalyzer indirectly measures the amount of alcohol in the suspect's breath by measuring the absorption of light by potassium dichromate before and after its reaction with alcohol. The reference potassium dichromate solution absorbs visible light with a maximum absorption at a wavelength of 420 nm. Beer's law establishes a linear relationship between this absorption, A, and the concentration of alcohol in the suspect's breath, c; k is a constant for the specific instrument.

$$A = kc$$

As the reaction between potassium dichromate and alcohol proceeds in the Breathalyzer, the concentration of potassium dichromate is reduced in the suspect's sample; likewise, the amount of light absorbed by the solution is reduced in direct proportion to the amount of alcohol in the sample cell.

The comparison of the suspect's sample and the reference sample results in an electric current that causes the needle in the Breathalyzer's indicator meter to move from its resting place. The Breathalyzer operator rotates a knob to bring the needle back to the resting place and reads the level of alcohol from the knob.

Because the Breathalyzer consumes chemicals, the officer administering the test must make sure that this instrument is continually supplied with a sufficient amount of fresh chemical reagents. If it is used infrequently, police officers may fail to replenish the chemicals before testing a DUI suspect. In such a case, the suspect's defense lawyers might argue that the Breathalyzer gave false readings due to the use of the outdated chemicals.

Recent advances in optical instruments and computers have spurred the development of new breath testing instruments that consume no chemical reagents and, therefore, avoid the limitations of the Breathalyzer. Today, breath testers based on the absorption of infrared light (Intoxilyzer) or the reaction of ethyl alcohol in a fuel cell (Alcosensor) have largely replaced the Breathalyzer.

Intoxilyzer

Various models of the Intoxilyzer are utilized by more than 35 states, and nearly 30 states use it as

FIGURE 12-8 The Intoxilyzer has replaced the Breathalyzer as the standard breath test in the United States.

their sole breath-testing device. The Intoxilyzer measures the infrared radiation (IR) absorption in a specific wavelength to confirm the presence of certain organic chemicals, such as alcohols.

IR is not as energetic as visible light, but it can change the vibrational or rotational motion of a molecule. The amount of IR absorbed by a compound corresponds to the energy it takes to stretch or bend bonds in a specific organic molecule. Thus each organic molecule has a unique IR spectrum (i.e., an IR fingerprint).

As with the Breathalyzer, the suspect blows into the Intoxilyzer's mouthpiece to give a breath sample for analysis. Unlike with the Breathalyzer, however, the tester is assured of analyzing the subject's "deep lung" air. The tube leading to the testing chamber has a spring valve that can be opened only by a breath sample with a high pressure. The person being tested must blow hard to open the valve, ensuring that deep lung air is sampled. As a consequence, the Intoxilyzer should not give false-positive results if alcohol is present only in the subject's mouth (e.g., if he or she recently used alcohol-containing mouthwash or ate alcohol-containing confectionery products).

The Intoxilyzer's measurement unit contains a quartz lamp that generates IR energy. This energy travels through a sample chamber containing the subject's breath (FIGURE 12-9). Next, a lens focuses the energy onto a wheel containing up to five IR filters. Each of these filters allows only a narrow wavelength band of IR to reach the detector. These wavelengths are chosen so that the Intoxilyzer will detect only ethyl alcohol. In particular, acetone, acetaldehyde, and toluene absorb light in the IR region and can give a false-positive result for ethyl alcohol. Acetone is a common solvent used in fingernail polish remover, toluene is a common solvent in oil-based paints, and acetaldehyde is found in the breath of cigarette smokers.

Once the IR energy passes through the narrowband filters, it is focused on a highly sensitive photo detector that converts this energy into electrical pulses. A microprocessor then interprets the pulses and calculates the BAC, which is reported on a digital display.

Alcosensor

The Alcosensor is an alcohol **fuel cell** that consists of a porous, chemically inert layer that is coated on both sides with a thin platinum layer (FIGURE 12-10). This platinum layer is impregnated with an acidic salt solution and has electrical wires attached to the platinum. The entire assembly is mounted in a plastic case, to which a plastic tube is attached. For breath testing purposes, the suspect blows into the plastic tube and the breath sample then passes over the platinum layer.

On the upper surface of the Alcosensor's fuel cell, any ethyl alcohol present in the breath sample is converted to acetic acid. This oxidation reaction produces two free electrons per molecule of

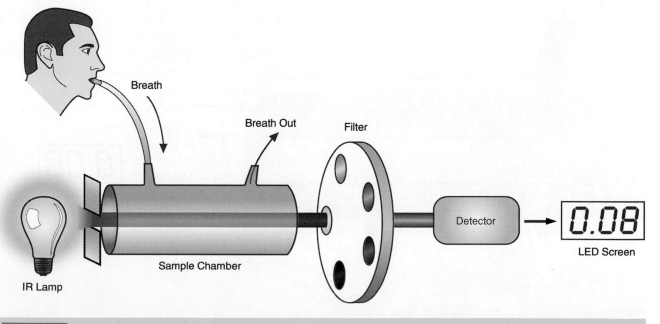

FIGURE 12-9 The Intoxilyzer uses infrared radiation to measure the amount of alcohol present in one's breath.

FIGURE 12-10 The BACTrack™ is a handheld device for measuring BAC.

duction process, in turn, consumes one electron per H$^+$ ion in the process. Thus the upper surface of the fuel cell has an excess of electrons, and the lower surface has a corresponding deficiency of electrons.

If the two platinum surfaces are connected electrically, the excess electrons in the upper layer flow as a current through the external circuit to neutralize the charge. This current is a direct indication of the amount of alcohol consumed by the fuel cell. The more alcohol oxidized, the greater the electrical current. A microprocessor measures the electrical current and converts the measured electrical current into a BAC, which is displayed as a digital readout on the Intoxilyzer.

Blood Tests for Alcohol

alcohol (**FIGURE 12-11**). In addition, the hydrogen ions (H$^+$) produced in this reaction migrate to the lower surface of the fuel cell, where they combine with atmospheric oxygen to form water. This re-

In Chapter 10, we learned that GC is a useful technique for separating organic compounds; the GC separates the gaseous components of a mixture by partitioning them between a mobile phase and a stationary phase. In Chapter 11, we explored this technology's use in conjunction with mass spectrometry (MS) to identify drugs of abuse. Now we will see that GC is the toxicologist's method of choice for identifying a suspect's BAC. When GC is used, the BAC level determined is easily admitted

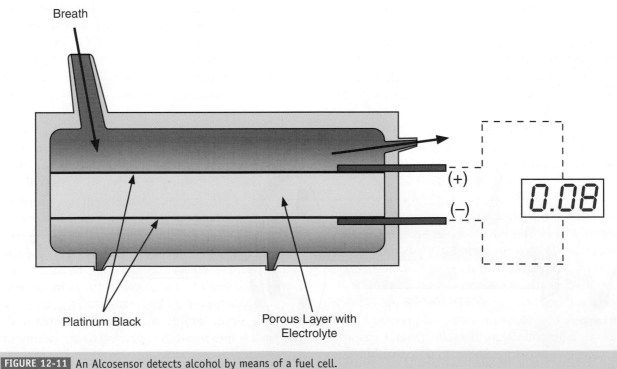

Breath

(+)

(−)

0.08

Platinum Black

Porous Layer with Electrolyte

FIGURE 12-11 An Alcosensor detects alcohol by means of a fuel cell.

as evidence and very rarely challenged by defense lawyers. The major drawback of this method is that it requires blood to be drawn from the suspect. GC analysis may also be used to determine the BAC of blood and tissues taken from a corpse.

Collection and Preservation of Blood

When collecting blood, the suspect's skin is first disinfected with a *nonalcoholic disinfectant*, such as Betadine. This practice prevents the suspect from claiming that alcohol in the disinfectant caused the BAC calculated by the lab to be abnormally high.

Next, the blood is removed by use of a sterile needle. The blood sample collected in this way is placed in an airtight container. Blood containers typically include an **anticoagulant** (e.g., EDTA, potassium oxalate), which is intended to prevent clotting of the blood, as well as a preservative (e.g., sodium fluoride), which inhibits microorganisms from growing and contaminating the sample. Blood containers must be stored in the refrigerator until they are delivered to the testing laboratory. Failure to refrigerate the sample or to add the preservative may cause the BAC values reported by GC to be abnormally low; failure to properly preserve the blood sample will never give a BAC that is falsely high.

After a person dies, bacteria can grow in the cadaver; these microorganisms produce ethyl alcohol, which may lead to a falsely high BAC reading. For this reason, pathologists prefer to collect blood samples from multiple sites on the body. This practice minimizes the chance of a false-positive result owing to a localized bacterial colony. For example, separate samples collected from the heart, the femoral (leg) vein, and the cubital (arm) vein should produce similar BAC results if the alcohol content was the result of alcohol consumption by the person before death.

Alternately, pathologists may collect samples other than blood from the cadaver and analyze them for alcohol content. For example, urine and vitreous humor do not support the growth of microorganisms that produce ethyl alcohol after the person's death. As a consequence, they may be sampled and analyzed via GC to determine whether a person consumed alcohol prior to his or her death. The collection of blood from a suspect must always be performed by a qualified technician under medically accepted conditions.

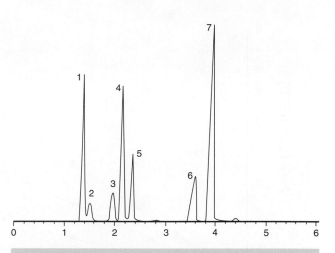

FIGURE 12-12 Gas chromatography (GC) provides a definitive measure of BAC. Here, the peaks correspond to (1) acetaldehyde, (2) methanol, (3) ethanol, (4) acetone, (5) isopropanol, (6) *n*-propanol, and (7) methyl ethyl ketone (MEK).

Source: © Agilent Technologies, Inc. 2010. Reproduced with permission, courtesy of Agilent Technologies, Inc.

Gas Chromatography Analysis of BAC

Gas chromatography (GC) is a dynamic method of separation and detection of volatile organic compounds. When a standard solution, which contains a known concentration of ethyl alcohol and other possible interfering chemicals, is prepared and injected into the GC, each component gives a Gaussian-shaped peak (**FIGURE 12-12**). During the GC separation, ethanol has a retention time of 1.9 minutes. This characteristic allows it to be separated from acetaldehyde and acetone (two potential sources of interference in breath alcohol tests), which reach the detector at a different time than ethanol. To ensure consistency of results, all of the GC's experimental conditions—including temperatures, flow rate, column composition, and length—must be kept constant for all following analyses.

When GC is used to test for BAC in a person's breath, standards containing different concentrations of each chemical that might be found in a person's breath are sequentially injected into the instrument. The time it takes for each compound to reach the detector will be unique for each compound (e.g., 1.9 minutes for ethyl alcohol). A calibration curve is constructed that relates the area under each peak in the curve to the concentration of each component in the standard (**FIGURE 12-13**). When a blood sample is analyzed, the area under the peak is measured. By referring to the calibration curve, the forensic examiner can calculate the concentration of alcohol in the sample (**FIGURE 12-14**).

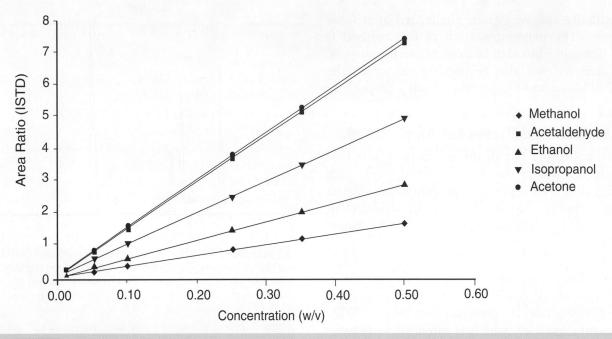

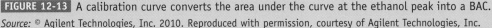

FIGURE 12-13 A calibration curve converts the area under the curve at the ethanol peak into a BAC.

Source: © Agilent Technologies, Inc. 2010. Reproduced with permission, courtesy of Agilent Technologies, Inc.

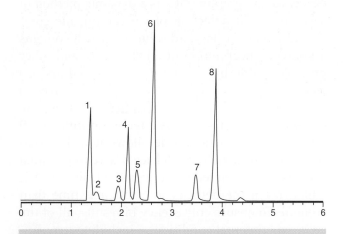

FIGURE 12-14 The concentration of alcohol in a blood sample is determined by GC. The calibration curve in Figure 12-13 is used to convert the peak area of ethanol (3) into concentration. Peak identification: (1) acetaldehyde, (2) methanol, (3) ethanol, (4) acetone, (5) isopropanol, (6) t-butanol, (7) *n*-propanol, and (8) methyl ethyl ketone (MEK).

Source: © Agilent Technologies, Inc. 2010. Reproduced with permission, courtesy of Agilent Technologies, Inc.

Noninvasive Alcohol Testing

To this point, this chapter has focused on the use of alcohol testing for drunk-driving infractions. In addition to this application, other uses for such

testing have emerged in recent years. For example, people who are on probation or under house arrest must comply with limitations on their alcohol use. Certain jobs that involve the use of heavy machinery—for example, bus driving—require alcohol testing on a regular basis. Finally, emergency medical technicians need to know if an unconscious patient has consumed alcohol and, if so, how much, before administering treatment.

In these circumstances, standard techniques that measure alcohol levels in the blood or the breath have two limitations. First, they require the handling of body fluids or gases, which poses a potential biohazard to the person administering the test. Second, the subject must willingly comply with testing. Furthermore, blood testing takes hours for results to be returned, and breath testing requires that the subject be conscious enough to blow into an instrument.

A touch-based alcohol monitoring instrument has recently been developed that overcomes these limitations. This new technology uses near-infrared reflectance spectroscopy to measure alcohol on the subject's skin. The device measures alcohol levels in 90 seconds by shining a harmless beam of near-IR light on the skin of the forearm. Any alcohol in the

SEE YOU IN COURT

The Daubert criteria apply to all forensic evidence presented in court—not just data gathered through laboratory tests, but also results from tests such as Breathalyzer tests conducted in the field. Breathalyzer tests for DUI cases have been the subject of an ongoing scientific dialog concerning their scientific usefulness (issues of temperature affecting accuracy) and validity in view of the Daubert criteria and the legislative approaches that seek to expedite the legal process.

For example, the case of *United States v. Kevin Vincent Brannon* turned on whether a reading from an uncompleted Breathalyzer test was admissible evidence. The defense stipulated to the government witness's "qualifications as an expert with the one exception of the contested evidence, the partial Breathalyzer results." The defense objected to the admission of the Breathalyzer evidence based on the argument that the witness testimony did not show that an uncompleted test had reached "the level of general acceptance within the scientific community," that the witness did not cite any studies to the effect that a partial test was reliable, and that the witness did not offer any proof that the machine was accurate when the test was not completed. The defense also argued on appeal that the witness did not satisfy the criteria laid out in the *Daubert v. Merrell Dow Pharmaceuticals, Inc.* (1993) ruling.

tissue absorbs this radiation, such that the amount of IR light reflected back to the detector in the instrument indicates how much alcohol is present.

This small instrument will be sold first to law enforcement agencies and later to employers for workplace safety. Eventually, it might be installed in cars as a starting "interlock" device mandated by court order as punishment for a DUI conviction. In this application, a person convicted of an alcohol-related charge would have to place his or her forearm against the instrument as proof that he or she is sober before the car would start.

WRAP UP

1. A temperature of 88°F when converted to the Celsius scale is 31°C. The body temperature dropped 6°C (37°C – 31°C). Given that body temperature drops 1.5°C per hour after death, this person has been dead approximately 4 hours.
2. The rigor observed in the body is consistent with a 4-hour period since death.

Chapter Spotlight

- Postmortem toxicology analysis focuses on four areas: the case history of the decedent, a survey of the area in which the body was found, collection of bodily fluids and organs (e.g., gastric contents, blood, liver, kidney), and a reasonably rapid analysis of samples.

- Samples for analysis should be taken quickly, before degradation occurs and before the body is embalmed.

- Unusual but useful samples may be obtained from hair, bone marrow, and vitreous humor in the eye.

- Interpretation of the data from a postmortem toxicology analysis seeks to answer two questions:

  - How did the substance enter the body?

  - Was the substance directly or indirectly responsible for the person's death?

- Field sobriety tests include a structured process that may include an officer interview, psychophysiological tests (eye exam and coordination tests), a physical examination of the suspect (for injection sites if DUID is suspected), any suspect statement, and a toxicology exam.

- Increased BAC is directly linked to a significantly higher risk of automobile crashes. The increase

in risk is exponential, meaning that the chances of getting in an accident more than double for every doubling of the individual's BAC.

- Alcohol is absorbed at each step in the alcohol's movement through the body: from the mouth to the esophagus to the stomach to the intestine.

- The amount of alcohol in any component of the body depends on the water content of that component.

- The amount of alcohol in the brain (indicative of the human response) is directly proportional to the BAC.

- Calculating a person's BAC at the time of an accident requires having a blood sample and knowing the exact time at which the blood sample was taken relative to the time of the accident.

- Henry's law can be used to determine the relative amount of a gas that will stay in a liquid phase (blood) versus a vapor phase (gas within the lungs) at a known temperature. This information is used to determine BAC through breath analysis.

- Types of breath alcohol tests include chemical tests (common), IR-based tests (most common), and alcohol fuel cells.

Key Terms

Alveoli Tiny air sacs within the lungs where the exchange of oxygen and carbon dioxide takes place.

Anticoagulant A chemical that inhibits the coagulation of blood.

Blood alcohol concentration (BAC) The amount of alcohol in a person's blood.

Catalyst A substance that increases the rate of a chemical reaction without being consumed during that reaction.

DUI Driving under the influence of alcohol.

DUID Driving under the influence of drugs.

Excretion The act or process of discharging waste matter from the blood, tissues, or organs.

Fuel cell A device that directly converts chemical energy into electrical energy through chemical reactions.

Metabolites The metabolic breakdown products of drugs.

Oxidation A chemical reaction in which oxygen is combined with other substances.

Retention time The amount of time it takes a compound to be separated via a gas chromatograph.

Vitreous humor Liquid inside the eye.

Putting It All Together

Fill in the Blank

1. Because there are so many possible poisons, the postmortem toxicological examination begins with a(n) _____ _____ of the deceased.

2. The ____, _____, and _____ are organs that should be collected for postmortem toxicological analysis.

3. If it is suspected that a poison was administered orally, the _____ should be analyzed first.

4. A positive immunoassay test leads to a confirmation test that is usually performed by _____.

5. If the poison is a metal, a(n) _____ test is used to identify which one.

6. Postmortem blood from the _____ has been found to contain higher concentrations of drugs than other parts of the body.

7. Arsenic concentrates in the _____ of the person who was poisoned.

8. The most widely abused drug in Western countries is _____.

9. A BAC of 0.08% is equivalent to _____ grams of alcohol per 100 mL of blood.

10. Is the field sobriety test a presumptive or confirmatory test?

11. Legislation that triggers sanctions at a certain BAC level is called "_____" legislation.

12. The laws that require motorists to undergo an alcohol breath test or face the loss of their driver's license are known as _____ _____ laws.

13. Alcohol acts as a(n) _____ on the central nervous system.

14. Most of the alcohol a person drinks is absorbed into the bloodstream through the _____ _____.

15. The higher the proportion of water in an organ, the _____ (higher/lower) the concentration of alcohol.

16. Alcohol in the body is oxidized in this organ: _____.

17. More than 90% of the ethyl alcohol that enters the body is completely oxidized to _____.

18. Alcohol that is not metabolized is excreted in the sweat, the urine, or the _____.

19. If alcohol is taken on a full stomach, the rate at which the BAC increases is _____ (slower/higher) than when alcohol is taken on an empty stomach.

20. Blood vessels called _____ carry blood away from the heart.

21. Blood vessels called _____ carry blood back to the heart.

22. The distribution of alcohol between the blood and the air in the lungs can be determined by applying _____ law.

23. One milliliter of blood will contain the same amount of alcohol as _____ milliliters of expired breath.

24. In the Breathalyzer, ethyl alcohol reacts with _____; this reaction is accompanied by a color change.

25. In the Breathalyzer, _____ ethyl alcohol molecules react with _____ potassium dichromate molecules.

26. The silver nitrate that is present in the Breathalyzer solution is not written in the

balanced chemical equation because it is a(n) _____ in this reaction.

27. The more alcohol present in the breath sample, the _____ (greater/smaller) the absorbance observed by the Intoxilyzer.

28. The Intoxilyzer contains _____ filters, each of which allows only a narrow-wavelength band of IR to reach the detector.

29. Common interferences in breath alcohol tests are _____, from fingernail polish, and _____, a common solvent in oil-based paints.

30. The reaction that takes place in the fuel cell of the Alcosensor converts ethyl alcohol to _____.

31. The Alcosensor fuel cell measures the flow of _____ as demonstrated via an electric current.

32. The most reliable method of determining BAC is by _____ _____.

True or False

1. In badly decomposed bodies, poisons are more likely to be found in the vitreous humor.

2. Finding a poison in the GI tract definitely indicates that this poison was the cause of death.

3. Gastric emptying is the most important determinant of the rate at which ethyl alcohol is absorbed into the bloodstream.

4. The rate at which the body removes alcohol from the blood is constant for any one individual.

5. The rate at which the body removes alcohol from the blood varies significantly from one person to the next.

6. IR absorbed by a molecule causes its bonds to break.

7. The Intoxilyzer uses fuel cells to determine the concentration of alcohol in a person's breath.

8. The Alcosensor uses IR to measure alcohol in a person's breath.

9. Failure to properly preserve the blood sample will always give a BAC that is falsely high.

10. When sampling blood from a corpse, it is best practice to collect blood samples from multiple sites on the body.

Review Problems

1. A 150-lb man goes to a bar and consumes 2 oz of 100-proof whiskey every hour. His 150-lb friend consumes 1 oz of the same whiskey per hour. Refer to Figures 12-4 and 12-5 and estimate the BAC of each man after 3 hours. Would either be "over the limit" for operating a vehicle?

2. Consider a traffic accident that occurs at 2:00 A.M. Reliable eyewitnesses place the driver in a bar and state that he consumed his last drink 2 hours before the accident at 12:00 A.M. After the accident, the driver is taken to a local hospital, where a blood sample is drawn at 4:00 A.M. for BAC determination. The GC analysis reports a BAC of 0.07% (i.e., 70 mg/100 mL). What was the driver's BAC at the time of the accident? Was the driver's BAC above the legal limit?

3. Consider a traffic accident that occurs at 2:00 A.M. Reliable eyewitnesses place the driver in a bar and state that he consumed his last drink 2 hours before the accident, at 12:00 A.M. After the accident, the driver is taken to a local hospital, where a blood sample is drawn at 5:00 A.M. for BAC determination. The GC analysis reports a BAC of 0.06% (i.e., 60 mg/100 mL). What was the driver's BAC at the time of the accident? Was the driver's BAC above the legal limit?

4. A driver who is involved in an auto accident has her breath tested for alcohol by an officer using the Breathalyzer. The Breathalyzer reports a BAC result of 0.09%. The driver is charged with driving under the influence and given a court date. During her court appearance, she presents witnesses who testify that

she is employed as a manicurist, worked late with them the night of the accident, and didn't have any alcohol to drink before the accident. Explain why the Breathalyzer gave such a result.

5. A person who is involved in an accident is given a breath test by an officer using an Intoxilyzer breath tester. The Intoxilyzer reports a BAC of 0.11%. At the driver's trial, he contests the results of the breath test. He testifies that he attached the dental bridge in his mouth with Super Glue just before getting in the car. The defendant claims that the Super Glue released solvents that interfered with the Intoxilyzer test. Use the web to investigate the composition of Super Glue and determine whether his claim is supported by science. (Note: This really happened; the person came to the free clinic at our law school!)

6. A person who is involved in an accident is asked to take a BAC test by breathing into an Alcosensor. Just before releasing the breath, the suspect burps and then blows into the instrument. The Alcosensor reports an alcohol level of 0.20%. Is this an accurate reading? Does this breath analysis reflect the BAC of the person? Explain. Should the test be repeated?

Further Reading

Caplan, Y. H., and Zettl, J. R. The determination of alcohol in blood and breath. In R. Saferstein (Ed.), *Forensic Science Handbook, vol. 1* (2nd ed), pp. 635–695. Upper Saddle River, NJ: Prentice-Hall, 2002.

Fenton, J. J. *Toxicology: A Case-Oriented Approach.* Boca Raton, FL: CRC Press, 2002.

Ferner, R. E. *Forensic Pharmacology: Medicines, Mayhem and Malpractice.* Oxford, UK: Oxford University Press, 1996.

Forney, R. B., and Hughes, F. W. Alcohol accumulations in humans after prolonged drinking. *Clinical Pharmacology and Therapeutics.* 1963; 4: 619.

Jachau, K., Sauer, S., Krause, D., and Wittig, H. Comparative regression analysis of concurrent elimination-phase blood and breath alcohol concentration measurements to determine hourly degradation rates. *Forensic Science International.* 2004; 143(2-3): 115–120.

Jones, A. W., and Jonsson, K. A. Food-induced lowering of blood-ethanol profiles and increased rate of elimination immediately after a meal. *Journal of Forensic Science.* 1994; 39: 1084–1093.

Klassen, C. D. *Casarett and Doull's Toxicology: The Basic Science of Poisons,* 7th ed. New York: McGraw-Hill, 2007.

Manahan, S. E. *Toxicological Chemistry and Biochemistry.* Boca Raton, FL: Lewis, 2002.

http://criminaljustice.jblearning.com/criminalistics

Answers to Review Problems

Interactive Questions

Key Term Explorer

Web Links

http://criminaljustice.jblearning.com/criminalistics

Biological Fluids: Blood, Semen, Saliva, and an Introduction to DNA

OBJECTIVES

In this chapter you should gain an understanding of:

- Tests for the presence of blood
- Serological blood typing
- Tests for the presence of saliva
- Tests for the presence of semen
- The principles of paternity
- DNA, genes, and chromosomes
- Mitochondrial DNA

FEATURES

On the Crime Scene

Back at the Crime Lab

See You in Court

WRAP UP

Chapter Spotlight

Key Terms

Putting It All Together

YOU ARE THE FORENSIC SCIENTIST

As we have learned, physical evidence recovered from the crime scene is most often used to identify a suspect. Trace evidence, such as hairs and fibers, and fingerprints are used to link an individual with the victim or the crime scene. Bloodstain patterns, by contrast, are a type of physical evidence that may indicate the sequence of events that took place at the crime scene.

Large amounts of blood are often at the site of a violent crime. The adult body contains 5 to 6 quarts of blood that are held within a pressurized vascular system. In many indoor crime scenes, bloodstains are found on the walls, floor, and ceiling. It is essential that crime scene investigators record all bloodstain patterns by making sketches of them and taking photographs.

Passive bloodstains are formed by the influence of gravity. Active bloodstains are formed by blood that has traveled by a force other than gravity. Transfer bloodstains result from contact with wet blood. (Chapter 2 describes these patterns in detail.)

Bloodstain pattern analysis is used to establish what occurred during a violent crime and to elucidate the probable sequence of events. A compete analysis will include the location of each bloodstain and the amount of blood found there. The investigation of bloodstain evidence requires considerable experience because the texture and geometry of the surface will inevitably influence the appearance of the bloodstain. Bloodstain evidence is used to verify or refute a suspect's recollection of events.

In violent crimes, some of the blood may have come from the assailant as well as from the victim. In these cases, DNA profiles of the bloodstains can be used to identify who was present at the crime scene.

1. The scene of a violent death in a dormitory room reveals bloodstain pattern evidence in many locations. Investigators find a large pool of blood on the floor near the dead victim, drops of blood on the floor leading out of the room, bloodstains with satellites on the inner wall, and even bloodstains on the ceiling. What type of bloodstain is each, and how were these four stains produced?

2. The suspect apprehended in the dormroom death claims that no physical contact occurred between himself and the victim. Investigation of his clothing reveals no bloodstains on his clothes. Does the bloodstain evidence in question 1 suggest a violent encounter between the victim and his murderer? Does it suggest that this suspect is the murderer?

Introduction

Forensic examiners search the crime scene for evidence that will establish a link between the victim and the perpetrator of the crime. Body fluids such as blood, saliva, and semen often are found at the crime scene. They can provide a direct link to the perpetrator. The popular *CSI: Crime Scene Investigation* television programs depict crime scenes in which forensic examiners carefully search for biological evidence. In various episodes of *CSI*, the crime lab "techs" often are able to place a perpetrator at the scene of the crime by analysis of DNA in a biological sample. (We will learn about DNA typing in Chapter 14.) For now, however, a student of criminology might ask this question: Which techniques did crime labs use prior to

the development of DNA typing? The answer: forensic serology.

You might wonder how important forensic serology is today given that DNA analysis is available in most jurisdictions in the United States. Serology was the preferred test for body fluids from the 1950s until the late 1980s, when the technology to analyze DNA became widely available. The emergence of DNA technology resulted in a steady growth in the number of forensic laboratories that were trained to perform the typing assays. However, although DNA typing remains the gold standard among body fluid tests, most forensic laboratories still routinely use many of the basic serological testing procedures. Some crime labs simply do not have a DNA typing facility—DNA typing requires expensive instrumentation and well-trained tech-

CRIME SCENE National DNA Databases for Sex Offenders

Sex offenders typically leave behind a trail of biological fluids containing DNA from which the investigator can make a sure claim of the perpetrator's guilt. Unfortunately, sex offenders often are repeat criminals. So, how can law enforcement personnel link these criminals with all of their crimes?

In July 2006 the federal Adam Walsh Child Protection and Safety Act was signed into law, making it the most comprehensive legislation targeting sex offenders in the United States. This law provides for the creation of a DNA database holding profiles of convicted molesters that will be made available to law enforcement officials. While all 50 states have laws requiring the registry of sex offenders, the new database will help make information uniform and available throughout the country. The law requires all convicted molesters to provide DNA samples for the registry; it is believed that the perpetrators of other unsolved past crimes will be more readily found.

The new legislation outlines protocols for the proper collection of samples from both perpetrators and victims. It also specifies methods for doing DNA comparisons with the database to help ensure proper matches in local jurisdictions. Furthermore, the law requires that convicted pedophiles wear global positioning system devices to maintain information about their locations. It stipulates a mandatory 30-year sentence for a convicted child rapist and makes failure to update authorities of the convicted molester's address a felony. It is estimated that nearly 150,000 convicted sex offenders may not be registered at present, or more than one-fourth of the nearly 500,000 known U.S. offenders. The new law also gives federal marshals jurisdiction to seek such offenders nationwide.

Individual states have gathered other criminal DNA samples from which other crimes have been solved. Virginia, for example, was the first to require all individuals arrested of a violent crime to provide a sample, taken with a buccal swab, for comparison to the Virginia state DNA database; it also was the first state to maintain a database of DNA samples from all convicted felons.

nicians. Some jurisdictions have too small a caseload to justify the cost of a DNA facility, while others simply cannot justify its cost when given their needs in other areas of law enforcement.

In this chapter, we review the more commonly used forensic serology tests involving body fluids. We also introduce the underlying biological concepts that are the foundation for not only serology but also DNA typing.

Blood

Blood is a complex fluid that is found in a human's cardiovascular system. It is a mixture of cells, proteins, enzymes, and inorganic salts. The liquid portion of blood is called **plasma**. Plasma accounts for 55% of the total blood volume; the other 45% consists of cellular material such as blood cells and platelets.

Blood plasma is a pale yellow liquid that is 90% water and 10% dissolved materials, such as proteins, waste products (carbon dioxide), and nutrients (amino acids and glucose). Other substances, such as drugs and alcohol, can be found in blood plasma, and plasma often is tested by a crime lab for its presence in a criminal investigation. **Blood serum** is blood plasma with its protein content removed. Blood clots when the protein fibrinogen reacts to form an entangled polymer. As it forms the clot, it forces a clear liquid (blood serum) out of the plasma.

The cellular components of blood are divided into three main types:

- Erythrocytes: Red blood cells. Erythrocytes are the most common type of blood cell and account for more than 44% of the total blood volume. They contain hemoglobin, the iron-containing protein that carries oxygen (and carbon dioxide). Erythrocytes are the only mammalian cells that do not have a nucleus.

- Leucocytes: White blood cells. Leucocytes make up a small fraction (less than 1%) of the total blood volume. They protect the body from infection.

- Thrombocytes: Platelets. Platelets are non-nucleated cell fragments that form in the bone marrow. They are involved in the blood clotting mechanism.

Tests for the Presence of Blood

At the scene of a violent crime, investigators often discover large quantities of blood on the victim, the walls, the floors, and the furniture. Police at the crime scene want to answer three questions:

- Is this blood?
- Is it from a human?
- How closely does it match the blood of the victim or suspect?

Sometimes the blood sample is small, perhaps because the perpetrator attempted to clean up the blood at the crime scene. In this case the forensic examiner needs to detect the presence of blood that might otherwise be overlooked, either because it is a very small drop or because the background it has landed on hides the blood from the naked eye.

Presumptive Tests for Blood

To find blood, forensic examiners use a variety of chemical tests known as **presumptive tests**. They are considered "presumptive" because they detect not only blood but also substances other than blood that can give a false-positive result. Most presumptive tests are not carried out directly on the bloodstained object. Instead, the suspected blood on the object is transferred to another absorbent material, such as filter paper or a cotton swab, and the test is completed on the adsorbent material. Luminol spray is the one presumptive test that is performed directly on the bloodstained object.

The presumptive tests for blood rely on hemoglobin's ability to catalyze the oxidation of certain reagents. In most cases the oxidizing agent is a 3% solution of hydrogen peroxide (H_2O_2). Five different color-change tests are routinely carried in a forensic examiner's field kit (FIGURE 13-1):

- In the Kastle-Meyer test, a drop of a solution of phenolphthalein, which is colorless, is applied to the stain. Next, a drop of the hydrogen peroxide solution is added. The formation of a bright pink color indicates the presence of blood.
- Another reagent that is commonly used is a solution of leucomalachite green (LMG), which is a colorless reagent. This test is performed in the same way as the Kastle-Meyer test. If blood is present, it produces a blue-green color when the hydrogen peroxide is added.

Blood is not the only material that will cause these reagents to change color, however. Vegetable matter, such as horseradish and potatoes, can give false-positive results (although they are unlikely to be present at a crime scene, of course).

Often, the presence of blood at a crime scene is expected based on statements made by witnesses but—under normal lighting—no blood can be found. For example, a spatter pattern on the wall that has been carefully cleaned may go completely unnoticed by the naked eye. In these cases the application of a luminol spray to the bloodstained area may be more appropriate than the color-change tests described previously. Luminol has been reported to detect bloodstains that have been diluted up to 300,000 times.

In this test, an alkaline solution containing both the luminol and an oxidizing agent, such as hydrogen peroxide, is sprayed onto the area being investigated. If blood is present, the luminol will be catalytically oxidized and, as a consequence, will produce a distinct glow (FIGURE 13-2). This luminescence can be viewed only in the dark. Thus the crime scene is darkened and luminol is applied until the luminescence appears and the geometry of the bloodstain becomes apparent. Sometimes an alternate light source that emits a red light is used to help illuminate the dark area.

Like the color-change tests, luminol is a presumptive test. It can produce false-positive results when exposed to household bleach, some metals, and some vegetables. It is important to note that luminol does not degrade DNA, nor does it interfere with any subsequent DNA testing that may follow when the evidence is submitted to the forensic laboratory.

Serological Tests for Blood

Once the presumptive test has indicated that the stain may be blood, the serologist will ascertain whether the blood is of human origin. This can be accomplished by the **precipitin** serological test, which identifies the presence of proteins that are specific to humans. The serologist could also analyze the DNA to show that the sequences found were specific to humans.

The primary function of red blood cells is to transport oxygen from the lungs to the body's organs, and to pick up carbon dioxide waste products from the organs and transport it back to the lungs. On their surface, red blood cells contain car-

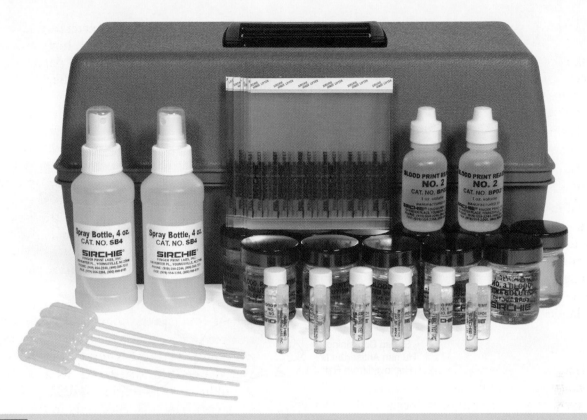

FIGURE 13-1 A field test kit contains presumptive color-change tests for blood.

bohydrate (sugar) molecules called **antigens**. Antigens are key components of the immune system's response to "invaders" such as infectious organisms and allergens (i.e., substances such as pollen and cat dander that cause allergic reactions). Human red blood cells have four specific antigens, which differ from the antigens found on other animals' red blood cells. The precipitin test takes ad-

vantage of the specific binding of these antigens for purposes of forensic testing.

The precipitin test (**FIGURE 13-3**) determines the species of origin for the blood sample by observing the formation of an antigen–antibody complex, which appears as a cloudy precipitate. When an animal is injected with human blood, the animal forms **antibodies**, which react to the human blood

(A)

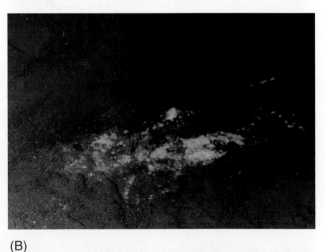

(B)

FIGURE 13-2 Luminol makes prints visible on (A) carpet and (B) dirt.

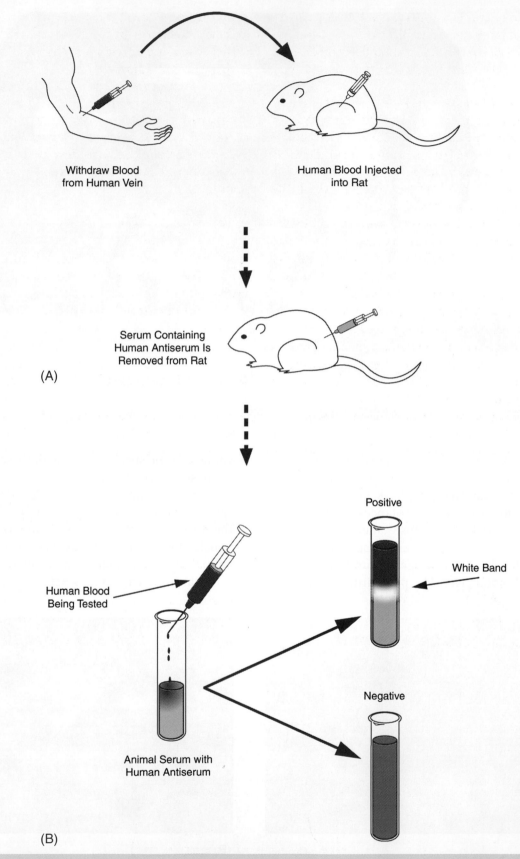

FIGURE 13-3 (A) An animal will form antibodies when injected with human blood. (B) The precipitin test may be completed in a capillary test tube. A layer of human antiserum is placed in the bottom of the tube, and an extract of the bloodstain is carefully added to the top. If the sample contains human blood, a cloudy precipitate forms at the interface of the two layers.

by treating it as a foreign invader (Figure 13-3A). These antibodies can be harvested by isolating the animal's antibody-containing blood serum from its blood; the resulting product is known as human **antiserum**.

The precipitin test may be used to identify human bloodstains in a number of different ways. For example, the test may be done in a capillary test tube. A layer of human antiserum is placed in the bottom of the tube, and an extract of the bloodstain is carefully added on top of it (Figure 13-3B). If a cloudy precipitate forms at the interface of the two liquids, the test is positive for human blood.

Another method, called cross-over electrophoresis, uses a gel-coated slide with two wells. A liquid extract of the bloodstain is placed in one well, and the human antiserum is placed in the other well (FIGURE 13-4). When an electric current is applied, both the antibodies and the antigens move to the middle of the plate. If a precipitate forms where the two meet, then the bloodstain is positive for human blood.

The precipitin test has several advantages. First, it requires only a small blood sample, so it can be used to test even minute bloodstains. Even if bloodstains have been washed to a point that only a tiny amount remains, this test may still produce a positive result. Second, the precipitin test is highly sensitive even when the bloodstains are old (e.g., 10–15 years).

If the precipitin test gives a negative result for human blood, it can be repeated using antiserum prepared for other animals. In this way, it becomes possible to determine the species of animal that produced the bloodstain. Test kits are available for a number of animals, including farm and domestic animals.

Serological Blood Typing

After the bloodstain has been identified as having a human origin, further forensic analysis is used to determine whether it can be associated with a particular individual. In the past, this process involved the use of blood typing systems that were based on serological techniques. Today, this approach has been superseded by DNA profiling, which can distinguish biological traits much more precisely and, in so doing, better individualize evidence. For this reason, only a brief description of serological blood typing is given here; techniques of DNA typing are presented in Chapter 14.

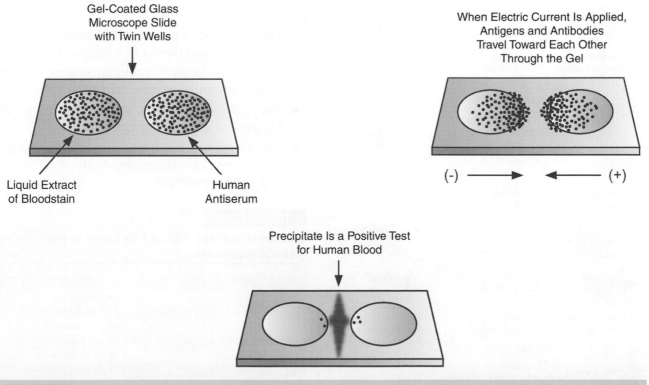

FIGURE 13-4 In cross-over electrophoresis, the formation of a precipitate (the solid line) shows that the bloodstain is positive for human blood.

ABO System

The ABO system of blood typing separates human blood into four broad classifications based on the presence or absence of either or both antigen A or antigen B on the surface of red blood cells. Individuals with type A blood have an A antigen on the surface of their red blood cells. Similarly, individuals with type B blood have B antigens. Individuals with type AB blood have both A and B antigens. People with type O blood have neither A nor B antigens.

A basic principle of **serology** is this: For every antigen, there exists a specific antibody. An antibody will react only with its specific antigen, and no other. As can be seen in **FIGURE 13-5**, if blood serum containing antibody A (anti-A) is added to red blood cells with the A antigen, the antibody will attach itself to the antigen on the cell. Antibodies have more than one reactive site, which means that they can link red blood cells together in a network of cross-linked cells.

For routine blood typing, only two antiserums are needed: anti-A and anti-B. To perform this test, the forensic examiner sequentially places a droplet of antiserum in samples of blood, and then observes them under a microscope. The key information sought in this case is whether the sample begins to clot in response to the antiserum—that is, whether it becomes **agglutinated**. Blood of type A will be agglutinated by anti-A serum; blood of type B will be agglutinated by anti-B serum; AB blood will be agglutinated by both anti-A and anti-B serum; and O blood will not be agglutinated by either type of serum.

Another useful serological tool is the measurement of the Rh (Rhesus disease) factor. If a person has a positive Rh factor, it means that his or her blood contains a protein that is also found in Rhesus monkeys. Most people (about 85%) have a positive Rh factor. The Rh system is actually much more complicated than the ABO system (approximately 30 subgroups have been identified), but for the sake of simplicity, Rh is usually expressed as either positive or negative. The Rh factor, like other antigens, is found on the surface of red blood cells. Forensic scientists typically express the distribution of the Rh factor, which is expressed as plus or minus, in terms of odds ratios (**TABLE 13-1**).

TABLE 13-2 shows the distribution of blood types across the United States. Because blood type is inherited from a person's parents, blood types tend to become concentrated among certain ethnic groups over long periods of time:

- The A blood type is most common among Caucasians and those of European descent.
- The B type is most common among African Americans and certain Asians (Thai).
- The AB type is most common among the Japanese and Chinese populations.
- The O type is most common among indigenous people (Aborigines and Native Americans) and Latin Americans.

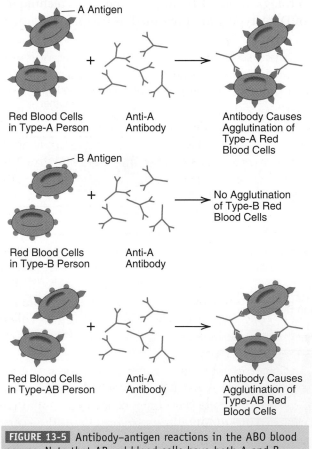

FIGURE 13-5 Antibody–antigen reactions in the ABO blood groups. Note that AB red blood cells have both A and B antigens.

A Antigen

Red Blood Cells in Type-A Person + Anti-A Antibody → Antibody Causes Agglutination of Type-A Red Blood Cells

B Antigen

Red Blood Cells in Type-B Person + Anti-A Antibody → No Agglutination of Type-B Red Blood Cells

Red Blood Cells in Type-AB Person + Anti-A Antibody → Antibody Causes Agglutination of Type-AB Red Blood Cells

TABLE 13-1	
Distribution of the ABO and Rh Factor in the General Population	
O+	1 in 3 persons
O-	1 in 15 persons
A+	1 in 3 persons
A-	1 in 16 persons
B+	1 in 12 persons
B-	1 in 67 persons
AB+	1 in 29 persons
AB-	1 in 167 persons

Source: American Association of Blood Banks.

TABLE 13-2

Distribution of Blood Types

O	A	B	AB
43–45%	40–42%	10–12%	3–5%
O+: 39%	A+: 35%	B+: 8%	AB+: 4%
O–: 6%	A–: 5%	B–: 2%	AB–: 1%

Source: American Association of Blood Banks.

Approximately 80% of the population is classed as **secretors**—that is, individuals who have significant concentrations of their antigens not only in their blood but also in other body fluids, such as perspiration, saliva, semen, and urine. The blood type of secretors can be determined from nonblood body fluids that are left at the scene of a crime by using standard serological techniques.

Other Blood Typing Systems

In addition to the use of blood groups, other blood typing systems have been developed based on the presence of certain proteins in the red blood cells. These proteins are polymorphic—that is, they occur in more than one form. The different forms of a particular protein can be identified and their statistical occurrence in the population calculated. This information can then be used to further characterize bloodstains.

The larger the number of independent factors that can be identified in a blood sample, including the blood type and presence of polymorphic proteins, the smaller the percentage of the population possessing that particular combination of blood traits. As we learned in Chapter 1, the probability of a match can be determined by multiplying the occurrence of each independent factor.

Forensic Characterization of Saliva

Saliva is more than 99% water, has a pH in the range of 6.8 to 7.0, and contains the digestive enzyme salivary amylase (α-amylase), which breaks down starch into smaller sugars. Adults produce approximately 1.0 to 1.5 L of saliva per day. Saliva is produced in three main pairs of salivary glands —the parotid, submaxillary, and sublingual glands— from which it flows into the mouth through ducts. Saliva cleanses the mouth and provides necessary lubrication.

It is not unusual for saliva to be present at a violent crime scene; it is always present if there are bite marks left on the victim. The presence of salivary amylase is a positive presumptive test for saliva. To perform this test, a sample of the suspected saliva stain is extracted or removed and then placed in a solution containing soluble starch. Next, the investigator adds a few drops of an iodine solution to this sample. Iodine reacts with starch to form a blue-black color. Thus, if the sample solution turns blue-black, then it contains starch and the test is negative for saliva. Conversely, if the solution becomes yellow, then this is a positive test for saliva: Starch in the solution has been broken down by the salivary amylase in the saliva into smaller sugars, so there is no starch left in the solution that will react with the added iodine.

As is the case with blood and semen, saliva can be used to identify an individual by the use of DNA profiling. The DNA in saliva will be extracted from cheek cells that have been shed into the saliva inside the mouth.

Forensic Characterization of Semen

Because so many crimes involve sexual offenses, forensic examiners often are called on to test for the presence of seminal stains. Semen is fluid produced by the testes and accessory sex glands, which include the prostate gland, bulbourethral glands, and seminal vesicles. Semen is more than 90% water and has a pH in the range of 7.2 to 7.4. One milliliter of semen contains 20 million to 100 million sperm cells (i.e., 2×10^7 to 10×10^7 sperm/mL). A normal male ejaculates 2 to 6 mL of seminal fluid. The average ejaculate is 3 mL in volume and contains about 300 million sperm. Some men have an abnormally low sperm count (i.e., less than 2×10^7 sperm/mL), a condition known as oligospermia. In other cases the seminal fluid contains no sperm cells, a condition called azoospermia; it is seen in males who have undergone a vasectomy.

The forensic examination of a crime scene for seminal stains is usually approached as a two-step process. First, the stain must be located. Only then can any presumptive tests begin. Often, the scene may include a large number of items that might have been stained by seminal fluid—for example, multiple garments, bed clothing, rugs, drapes, and solid surfaces. The forensic examiner often searches the crime scene with an ultraviolet (UV) light to detect the presence of seminal stains; the UV light makes the search of the scene go quickly, and semen fluoresces under UV light, making stains easy to locate. Other body fluids such as urine also fluoresce in UV light, however, so this test is always considered presumptive for semen.

Another presumptive test often used at a crime scene to detect seminal fluid is the acid phosphatase test. The prostate gland secretes the enzyme acid phosphatase, and its concentration is much higher in seminal fluid than in any other body fluid. This compound can be easily detected with a simple color test: In the presence of diazotized o-dianisidine, acid phosphatase will react with α-naphthyl phosphate to produce a purple color. If the stain is semen, the reagent will produce a rapid color change, with the purple appearing in less than 1 minute. Although some fruit juices, contraceptive creams, and vaginal secretions do produce a purple color with this test, the color change is much slower with these substances, because they contain a lower concentration of acid phosphatase. A stain that develops a strong purple color in less than 30 seconds is a strong indication of the presence of seminal fluid.

The acid phosphatase test is often used when many garments need to be tested for seminal fluid. A filter paper is moistened and rubbed lightly over the garment; if seminal fluid is present, some will likely be transferred to the filter paper. Next, a drop of the test solution is applied to the filter paper, and the examiner looks for a color change in 30 seconds. If a large object, such as a bed sheet, needs to be tested, the filter paper test can be carried out on sections of the object. In this way any fabric or surface can be systematically searched for seminal fluid.

Another presumptive test used for the identification of seminal fluid that is not dependent on the presence of sperm is the p30 test. This test uses serological methods similar to those described for the ABO serological tests to detect the presence of p30, a protein produced by the prostate gland. The p30 enzyme is found exclusively in seminal fluid, so vaginal stains will not give a positive p30 test result, unlike with the acid phosphatase color test.

One of the best and simplest tests for the presence of seminal fluid is the direct observation of sperm under a high-powered microscope. Sperm have a unique shape: They are 50–70 microns long and contain a head and a thin tail (FIGURE 13-6). To

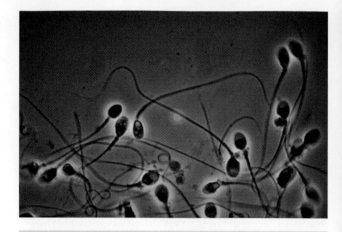

FIGURE 13-6 Human sperm as viewed under a microscope.

use microscopy to identify seminal fluid, the semen-stained object is wetted with water and a small sample of the water transferred to a microscope slide. After drying, the slide is stained and observed under a compound microscope.

Once a stain has been identified as being a seminal stain, its sperm cells can be used to link a particular suspect to the crime scene through the use of DNA profiling. We will learn about DNA profiling in Chapter 14.

Rape Evidence Collection

In many cases of rape, there are no witnesses to the act other than the victim and the perpetrator. While the victim's physical injuries may indicate that he or she suffered a violent assault, winning a conviction in the case often hinges on the prosecutor's ability to definitively link the perpetrator to the victim and his or her injuries. To do so, the prosecutor may cite both a variety of physical evidence—such as blood, semen, hairs, saliva, and fibers—that was transferred from victim to perpetrator, or vice versa, during the commission of the crime. This kind of physical evidence may also help investigators reconstruct the sequence of events as this violent crime was being committed.

To collect evidence from the scene of a sexual assault, the investigator first surveys the scene. Did it take place outdoors or indoors? If the assault took place on a bed, all sheets and beddings should be placed in paper bags and submitted to the crime lab. If a stain is found on an object that is too big to ship to the lab, then the stain should be cut out and packaged. An unstained sample of the same object should also be cut out and packaged as a substrate control.

In addition to gathering evidence from the surroundings, the investigator must collect evidence from the victim, if he or she is still present at the scene. To do so, the investigator places a clean bed sheet on the floor close to the scene of the rape and puts a clean sheet of paper on top of it. All persons, other than the investigator, should be asked to leave to make this area as private as possible. The victim then is asked to remove his or her shoes, stand on the paper, and remove each piece of clothing. As he or she removes each piece of clothing, an examiner should collect any objects or fibers that fall off the clothes and onto the paper. The forensic examiner must wear disposable gloves and minimize direct personal contact with the evidence and the paper. Each piece of clothing should be placed in a separate paper bag to avoid cross-contamination. When this process is complete, the victim should step off the paper; this paper is then folded and—along with any evidence on it—preserved. Clean clothing should be immediately provided to the victim with privacy to redress. When the victim is dressed, he or she should be immediately transported to the hospital.

The victim of a sexual assault should be examined by a physician as soon as possible after the assault, both to ensure that the victim receives any needed medical care and to enable the collection of any remaining evidence by trained law enforcement personnel. The following physical evidence is collected from the victim:

1. Blood sample
2. Combings from pubic hair
3. Pubic hair reference samples
4. Vaginal swab and smear
5. Rectal swab and smear
6. Head hairs
7. Fingernail scrapings
8. Oral swab
9. All clothing (if not collected at the scene as previously described)
10. Urine specimen

In addition, if the victim reports that the assailant bit, sucked, or licked an area of the victim's body, then saliva residues should be collected off the victim's skin. Most commonly, the person gathering this rape-related evidence uses a standardized col-

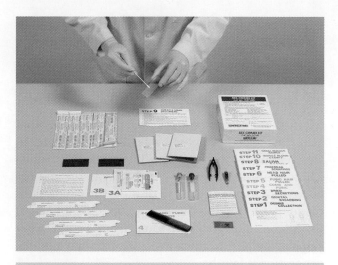

FIGURE 13-7 A rape evidence kit is used to collect biological and other physical evidence from a victim of sexual assault.

lection kit that has been approved by local authorities (**FIGURE 13-7**).

The vaginal swab and smear are especially important when dealing with female rape victims. The state of any sperm found in these samples—that is, whether the sperm are living (motile) or dead (nonmotile), intact (with tails) or degraded—may allow investigators to establish a timeline for the crime. In a rape victim who survives the attack, intact motile sperm may remain in the vagina for 4 to 6 hours after intercourse. After this point, the sperm die. They may remain intact for 16 hours or more after the assault, but eventually they disappear from the vagina, usually 3 days following the attack.

The following pieces of evidence are routinely collected if a subject is apprehended:

1. All clothing (with each article packaged in a separate paper bag, as described in Chapter 1)
2. Combings of pubic hair
3. Head hair and pubic hair standards
4. Penile swab (to collect skin cells and any seminal fluid)
5. Blood sample (collected and preserved in the manner described in Chapter 12)

At this point, the forensic examiner's job is to determine whether the evidence collected from the suspect can be definitively linked to the evidence collected from the victim. In the past, a variety of techniques were used to identify such a relationship. More recently, the emergence of DNA profiling has changed the way that evidence from a suspect is processed (as described in detail in Chapter 14).

Principles of Paternity

All of the antigens and polymorphic proteins that coat the surface of a blood cell (described earlier in this chapter) are genetically controlled traits. These traits are inherited from our parents and become a permanent feature of a person's biological makeup. Identifying these traits not only allows us to differentiate one individual from another, but can also help us to track the traits through a multigenerational family.

Human cells consist of many parts, of which we will consider only a few here (FIGURE 13-8). Each cell is enclosed in a cell membrane through which the cell receives nutrients and eliminates waste. Inside the cell are a number of structures that serve a variety of purposes. The largest is usually the cell nucleus, which controls heredity; the ribosomes, where protein synthesis occurs; and the mitochondria, where energy is produced for the cell.

The hereditary material found in the nucleus of the cell consists of elongated, threadlike bodies called chromosomes. The number of chromosomes varies with the species. Humans have 46 chromosomes: 22 matched pairs and 2 sex-determining chromosomes (XX = female, XY = male). All body (somatic) cells are diploid, meaning that they contain two sets of each chromosome. By contrast, gamete cells (sperm or egg) are haploid—they contain only one set of chromosomes. During fertilization, an egg and a sperm combine, such that the resulting zygote becomes diploid again. In this process, one chromosome from each parent is donated to the new zygote.

The egg cell always contains an X chromosome. The sperm may contain either a Y chromosome or an X chromosome. When a sperm possessing an X chromosome (which is much longer than a Y chromosome) fertilizes an egg, the new cell is an XX and develops into a female. A sperm possessing a Y chromosome produces an XY fertilized egg, which develops into a male. Because the sperm determines whether the zygote is an XX or an XY, the father biologically determines the sex of the offspring. Within a multigenerational family, sons always inherit their Y chromosome from their father, so paternity can often be determined by comparison of the Y chromosomes from father and son.

Recently, some interesting studies have used DNA technology to probe the Y chromosome with the goal of showing paternity. A year after becoming president of the United States, Thomas Jefferson was accused by a Richmond, Virginia, newspaper of being the father of a child born of his slave, Sally Hemings. For almost 200 years this allegation has been a matter of controversy. To resolve this question, Dr. Eugene Foster began a study that tested Y-chromosome DNA markers to trace the Jefferson male line to a descendant of Sally Hemings's youngest son, Eston Hemings. The study collected samples from living individuals, 15 generations later, who possessed the Jefferson Y chromosome, including the descendants of Eston Hemings. The prestigious scientific journal *Nature* published a report of Foster's study in 1998 that concluded that Jefferson had fathered Eston Hemings.

Chromosomes are made of nucleic acids. The nucleic acid in chromosomes is deoxyribonucleic acid (DNA), the primary hereditary material. The DNA material is composed of a series of **coding regions** and **noncoding regions** that are arranged along the chromosomes. The coding regions, or *genes*, contain the information that a cell needs to make proteins. Structurally, genes are sections of the DNA molecule. The position of a gene along the chromosome is referred to as a **locus**. Pairs of chromosomes are considered homologous because they are the same size and contain the same infor-

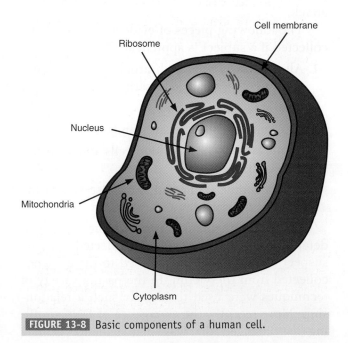

Cell membrane

Ribosome

Nucleus

Mitochondria

Cytoplasm

FIGURE 13-8 Basic components of a human cell.

mation. Each gene is located at the same position (locus) on each chromosome of a pair of homologous chromosomes.

When fertilization occurs, one chromosome is inherited from the mother and one chromosome is inherited from the father. It is possible that the DNA sequence in each chromosome in the homologous pair might have been altered by mutations so that the two pairs will not be identical. If the genetic locus is located at a different position, it is called an **allele**. If two alleles at a specific genetic locus on homologous chromosomes are different, they are called **heterozygous**; if the alleles are identical at a specific locus, they are called **homozygous**.

The ABO blood typing system is a simple example of how alleles are positioned on the chromosome. In the ABO system, a chromosome pair made of AA and BB is homozygous; a pair made of two dissimilar types, such as AO, is heterozygous. If the chromosome inherited from the father carried the B gene and the chromosome inherited from the mother carried the B gene as well, then the offspring would have a BB blood type. Similarly, if one chromosome contained the A gene and the other an O gene, the offspring would have an AO blood type.

Introduction to DNA

The complex compounds called **nucleic acids** are found in all living cells except mammalian red blood cells. Two kinds of nucleic acids play crucial roles in inheritance: **deoxyribonucleic acid (DNA)** and **ribonucleic acid (RNA)**. DNA is found primarily in the nucleus of cells; hence it is referred to as nuclear DNA (nDNA). RNA is found primarily in the **cytoplasm**, the part of the cell surrounding the nucleus.

Functions of Nucleic Acids

DNA and RNA are responsible for the storage and transmission of genetic information in all living organisms. They hold the key to how genetic information is transferred from one cell to another and how genetic traits are transmitted, via sperm and eggs, from parents to offspring. The major function of DNA—and one in which RNA is also involved—is the control and direction of protein synthesis in body cells. Chemical information stored in the DNA of genes specifies the exact nature of the protein to be made and, therefore, dictates the character of the organism.

SEE YOU IN COURT

Due to the reliability of DNA evidence, DNA (when tested properly) is widely used in criminal investigations and is widely accepted as evidence in the judicial system. DNA evidence can secure convictions, and it is also a way of exonerating those who have been wrongly accused. Nevertheless, there have been cases where the admissibility of DNA evidence has been challenged.

The first case that challenged the use of DNA was *People v. Castro*. In 1989 the New York Supreme Court, in a 12-week pretrial hearing in this case, exhaustively examined numerous issues relating to the admissibility of DNA evidence. Jose Castro was accused of murdering a female neighbor and her 2-year-old daughter. Blood collected from Castro's wristwatch was sent for DNA analysis and matched the neighbor's DNA profile. During the trial, however, the defense focused on the gray areas of DNA profiling and challenged the reliability of DNA testing by forensic scientists.

In its decision, the court made the following points:

- DNA identification theory and practice are generally accepted among the scientific community.
- DNA forensic identification techniques are generally accepted by the scientific community.
- Pretrial hearings are required to determine whether the testing laboratory's methodology was substantially in accord with scientific standards and produced reliable results for jury consideration.

The court ruled that the DNA test results could be used to show that the bloodstain on the watch was not Castro's, but the tests could not be used to show that the blood was not the neighbor's or the daughter's.

Since this case, all court hearings dealing with DNA profiling have been obligated to include statements on all laboratory procedures, including statistical probability calculations, a list of all possible laboratory errors, and all chain of custody documents. All laboratory practices must be in accordance with scientific standards to be considered reliable in court.

Structure of Nucleic Acids

In much the same way that polysaccharides are composed of monosaccharides and proteins are composed of amino acids, nucleic acids are composed of long chains of repeating units called **nucleotides**. DNA molecules are the largest of the naturally occurring organic molecules.

Each nucleotide in a chain is made up of three components: (1) a sugar, (2) a nitrogen-containing heterocyclic base, and (3) a phosphoric acid unit (FIGURE 13-9). The sugar is a pentose, either **ribose** or **deoxyribose** (FIGURE 13-10). The only difference between these two sugars occurs at carbon 2, where ribose has a hydrogen atom and an —OH group, and deoxyribose has two hydrogen atoms. As their names indicate, the sugar in DNA is deoxyribose while that in RNA is ribose.

Nucleic acids may contain any of five different bases (FIGURE 13-11). Two—**adenine** and **guanine** —are double-ring bases that are classified as purines. The other three bases—**cytosine**, **thymine**, and **uracil**—are single-ring bases that belong to the class of compounds called pyrimidines. Purines and pyrimidines are bases because their nitrogen atoms can accept protons. Adenine, guanine, and cytosine are found in both DNA and RNA. Thymine is found in DNA, while uracil is found in RNA.

FIGURE 13-12 depicts the formation of a representative nucleotide from its three components. In

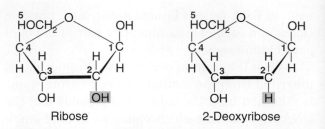

FIGURE 13-10 The pentoses ribose and deoxyribose differ at carbon 2, where ribose has an H atom and an —OH group, and deoxyribose has two H atoms. Ribose is present in RNA; deoxyribose is present in DNA.

this case, the pentose forms an ester bond with a phosphoric acid unit through the —OH on carbon 5 and forms another bond with the base through the —OH group on carbon 1. Water is eliminated as the bonds are formed. When successive nucleotides bond together to form long-chain nucleic acids, the phosphoric acid units form a second ester bond through the —OH group on carbon 3 of a second pentose molecule (FIGURE 13-13). The alternating sugar–phosphoric acid units form the backbone of the nucleic acids.

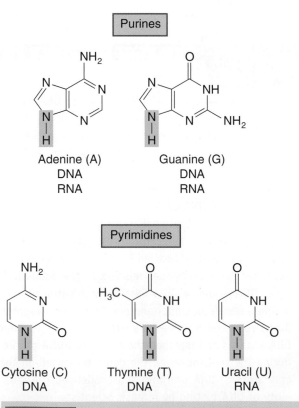

FIGURE 13-11 Of the five bases found in nucleic acids, adenine and guanine are purines, while cytosine, thymine, and uracil are pyrimidines.

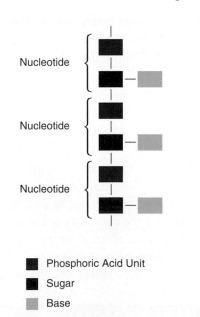

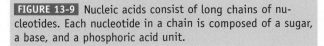

FIGURE 13-9 Nucleic acids consist of long chains of nucleotides. Each nucleotide in a chain is composed of a sugar, a base, and a phosphoric acid unit.

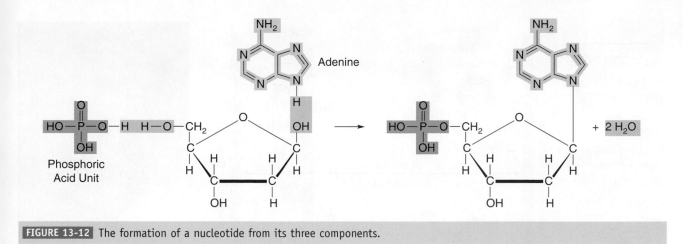

FIGURE 13-12 The formation of a nucleotide from its three components.

The Double Helix

The sequence of the four bases in a nucleic acid is referred to as its primary structure. Nucleic acids also have a secondary structure. The secondary structure of DNA was determined in 1953 by James D. Watson and Francis H. C. Crick, a feat for which they won the Nobel Prize in 1962. Their discovery, with all its biological implications, is one of the most significant scientific achievements of the twentieth century.

On the basis of the available information, Watson and Crick concluded that a DNA molecule must consist of two polynucleotide chains wound around each other to form a **double helix**, a structure that can be compared to a spiral staircase (**FIGURE 13-14**). The phosphate–sugar backbone represents the hand-rails, and the pairs of bases linked together by hydrogen bonds represent the steps (Figure 13-14A). Working with molecular-scale models, Watson and Crick realized that, because of the dimensions of the space available inside the two strands of the helix and the relative sizes of the bases, base G (guanine) on one chain must always be bonded to base C (cytosine) on the other chain, and base A (adenine) must always be bonded to base T (thymine). This arrangement gives "steps" of almost equal lengths and explains why amounts of C and G and amounts of A and T in DNA molecules are always equal (Figure 13-14B). Although hydrogen bonds are relatively weak, so many are present in a DNA molecule that, under normal physiological conditions, they hold the two chains together. The C-G (and G-C) base pairs are held together by three hydrogen bonds, whereas the A-T (and T-A) base pairs are held together by two hydrogen bonds. Figure 13-14C shows a space-filling model of DNA.

DNA, Genes, and Chromosomes

The DNA in the nuclei of cells is coiled around proteins called histone molecules to form structures known as **chromosomes** (**FIGURE 13-15**). The

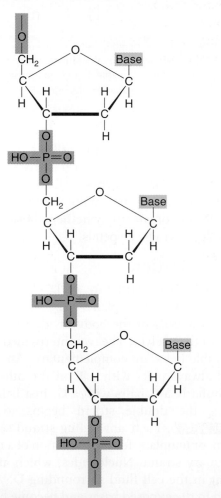

FIGURE 13-13 A three-nucleotide segment of a DNA strand, showing the alternating phosphate–sugar units that form the backbone of the molecule.

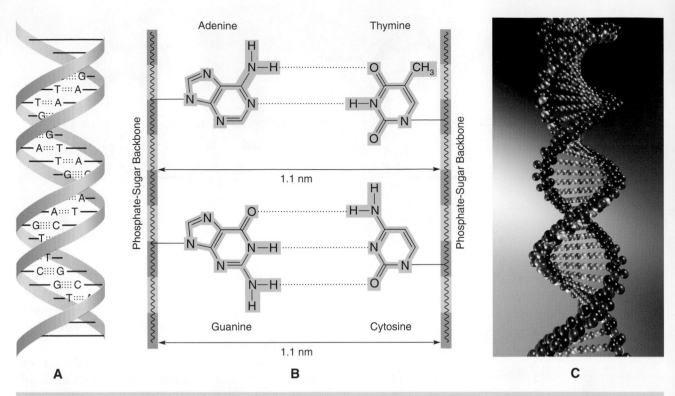

FIGURE 13-14 (A) A schematic representation of the double-helix model of DNA. Hydrogen bonding between adenine and thymine and between cytosine and guanine is represented as dotted lines. (B) The steps in the spiral staircase structure of the DNA molecule are formed by two kinds of base pairs: Adenine (a purine) pairs with thymine (a pyrimidine) through two hydrogen bonds; guanine (a purine) pairs with cytosine (a pyrimidine) through three hydrogen bonds. These pairings give steps of equal width (1.1 nm) separating the two strands. (C) A space-filling model of DNA.

number of chromosomes varies with species. Humans have 46; each parent contributes 23 to his or her offspring.

Long before the structure of DNA was understood, genes were defined as sections of chromosomes that determined inherited characteristics, such as blue eyes and dark hair in humans. We now define genes as those segments of DNA molecules that control the production of the different proteins in an organism. Proteins, in turn, control the chemical reactions of life that occur in an organism, including the development of blue eyes and dark hair. Ten years ago scientists thought humans had 50,000 to 100,000 genes. One surprising result of the Human Genome Project is the finding that humans seem to have fewer than 30,000 genes.

DNA molecules vary in terms of the number and sequence of the base pairs they contain. The precise sequence of base pairs in the DNA molecule is the key to the genetic information that is passed on from one generation to the next; it is this sequence that directs and controls protein synthesis in all living cells. Each organism begins life as a single cell. The unique DNA in the nucleus of that cell determines whether the cell, as it multiplies, develops into a human, a bird, a rose, or a bacterium. The DNA carries all the information needed for making and maintaining the different parts of the organism, whether those parts are hearts, legs, wings, or petals.

Cell Replication

The double-helix structure of DNA proposed by Crick and Watson explains very simply and elegantly how cells in the body divide to form exact copies of themselves. The two intertwined chains of the double helix are complementary: An A on one strand always pairs with a T on the other strand, and similarly a C pairs with a G. Just before a cell divides, the double strand begins to unwind (FIGURE 13-16). Each unwinding strand serves as a pattern, or template, for the formation of a new complementary strand. Nucleotides, which are always present in the cell fluid surrounding DNA, are attracted to the exposed bases and become hydrogen-bonded to them: A to T, T to A, C to G, and G to C. In this way, two identical DNA molecules are formed—one for each of the two daughter cells.

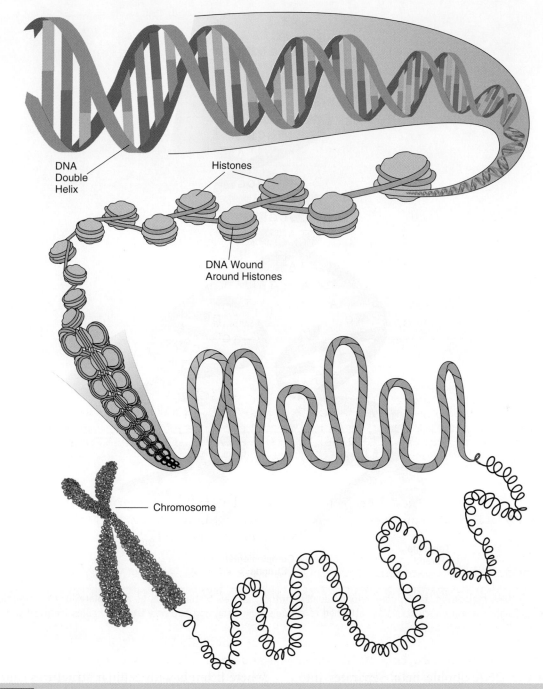

DNA
Double
Helix

Histones

DNA Wound
Around Histones

Chromosome

FIGURE 13-15 Chromosomes, which are found in cell nuclei, consist of tightly coiled strands of DNA.

Protein Synthesis

Protein synthesis is carried out in a series of complex steps involving RNA. As we noted earlier, the sugar in RNA is ribose, and the bases are uracil (U), adenine (A), guanine (G), and cytosine (C). The primary structure of RNA is similar to that of DNA: ribose–phosphoric acid units form the backbone, and each ribose unit is bonded to one of the four bases. RNA molecules are much smaller than DNA molecules and contain from 75 to a few thousand base pairs. RNA molecules exist primarily as single strands rather than in a double-helix form.

Protein synthesis proceeds in two main steps:

$$\text{DNA} \xrightarrow{\text{Transcription}} \text{mRNA} \xrightarrow[\text{Translation}]{\text{tRNA}} \text{Protein}$$

In the first step, transcription, a single strand of RNA is synthesized inside the cell nucleus. A

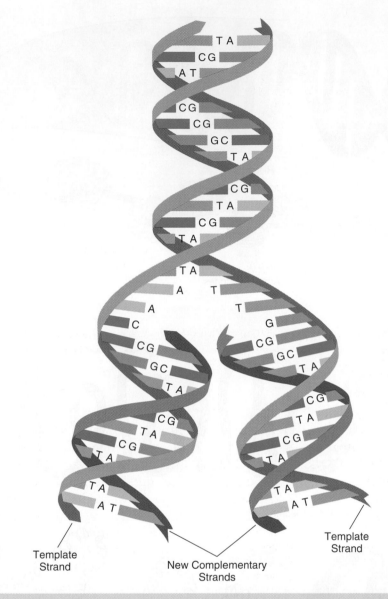

Template
Strand

New Complementary
Strands

Template
Strand

FIGURE 13-16 In DNA replication, the double helix gradually unwinds, and each strand acts as a template for the formation of its complement. Complementary nucleotides are attached to single strands, forming two new DNA molecules identical to the original one.

segment of the DNA double helix separates into single strands, and the exposed bases on one strand act as the template for the synthesis of a molecule of RNA. The base sequence in the RNA, which is known as messenger RNA (mRNA), complements the base sequence on the DNA strand with one exception: Wherever there is an adenine in the DNA, the RNA transcribes a uracil instead of thymine (**FIGURE 13-17**).

The second step in protein synthesis is the translation of the code that has been copied from the DNA strand into the new protein. To direct this synthesis, the mRNA leaves the nucleus and takes its chemical message to the cytoplasm of the cell,

where it binds with cellular structures called ribosomes. Taking part in this translation are molecules of transfer RNA (tRNA; another kind of RNA), which are responsible for delivering amino acids one by one to the mRNA. Each tRNA molecule carries a three-base sequence called an **anticodon**, which determines which specific amino acid it will deliver. Three-base sequences along the mRNA strand, called **codons**, determine the order in which tRNA molecules bring the amino acids to the mRNA.

Guided by the first codon on the mRNA strand, a tRNA molecule with an anticodon that is complementary to this codon transports a specific

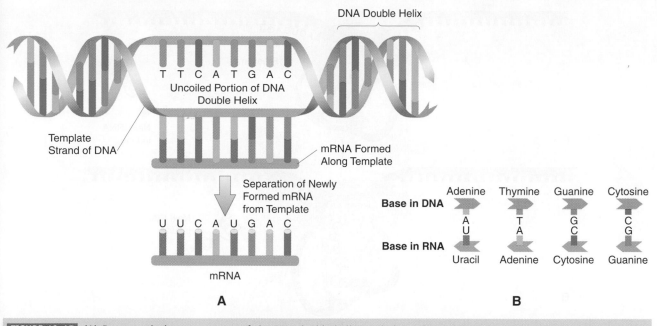

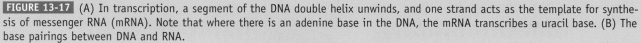 **FIGURE 13-17** (A) In transcription, a segment of the DNA double helix unwinds, and one strand acts as the template for synthesis of messenger RNA (mRNA). Note that where there is an adenine base in the DNA, the mRNA transcribes a uracil base. (B) The base pairings between DNA and RNA.

amino acid to the mRNA codon (**FIGURE 13-18**). For example, if the bases are lined up as shown in Figure 13-18A, where the first codon is A-U-G, a tRNA molecule with a U-A-C anticodon will transport the amino acid methionine (Met) to the mRNA, where the tRNA anticodon will pair up

through hydrogen bonding with the complementary codon. Similarly, a second tRNA molecule will bring the amino acid phenylalanine (Phe) to the next mRNA codon, U-U-U. A peptide bond then forms between the two amino acids, and the methionine separates from the tRNA (Figure 13-18B). When a third amino acid, glutamic acid (Glu), has been added, the first tRNA molecule is released from the mRNA (Figure 13.18C).

The actual protein synthesis occurs in the ribosomes, which move along the mRNA one codon at a time as the amino acid chain grows. In this way, the mRNA is read codon by codon, and the protein is built up one amino acid at a time in the correct sequence. When the synthesis is completed, the protein separates from the mRNA (Figure 13-18D).

The Genetic Code

After years of work, researchers have established the specific amino acid that each of the three-base sequences in mRNA codes for. This information is known as the **genetic code** (**TABLE 13-3**). Sixty-four possible three-base codons can be formed from the four bases in mRNA ($4^3 = 64$). Given that all proteins are made from only 20 amino acids, there is some redundancy in this system. For example, the codon A-U-G not only encodes for methionine, but

Example

Show the sequence of bases in the mRNA that would be synthesized from the lower strand of DNA with the following sequence of bases:

-T-A-C-G-G-T-T-C-A-C-
-A-T-G-C-C-A-A-G-T-G- (Template strand)

Solution

The bases pair as follows: A-U, T-A, G-C, C-G. Therefore, the mRNA would have the following sequence of bases:

-U-A-C-G-G-U-U-C-A-C-

Notice how the sequence in the top strand of DNA corresponds to the sequence in the mRNA.

Practice Exercise

Show the sequence of bases in the DNA segment that acted as the template for the synthesis of the following section of mRNA:

-C-G-G-U-A-U-C-U-A-

Solution

DNA: -G-C-C-A-T-A-G-A-T-

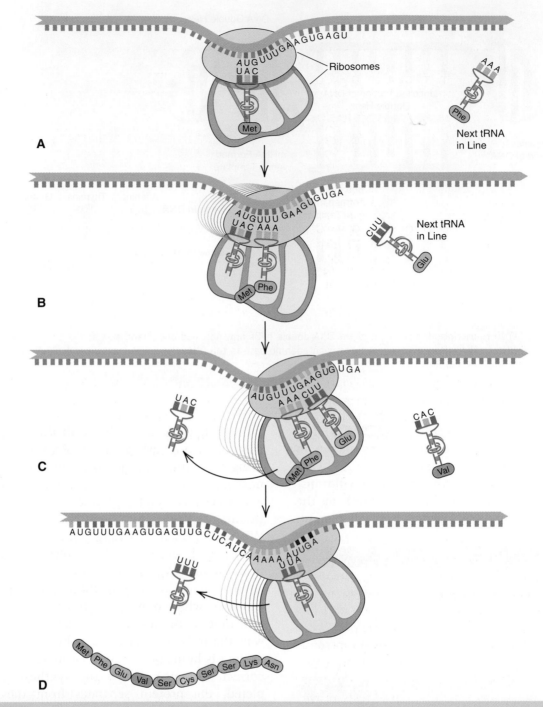

FIGURE 13-18 Translation and protein synthesis. (A) A transfer RNA (tRNA) molecule carrying the anticodon U-A-C delivers the amino acid methionine (Met) to the mRNA strand, where the anticodon bonds with the codon A-U-G. Another tRNA molecule carrying the anticodon A-A-A brings the amino acid phenylalanine (Phe) to the next mRNA codon, U-U-U. (B) A peptide bond forms between Met and Phe, and Met separates from the tRNA. A third amino acid, glutamic acid (Glu), is brought to the next mRNA codon, G-A-A. (C) The tRNA molecule carrying the U-A-C anticodon is released for the mRNA. (D) When synthesis of the protein is complete, the protein molecule separates from the mRNA.

TABLE 13-3

The Human Genetic Code

First Base	Second Base U		Second Base C		Second Base A		Second Base G		Third Base
U	U-U-U	Phe	U-C-U	Ser	U-A-U	Tyr	U-G-U	Cys	U
	U-U-C	Phe	U-C-C	Ser	U-A-C	Tyr	U-G-C	Cys	C
	U-U-A	Leu	U-C-A	Ser	U-A-A	Stop	U-G-A	Stop	A
	U-U-G	Leu	U-C-G	Ser	U-A-G	Stop	U-G-G	Trp	G
C	C-U-U	Leu	C-C-U	Pro	C-A-U	His	C-G-U	Arg	U
	C-U-C	Leu	C-C-C	Pro	C-A-C	His	C-G-C	Arg	C
	C-U-A	Leu	C-C-A	Pro	C-A-A	Gln	C-G-A	Arg	A
	C-U-G	Leu	C-C-G	Pro	C-A-G	Gln	C-G-G	Arg	G
A	A-U-U	Ile	A-C-U	Thr	A-A-U	Asn	A-G-U	Ser	U
	A-U-C	Ile	A-C-C	Thr	A-A-C	Asn	A-G-C	Ser	C
	A-U-A	Ile	A-C-A	Thr	A-A-A	Lys	A-G-A	Art	A
	A-U-G	Met	A-C-G	Thr	A-A-G	Lys	A-G-G	Arg	G
G	G-U-U	Val	G-C-U	Ala	G-A-U	Asp	G-G-U	Gly	U
	G-U-C	Val	G-C-C	Ala	G-A-C	Asp	G-G-C	Gly	C
	G-U-A	Val	G-C-A	Ala	G-A-A	Glu	G-G-A	Gly	A
	G-U-G	Val	G-C-G	Ala	G-A-G	Glu	G-G-G	Gly	G

it also is a "start" signal for protein synthesis. Other codons act as stop signals to terminate protein synthesis. Most of the approximately 30,000 genes in each cell of our bodies encode for a specific protein.

Nuclear DNA and the Law

In 1985 a routine analysis of the structure of a human gene led Dr. Alec Jeffreys of Leicester University (United Kingdom) to the discovery that portions of the DNA structure are as unique to each individual as fingerprints. Forensic use of DNA evidence in criminal cases began in 1986, when police asked Jeffreys to verify a suspect's confession that he was responsible for two rape/murders in the English Midlands. As in many cases, this forensic examination actually led to the exoneration of the suspect: Jeffreys' DNA tests proved that he had not committed the crimes despite his confession. Police then began a dragnet to obtain blood samples from several thousand male inhabitants in the area in an effort to identify a new suspect. One soon turned up: Robert Melias. In 1987 Melias became the first person convicted of a crime (rape) on the basis of DNA evidence.

Use of DNA evidence in criminal cases in the United States began around the same time as in the United Kingdom. In 1987 an Orange County, Florida, court convicted Tommy Lee Andrews of rape after tests matched his DNA (obtained from a blood sample) with the DNA found in semen traces taken from a rape victim.

Two other important early cases involving DNA testing were *State v. Woodall* and *Spencer v. Commonwealth*. In 1989 the West Virginia Supreme Court became the first state high court to rule on the admissibility of DNA evidence when it took up the *Woodall* case. In its decision, the court accepted the results of DNA testing performed by the defense team but ruled that its inconclusive results failed to exculpate Woodall. As a consequence, the court upheld the defendant's conviction for rape, kidnapping, and robbery of two women. Subsequent DNA testing determined that Woodall was innocent, and he was eventually released from prison.

In *Spencer v. Commonwealth*, the multiple murder trials of Timothy Wilson Spencer became the first cases in the United States in which the admission of DNA evidence led to guilty verdicts resulting in the death penalty for the defendant (in this case, in Virginia). In 1990 the Virginia Supreme Court upheld the murder and rape convictions of Spencer, who had been convicted on the basis of DNA testing that matched his DNA with that of semen found in several victims.

In *People v. Castro* (1989), the New York Supreme Court examined numerous issues relating to the admissibility of DNA evidence. (The earlier "See You in Court" feature discusses this case in depth.) Although the court found the scientific theory underlying DNA to be convincing, it required

Chemical-based tests, while presumptive, are commonly used for initially establishing the presence of blood, saliva, and semen evidence at a crime scene.

Blood

- The Kastle-Meyer test and the LMG reagent are commonly used tests that rely on color changes to detect the presence of blood.
- Luminal sprays can detect very small amounts of blood when viewed in the dark.
- In both cases, these chemical-based tests can produce false-positive results, so serological tests such as a precipitin test and blood typing must be used for definitive identification of blood.

Saliva

- A color change test for saliva leverages the ability of an abundant enzyme in saliva (amylase) to digest starch, resulting in a yellow color when iodine is added to a saliva-containing solution.

Semen

- UV light, although not specific for semen, is used in the initial surveillance process.
- A chemical test that detects the ability of an enzyme present in seminal fluid to react with a specific substrate results in a rapid color change in the presence of seminal fluid.
- A more specific test for semen involves serological methods for detecting a protein that is present only in semen.
- Microscopic identification of sperm is a direct method of establishing the presence of seminal fluid.

Following this preliminary identification of the type of biological material at a crime scene, forensic samples are collected for DNA profiling to further characterize the samples at the individual level.

---

laboratories and personnel to follow appropriate practices and prove the validity of their procedures before the resulting evidence would be accepted in court.

In summary, defendants have successfully challenged the improper application of DNA scientific techniques in particular cases, especially when these methodologies are used to declare "matches" based on frequency estimates. However, DNA testing, when properly applied, is generally accepted as admissible under Frye or Daubert standards. As stated in the National Research Council's 1996 report on DNA evidence, "The state of the profiling technology and the methods for estimating frequencies and related statistics have progressed to the point where the admissibility of properly collected and analyzed DNA data should not be in doubt."

Mitochondrial DNA

The DNA profiling described in the previous cases was carried out using the DNA taken from the nucleus of the cell, which is known as nuclear DNA (nDNA). DNA is also present in the mitochondria of the cell and can be used for testing purposes as well.

Mitochondria are organelles that are found outside the nucleus of the cell in the cytoplasm (refer to Figure 13-8). In humans, the egg cell is about 5000 times the volume of the sperm head and contributes virtually all the cytoplasm to the zygote, including the mitochondria. Mitochondria are the power plants of the human body, providing approximately 90% of the energy we need. Each human cell contains hundreds—if not thousands—of mitochondria.

Although mitochondrial DNA (mtDNA) is chemically the same as nuclear DNA, mtDNA differs from nDNA in several important ways. Unlike nuclear DNA, which is linear and about 3 billion base pairs long, mtDNA is circular and much smaller, consisting of only 16,569 base pairs. Cells can contain hundreds or even thousands of copies of mtDNA, whereas each cell nucleus contains only two copies of each chromosome of nDNA.

Mitochondrial DNA is inherited from a person's mother, because the egg is the major contributor of cytoplasm to the zygote. Brothers and sisters will have the same mtDNA as their mother, maternal aunts and uncles, and maternal grandmother (FIGURE 13-19). Because of this pattern of maternal inheritance, mtDNA can be used to identify the remains of a dead person by its comparison to living relatives with the same mtDNA. Over the past 15 years, the identification of the remains of persons of historical interest, such as Jesse James

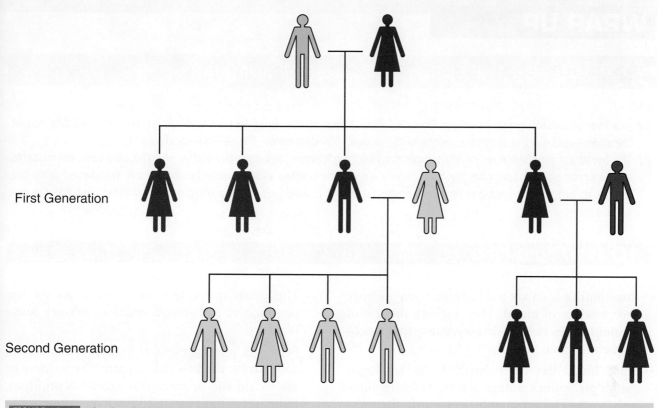

First Generation

Second Generation

FIGURE 13-19 Mitochondrial DNA is maternally inherited.

and Tsar Nicholas II of Russia, has been accomplished by analysis of their mtDNA. Likewise, many "missing in action" soldiers and the "Unknown Soldier" of the Vietnam War (who is interred in Arlington Cemetery) have been identified by comparing the mtDNA recovered from their bone fragments to the mtDNA of living maternally related relatives.

In the early 1990s forensic laboratories began to develop techniques that would allow mtDNA to be used as a tool for human identity testing in criminal cases. After the technology was validated, the first use of mtDNA as evidence in a court case occurred in 1996 in *State of Tennessee v. Ware*. In this case, only circumstantial evidence pointed to Paul Ware as having committed the rape and murder of a 4-year-old girl. The defense claimed that another man in the home, the babysitter, had framed Ware, who was found drunk and asleep next to the body of the child.

Surprisingly, the girl's blood was not found on the suspect, nor was his semen found on her. The medical examiner determined that the child had been strangled. During his examination, the doctor found a hair stuck to the victim's lip, another hair touching the mucosa of her rectum, and a third hair

in the victim's pharynx. The police recovered additional hairs off the victim's bed. A special agent with the Federal Bureau of Investigation's Hair and Fibers Unit testified regarding the testing he performed on the hairs found at the crime scene and on the victim's body. He concluded that the hairs from the bed and the hair in the victim's pharynx were consistent with having originated from the defendant. He pointed out, however, that hair comparison is not a means of making a positive identification and, as such, he could not conclusively say whether the hairs belonged to the defendant.

Mitochondrial DNA was extracted from two of the hairs recovered from the crime scene. DNA was also extracted from a sample of Ware's saliva and from the victim's blood. The mtDNA sequences from the hair in the pharynx and from the hair found on the sheet were compared and found to be exact mtDNA matches for each other. They were further compared with the saliva sample from Ware and found to match its mtDNA. Each of these three mtDNA samples was compared with the known mtDNA sequence of the victim but did not match it. Mitochondrial DNA had for the first time been used to conclusively identify the perpetrator of a murder.

WRAP UP

1. The four bloodstains can be categorized as follows: pool on the floor—passive; drops on the floor leading out of the room—passive; bloodstains on the wall—active; bloodstains on the ceiling—active.
2. The bloodstain evidence definitely suggests a violent encounter between the victim and the assailant. Bloodstains at the crime scene appear on the walls, and even on the ceiling. Blood probably would have transferred onto the assailant, but this suspect has no blood on his clothing. It would be hard to imagine that this suspect was the assailant.

Chapter Spotlight

- Presumptive tests are used to tentatively identify the presence of blood. They include the use of luminol spray, the Kastle-Meyer test, and leuco-malachite green (LMG).

- Once blood has been detected, the serologist uses a precipitin serological test, which requires only a small blood sample, to identify the presence of proteins that are specific to humans.

- Commercially available precipitin test kits are available for testing the blood of domestic animals and can be used if the first precipitin test gives a negative result for human blood.

- If human blood has been identified, further forensic analysis is performed to determine whether it can be associated with a particular individual. Serological techniques and DNA profiling are two techniques that distinguish individual biological traits.

- The ABO system of blood typing classifies human blood into four broad groups based on the presence or absence of either or both antigen A or antigen B on the surface of red blood cells.

- The measurement of the Rh (Rhesus disease) factor is another serological tool. If a person has a positive Rh factor, his or her blood contains a protein that also is found in Rhesus monkeys. These antigens are also found on the surface of red blood cells.

- Saliva is often present at violent crime scenes and can be used to identify an individual through DNA profiling. Saliva is always present if the perpetrator left bite marks on the victim.

The presumptive test for saliva relies on the presence of a digestive enzyme, salivary amylase.

- Semen is often present at violent crime scenes and can be used to link a particular suspect to the victim and/or scene through DNA profiling. Four techniques are used to detect semen: (1) UV light (since semen fluoresces under UV light), (2) the acid phosphatase test, (3) the p30 test, and (4) direct observation of sperm under a high-powered microscope.

- When collecting evidence relating to a sexual assault, physical and biological evidence may help to establish a link between the victim and possible perpetrators. Physical evidence such as blood, semen, hairs, saliva, and fibers may also help investigators reconstruct the sequence of events that took place during the commission of the crime.

- Portions of DNA are as unique to each individual as fingerprints. As a consequence, courts have allowed DNA profiling results to be submitted as evidence, as long as the laboratory and personnel who analyzed the DNA used the proper techniques.

- DNA technology can probe the Y chromosome to show paternity.

- Mitochondrial DNA (mtDNA) is maternally inherited and can be used to identify the remains of a dead person by comparing the decedent's mtDNA with the mtDNA of his or her living relatives.

Adenine One of the two double-ring bases found in nucleic acids. It belongs to the class of compounds called purines.

Agglutination A situation in which red blood cells stick together.

Allele The sequence of nucleotides on DNA that constitutes the form of a gene at a specific spot or a chromosome. There can be several variations of this sequence, each of which is called an allele. For example, in the gene determining a person's blood type, one allele may code for type A, whereas the other allele may code for type B.

Antibody A protein that inactivates a specific antigen.

Anticodon The three-base sequence carried by the tRNA molecule that determines which specific amino acid it will deliver.

Antigen Any substance (usually a protein) capable of triggering a response from the immune system.

Antiserum Blood serum containing antibodies against specific antigens.

Blood serum Blood plasma with its protein content removed.

Chromosomes The structures in the nuclei of cells that are coiled around histone molecules.

Coding region A section of DNA that contains the code for protein synthesis.

Codons Three-base sequences along the mRNA strand that determine the order in which tRNA molecules bring the amino acids to the mRNA.

Cytoplasm The part of the cell surrounding the nucleus.

Cytosine One of the three single-ring bases found in nucleic acids. It belongs to the class of compounds called pyrimidines.

Deoxyribonucleic acid (DNA) A double-stranded helix of nucleotides that carries the genetic information of a cell.

Deoxyribose The sugar in DNA; it has two hydrogen atoms.

Double helix A structure that can be compared to a spiral staircase, consisting of two polynucleotide chains wound around each other.

Genetic code The specific amino acid that each of the three-base sequences in mRNA codes has.

Guanine One of the two double-ring bases found in nucleic acids. It belongs to the class of compounds called purines.

Heterozygous Having different alleles at one or more corresponding chromosomal loci.

Homozygous Having identical alleles at corresponding chromosomal loci.

Locus A specific location on the DNA chain.

Noncoding region A section of DNA that does not code for protein synthesis.

Nucleic acids Complex compounds that are found in all living cells except mammalian red blood cells.

Nucleotides Long chains of repeating units that make up nucleic acids.

Plasma The colorless fluid of the blood in which the red and white blood cells are suspended.

Precipitin An antibody that reacts with a specific soluble antigen to produce a precipitate.

Presumptive test A test that provides a reasonable basis for belief.

Ribonucleic acid (RNA) A nucleic acid found primarily in the cytoplasm.

Ribose A sugar that contains a hydrogen atom and an —OH group.

Secretor An individual who has significant concentrations of antigens not only in his or her blood but also in other body fluids, such as perspiration, saliva, semen, and urine.

Serology The science that deals with the properties and reactions of blood serum.

Thymine One of the three single-ring bases found in nucleic acids. It belongs to the class of compounds called pyrimidines.

Uracil One of the three single-ring bases found in nucleic acids. It belongs to the class of compounds called pyrimidines.

Putting It All Together

Fill in the Blank

1. The liquid portion of blood is called _____.

2. Blood plasma is _____% water.

3. Red blood cells are called _____.

4. White blood cells are called _____.

5. Platelets are called _____.

6. The only mammalian cells that do not have a nucleus are _____.

7. The one presumptive blood test that is performed directly on the bloodstained object is the _____ test.

8. Most presumptive tests for blood rely on _____ ability to catalyze the oxidation of certain reagents.

9. In most cases, the presumptive tests for blood apply a dye solution that is followed by the application of _____ _____.

10. Red blood cells have specific carbohydrates on their surface that are called _____.

11. When animals are injected with human blood, they form _____.

12. A person who has a significant concentration of blood antigens in his or her perspiration and saliva is referred to as a(n) _____.

13. The two most common blood types in the United States are __and ___.

14. The presumptive test for saliva searches for the enzyme called _____ _____.

15. A(n) _____ light is used to search the crime scene for semen stains.

16. A presumptive test for seminal fluid that is not dependent on the presence of sperm checks for the presence of _____, a protein produced in the prostate.

17. Semen can be unambiguously identified by the presence of _____ when observed under a microscope.

18. In cases of rape, the violent contact between the victim and the assailant may result in a(n) _____ of evidence.

19. Inside the cell are _____, where protein synthesis occurs.

20. The cell's energy is produced by the _____.

21. Hereditary material is found in the nucleus of the cell in elongated bodies called _____.

22. Chromosomes are made of _____.

23. The position of a gene along a DNA strand is known as a(n) _____.

24. If a genetic locus is located at a different position, it is termed a(n) _____.

25. If two alleles at a specific genetic locus on homologous chromosomes are different, they are _____ (heterozygous/homozygous).

26. If the mother has type AA blood and the father has type BB blood, their offspring will have type ____ blood.

27. Nucleic acids are found in all living human cells except _____ _____ _____ cells.

28. DNA is found in the _____ of the cell.

29. RNA is found in the _____ of the cell.

30. The base uracil is found only in _____.

31. The sequence of the four bases in DNA is referred to as its _____ structure.

32. DNA consists of two strands of bases wound around each other to form a(n) _____ _____.

33. The base G on one strand of DNA always hydrogen-bonds to a(n) _____ base on the other strand.

34. The base A on one strand of DNA always hydrogen-bonds to a(n) _____ base on the other strand.

35. Genes are the segments of DNA that control the production of _____.

36. Each transfer RNA molecule carries a three-base sequence that is called a(n) _____.

37. Mitochondria produce _____ for the cell.

True or False

1. Luminol interferes with DNA testing and, therefore, should not be applied directly to a bloodstain.
2. People who have type O blood have both an A antigen and a B antigen.
3. When antiserum A is added to type O blood, the blood clots.
4. Most people have a positive Rh factor.
5. Mitochondrial DNA is designated by the shorthand form "nDNA."
6. The DNA in the mitochondria is inherited equally from the person's mother and father.
7. Cells can contain thousands of copies of mitochondrial DNA.

Review Problems

1. A stain found at the scene of a crime is suspected of being a bloodstain.
   a. Describe in detail two presumptive tests for blood.
   b. The presumptive tests indicate that this stain is blood. Describe two serological tests that will confirm that the blood is human.
2. You are asked to find the ABO type of a bloodstain on a pair of jeans.
   a. Your lab has two types of commercial antiserum available: anti-A and anti-B. Describe how these materials are used.
   b. You add anti-A to the blood sample and the blood clots. You take another sample of blood and add anti-B to it; it also clots. Which ABO blood type is the blood sample?
3. A stain found on the clothing of a rape victim is suspected to be semen.
   a. Describe two presumptive tests for semen.
   b. Describe a test that would confirm the results of the presumptive tests.
4. A victim of a sexual assault tells you that the assailant licked her. Describe tests you could perform to test for saliva.

5. The base sequences of four DNA fragments are listed below. What would be the sequence of each of the complementary strands of DNA?
   a. A-C-T-G-T-A
   b. A-G-G-A-C-T
   c. G-A-T-A-C-A
   d. T-G-C-T-T-A
6. Describe why cytosine binds only with guanine and tyrosine binds only with adenine.
7. Describe the difference between coding regions and noncoding regions in DNA.
8. We know that DNA controls protein synthesis and that nuclear DNA is held inside the nucleus of the cell. Protein synthesis takes place outside the nucleus. Explain how DNA directs protein synthesis.
9. A forensic lab recovers a badly decomposed body. Based on the size of the skeleton and the location where it was found, investigators think it may be the remains of a missing woman. Describe in detail how the investigators should approach the DNA analysis and from whom the reference DNA samples should be taken.

Further Reading

Bevel, T., and Gardner, R. M. *Bloodstain Pattern Analysis, with an Introduction to Crime Scene Reconstruction*, 3rd ed. Boca Raton, FL: Taylor & Francis, 2008.

Butler, J. M. *Forensic DNA Typing*. London: Elsevier Academic, 2005.

Foster, E. A., Jobling, M. A., Taylor, P. G., Donnelley, P., deKnijff, P, Mieremet, R., Zerjal, T., and Tyler-Smith, C. Jefferson fathered slave's last child. *Nature*. 1998; 396: 27–28.

James, S. H., P. E. Kish, and T. P. Suton, *Principles of Bloodstain Pattern Analysis: Theory and Practice*, Boca Raton, FL: Taylor & Francis, 2005.

Jones, E. L. The identification of semen and other body fluids. In R. Saferstein (Ed.), *Forensic Science Handbook, vol. 2* (ed 2), pp. 329–399, Upper Saddle River, NJ: Prentice-Hall, 2005.

Sutton, T. P. Presumptive blood testing. In S. H. James (ed.), *Scientific and Legal Applications of Bloodstain Pattern Interpretation*. Boca Raton, FL: CRC Press, 1999.

http://criminaljustice.jblearning.com/criminalistics

Answers to Review Problems

Interactive Questions

Key Term Explorer

Web Links

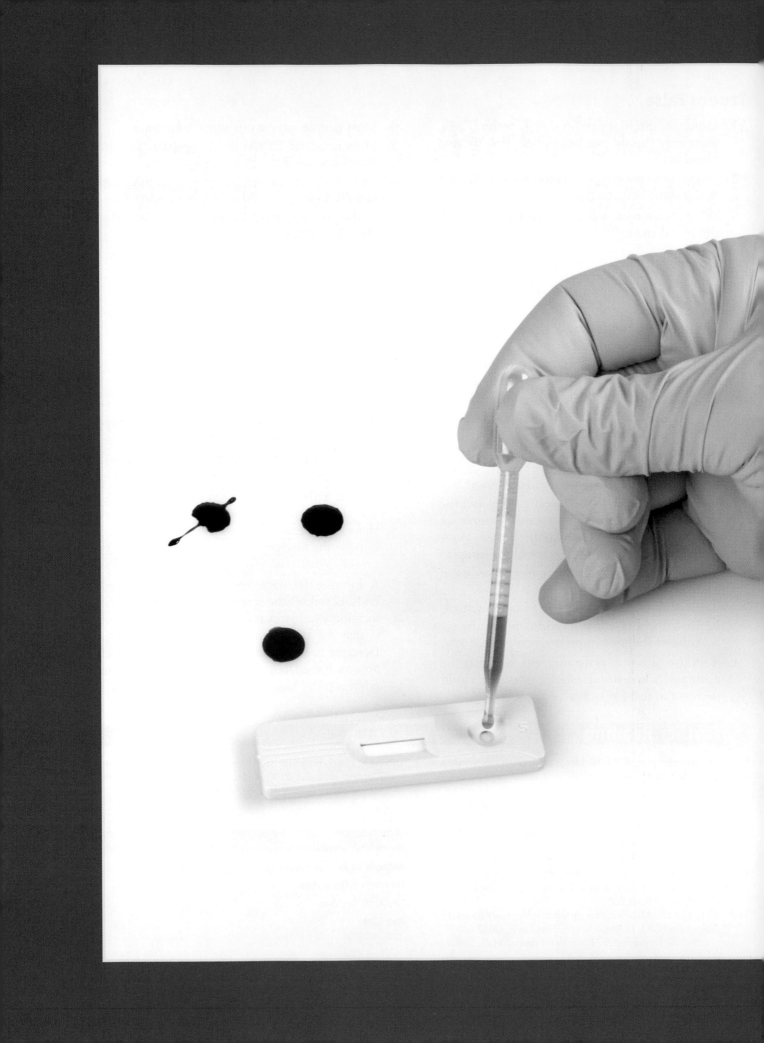

Forensic DNA Typing

OBJECTIVES

In this chapter you should gain an understanding of:

- The use of the polymerase chain reaction (PCR) to make many copies of a DNA sequence
- Short tandem repeats (STRs) and their forensic importance
- The use of electrophoresis to analyze STRs
- The Combined DNA Index System (CODIS)
- DNA paternity testing
- Mitochondrial DNA testing

FEATURES

On the Crime Scene

Back at the Crime Lab

See You in Court

WRAP UP

Chapter Spotlight

Key Terms

Putting It All Together

Depending on the crime, biological evidence could consist of blood, semen, saliva, fingerprints, tissue, bone, urine, hair, or feces. By extracting and analyzing the DNA from these samples, the forensic scientist can produce a profile that can identify who was present at the crime scene. DNA typing has become so reliable that if a suspect's DNA profile matches that of the evidence found at the crime scene, it is almost a certainty that he or she was there. By contrast, serological blood typing allows for only the elimination of a suspect; it hardly ever leads to certain identification of a suspect.

Before the DNA present in the evidence can be typed, the evidence sample must be processed to successfully isolate the DNA from all other material present in the biological sample. Because DNA is located inside cells, first the cell wall (plasmid membrane) and then the nuclear membrane that surrounds the nucleus must be opened for the DNA to be released.

Blood samples, for instance, are mixed with a solution containing detergent, salt, and an enzyme that breaks proteins (the components of cell walls) apart. Proteinases are enzymes that digest proteins by breaking them into smaller units; through their action, DNA is released from the cell and travels into the solution. The salt is added to stabilize the DNA, and the detergent breaks down lipids, another cellular component.

After the enzyme has finished working, an organic solvent, such as chloroform, is added to the DNA-containing mixture. Chloroform is not soluble in water, so it forms a second liquid phase at the bottom of the container, much like the way in which vinegar and oil form two phases in salad dressing. The chloroform dissolves cellular components. Because DNA is soluble in water, it stays in the aqueous phase. When the water layer is removed, the cellular components are left behind in the chloroform layer.

This separation step isolates about 90% of the DNA in the sample in the water layer and can be repeated to recover even more of the DNA present. The DNA in the sample is now ready to be typed.

1. Enzymes called nucleases are responsible for the breakdown of DNA. Nucleases are present on many common surfaces—even our skin. A blood sample collected at a crime scene cannot be analyzed immediately due to problems in the laboratory. Suggest how the wet blood sample should be stored until it can be analyzed.
2. Another bloodstain is found on a suspect's denim jeans. How should it be stored?

Introduction

The landmark discovery of the structure of DNA by Watson and Crick in 1953 set in motion the development of one of the most useful tools in forensic science. Watson and Crick were the first to describe the basic structure of DNA and its method of replication. The DNA in all of the cells of a single individual is the same throughout the entire body because humans evolve from a single fertilized egg—that is, a single source of DNA.

By the early 1980s, after studying the composition and organization of the basic genetic material, scientists had concluded that most human DNA varies very little from one individual to the next. However, a British researcher, Alec Jeffreys at the University of Leicester, discovered a region in human DNA that showed enormous variations from one person to the next. In 1983 Jeffreys found that DNA contained repeated sequences of genetic codes within "mini-satellites"; these mini-satellites, in turn, contained core sequences that are unique to particular individuals. Probes can now be designed to detect a person's individual type in these regions of variable DNA. Jeffreys' discovery was a crucial element in the development of the technology now known as DNA fingerprinting.

A person's genetic code can now be determined from very small amounts of DNA. Therefore, bloodstains, saliva, bone, hair, semen, or any other biological material found at a crime scene can be a ready source of DNA for forensic investigators.

DNA fingerprinting was first used in the Narborough area of Leicester, England, to identify the criminal who raped and murdered two girls in 1983 and 1986. The most likely suspect was a local youth, Richard Buckland, who gave police information regarding the location of the body of one of the girls. Using Jeffreys' technique, police determined that the DNA in samples taken from Buckland and from the body of the girl did not match. In an attempt to identify the actual perpetrator, the local police took DNA samples from 5000 local men over the course of 6 months, but no matches were found. Some time later, a man named Ian Kelly was overheard bragging that he gave his sample while masquerading as a friend, Colin Pitchfork. Pitchfork, a local baker, was taken into custody, and an analysis of his sample did match that of the killer. Pitchfork later confessed to both murders. This series of events marked the first forensic use of DNA fingerprinting both to exonerate a suspect and to identify the killer.

Restriction Fragment Length Polymorphisms

Our DNA contains a series of genes, which are sections of the DNA molecules that carry the code that controls the production of proteins within the human body. These proteins, in turn, control all of the chemical reactions that occur within our bodies, including those that produce characteristics such as blue eyes and blond hair.

Other DNA sections do not contain a code for synthesizing proteins but rather act primarily as spacers between the coding areas. The sequence of bases in these noncoding DNA regions varies considerably from person to person; as a consequence, these regions can be used for DNA profiling.

As discussed in the "You Are the Forensic Scientist" section earlier in this chapter, the first step in the analysis of a sample is to extract DNA from a particular chromosome. Next, restriction enzymes—which, in essence, are "molecular scissors"—cut the DNA strands into fragments at specific base sequences. This technique, which relies on **restriction fragment length polymorphisms** (**RFLPs**), allows for the individualization of DNA evidence.

For example, one restriction enzyme will cleave DNA wherever the sequence G-A-A-T-T-C occurs (FIGURE 14-1). When this enzyme reacts with the DNA, it cuts at wherever it finds this sequence, and only at this sequence. Given that different people will have different sequences in the noncoding regions of their DNA, the number and lengths of the fragments that are produced by the

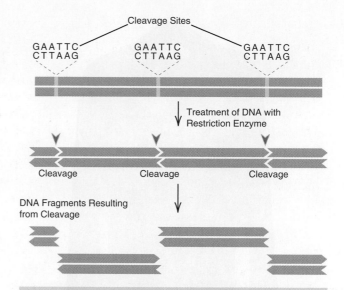

FIGURE 14-1 When DNA is treated with a restriction enzyme, the two strands are cut into fragments at specific sites. One restriction enzyme cleaves the DNA strands wherever the base sequence G-A-A-T-T-C occurs, producing a number of fragments of different length.

restriction enzymes can be matched to a person's reference DNA. Typically RFLP sequences are 15 to 34 bases in length; these sequences may be repeated as many as 1000 times in the strand that is cut out. RFLP produces large DNA fragments that are 15,000 to 30,000 bases long.

Ten years ago RFLP was the primary technique used for DNA typing and was the first scientifically accepted protocol in the United States for typing DNA. RFLP evidence was used in the impeachment proceedings for President Bill Clinton in 1998. During the investigation of the relationship between President Clinton and White House intern Monica Lewinsky, it was revealed that Lewinsky possessed a dress that she claimed was stained with the president's semen. The FBI laboratory compared the DNA extracted from the stain on her dress with DNA from the president's blood sample. A seven-probe RFLP match was obtained between the semen in the stain and the president's blood sample. The Federal Bureau of Investigation (FBI) concluded that the frequency with which the seven RFLP matches would occur randomly was about one in 8 trillion—there was no doubt that the semen on the dress came from President Clinton (FIGURE 14-2).

RFLP has many disadvantages, however. The RFLP method takes 6 to 8 weeks to obtain results; it requires a large sample of intact, nondegraded DNA; and it is not amenable to high-volume sam-

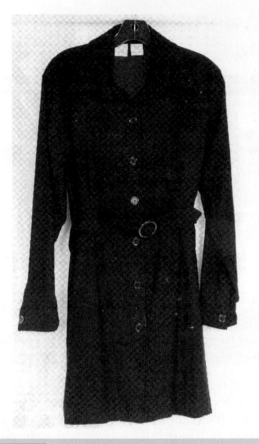

FIGURE 14-2 Restriction fragment length polymorphism (RFLP) evidence was used when examining Monica Lewinsky's dress.

ple processing. The DNA strands that are produced by RFLP methods are very large (30,000 base pairs); in forensic cases where the recovered DNA is partially degraded, these large sections are often damaged. When the restriction enzyme cuts out the damaged DNA sequence, the RFLP technique may therefore be unable to match it to an undamaged reference DNA sample.

In recent years, RFLP methods have been replaced by methods based on PCR, largely because of PCR's ability to process forensic samples that are either present in minute quantities or partly degraded.

Polymerase Chain Reaction: A DNA Copy Machine

Today, many copies of a particular portion of the DNA sequence are manufactured in the DNA lab by using the **polymerase chain reaction (PCR)**. When used for forensic purposes, the PCR amplifies only those DNA regions that are of interest; it ignores all other regions in the DNA strand. PCR is fast and, because it can make copy after copy, extremely sensitive.

It is well known that two complementary DNA strands will **denature** (separate) if they are heated above a certain temperature (i.e., the melting point); when cooled, the strands will **anneal** (come back together). At temperatures above the melting point, the hydrogen bonds that hold the two strands together are broken, and the DNA becomes two single strands (ssDNA). Below the melting point, the complementary hydrogen bonds reform. This process can be repeated time and time again without degrading the DNA. In the PCR amplification process, DNA is repeatedly heated and cooled in a process that is known as **thermocycling** (**FIGURE 14-3**).

FIGURE 14-4 shows how the PCR reaction is carried out. The DNA sample is placed in a small sample tube, and **primers** and an enzyme, **DNA polymerase**, are then added. The primers are short sequences of DNA that are specifically designed and synthesized to be complementary to the ends of the region to be amplified. Two primers are used for each DNA locus (region): a forward primer and a reverse primer. The specific sequence (of bases) at the ends of each region is well known (and is called the flanking region), so the primer can be designed to attach there and no other location on the DNA strand.

FIGURE 14-3 A thermocycler repeatedly heats and cools DNA as part of the polymerase chain reaction (PCR) amplification process.

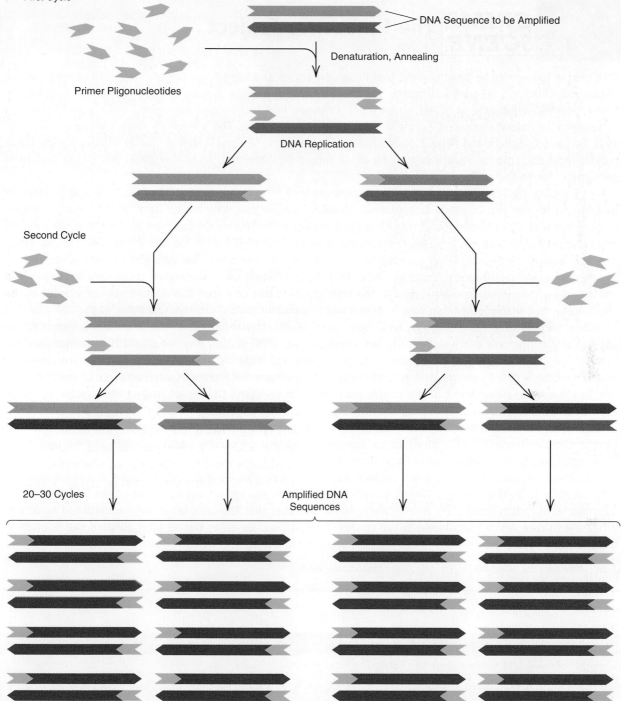

First Cycle

DNA Sequence to be Amplified

Denaturation, Annealing

Primer Pligonucleotides

DNA Replication

Second Cycle

20–30 Cycles

Amplified DNA Sequences

FIGURE 14-4 The polymerase chain reaction (PCR) is used to amplify a particular DNA sequence. Only the DNA region to be amplified is shown here. A primer (color) that is complementary to the ends of the target sequence is used in repeated rounds of denaturing, annealing, and DNA replication. The number of copies of the target sequence doubles with each round of replication.

on the CRIME SCENE — The Innocence Project

DNA testing has proven to be a powerful tool in solving crimes and identifying perpetrators. DNA sampling of arrested individuals is pro forma in Virginia, and this analytical technique is used as the basis for the CODIS database of the Federal Bureau of Investigation (FBI). However, DNA testing also has proved to be equally powerful in helping to exonerate innocent people who have been incarcerated. The Innocence Project was established in 1992 by Barry C. Scheck and Peter J. Neufeld at the Benjamin N. Cardozo School of Law at Yeshiva University to provide legal and criminal justice support for those wrongfully convicted. As of May 2007, 200 people had been exonerated due to their efforts.

In February 2001 Earl Washington, Jr., was released from prison after serving 18 years for a rape and murder that he did not commit. In 1983, facing intense pressure, Washington, who has an IQ of 69, confessed to police that he committed the crime. Even though his story did not match the evidence found at the scene, Washington was convicted of the murder and faced the death penalty. Just days before his scheduled execution in 1994, Virginia Governor L. Douglas Wilder commuted his sentence to life in prison because of questions raised regarding the DNA evidence. However, it was not until 2000 that Washington was exonerated completely by further DNA testing supported by the Innocence Project; the testing proved not only that Washington was not present at the crime scene, but that another man held in prison was responsible for the rape and murder.

Over the past 10 years, DNA typing techniques have significantly improved. At the time when Governor Wilder commuted Washington's death sentence to life imprisonment, RFLP testing was the standard technique used for DNA testing. This sampling method relies on large fragments of DNA; old samples may have degraded, however, so this technique cannot always yield accurate results. Furthermore, RFLP requires large amounts of the DNA sample. In Washington's case, most of the sample consisted of the victim's fluid—not that of her attacker. Thus the RFLP tests done in 1994 were unable to conclusively rule out Washington as the perpetrator.

By 2000, however, a new DNA test had emerged: STR typing. STR typing uses amplification to enhance the sample DNA, and much smaller amounts are needed to produce conclusive results. Thanks to this technique, Washington was exonerated and ultimately released from prison.

Washington's case also brought out the fact that the Virginia DNA lab had made several errors in his case. An external audit in 2004 ordered by Virginia Governor Mark Warner and carried out by the American Society of Crime Laboratories Directors/Laboratory Accreditation Board indicated that the lab's senior DNA analyst had made significant mistakes, which contributed to the erroneous conviction. Governor Warner later adopted the recommendations made by the auditors to remedy the problems highlighted by this investigation.

Source: Data from the Innocence Project. http://www.innocenceproject.org.

EXAMPLE

Suppose the following DNA segment must be amplified by PCR:

GTCCTAATGAATGAATGAATGAATGTCCAGCAGAGTTACTTACTTACTTACTTAC**AGGTC**

The end of each sequence of interest (flanking region) must be known before a primer can be made that will bind to each end of this sequence. In this example, the ends are designated in bold type. The complementary primer for the upper strand is CAGGA; it will bind at the left on the first five bases. The complementary primer for the lower strand is TCCAG; it will bind at the right on the last five bases.

Step 1: When the DNA is heated to 194°F (90°C), the DNA denatures and separates into two single strands (ssDNA).

Step 2: The primers are added, and the temperature is lowered to 140°F (60°C). The primers bind to the complementary region on the sample DNA.

CAGAG

GTCCTAATGAATGAATGAATGAATGTCCA**G**CAGAGTTACTTACTTACTTACTTAC**AGGTC**
                                               TCCAG

Step 3: The DNA polymerase and a mixture of the free nucleotides (G, T, A, C) are added, and the temperature is raised to 162°F (72°C). The top strand is extended to the right, and the bottom strand is extended to the left.

```
CAGAG →
GTCTCAATGAATGAATGAATGAATGTCCAG
CAGAGTTACTTACTTACTTACAGGTC
← TCCAG
```

This produces two identical copies:

```
GTCTCAATGAATGAATGAATGAATGTCCAGCAGAGTTACTTACTTACTTACAGGTC
```

and

```
GTCTCAATGAATGAATGAATGAATGTCCAGCAGAGTTACTTACTTACTTACAGGTC
```

This completes the first cycle of the PCR reaction.

When the temperature reaches 194°F (90°C), the two complementary DNA strands denature in the first step of the PCR amplification. In the second step, as the temperature is lowered to 140°F (60°C), each primer finds and binds to its complementary sequence on the DNA strand as the hydrogen bonds re-form. In the third step, the temperature is raised to 162°F (72°C) so that the DNA polymerase enzyme starts to add bases to extend the primer and to build a DNA strand that is complementary to the sample DNA. At this point, an exact complementary copy of the DNA sequence has been synthesized.

This heating and cooling process is then repeated to make more copies. Each cycle doubles the number of copies; 30 cycles, which takes about 21.5 hours, will produce about 1 billion copies of the target region of the DNA. The concentration of the DNA copies, which are sometimes collectively referred to as an **amplicon**, is now high enough that the DNA can be analyzed by a variety of techniques.

PCR has been simplified by the availability of reagent kits that allow a DNA technician to simply add a DNA sample to a commercial PCR mixture that contains the primers and all the necessary reagents for the amplification reaction at the specific DNA loci of interest.

A DNA profile can be produced for a sample as small as 0.000000002 g (0.2 ng, or 0.2×10^{-9} g) after PCR amplification. This ability to amplify small amounts of DNA comes with a price, however. The DNA laboratory and DNA technicians must take extreme care to eliminate extraneous DNA from the PCR amplification area of the laboratory. Most forensic DNA laboratories restrict entry to these areas to only those individuals who are performing the PCR reactions. They also require each individual working there to be typed to ensure the technician does not contaminate the forensic sample and that the PCR product is a result of the amplification of the sample DNA, not the technician's DNA.

Short Tandem Repeats

One of the newer methods of DNA typing uses **short tandem repeat (STR)** analysis for identification. STRs are loci (locations) on the sample DNA that contain a short sequence of bases that is repeated over and over. STR sequences are named based on the length of the repeating unit. Although STRs can have anywhere from three to seven repeating units, most STRs used for forensic purposes are tetranucleotide repeats—that is, four bases repeat over and over. Examples of forensic STRs include the FGA locus, which has a TTTC repeat and is found in the human alpha fibrinogen locus; the THO1 locus, which has an AATG repeat and is found in the human tyrosine hydroxylase gene; and the TPOX locus, which also has an AATG repeat and is found in the human thyroid peroxidase locus.

The entire length of an STR is very short (fewer than 400 bases). Because the STR is so short, it is less susceptible to degradation than are the fragments used for RFLP. Thus STRs are often recovered from bodies or stains that have started to decompose. The small size of the STR also means that it can be amplified very quickly in PCR amplification reactions, which is a huge advantage when analyzing crime scene evidence that consists of only small samples.

The number of repeats in STRs varies widely among individuals, which makes them very useful for human identification purposes. For the THO1 STR, for example, the most common repeats are 6 through 10 (FIGURE 14-5). Notice also that each

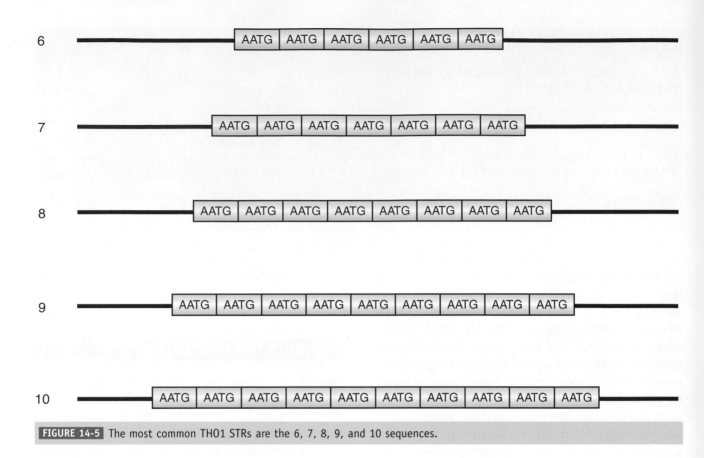

FIGURE 14-5 The most common TH01 STRs are the 6, 7, 8, 9, and 10 sequences.

individual STR is longer by 4 bases: The 6 allele is 24 bases long; the 7 allele is 28 bases long; the 8 allele is 32 bases long; and so on. This fact becomes important when we discuss the separation and identification of each STR.

STRs are scattered throughout the genome and occur, on average, every 10,000 bases. Searches of the recently completed human genome reference sequence have attempted to comprehensively catalog the number of STRs in this genome. The STR markers that are used in forensic DNA typing are typically chosen from separate chromosomes to avoid any problems with linkage between the markers. We will describe those STRs in more detail later in this chapter.

DNA Sequence Variations Among Individuals

Individuals differ genetically because they possess different combinations of alleles at numerous loci in their genomes. It is currently estimated that the DNA sequences of two randomly selected individuals will differ at about 1 million different posi-

tions. The Human Genome Project has yielded a number of unexpected findings, one of the most surprising of which is that most of the genome is not involved in coding for proteins. It is now estimated that only 3% of a person's total DNA is involved in coding directly for proteins. This small percentage has led the popular press to use the term "junk DNA" to describe the parts of the genome whose function is now unknown. More precise language would describe the portion that carries the code for protein synthesis as the "coding region" and the rest of the genome as the "noncoding region." Thus, if the DNA from two individuals differs at 1 million different positions, then statistics tell us that most of those differences will fall in the noncoding regions and will not affect protein function.

If the differences in DNA between two individuals are the result of a mutation, and that discrepancy appears in a noncoding region, it will not affect the ability of that person to survive. Mutations in the noncoding regions have no effect on the **phenotype** of the person and can be passed on to his or her children. The phenotype comprises the observable characteristics of a person (e.g., blue eyes and blond hair). Conversely, a mutation that affects an allele in a coding gene may prove fatal to that person in childhood, and it is unlikely such an individual would live long enough to pass it on to his or her offspring. For this reason, the most serious genetic diseases are rare.

The loci selected for DNA typing are **selectively neutral loci**; that is, they are genetically inherited characteristics that confer neither any benefit nor any harm to the individual's ability to reproduce successively. A mutation in a noncoding region is considered to be selectively neutral because the survival of the individual carrying it is not threatened by its presence. Because they are selectively neutral, these loci become more commonly found in the population.

Inheritance of Alleles

The alleles discussed in the previous section are inherited from the individual's parents following the fundamental rules of genetics. For example, FIGURE 14-6 shows fertilization using a single chromosome as an example. This particular chromosome has alleles that are STRs and differ in the number of repeats. The father has 2 repeats on one of the chromosomes and 5 repeats at the same locus on the other, so he is heterozygous and has a 2, 5 genotype for this locus. The mother also is heterozygous and has an 8, 3 genotype for the same locus.

During meiosis, the chromosomes separate, such that each gamete receives only one of the two chromosomes from the parent. In the example illustrated in Figure 14-6, therefore, the father's seminal fluid will contain a mixture of sperm cells carrying either the 2 repeat or the 5 repeat. The mother's eggs will carry either a 3 repeat or an 8 repeat allele.

At conception, the haploid sperm fuses with the haploid egg, forming a diploid fertilized egg. This egg will have one of four possible combinations of alleles—in the case of Figure 14-6, either 2, 8; 2, 3; 5, 8; or 5, 3. Since these alleles are in the noncoding region and are not dominant or recessive, they have no bearing on the phenotype of the child.

Individuals will possess different alleles in numerous loci in their genome. To determine what makes a person an individual, genetically speaking, we need to be able to measure the length of an STR at different locations.

Analyzing the STR by Electrophoresis

Electrophoresis is a technology used to cause ions in solution to migrate under the influence of an electric field. DNA molecules, which contain many phosphates, form negative ions and are attracted to a positive electrode—a fact that can be used to separate them out in a solution. In this way, electrophoresis can be used to analyze DNA.

Because the smaller DNA molecules move faster, electrophoresis is used to separate STRs according to their length (i.e., shorter STRs will reach the positive electrode sooner than longer STRs). Once separated, the actual STR length (measured in base pairs) is determined by comparison with STR standards. This length can then be used to establish the number of repeats and hence to elucidate the genotype of the individual at each amplified locus.

Gel Electrophoresis

Fifteen years ago, **gel electrophoresis** was the only DNA separation technique that was readily available. In this type of electrophoresis, a polyacrylamide gel is molded into a semisolid slab and placed in a tray containing a buffer solution; this solution maintains the pH within a relatively narrow range and carries the applied electric current. The technician then uses a pipette to place the DNA samples into wells located in the top of the slab (FIGURE 14-7). When a voltage (30 volts) is applied, the DNA in the samples starts to migrate through the gel. Because DNA has a negative charge on its phosphate groups, it migrates toward the positive electrode. The gel is an entangled polymer that acts as a sieve. That is, the shorter STRs move more easily in the gel and travel farther from the starting point; the longest STRs will travel the shortest distance. By observing the distance the STR travels on the gel slab, the technician can determine its number of repeats.

After the voltage has been applied for 2 to 3 hours, the electrophoresis is stopped and the DNA, which is not visible to the naked eye, must be made visible. Each group of similar-length DNA molecules should appear as a narrow band in the gel. Finding the DNA bands requires an additional step, which takes more time and subjects the technician to extra risks. To highlight the lines of DNA, a stain that binds to DNA (e.g., ethidium bromide)

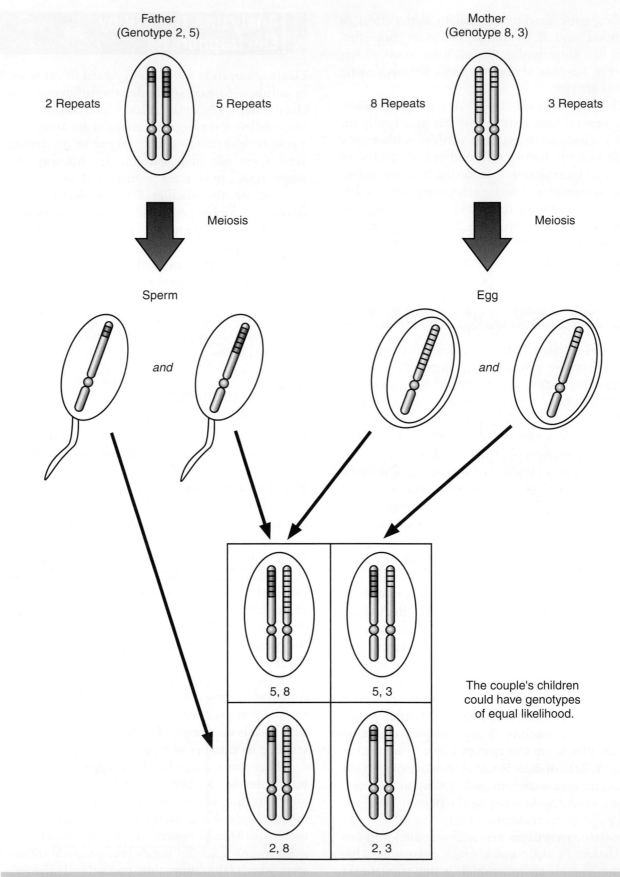

FIGURE 14-6 The inheritance of one allele at a single locus. This couple's offspring could have four possible combinations at this locus: 5, 8; 5, 3; 2, 8; or 2, 3.

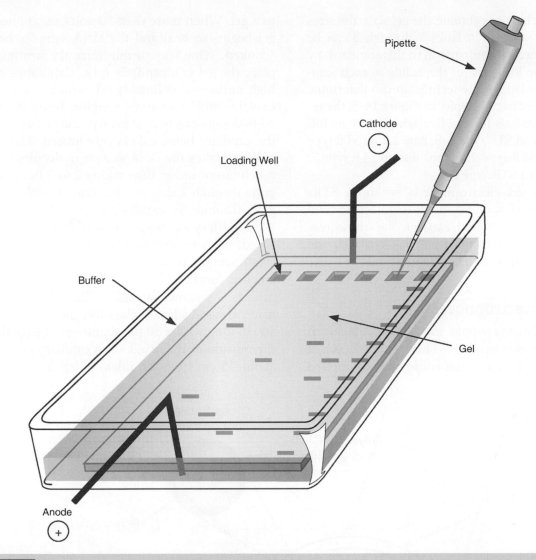

FIGURE 14-7 Gel electrophoresis. First, the DNA samples are placed in wells in the gel. After the voltage has been applied for a few hours, the DNA moves through the gel. The shortest STR moves the farthest while the longest STR moves the shortest distance.

is often added to the gel: under ultraviolet light, this dye fluoresces, causing the DNA to appear as bands on the gel. Unfortunately, ethidium bromide is a carcinogen, so anyone using it must exercise extreme caution. In the past, radioactive isotopes were also used to find the bands in the gel. Labs working with radioactive materials must be licensed by the Nuclear Regulatory Agency. Most DNA labs found that working with radioactive isotopes was too expensive and dangerous, which prompted many of them to abandon this technology as a DNA typing strategy.

The gel electrophoresis separation in **FIGURE 14-8** shows the separation of five different DNA samples and a standard DNA "ladder." The ladder, which appears in lane 6, is a standard that has been produced to contain DNA molecules with 2, 3, 4, 5, 6, 7, 8, 9, and 10 repeats of the STR being investigated. It is

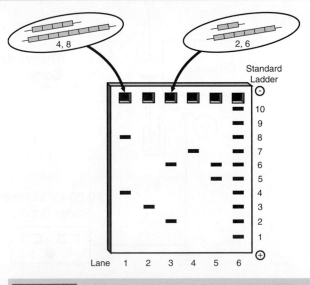

FIGURE 14-8 A gel electrophoresis separation of the STR samples shown in Figure 14-7.

placed in one lane to calibrate the gel slab; the sizes of the actual samples in lanes 1 through 5 can be then determined by comparison to this standard. By comparing the locations of the bands in each sample lane to the ladder, the technician can determine the STR type for each sample. In Figure 14-8, the results of the analysis for the five samples are as follows: lane 1 is an STR type 4, 8; lane 2 is an STR type 3; lane 3 is a STR type 2, 6; lane 4 is an STR type 7; and lane 5 is an STR type 5, 6.

Although gel electrophoresis separates STRs effectively, it is slow and difficult to automate (and dangerous, as mentioned earlier). For these reasons, capillary electrophoresis has replaced gel electrophoresis in most DNA laboratories.

Capillary Electrophoresis

The STR fragments would move more quickly if more voltage was applied. There is, however, a practical limit to how much voltage can be applied

to a gel. When more than 50 volts are applied, the gel begins to heat and the DNA samples become "cooked." One way to minimize the heating is to place the gel in a capillary tube. Capillaries have a high surface-to-volume ratio, which makes them easy to cool. Potentials ranging from 20,000 to 50,000 volts can be applied without the contents of the capillary being excessively heated. The higher voltage makes the DNA sample molecules migrate much faster; rather than taking 2 to 3 hours to migrate through a slab gel, the same sample will migrate through the capillary in 10 to 15 minutes.

Capillary electrophoresis (CE) produces high-speed, high-resolution separations on extremely small sample volumes, 1000 times less than the amount needed for a slab gel electrophoresis process. As shown in **FIGURE 14-9**, the capillary is made of fused silica (a glass-like material), and it is 50 cm long with a 50 μm diameter. The coating on the outside of the fused silica capillary is removed about 35 cm from the inlet so that a detector can

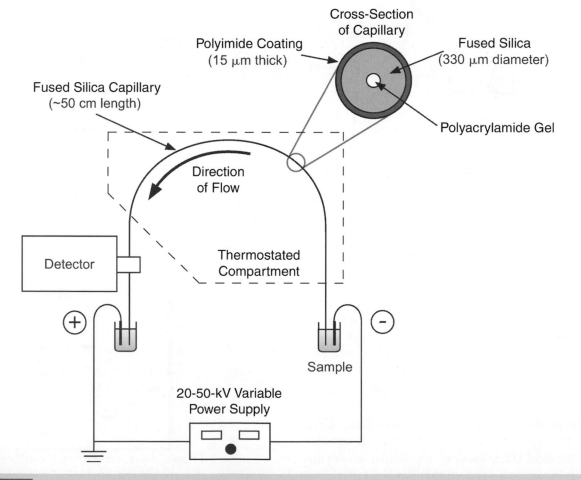

FIGURE 14-9 A capillary electrophoresis (CE) instrument may be used with much smaller samples than can be analyzed with gel electrophoresis.

measure the DNA as it flows by. As with slab gel electrophoresis, the smallest DNA molecules will arrive at the detector window first and the largest molecules last.

The CE used for DNA determination is a laser fluorescence detector (FIGURE 14-10). The argon laser, which produces light with a 488-nm wavelength, is placed so as to send its beam across the capillary. Just below the laser beam are four detectors that are tuned to detect one color each (red, yellow, green, and blue). The CE also can accept multiple samples; they are placed in the sample tray. As each DNA analysis is completed, the instrument robotically moves the tray to the next sample and begins the analysis anew.

As mentioned earlier, DNA molecules are not visible to the naked eye, so they must be labeled if the CE is to detect them. Fluorescence-based detection of STRs is now widely used in forensic laboratories. In this procedure, a fluorescent dye is attached to the PCR primer that is used to amplify the STR region of interest. The PCR amplification is carried out as usual, except that the amplicon produced will have a dye attached to it. The fluorescence measurement begins when the STR dye passes by the laser in the CE detector. The laser beam is absorbed, which causes the dye to enter an excited state. After losing some energy to its surroundings, the STR dye emits light and returns to its ground state.

Depending on the structure of the dye, the color of the light emitted at this point will be red, blue, yellow, or green. Four commonly used dyes are attached to primers that are sold as part of commercial DNA analysis kits. These dyes have complex organic structures and names, so they are more colloquially known by their nicknames: FAM emits blue light, JOE emits green light, TAMRA emits yellow light, and ROX emits red light. If ROX dye is attached to the amplicon, for example, the red detector will sense its emission after it has been excited by the laser beam.

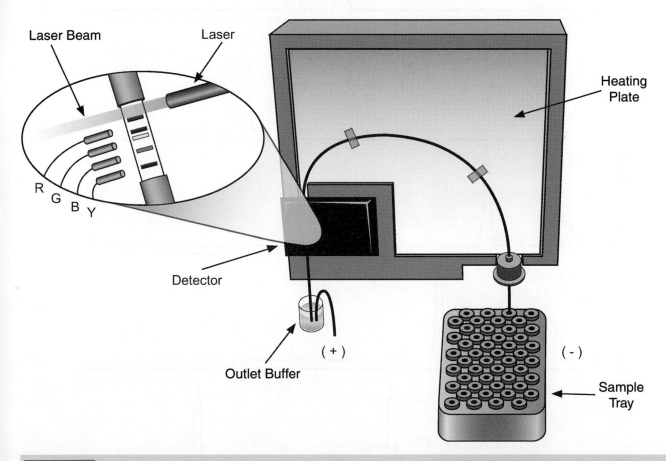

FIGURE 14-10 In CE, the laser in the detector irradiates the DNA as it passes. The type of dye attached to the STR determines what color light is emitted from the sample and then picked up by the detector: STRs with the FAM dye attached emit light with a blue color; those with the JOE dye attached emit green; those with the TAMRA dye attached emit yellow; and those with the ROX dye attached emit red.

If the same DNA samples that were analyzed by slab gel electrophoresis in Figure 14-8 were analyzed by CE, the results would appear as a series of Gaussian-shaped peaks (FIGURE 14-11). This CE separation, which takes only 10 to 15 minutes, produces computer-readable information about the location of the peaks along the *x*-axis. The results can then be compared to the standard ladder, allowing the STR genotype to be determined. The position of the peak on the *x*-axis indicates which STR allele is present. Unlike in gel electrophoresis, where the bands are developed in a separate step after the separation is complete, the STRs are detected instantaneously as they move past the detector.

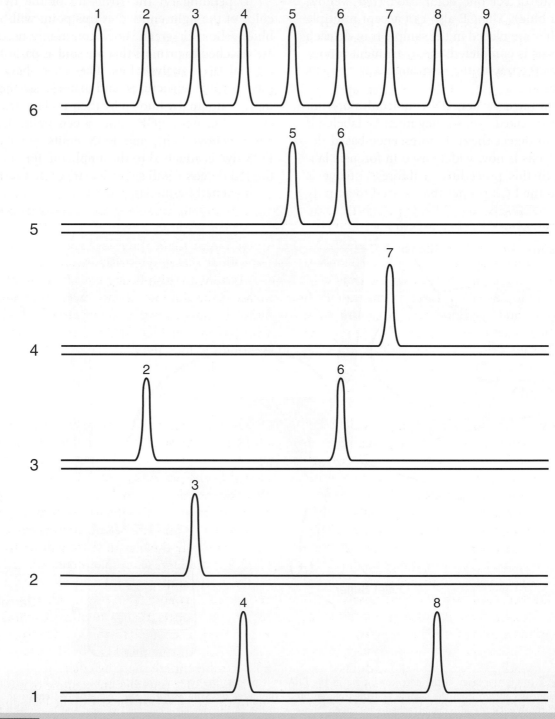

FIGURE 14-11 A CE separation of the STR products shown in Figure 14-8.

Forensic DNA analysis has represented a quantum leap in technology for the criminal justice system. The power of DNA analysis as evidence follows from its theoretical power to exclude large proportions of the population as potential contributors of genetic samples (e.g., blood, hair, semen). Thus, when a suspect's DNA matches a DNA sample that is recovered from the scene of a violent crime, a prosecutor may suggest that the suspect is a likely source of that sample and was present at the crime scene because the suspect was *not excluded* as a potential contributor.

In most cases, the probative value of a DNA match is presented in terms of frequency statistics that describe how common the particular DNA profiles are in a given population. For example, a prosecutor would make the argument that a DNA match statistic of 1 in 1,000,000,000 means that approximately one person out of every 1 billion members of a population will match that DNA profile. Put another way, the chance that a randomly selected person will match this DNA profile is one in 1 billion. If this DNA match statistic is low (one in many thousands, millions, or billions), the prosecutor would argue that it is *unlikely* that the match between a sample taken from the suspect and a sample recovered from the crime scene is purely coincidental.

Multiplex DNA Analysis

As previously discussed, CE can reduce the time needed to analyze a PCR-amplified STR. Forensic DNA laboratories need to process many samples quickly, and they need to amplify and identify a number of DNA loci simultaneously. By using **multiplexing**, multiple STR loci may undergo PCR amplification and be analyzed simultaneously. Multiplexing is accomplished by placing each of the four dyes on specific primers and by adjusting the size of the STR amplicon produced.

Multiplexing by Size

The number of repeats in STRs varies widely. For instance, THO1 contains 5 to 11 repeats, whereas FGA has 18 to 30 repeats. As a consequence, the THO1 amplicon will be 20 to 44 base pairs long, whereas the FGA amplicon will be 72 to 120 base pairs long. Since the CE separates STRs on the basis of their length (shorter = faster arrival at the electrode), the peaks for THO1 will appear much sooner on the *x*-axis than will the peaks for FGA. Given that some of the amplicons from different loci are different sizes, they will be clearly separated from one another and will appear at different locations on the *x*-axis of the CE analysis.

FIGURE 14-12 shows an example of multiplexing by size, in which PCR-amplified products from three different loci (D3S1358, VWA, and FGA) are simultaneously separated by CE. The D3S1358 alleles are the smallest STRs (105 to 140 base pairs), the VWA alleles are an intermediate size (150 to 195 base pairs), and the FGA alleles are the largest (215 to 270 base pairs). In this example, the primers for the three loci are simultaneously amplified via PCR and separated via CE. The STR genotypes at these three loci are easily identified for both of the samples that are analyzed.

Although multiplexing by size is useful for increasing the number of loci that can be analyzed simultaneously, it is not possible to multiplex more than five or six loci. To increase the number of multiplexed loci, the placement of the fluorescent dyes can be adjusted.

Multiplexing by Dye Color

To multiplex samples on the basis of dye color, one of the four fluorescent dyes can be attached to the primers for one specific STR locus. If ROX is attached to the primers for one locus, for example, only the red detector will see the PCR-amplified products from that STR locus; by contrast, if FAM is attached to another set of primers for another locus, then only the blue detector will see the PCR-amplified products from that locus. For those loci that produce amplicons that are the same size and cannot be separated from one another by CE, different dyes may be used to amplify each STR locus. In this way, even though the CE cannot separate the STRs from one another, the detectors will be able to differentiate them by color.

FIGURE 14-13 provides an example of multiplexing by dye color, where the size of the PCR-amplified product is shown horizontally and the color of the dye vertically. Inspection of the figure shows that

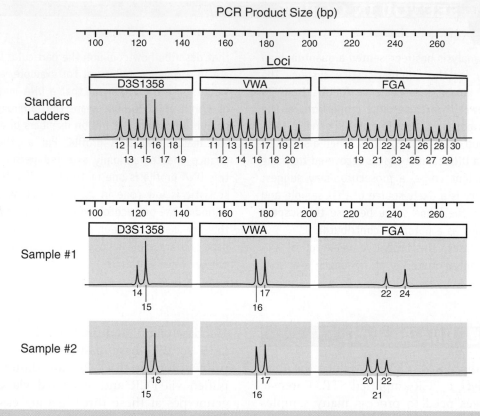

FIGURE 14-12 The sizes of the PCR-amplified products of the D3S1358, VWA, and FGA STRs are different enough that the three loci can be simultaneously separated (multiplexed) by CE.

some loci, such as the D13S317, TPOX, and FGA loci, give PCR products that are approximately the same size. The CE would be unable to separate the PCR-amplified products from these three loci so as to accurately type each locus. This problem can be overcome by attaching a green dye to the D13S317 primers, a yellow dye to the TPOX primers, and a red dye to the FGA primers; now the PCR-amplified products from each locus will be sensed by a different detector. Even though the CE cannot separate the products, the multiplexing of detectors will allow all of them to be identified simultaneously, each by its own detector.

Further inspection of Figure 14-13 reveals that by using size and color criteria, 15 STR loci can be simultaneously determined. Forensic labs routinely use a commercial PCR kit, the AmpFISTR, to multiplex these loci.

Multiplexing with Multiple Capillaries

All of the CE applications discussed so far have used one capillary, with samples being analyzed sequentially. Parallel capillaries that lie next to one another can process multiple samples simultane-

ously and are called capillary array electrophoresis (CAE) systems (**FIGURE 14-14**). Several commercial instruments are available that place 16 parallel capillaries into a single array. The sample tray is, therefore, able to inject 16 samples simultaneously, and the detector scans and records the data from all 16 capillaries. By comparison, an array containing 96 parallel capillaries was developed to meet the high-capacity needs of the Human Genome Project. CAE instruments are helping to reduce the time between when a DNA sample is collected at the crime scene and when it is actually analyzed.

Forensic STRs

Forensic scientists around the world have debated which STR markers are the best to use for human identification and how many STRs are needed to make a definitive identification. In choosing forensic STRs, care has to be taken to identify those that are not genetically linked. In forensic DNA testing, the goal is to distinguish unrelated individuals from one another. To have high discrimination power, the DNA databases used in most developed

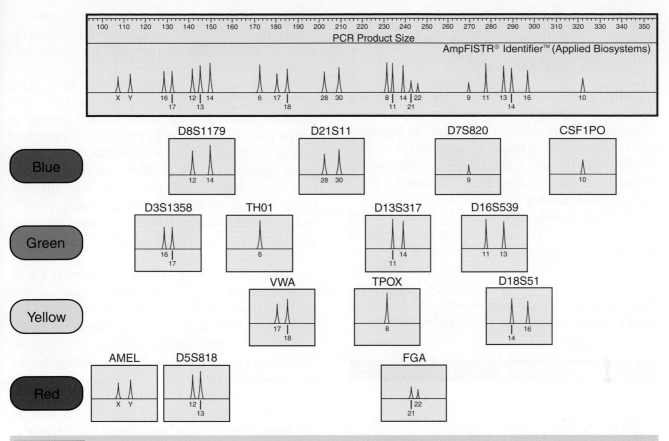

FIGURE 14-13 By carefully labeling the primers with fluorescent dyes, the STRs that cannot be separated by size may be identified by color.

countries rely on 10 or more STR loci, each of which is found on a different chromosome. This practice is meant to ensure that the alleles of a particular locus are inherited independently (i.e., independently of other alleles on other chromosomes). European forensic laboratories use 10 STR loci,

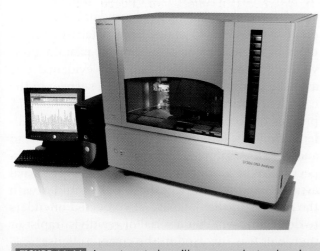

FIGURE 14-14 An automated capillary array electrophoesis (CAE) analyzer.

and U.S. laboratories use the 13 CODIS loci. Eight STR loci are common to the European and U.S. DNA tests: FGA, THO1, VWA, D3S1358, D8S1179, D16S539, D18S51, and D21S11.

A standard nomenclature is used to designate the location of a DNA marker depending on where it occurs in the genome. If the marker is part of a gene or falls within it, the gene name is used. For example, the THO1 forensic STR, which falls within a gene, is designated "TH" from tyrosine hydroxylase, the name of the gene. It is designated "01" because it is in the first intron of the gene. If the STR falls outside a gene region, its name indicates the chromosome and locus on which it is found. For example, for the STRs D5S818 and DYS19, which fall outside of gene regions:

D: Stands for DNA.

5: Chromosome number.

S: Indicates the DNA marker is a single copy sequence.

818: Indicates the order in which the marker was discovered—that is, the 818th locus.

D: Stands for DNA.

Y: Y chromosome.

S: Indicates the DNA marker is a single copy sequence.

19: 19th locus described.

Alleles are generally named according to the number of repeats in the STR. For example, for THO1, allele 6 contains six repeats of the sequence AATG, and allele 7 contains seven repeats. The THO1 STR also has a 5.3 allele; this designation means there are five repeats of AATG and a sixth repeat that is missing one of the four bases.

The amelogenin gene is not an STR locus and has two alleles. This gene, which encodes for a pulp tooth protein, is unusual in that it is present on both the X and Y chromosomes. Curiously, the gene on the Y chromosome is 6 base pairs longer than the gene on the X chromosome. This locus is used to determine the sex of the individual.

CODIS

As part of the DNA Identification Act of 1994, Congress set aside $40 million to help develop the DNA-related capabilities in state and local labs and to create the **Combined DNA Index System (CODIS)**, a national database containing the DNA of individuals who are convicted of committing sexual and violent crimes. In developing CODIS, the FBI laboratory initiated a project to standardize the number of STR loci used to analyze samples in the United States. **TABLE 14-1** contains a list of the CODIS core

TABLE 14-1

CODIS Core STR Loci

Locus	Repeat Sequence	Known STR Repeats	Chromosome
TPOX	AAGT	6–14	2
D3S1358	GATA	9–20	3
FGA	CTTT	15–30	4
D5S818	AGAT	5–15	5
CSF1PO	AGAT	6–15	5
D7S820	GATA	6–14	7
D8S1179	TCTA	8–18	8
TH01	TCAT	3–13.3	11
VWA	TCTA	11–22	12
D13S317	TATC	7–15	13
D16S539	GATA	5.8–15	16
D18S51	AGAA	9–27	18
D21S11	TCTA	24–38	21
Amel-X			X
Amel-Y			Y

STR loci, their repeat sequences, the number of known repeats, and the chromosomes on which the STRs lie. **FIGURE 14-15** shows where each locus falls on its chromosome.

CODIS has three tiers (i.e., levels): local, state, and national. Each lab participating in the CODIS program maintains a Local DNA Index System, in which DNA profiles can be entered and compared within the lab itself. Each state has a State DNA Index System, in which profiles from all of the local labs in the state as well as profiles from the convicted offender database may be compared. The FBI maintains the National DNA Index System, in which profiles from all of the individual state systems can be compared.

One of the goals of CODIS was to create a database of a state's convicted offender profiles that could be used to solve crimes for which police had no suspects. The first convicted offender database was the sex crimes database. Today, all 50 states have a sex crimes database law mandating that individuals who are convicted of a sex crime must submit a DNA sample that will be typed by CODIS, with the resulting profile being entered into the database. This database has proved to be very useful to law enforcement officers. Today, when a biological sample is found at the scene of a sex crime, in one-third of the cases it will match a DNA sequence already in the database, allowing the crime to be solved immediately. In these cases the offender served his or her sentence for the first offense and, after being released from prison, relapsed into committing sex crimes.

The success of the sex crimes database has motivated states to enact additional database laws. All 50 states have created DNA-profile databases for convicted murders, 49 states have databases for violent felons, and 43 states have databases for all types of felons. Historically, forensic examinations were performed by laboratories if evidence was available and police had identified a suspect in the case. By creating a database that contains the DNA profiles of convicted sex offenders and other violent criminals, forensic laboratories are now able to analyze DNA samples from cases in which no suspects have been identified (e.g., as stranger sexual assaults) and to compare those DNA profiles with the profiles in the database of convicted offenders to determine whether a serial or recidivist rapist or murderer was involved in the current offense.

CODIS began as a pilot project with 12 state and local forensic laboratories; today, it relies on

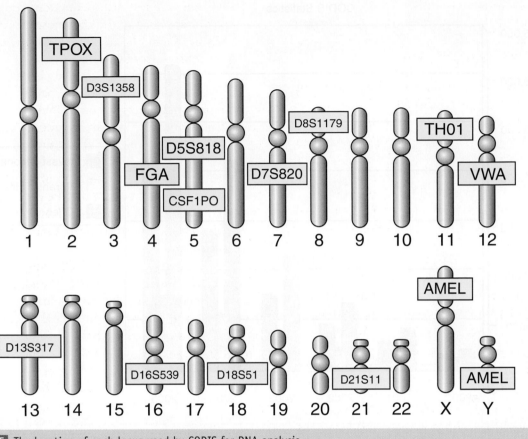

FIGURE 14-15 The location of each locus used by CODIS for DNA analysis.

laboratories representing 49 states and the District of Columbia. The FBI's primary method of measuring the effectiveness of the CODIS program is to ascertain the number of investigations it assists either by identifying a suspected perpetrator or by linking serial crimes. In 2005 more than 30,000 investigations were aided by CODIS (FIGURE 14-16).

The CODIS software compares the DNA profile submitted by the local or state lab with the DNA profiles obtained from convicted offenders and other crime scenes, which are submitted by other members of the CODIS network. For example, in a rape case, a DNA profile of the perpetrator is obtained from the sexual assault evidence kit. If there is no suspect in the case or if the suspect's DNA profile does not match the profile obtained from the evidence, the laboratory will search the DNA profile against the Convicted Offender Index. If the CODIS software finds a match in the Convicted Offender Index, the laboratory will be sent the identity of the suspected perpetrator. If it does not find a match in the Convicted Offender Index, the DNA profile is searched against the crime scene DNA profiles contained in the

Forensic Index (a database of DNA evidence from unsolved cases). If the CODIS software finds a match in the Forensic Index, the laboratory has linked two or more crimes together, and the law enforcement agencies involved in the cases may then pool the information obtained on each of the cases. Matches made by CODIS and confirmed by the participating laboratories are often referred to as CODIS "hits."

The CODIS database contains only the information required for performing DNA matches. It does not contain information such as criminals' personal histories, case-related histories, or other identifying information, such as Social Security numbers. When CODIS identifies a potential match, the laboratories responsible for originally submitting the matching profiles contact each other to verify the match. If the match is confirmed by qualified analysts, the laboratories may exchange additional information, including the names and phone numbers of criminal investigators. In the event of a match with the Convicted Offender Index, the identity and location of the offender are released.

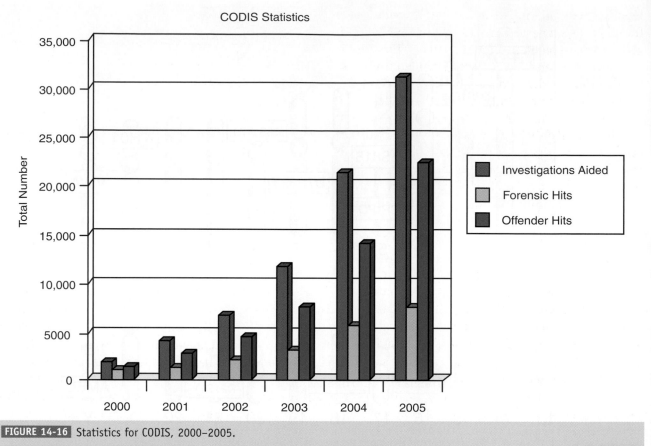

FIGURE 14-16 Statistics for CODIS, 2000–2005.

Source: Data from Federal Bureau of Investigation (FBI). CODIS Combined DNA Index System. http://www.fbi.gov/hq/lab/html/codisbrochure_text.htm.

Interpretation of DNA Profiles

As we will later see in this chapter, it is somewhat easier to exclude a person from suspicion based on DNA evidence than to prove that an individual is the lone possible suspect. In the United States, the Innocence Project reports that three times more suspects are proven innocent by DNA analysis than are proven guilty. But what if the DNA profile from the crime scene and the DNA profile of the suspect appear to be identical? What if two individuals have fairly common DNA profiles and by coincidence one is identified as the other? Forensic DNA loci must be chosen so as to minimize the chance that two people will have the same profile. To understand how this is done, we need to discuss population genetics.

Simple Population Genetics

Earlier in this chapter, we learned how a PCR-amplified locus could be separated by CE and its allele number (length) determined. Figure 14-11 showed a CE analysis of five individuals with dif-

BACK AT THE CRIME LAB

Cheek swabs can be collected from individuals and may, in fact, result in the discovery of some of the most highly concentrated DNA cells. The cheek swab is nonthreatening in that the individual feels less intimidated by the sample-collection process. The procedure is quick and simple: A cotton-tipped swab is scrubbed on the inside of the cheek. No food or drink is allowed 20 minutes prior to the collection. It is preferred that this technique not be used if the mouth is bleeding.

ferent alleles at one locus. What if evidence is found at the scene of a crime that, when analyzed, gives a genotype of 2, 6 for this allele? First, individuals 1, 2, 4, and 5 can all be excluded from suspicion because they do not have the 2, 6 alleles. Individual 3, however, has both of these alleles. On the basis of this match, can we definitively say that this person is guilty? The answer to the question can be found in population genetics. If the 2, 6 allele is a very rare combination in the general population, then this evidence will carry more weight. If it is very common, then the match itself might not mean much.

In the 1990s the forensic community carried out studies that tried to determine just how common certain alleles were in the population. To do so, researchers took DNA samples from many individuals and performed genotyping to determine the allelic frequency. For our hypothetical example, the allele frequencies are found in **TABLE 14-2**.

From Table 14-2, we can determine that the 2 allele has a frequency of 0.23, which means that if 100 people were sampled randomly (200 alleles, with 2 alleles for each person), 23 people would be expected to have the 2 allele. The sum of all the allele frequencies adds up to 1.0. For the example locus, there is no allele that is bigger than an 8 repeat (the allele frequency is zero).

To determine how frequently the 2, 6 genotype occurs in the population, we must apply the Hardy–Weinberg principle. The Hardy–Weinberg principle states that allele frequencies remain constant from generation to generation and that the allele frequencies can be easily calculated. The simplest case is one with an STR with two alleles, P and Q, that have allele frequencies p and q. The Hardy–Weinberg principle states that for this example the genotype frequencies will be

$$(p + q)^2 = p^2 + 2\,pq + q^2$$

where

p^2 is the frequency of the homozygous genotype PP.

TABLE 14-2	
Frequency of Occurrence	
Allele	**Allele Frequency**
2	0.23179
3	0.00170
4	0.19040
5	0.08400
6	0.36000
7	0.00961
8	0.12250

q^2 is the frequency of the homozygous genotype QQ.

$2\,pq$ is the frequency of the heterozygous genotype PQ.

The Hardy–Weinberg principle can be expanded to include any number of alleles for a specific locus. **TABLE 14-3** shows how the mathematical equations are expanded when multiple alleles are considered. When these equations are reduced to their simplest terms, you will notice that the frequency of a particular homozygote is that allele frequency squared. For a heterozygote, the expected frequency is two times the product of the two allele frequencies.

Now we can answer our original question: What is the frequency of the 2, 6 allele in our example? Using the information in Table 14-3, we can calculate how common the 2, 6 genotype is in the population. From Table 14-3, the frequency of the 2 allele (p) is 0.23, and the frequency of the 6 allele (q) is 0.36. The Hardy–Weinberg principle tells us that the expected frequency of the 2, 6 heterozygote is 2 pq or 2(0.23)(0.36) = 0.166. That means that 166 people out of 1000 would have the 2, 6 genotype. This is obviously a very common genotype. If the evidence indicated that the suspect had a 3, 5 genotype, however, the frequency would be 2(0.0017) (0.084) = 0.000143. Thus only 1 person out of 10,000 would have this rare genotype. When only one allele is analyzed, the discriminating power of the DNA, although good, is not maximized. By

TABLE 14-3			
Allele Frequencies Calculated by the Hardy–Weinberg Principle			
Number of Alleles	**Names of Alleles**	**Frequencies of Alleles**	**Genotype Frequencies**
2	P, Q	p, q	$(p + q)^2 = p^2 + 2\,pq + q^2$
3	P, Q, R	p, q, r	$(p + q + r)^2 = p^2 + 2\,pq + 2pr + 2qr + q^2 + r^2$
4	P, Q, R, S	p, q, r, s	$(p + q + r + s)^2 = p^2 + 2\,pq + 2pr + 2qr + 2\,ps + 2\,rs + 2\,qs + q^2 + r^2 + s^2$

multiplexing the 13 CODIS loci, we can use the Hardy–Weinberg principle to greatly increase the discriminating power of DNA typing.

Interpreting Multiplex DNA Profiles

Although individual genotypes might not be sufficient to convict a criminal, the use of multiple DNA loci and the Hardy–Weinberg principle can tremendously increase the discriminating power of the DNA analysis. As mentioned earlier, researchers have determined the expected genotype frequencies for each locus—that is, how common a particular genotype is expected to be in the general population. To estimate how common a particular CODIS profile is likely to be in the general population, the genotype frequencies for all the loci are multiplied together. This is a sensible approach because the loci are found on different chromosomes, so the inheritance of one genotype has no influence on the inheritance of the other alleles at different loci.

To see how this works, let's consider an example. Suppose a bloodstain found at the scene of a crime was DNA typed using a standard CODIS protocol. A suspect was ordered by the court to submit a sample for CODIS DNA typing. The sample from the bloodstain and the suspect's blood gave the same allelic pattern. The STR alleles that were identified, and their frequency of occurrence, can be seen in **TABLE 14-4**.

For this particular DNA profile, its frequency of occurrence in the overall population can be calculated by multiplying all of the individual STR frequencies: $0.098 \times 0.177 \times 0.066 \times 0.059 \times 0.077 \times 0.037 \times 0.103 \times 0.023 \times 0.022 \times 0.089 \times 0.051 =$

8.62×10^{-16}. This is an incredibly small frequency. The chance that two people selected at random would have the same profile is $1/8.62 \times 10^{-16} = 1.15 \times 10^{15}$. This 1.15×10^{15} match probability is the reciprocal of the frequency. It is larger than the population of the Earth, so there is no doubt that this DNA sample must match the DNA profile of the suspect—and only the DNA profile of the suspect.

The probability also can be expressed as a likelihood ratio where the probabilities of alternative events are compared. The likelihood ratio has the same numerical value as the match probability (for our example, 1.15×10^{15}). Although the numerical value is the same, jurors often find it easier to understand a likelihood ratio. For this example, the DNA evidence shows that it is 1.15×10^{15} times more likely that the sample came from the suspect than from anyone else.

As you can see in Table 14-4, some of the CODIS loci are more common than others. The true discriminating power of CODIS is achieved by multiplying the individual frequencies of the 13 loci.

Paternity Testing

DNA profiling has proved to be a powerful tool in establishing familial relationships between individuals. Many civil cases involving paternity or inheritance are resolved by DNA testing, and it isn't uncommon to see billboards in some states advertising DNA paternity testing services. DNA paternity testing is also important in criminal cases where the body of a deceased, missing child can be identified by comparing his or her DNA to that of the parents.

TABLE 14-4

Genotype Frequencies of the CODIS Loci in Figure 14-12

Locus	Allele (p)	Frequency	Allele (q)	Frequency	2pq	p^2	Likelihood Ratio
D3S1358	16	0.2315	17	0.2118	0.098		1 in 10.20
VWA	17	0.2628	18	0.2219	0.117		1 in 8.57
FGA	21	0.1735	22	0.1888	0.066		1 in 15.26
D8S1179	12	0.1454	14	0.2015	0.059		1 in 17.07
D21S11	28	0.1658	30	0.2321	0.077		1 in 12.99
D18S51	14	0.1735	16	0.1071	0.037		1 in 26.91
D5S818	12	0.3539	13	0.1462	0.103		1 in 9.66
D13S317	11	0.3189	14	0.0357	0.023		1 in 43.92
D7S820	9	0.1478				0.022	1 in 43.28
D16S539	11	0.2723	13	0.1634	0.089		1 in 11.24
TH01	6	0.2266				0.051	1 in 19.50
TPOX	8	0.5443				0.296	1 in 3.38
CSF1PO	10	0.2537				0.064	1 in 15.53

Currently, four tests are used to profile DNA evidence.

RFLP Analysis: Restriction Fragment Length Polymorphisms

- Large amounts of intact DNA are required.
- DNA is cut to generate large fragments that are sorted based on size by gel electrophoresis.
- This time-intensive process involves transferring the DNA to a membrane and probing with a radio-labeled DNA probe to look for pattern differences between individuals.
- Replaced today almost entirely with PCR-based technologies.

PCR: Polymerase Chain Reaction

- Very powerful technology that revolutionized forensics identification.
- PCR itself doesn't accomplish DNA typing; it merely increases the amount of DNA available for typing.
- This technique, in theory, can make millions of copies of a specific region of DNA from a single DNA molecule.

STR Analysis: Short Tandem Repeats

- Type of DNA targeted for most of the currently popular forensic DNA tests.
- With PCR, only small amounts of partially intact DNA are required for analysis.

- STR is a generic term that describes any short, repeating DNA sequence.
- Human DNA has a variety of STRs scattered among DNA sequences; those in noncoding areas of the DNA are used in forensic identification.
- PCR that uses constant regions of DNA sequence as primers allows the copying of variable STR regions of DNA sequence.
- Thirteen different STRs in nuclear DNA (CODIS 13) are typically screened for forensic identification in a multiplex DNA analysis.

Mitochondrial DNA Analysis

- Can be used when the availability of nuclear DNA is limited due to degradation.
- For every copy of nuclear DNA, there can be hundreds or even thousands of copies of mtDNA.
- mtDNA analysis is used in conjunction with decomposed or burned biological evidence.
- mtDNA has played a significant role in disaster cases, post-conviction exoneration, cold cases, genealogical studies, and missing persons investigations where the quality of DNA available for analysis was poor.
- Not as accurate as nDNA-based analysis.

The inheritance of alleles follows the basic laws of genetics that were discussed earlier. A child can receive only one of the father's alleles and only one of the mother's alleles. By comparing the DNA profiles of the mother, father, and child, a pattern should be obvious.

TABLE 14-5 shows results of DNA typing for a hypothetical paternity case where there are two potential fathers. Comparing the child's genotype for each locus with the genotypes for the mother and the suspected fathers allows a determination to be made. Consider locus D13S317, where the child has a genotype of 10, 14. Allele 14 must come from the child's mother, because neither of the potential fathers has this allele. The child's 10 allele must therefore come from the father, and only suspected father 2 has the 10 allele. This excludes suspected

TABLE 14-5

Comparison of DNA Profiles from a Child and Suspected Fathers

Locus	Mother	Child	Suspected Father 1	Suspected Father 2
D3S1358	16, 17	15, 17	16, 19	15, 16
VWA	17, 18	18, 18	14, 16	17, 18
FGA	21, 22	22, 21	20, 21	21, 23
D8S1179	12, 14	12, 14	12, 13	14, 18
D21S11	28, 30	28, 30	28, 31	30, 33
D18S51	14, 16	13, 14	12, 15	13, 14
D5S818	12, 13	10, 12	11, 13	13, 15
D13S317	11, 14	10, 14	9, 11	10, 15
D7S820	9, 9	9, 7	8, 12	7, 10
D16S539	11, 13	9, 11	9, 10	9, 12
TH01	6, 6	6, 8	6, 9.3	6, 8
TPOX	8, 8	8, 9	8, 10	8, 9
CSF1PO	10, 10	10, 12	12, 14	12, 13

father 1 from being the father. Applying the same logic to the other loci confirms this conclusion. For each of the child's alleles that does not have a maternal origin, suspected father 2 can be the source. These DNA data strongly favor suspected father 2 as being the father of the child.

When presented as testing results, the data from the paternity analysis in Table 14-5 would be converted into a likelihood that suspect 2 is the father. The paternity index (PI) is the likelihood that an allele from the child supports the assumption that the tested man is the true biological father rather than a randomly selected unrelated man. The combined paternity index (CPI) is determined by multiplying the individual PIs for each locus tested. For the child represented in Table 14-5, suspected father 2 is 500,000 times more likely to be the child's father than a male picked out from the general population at random. Although we will not learn how to compute the CPI, you should realize that courts use the CPI to introduce an objective statistical analysis of the DNA evidence.

Mitochondrial DNA Analysis

The forensic DNA analysis that preceded this section was based on the presence of STR loci in nuclear DNA (nDNA). In some cases, however, it is not possible to obtain an STR profile from nDNA because of degradation of the sample. Badly decomposed or burned bodies, old bones, and human hair without follicular tags, for example, may provide badly decomposed DNA that is present in an extremely low concentration. For the majority of the population, DNA reference samples are not generally taken nor held in a data bank for later use. Thus, it is difficult to identify individuals found dead because no DNA reference sample exists. Missing persons identification often is hampered because of this.

The U.S. military was faced with this problem in the early 1990s when it began using DNA techniques to identify soldiers who were missing in action from the Vietnam War. In these cases, the soldiers had been dead 20 years before their remains were located, and there were no longer any of their biological samples left that could be used for comparison. This problem prompted the military to begin in the early 1990s to collect and retain a blood sample from every recruit as he or she entered basic training.

For remains recovered in Vietnam, the U.S. military found examination of recovered mitochondrial DNA (mtDNA) to be a useful forensic tool. As mentioned in Chapter 13, although mtDNA is not resistant to degradation, a cell may contain thousands of copies of mtDNA, which increases the probability that at least some copies will survive in extreme cases where the DNA is exposed to environmental factors that degrade it.

Human mtDNA is a circular DNA molecule that is only 16,569 base pairs in circumference (FIGURE 14-17). Unlike nDNA, it has no noncoding

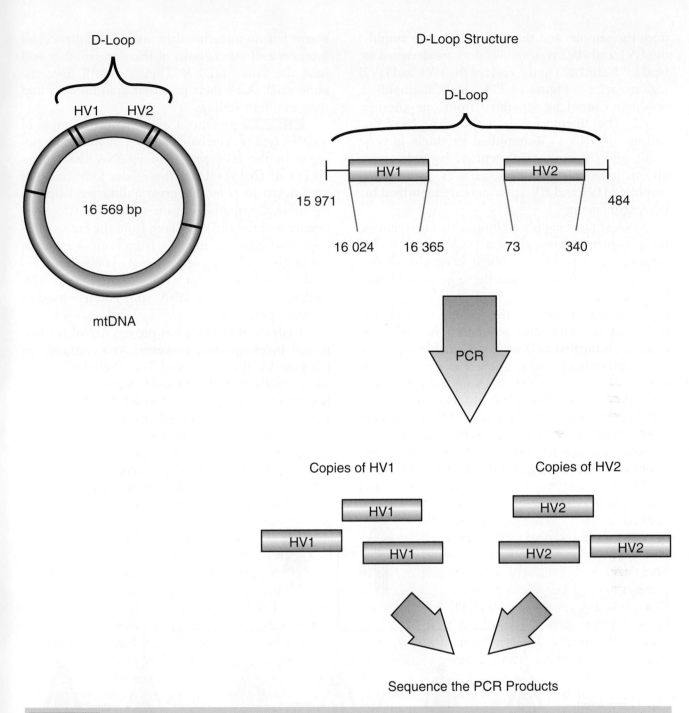

FIGURE 14-17 Mitochondrial DNA has a small circular structure. Primers are designed to amplify the HV1 and HV2 regions.

or "junk" DNA; almost every base in mtDNA has a function. The genetic information that mtDNA carries is essential, and most of it is the same between individuals. Geneticists refer to this characteristic as being "highly conserved." Most mtDNA has no forensic value because it varies so little between individuals. Nevertheless, one region, called the D-loop, is about 1000 base pairs long and is forensically useful. There are no STRs in mtDNA, but the D-loop contains two hypervariable regions,

HV1 and HV2, whose sequences vary. The sequence variation in these regions typically consists of a change in a single base (i.e., a point mutation), which does not alter the length of the DNA. In the hypervariable regions, a difference of less than 3% is expected between unrelated individuals. Because of this narrow window, analysis of mtDNA is much harder than analysis of the STRs on nDNA.

Analysis of mtDNA is similar in its beginnings to the analysis of nDNA. The mtDNA is extracted from the sample, and then PCR is used to amplify the HV1 and HV2 regions. Primers are designed to bind to the mtDNA at the ends of the HV1 and HV2 regions (refer to Figure 14-17). The PCR-amplified products cannot be separated from one another by CE. The difference between the lengths of the various mtDNA PCR-amplified products is very small, and CE cannot differentiate between them on this basis. Instead, the sequences of the PCR-amplified HV1 and HV2 products are identified by DNA sequencing.

DNA sequencing is a technique that determines the sequence of bases along a DNA strand. It is a slow, complex technique, but it is capable of detecting point mutations. Once the sequences of bases have been elucidated, the DNA sequences of the hypervariable regions of the sample mtDNA are compared with a reference sequence. The Anderson sequence is the first mtDNA hypervariable sequence to be determined, and it is used as the reference sequence.

When comparing mtDNA, the analyst must remember how mtDNA is inherited. As can be seen in **FIGURE 14-18**, mtDNA is inherited from a person's mother, a phenomenon referred to as maternal inheritance. As discussed in Chapter 13, the father's

sperm has no mitochondria. As a consequence, all brothers and sisters born of the same mother will have the same mtDNA. They also will have the same mtDNA as their maternal grandmother and their mother's siblings.

TABLE 14-6 presents a hypothetical example of mtDNA typing. The base pairs are identified by position in the HV1 region of mtDNA (base pairs 16,111 to 16,357). Only those base locations that are known to show hypervariability are listed in the table. Comparison of the mother's mtDNA sequence and the mtDNA taken from the bones indicates that only the mtDNA from bone 3 matches the mother's mtDNA sequence. As you will notice, mtDNA analysis is not as discriminating as STR analysis using nuclear DNA, so it is rarely used in criminal proceedings.

Analysis of mtDNA has proven useful for historical investigations, however. As mentioned in Chapter 13, the remains of Tsar Nicholas II were successfully identified by this means and are now buried in a cathedral in St. Petersburg, Russia. The U.S. military has identified more than 1800 soldiers who were missing in action from the Vietnam War using this technology. The soldier who was interred in the Tomb of the Unknown Soldier of the

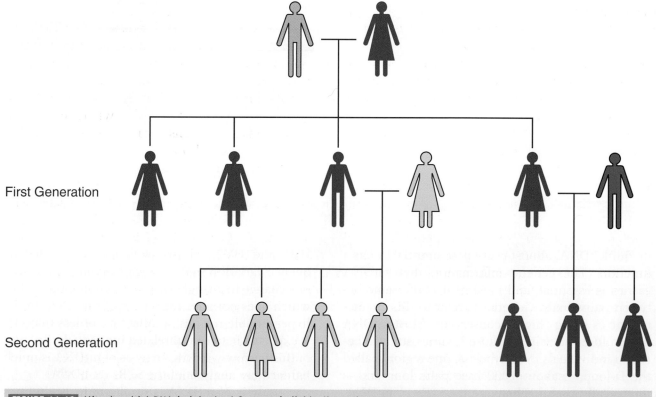

First Generation

Second Generation

FIGURE 14-18 Mitochondrial DNA is inherited from an individual's mother.

TABLE 14-6

mtDNA Sequences from Soldiers Who Were Missing in Action

Sample	mtDNA Base Pair Positions HV1											
	16111	16126	16169	16261	16264	16278	16293	16294	16296	16304	16311	16357
Anderson	C	T	C	C	C	C	A	C	C	T	T	T
Bones 1		T					G					
Bones 2	T									T		
Bones 3	T											C
Bones 4			C			G						
Mother	T											C

If a position is blank, the base is identical to the Anderson sequence.

Vietnam War at Arlington Cemetery is no longer unknown, thanks to mtDNA analysis; Michael Blassie was identified in 1998 after investigators compared the mtDNA from the bones in the tomb with the mtDNA from his mother.

The Y Chromosome: STRs and SNPs

Although the CODIS loci have all been chosen to be on different chromosomes, there is no locus on the Y chromosome. This omission can sometimes create problems when PCR is used. In cases of rape, for example, a vaginal swab is taken. After the DNA is extracted from this sample, it is amplified by PCR. Unfortunately, PCR most readily amplifies the DNA that is present in highest concentration. Given that there are 300 times more female epithelial cells than sperm cells in such samples, the female component of the sample is inevitably amplified to a greater degree.

To avoid this problem, STR analysis is directed at the Y chromosome. Because females do not have a Y chromosome, PCR amplification of the Y-chromosome STRs will produce copies of the male's DNA. Rape kits containing primers that will specifically bind and amplify only the Y-chromosome STRs are now available to law enforcement.

Only recently has the complete sequence of the Y chromosome been deduced. Besides STRs, the Y chromosome contains many polymorphisms that show promise as forensic markers. In **single-nucleotide polymorphisms (SNPs)**, for example, the base difference occurs at only one specific site. Using a technique known as minisequencing, the forensic scientist can determine the base at a specific SNP location. The maximum number of alleles for each SNP is four (one for each A, C, T, and G).

SNPs at different loci can be identified simultaneously, producing an SNP DNA profile. That is, the base at each SNP loci is determined, leading to the identification of a genotype for each SNP locus. SNP allele frequencies have been calculated, and likelihood ratios can be determined as well. Because there are only four alleles for each SNP, a large number of SNPs must be analyzed to achieve the same level of discrimination as is provided by current STR profiling, however.

Low-Copy-Number DNA Typing

To analyze a wider range of evidence, an extremely sensitive technique called low-copy-number (LCN) DNA typing has recently been developed. LCN analysis is used when the quantifying test indicates there is too little DNA available to perform a regular DNA analysis. It has been used successfully to type DNA from a doorknob that was touched by a suspect, a few skin cells left in a fingerprint, and a single flake of dandruff.

LCN analysis is used only in cases where standard typing protocols have already failed. This technique increases the number of PCR cycles performed. That is, because there is a smaller amount of DNA, the PCR is cycled longer to make more copies. Of course, as the amount of suspect DNA decreases, the chance of contamination by DNA from other sources increases; in such a case, the DNA contamination will be amplified by the PCR along with the suspect's DNA. As a result, DNA labs limit access to PCR areas to only a few technicians who have already been typed. If samples of evidence are contaminated with their DNA, this problem will then be obvious when the typing results are analyzed.

WRAP UP

YOU ARE THE FORENSIC SCIENTIST SUMMARY

1. This sample could be stored in two ways until it is analyzed. First, a small portion of the sample could be dried for storage. The nucleases will be rendered inactive in dry conditions. DNA labs have special adsorbent paper that is used to store blood drops. Once dry, the paper with the bloodstain is put into a plastic bag, properly labeled, and placed into the freezer. Second, the remainder of the sample could be stabilized by adding an anticoagulant (sodium oxalate) and a blood preservative (sodium fluoride). It is then properly labeled and placed in a lab refrigerator.

2. The bloodstain on the jeans should be allowed to dry. Once dry, the jeans should be placed in a properly labeled evidence bag.

Chapter Spotlight

- The RFLP technique uses restriction enzymes that cut DNA strands into fragments at specific base sequences.

- PCR is a fast method that allows many copies of a particular portion of the DNA sequence to be manufactured in the laboratory. PCR amplifies only the DNA regions that are of interest; all other regions in the DNA strand are ignored.

- STR analysis is a more recent method for DNA identification.

- Gel electrophoresis is a technique used for DNA separation.

- CE systems can be used to help reduce the time span between when a DNA sample is collected at a crime scene and when it is actually analyzed.

- Multiplex DNA analysis allows multiple STR loci to be amplified by PCR and analyzed simultaneously.

- CODIS is the United States' national database that holds the DNA profiles of persons who are convicted of sexual and violent crimes. It was created to solve crimes in which police have not identified any suspects. Contributors to and users of CODIS come from three levels: local, state, and national.

- For DNA typing purposes, forensic DNA loci must be chosen so as to minimize the chance that two people will have the same profile. To do so, researchers have collected DNA samples from many individuals and performed genotyping to determine how common certain alleles are in the population. To determine how fre-

quently the genotype occurs in a population, they apply the Hardy–Weinberg principle.

- DNA profiling has proved to be a powerful tool in establishing familial relationships between individuals.

- Mitochondrial DNA (mtDNA) has proven to be a useful forensic tool in cases involving badly decomposed or burned bodies, old bones, and human hair without follicular tags, in which the nDNA may be badly decomposed or present in an extremely low concentration. mtDNA analysis is also useful in cases where an individual has been dead for a long time before his or her remains are located, such that there may be no nDNA reference sample to match.

- The U.S. military has used mtDNA analysis to identify soldiers who were missing in action from the Vietnam War.

- The CODIS loci have been chosen so they are found on different chromosomes. There is no locus on the Y chromosome, however. To avoid overamplification of a female victim's DNA in cases of rape, STR analysis is used to amplify only the male's DNA and is directed at the Y chromosome. Because females do not have a Y chromosome, PCR amplification of the Y-chromosome STRs will produce copies of the male's DNA.

- LCN DNA typing is used when the quantifying test indicates there is too little DNA (e.g., a single flake of dandruff) to perform a regular DNA analysis.

Key Terms

Amplicon A DNA sequence that has been amplified by PCR.

Anneal When two complementary strands of DNA bind together.

Capillary electrophoresis (CE) A method of separating DNA samples based on the rate of movement of each component through a gel-filled capillary while under the influence of an electric field.

Combined DNA Index System (CODIS) The national database containing the DNA profiles of individuals who are convicted of sexual and violent crimes.

Denature The separation of two complementary strands of DNA upon heating.

DNA polymerase An enzyme that is a catalyst in the polymerase chain reaction.

Gel electrophoresis A method of separating DNA samples based on the rate of movement of each component in a gel while under the influence of an electric field.

Multiplexing A technique in which multiple STR loci are amplified and analyzed simultaneously.

Phenotype The observable characteristics of an organism (e.g., blue eyes, blond hair).

Polymerase chain reaction (PCR) A reaction that is used to make millions of copies of a section of DNA.

Primer A short sequence of DNA that is specifically designed and synthesized to be complementary to the ends of the DNA region to be amplified.

Restriction fragment length polymorphism (RFLP) A DNA typing technique in which sections of the DNA strand are cut out by restriction enzymes and differentiated based on their length.

Selectively neutral loci Inherited characteristics that confer neither a benefit nor any harm to the individual's ability to reproduce successively.

Short tandem repeat (STR) A DNA locus where sequences of 3 to 7 bases repeat over and over.

Single-nucleotide polymorphism (SNP) A DNA sequence variation that occurs when a single nucleotide (A, T, C, or G) in the genome sequence is altered.

Thermocycle The heat–cool cycle used in the PCR reaction.

Putting It All Together

Fill in the Blank

1. A gene that has been manipulated by molecular biologists is called _____ DNA.

2. Most DNA typing techniques focus on the _____ (coding/noncoding) portions of DNA.

3. In RFLP, _____ enzymes are used to cut out the DNA sequence of interest.

4. PCR-based methods have largely replaced RFLP because of PCR's ability to process _____ samples or _____ samples.

5. When DNA is heated, the two complementary strands _____ (separate); when the strands are cooled, they _____ (come back together again).

6. In the PCR amplification process, the heating and cooling process is called _____.

7. The short sequences of DNA designed to be complementary to the ends of the DNA region to be amplified are called _____.

8. The DNA copies that are produced by PCR amplification are called a(n) _____.

9. Most forensic STRs have _____ (3, 4, 5, 6) bases that repeat over and over.

10. The length of an STR is _____ (longer/shorter) than an RFLP DNA fragment.

11. STR markers that are used for forensic DNA typing are chosen from separate _____ to avoid any linkage between markers.

12. Mutations in the noncoding region will have no effect on the _____ of the individual.

13. A mutation that does not threaten the individual's survival is called _____ _____ _____.

14. Observable characteristics of an individual, such as eye and hair color, describe the person's _____.

15. DNA moves in electrophoresis because of the influence of a(n) _____ _____.

16. As more voltage is applied to an electrophoresis gel, the temperature _____ (increases/decreases/stays the same).

17. In a CE separation, the _____ (largest/smallest) STRs arrive at the detector first.

18. In a CE separation, the STR is identified by comparison to a standard _____.

19. In a CE instrument, the STR is detected by a(n) _____ _____ detector.

20. The fluorescent dyes that are used for detecting STRs are attached to the _____ _____.

21. The process of amplifying and identifying multiple STR loci simultaneously is called _____.

22. In multiplexing, the _____ of the STR and the color of the _____ are adjusted to give an optimal separation of multiple STRs.

23. The DNA databases in most countries rely on _____ or more STR loci, each of which is found on a different chromosome.

24. CODIS matches _____ (10, 11, 12, 13, 14, 15) core loci.

25. CODIS has three tiers: _____, _____, and _____.

26. The match probability is the _____ of the frequency.

27. The likelihood ratio is the same numerical value as the _____ _____.

28. At a specific locus, a child can inherit only _____ allele from his or her father and _____ allele from his or her mother.

29. The genetic information carried by mtDNA is essential. Geneticists refer to this as being highly _____.

30. DNA _____ is a technique that determines the sequences of bases along a DNA strand.

31. The first mtDNA sequence to be determined is the _____ sequence, and it is used as a reference sequence.

32. A person inherits mtDNA from his or her _____.

33. A maximum of _____ alleles are possible for each SNP.

True or False

1. RFLP fragments are short, only 100 bases long.

2. Forensic PCR techniques make copies of the entire DNA strand.

3. STRs are named based on the number of repeating units they contain.

4. Larger STRs move faster in electrophoresis than do small STRs.

5. DNA molecules are colored and visible to the eye.

6. The CODIS database links the DNA typing information to the person's Social Security number.

7. The Hardy–Weinberg principle states that allele frequencies remain constant from generation to generation.

8. The combined paternity index is determined by multiplying the individual paternity index for each locus tested.

9. Human mtDNA is longer than nDNA.

10. Mitochondrial DNA has no noncoding DNA regions.

Review Problems

1. The gel electrophoresis separation in Figure 14-7 shows five DNA samples being separated. How would the separation be affected by the following changes?

    a. Decreasing the voltage from 30 volts to 10 volts.

    b. Increasing the voltage from 30 volts to 300 volts.

    c. Exchanging the buffer solution with distilled water.

2. Explain how the DNA bands in Figure 14-7 are made visible.

3. For sample 1 in Figure 14-12

    a. Which alleles were found for STR VWA?

    b. Which alleles were found for STR FGA?

4. For the sample that is analyzed in Figure 14-13

    a. Which detector will detect the STR that indicates the sex of the donor?

    b. Which detector will detect the CSF1PO STR?

    c. Is the DNA fragment for the D3S1358 STR larger or smaller than that for the TPOX STR?

5. The sample in lane 5 in Figure 14-7 is that of a child. Which of the other samples (4, 3, 2, 1) could have come from this child's father?

6. Do the mitochondrial HV1 sequences found for the recovered bones in Table 14-6 suggest that the bone samples were maternally related?

7. A sample of blood from a crime scene was compared with a sample of blood collected from a suspect. Analysis of these two blood samples by CODIS gave identical results.

Locus		Alleles
D3S1358	14	17
vWA	17	17
FGA	21	22
D8S1179	13	13
D21S11	29	33
D18S51	14	14
D5S818	12	12
D13S317	11	13
D7S820	9	10
D16S539	12	14
TH01	6	7
TPOX	8	11
CSF1PO	12	12

What is the probability that these two blood samples came from random Caucasian people? The Canadian Society of Forensic Science website provides a random match probability calculator (http://strwatch.com/STRcalc.aspx).

Further Reading

Budowle, B., and Moretti, T. R. Genotype profiles for six population groups at the 13 CODIS short tandem repeat core loci and other PCR-based loci. *Forensic Science Communications*. 1999; 1(2). http://www.fbi.gov/hq/lab/fsc/backissu/july1999/budowle.htm.

Butler, J. M. *Forensic DNA Typing*. London: Elsevier Academic Press, 2005.

Gill, P., Ivanov, P. L., Kimpton, C., Percy, R., Benson, N., Tully, G., Evett, I., Hagelberg, E., and Sullivan, K. Identification of the remains of the Romanov family by DNA analysis. *Nature Genetics*. 1999; 6:130–135.

Hartl, D. L., and Jones, E. W. *Genetics: Analysis of Genes and Genomes*, ed 7. Sudbury, MA: Jones and Bartlett, 2008.

Inman, K., and Rudin, N. *An Introduction to Forensic DNA Analysis*, ed 2. Boca Raton, FL: Taylor & Francis, 2002.

Isenberg, A. R., Forensic mitochondrial DNA analysis. In R. Saferstein (ed.), *Forensic Science Handbook, vol. 2*. Upper Saddle River, NJ: Prentice Hall.

Isenberg, A. R., and Moore, J. M. Mitochondrial DNA analysis at the FBI laboratory. *Forensic Science Communications*. 1999; 1(2). http://www.fbi.gov/hq/lab/fsc/backissu/july1999/dnalist.htm.

Wambaugh, J. *The Blooding*. New York: Bantam Books, 1989.

http://criminaljustice.jblearning.com/criminalistics

Answers to Review Problems
Interactive Questions
Key Term Explorer
Web Links

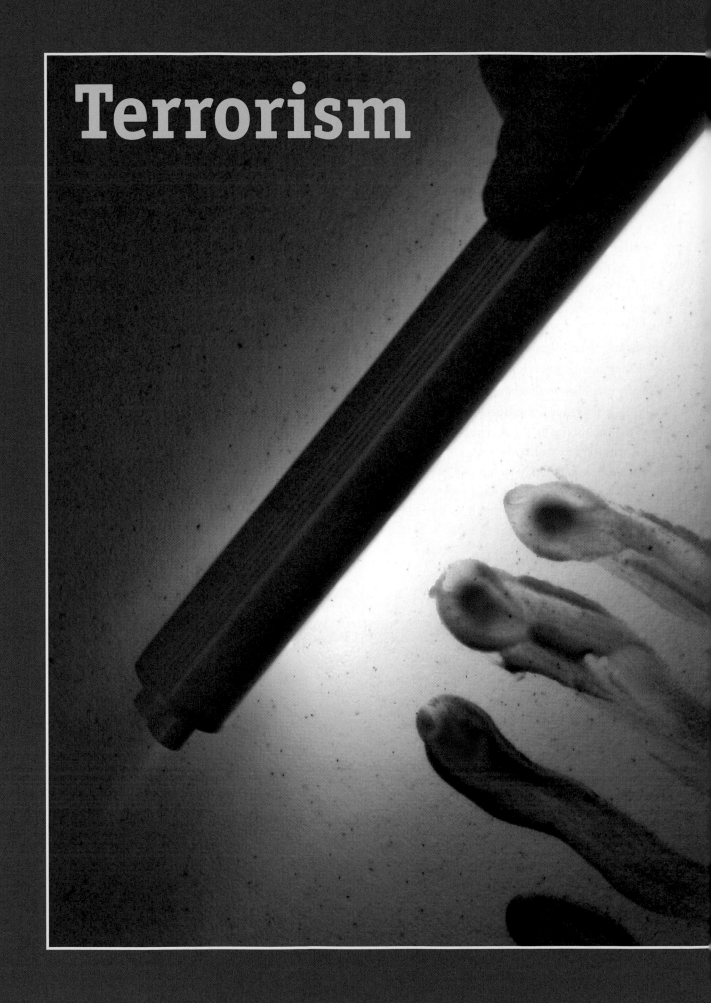

Terrorism

SECTION

6

Computer Forensics, Cybercrime, and Cyberterrorism

One of the world's leading industries in the production of electricity, natural gas, paper and pulp manufacturing, and communications during the 1990s and early 2000s was Enron Corporation of Houston, Texas. At its peak in 2000, Enron employed approximately 22,000 people and had revenues of nearly $101 billion. For six consecutive years, *Fortune* magazine named Enron as one of America's "Most Innovative Companies." However, by the end of 2001 Enron had become the symbol of corporate fraud and financial abuse. Through expert computer forensics, it was found that Enron's enormous financial holdings were sustained substantially by institutionalized, systematic, and creatively planned accounting fraud. The "Enron scandal" has since become a popular symbol of willful corporate fraud and corruption. The scandal brought into question the accounting practices of many corporations throughout the United States.

The guilty verdicts directed at the corporate managers of Enron were sealed by overwhelming computer forensic evidence. The government searched more than 400 computers and handheld devices, as well as over 10,000 backup storage devices. These searches involved the use of sophisticated forensic software from AccessData, Inc., that allowed investigators to open password-protected or encrypted files and to read email and other documents that had been deleted by Enron employees.

While the specifics of the case are complex and intricate, computer forensic investigators were able to help achieve convictions by using advanced software techniques to isolate key documents, including spreadsheets and emails, that were exculpatory to assist the government's case against the Enron corporate leaders. Investigators found what were considered to be "hot" documents, invoices, spreadsheets, contracts, memos, and notes that showed a pattern of premeditated and well-thought-through fraud. Unfortunately, with the downfall of Enron, many of its long-term employees lost not only their jobs but also their entire retirement funds.

1. After reading this chapter, describe the steps forensic investigators should take to secure forensic evidence of fraud as they enter a corporate office that contains a computer.
2. Once investigators have decided to seize the computer, what steps should be taken to protect it as it is shipped back to the crime lab?

Introduction

Computers and digital devices are ubiquitous in our modern society, so much so that they often provide the facts that can establish a motive for a crime, supply information that will contradict a supposed alibi, or be the scene of the crime itself. The modern forensic investigator must now understand what a cybercrime is and how to process a crime scene where computers are present. Furthermore, police, as first responders to a crime scene, must be well versed in preserving the evidence—not only for the physical crime scene but also for the computers that are present. Cybercops, forensic investigators who specialize in computer technology, need to understand what the evidence is indicating. Does the victim's cell phone contain a phone number that may suggest an extramarital relationship? Do financial records indicate that large withdrawals were

made from the victim's bank account? Was a sexual predator attempting to lure a teenage child out of his or her home via the Internet? Sometimes computers are pieces of evidence in prosecuting another crime, and sometimes computers are the major instrument used in the crime itself.

A computer is an electronic machine that performs complex calculations in a predetermined sequence. One of the earliest examples of non-electronic devices to allow complex calculations is the abacus, or counting frame. Evidence indicates that the ancient Chinese developed this counting tool as early as 3000 B.C. Similar devices were found in ancient Rome, Sumaria, and India.

The abacus and other types of devices used for calculations are different from a computer in that they are not programmable, whereas the computer, of course, is. Much of the underlying technology of computers is attributed to the British inventor Charles Babbage. Babbage designed a device in the mid-1830s that used a steam engine for power and could calculate and print logarithmic tables with great precision. Babbage's computer was able to read instructions from cards with holes punched in them (essentially the computer's memory), make calculations by following these instructions, and print out the results. While his design was clever, his machine was ultimately a failure. Its parts, which were made by hand, had to have perfect tolerances, and because such perfection was not possible, the machine made too many errors.

In the 1880s Herman Hollerith developed a device to record data on cards that could be read by a machine. He invented the punch card (**FIGURE 15-1**), the tabulator, and the key punch machine, which would later become the foundation of the modern information processing industry. Hollerith's technology was used to record the 1890 census in the United States and allowed the data to be collected months ahead of schedule and far under budget. Hollerith's technology was later used by IBM to build the data-processing industry, and by 1950, the punch card was the standard for entering data into a computer.

In the 1950s the invention of transistors offered the possibility of a smaller, faster computer that used much less power. As transistors became even smaller, the integrated circuit (IC) was developed, which placed entire circuits on a solid-state chip. Microprocessors were the next major invention in the data industry. Thousands of transistors

on a single silicon chip allowed the computer to be miniaturized to a point where it could sit on the top of a desk. Human interface devices, such as the computer keyboard and a pointing device, which we know as "the mouse," replaced the punch card as the only way to enter data into a computer. These developments led to the era of the personal computer.

Since the 1990s computer technology has advanced rapidly, and the price of high-powered computers has dropped dramatically. Electronic devices (e.g., laptop computers, cell phones, mp3 players, digital cameras) travel with us as we carry on our lives. It is not surprising, therefore, that these electronic devices have become increasingly important as sources of evidence in routine criminal investigations.

Additionally, computer systems now control much of our banking and commerce. Computer networks control how electricity is brought to our homes and businesses, and they are used in many air and land traffic control systems. These networks make tempting targets for criminals and terrorists.

In a criminal investigation, a computer forensic expert will seek to acquire, preserve, and analyze computer data. A computer may have been used in the commission of a crime of fraud and identity theft, and it can provide evidence that a crime was committed. Even if the computer was not directly used in the commission of a crime, it is a device that maintains a chronological record of the user's activity.

FIGURE 15-1 The 80-column IBM punch card. Computer users submitted their programs to the computer center as a stack of punch cards. Each card carried one line of the program.

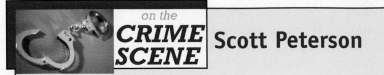
Personal Computer

Whether a desktop or a laptop, Macintosh or PC, all personal computers are made up of the same basic components. FIGURE 15-2 shows the internal components of the personal computer. Some of the components, such as the power supply, system bus, and motherboard, are required for the computer to function properly, but they are usually of little interest to the forensic investigator. The forensic examination most often focuses on the other parts of the computer.

Central Processing Unit (CPU)

The **central processing unit (CPU)**, or microprocessor, is the brain of the computer. The microprocessor is found on the motherboard, the main circuit board of the computer. The motherboard also contains the ROM and RAM memory expansion slots, slots to add other features, such as graphics cards (called PCI slots), and USB ports. It is the CPU's job to execute stored instructions.

The CPU repeats four basic steps; fetch, decode, execute, and writeback. In the first step, fetch, the CPU retrieves instructions that are stored in the read-only memory (ROM). In the decode step, the CPU breaks the instructions apart and sends each subset to a part of the CPU that will then process it. In the third step, the execute step, the CPU connects all of its parts together to complete the task found in the instructions. Finally, in the writeback step, the CPU writes the results of the execute step in the memory. Once these steps are successfully completed, the CPU goes to the next instruction.

Read-Only Memory (ROM)

Read-only memory (ROM) computer chips store firmware programs that hold the instructions to power up, or boot, the computer to control DVD drives, hard disk drives, and graphics cards. ROM

FIGURE 15-2 Hardware components of a personal computer: central processing unit (CPU), random-access memory (RAM), and hard disk drive (HDD).

also is known as flash memory. ROM is an important part of the BIOS (basic input/output system) operation. This BIOS process is often important in forensic investigations. When powering up the computer, the BIOS enables the CPU to communicate with the hard disk drive and the input/output devices that are attached to the computer. We will later see that booting a computer with the hard disk drive attached can cause the data held on the

disk drive to be altered, which would compromise the evidence the hard disk drive holds. If forensic examiners control the BIOS functions, they can secure the data held on the hard disk drive.

Random-Access Memory (RAM)

Random-access memory (RAM) is computer memory that holds data that can be accessed by the CPU in any order, no matter where it is physically located on the memory chip; thus the name random-access memory. The CPU knows that the hard drive retrieves data slowly; it takes information from the hard drive that it may need to complete a task and stores information in RAM to make it instantly available. RAM is considered to be volatile memory—that is, memory that loses its information when the power is turned off. The computer's RAM operates much faster than the hard drive, and programs are loaded into RAM to speed computer processing. Because each computer has a limited amount of RAM, the computer's operating system only keeps there applications that are currently in use. As shown in **FIGURE 15-3**, the computer often swaps other programs that were opened but are not currently in use out of the RAM and writes them onto the hard disk drive. As a computer swaps data back and forth it creates space on the hard drive called swap space. As

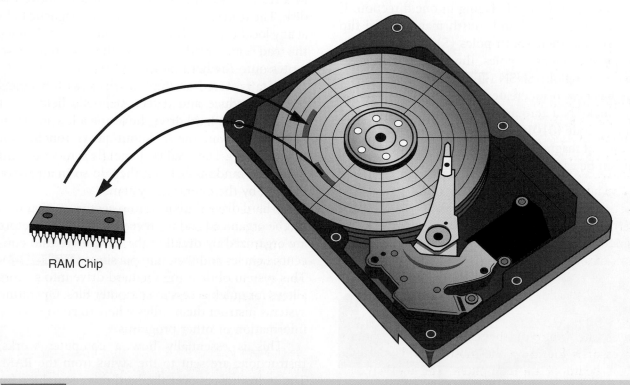

RAM Chip

FIGURE 15-3 As computer data swaps between the RAM and HDD, it creates swap space on the HDD.

the hard drive swaps information with RAM, some of the **swap data** can be left on the hard drive. This data may become forensically useful.

How Data Is Stored: The Hard Disk Drive (HDD)

The **hard disk drive (HDD)** is the only component of the computer that can store data while the machine is powered down. The HDD operates by running needles across a spinning disk that cause permanent, but alterable, changes in the magnetic field of the disk. This is similar to the grooves on a vinyl record but on a much smaller and more compact level.

The HDD is the computer's major storage device. It stores the operating system, all programs, and data files. As shown in **FIGURE 15-4**, the HDD is found in a sealed case that contains disks that are coated with magnetic material. In a desktop computer, where space is plentiful, the HDD disks are 3.5 inches in diameter. In a laptop computer the disks are 2.5 inches or smaller.

The hard drive contains a platter that is coated with a magnetic material. Imagine you have five bar magnets which, when you line them up in a row, have the north (N) pole facing in one direction. If you take the second and fourth magnets and flip them so that their south poles (S) are sitting next to their neighbor's N poles, the arrangement gives you a reading of NSNSN (**FIGURE 15-5**). A magnetic code has now been created using just N or S poles. If we make N a zero and S a one, we have created a binary code (01010). The hard disk drive coverts this line of magnets into a spinning disk of magnets that can be easily read by a small detecting magnet

FIGURE 15-5 Binary information is stored by magnetic material on the HDD by inducing a north (N) or south (S) magnetic pole.

in a read sensor that is held by an arm above the disk. The read sensor also can flip the magnetic field at any location from N to S. Therefore, not only does the read sensor read what is on the disk, but it also writes onto the hard disk.

Data is stored on the hard drive with magnets that can induce and detect magnetic fields as it flies across the hard drive, hovering a few micrometers away from the disk but never touching it (**FIGURE 15-6**). The needles do not fly across the hard drive in a random fashion; they do so in a pattern laid out by the operating system.

A hard drive is divided into **sectors** so that files can be organized and retrieved quickly. The sectors are organized by dividing the hard drive into concentric circles and then into pie slices (**FIGURE 15-7**). This system of dividing the hard drive into sectors allows for quick access to computer files. Operating systems instruct the needles where to go to retrieve information or other programs.

This is essentially how a computer works. Instructions are sent to the stylus from the RAM. The stylus hovers over the hard drive searching for the right sector on which to look for its instructions,

FIGURE 15-4 The magnetic disk of an HDD stores data.

FIGURE 15-6 The magnetic head travels across the disk without touching it.

and the magnetic field is altered, allowing the stylus to pick up the information from the hard drive sector and report back or write back the information to the screen. Understanding the inner working of a computer allows the forensic specialist to know how to treat a computer suspected of being used in a crime without corrupting information, thus preserving its use as evidence.

The Electronic Crime Scene

Electronic Equipment and the Fourth Amendment

The Fourth Amendment (refer to Chapter 1) prohibits law enforcement from accessing and viewing information stored in a computer without a search

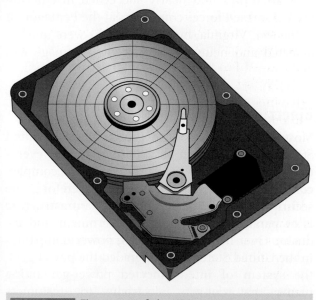

FIGURE 15-7 The sectors of the HDD are arranged as concentric circles that have been sliced into pie-shaped areas.

warrant. However, the U.S. Supreme Court held that a search of electronic equipment does not violate this amendment if it does not violate a person's reasonable expectation of privacy. This expectation of privacy is relinquished if a person gives a third party control of the information; this can be done by making a CD copy and giving it to a friend or by taking the computer to a repair shop. In a manual released in 2002, the U.S. Department of Justice stated that the privacy of information stored in a computer is determined by considering the computer as if it were a storage device such as a file cabinet. The Fourth Amendment's expectation of privacy applies only to law enforcement officers and does not apply to private individuals. Therefore, there is no violation if a private individual finds information and makes that data available to law enforcement. In *United States v. Hall* (1998) a pedophile took his computer to a computer repairman, who in the process of repairing the computer noticed files that contained child pornography. The repairman notified police, who obtained a warrant for the defendant's arrest.

There are exceptions to requiring a warrant that involve consent, the plain view doctrine, and searches that lead to an arrest. Cases that involve consent usually involve parents, other family members, roommates, or employers and whether these people have the authority to consent to the search of a computer that is being used by a suspect. The courts have held that parents can consent to searches of their minor children's computers. However, if an adult child is living with his parents, pays rent, and has denied access to his parents, the courts have held that parents cannot give consent to a search and a warrant must be issued (*United States v. Whitfield*, 1991).

Cybercrime

Cybercrime is a general term used to describe crimes intended to damage or destroy computer networks, obtain money fraudulently, disrupt business, or otherwise commit felonies through the use of computers. Although cybercrime is not a violent crime, it can be devastating to individuals who are victims of identity theft or to the larger society in the case of an act of terrorism. The global cost of cybercrime is estimated to exceed $500 billion per year; however, it is thought that only about 10% of cybercriminals are reported to police and fewer than

2% of cybercrime cases result in a conviction. In July 2008 former U.S. Attorney General Alberto Gonzales reported to Congress that cybercrime in the United States had resulted in annual losses of $250 billion and the loss of more than 750,000 jobs. Nearly 75% of U.S. businesses claim to have been the victims of cybercriminals. As shown in FIGURE 15-8, incidences of cybercrime have increased dramatically since 2000. Symantec, a computer security company, provided reports that in a single day it prevents 59 million cybercrime attempts. Because of these vulnerabilities, it is estimated that 2.6 million households must replace their computer systems annually. Moreover, about one in every five households does not have an antivirus **software** program installed to repel cyberattacks.

Cyberspace and Cyberterrorism

Cyberspace is the term used to describe the millions of interconnected computers, servers, routers, and switches that interconnect the digital infrastructure. Cyberspace often is compared to our nervous system. Just as the nervous system plays a major role in every part of the body, cyberspace plays a major role in every part of American life, including agriculture, water, public health, defense, industry, travel, commerce, and recreation. Thus, protecting cyberspace from attack is critical in protecting our country.

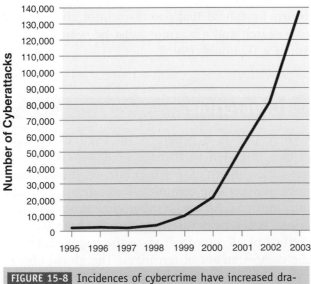

FIGURE 15-8 Incidences of cybercrime have increased dramatically since 2000 and are expected to continue to increase.

Cyberterrorism is defined as any type of attack on the computer infrastructure. Terrorists are keenly aware of the prominent role computers play in the United States and worldwide; a successful attack on the computer infrastructure could inflict major damage without bloodshed. The vast national and international computer network was specifically designed to be interactive and easily used, and, as such, it is ripe for exploitation. In 2003 the Department of Homeland Security developed a public/private plan called the National Strategy to secure cyberspace. Its overarching goal is to increase cyberspace security by:

- Establishing a public/private structure for responding to national-level cyberincidents
- Providing for the development of tactical and strategic analysis of any cyberattacks and assessing vulnerability
- Encouraging the development of a private-sector capability to protect cyberspace
- Supporting the role of the Department of Homeland Security in coordinating crisis management
- Coordinating processes for voluntary participation in cyberspace protection
- Improving public-private information sharing involving cyberattacks, threats, and vulnerability

To date, there are no recorded incidences of a cyberterrorist attack, per se; however, the potential threat exists. There have been a few cases of hackers who attempted and nearly succeeded in entering the U.S. armed forces computers at the Pentagon in Arlington, Virginia, but these attacks were quickly detected and neutralized before any damage was done.

Electric Power Grid

Most industrialized countries, especially the United States, rely heavily on electricity to operate everything from the lights in our houses to the complex computerized banking processes. Therefore, the security of America's electric power infrastructure is of paramount concern to government and industry. Over the past five years, power companies in the United States have upgraded the power grid, the system of interconnected power-generating plants that delivers electricity to customers throughout the country. Electric utilities have been implementing **smart grid** technology (adding computer-based technology to the U.S. power grid to manage the efficiency and reliability of electricity distribution). Thus, major electric markets will be supplied energy during peak demand times from other markets where demand is less.

Smart meters (FIGURE 15-9) are an important part of the smart grid technology. They are installed in individual homes or offices to provide real-time feedback on power consumption patterns. It is estimated that there are 8 million smart meters in use in the United States today, and more than 50 million more could be installed over the next five years, according to the Edison Institute.

Security researchers have found that smart meters are the weakest link in the smart grid technology. Smart meters allow electric power companies direct access to information about the power usage of individual locations. They also allow both the power company and individual users the ability to manage power consumption. For example, during a summer heat wave, the power company could automatically raise the household air conditioning temperature to lower demand through the smart meter. Similarly, individuals could change their own household heating or air conditioning temperature over the Web. It is this two-way communication system of the smart meter that makes the grid vulnerable to an attack.

Many smart meters require little authentication to carry out such commands as disconnecting customers from the power grid. In 2009, IOActive, a cybersecurity firm, demonstrated how a computer worm can spread by taking advantage of the software update feature built into some popular brands of smart meters. If multiplied by thousands of users, it is easy to see how such a worm, if launched by cybercriminals, could rapidly sever huge populations from the power grid. While simply an inconvenience for the average household, such an attack would cut power to industry, banking, commerce, and law enforcement. Gaining access to even a small portion of the power system could allow cybercriminals access to computer-controlled security devices. For example, in 2009, federal authorities reported that they had detected intruders, most likely from Russia or China, who had penetrated the computer systems that control the U.S. power grid. Although these intruders did no damage, government agents think they were mapping the computer network to plan another attack.

Taking a proactive approach, the U.S. Department of Energy created a new grant program in 2009 seeking proposals to develop technology to prevent cyberattacks on the U.S. power grid. The requirements from the Department of Energy came as a result of mounting concern from security experts that many existing smart grid efforts do not have sufficient built-in protections against computer **hacking**.

Other Cybercrimes

Thirty states have passed or are considering legislation targeting cybercriminals who use **spyware**. Virginia became the first state to pass an anti-phishing bill, which America Online used as a basis to file lawsuits. However, the Virginia Supreme Court declared the bill unconstitutional in 2008 on the grounds that it violated the First Amendment right to free speech. The U.S. Supreme Court declined to hear the case. Virginia's district attorney is preparing to draft new anti-spam legislation that should address any constitutional concerns. However, U.S. law will have very little effect on reducing **spam** (unsolicited bulk email) since much of the spam originates outside of the United States and is, therefore, not within this country's jurisdiction.

Identity Theft

One of the fastest-growing types of cybercrime is **identity theft**, the unauthorized use or attempted use of another person's financial records and/or

FIGURE 15-9 Smart electric meters are becoming an important part of smart grid technology.

identifying information without the owner's permission. In the 1990s cases of identity theft began to increase at a rate of more than 30% each year. Between 1996 and 2000 identity theft and fraud losses incurred by MasterCard and Visa increased by more than 43%. However, since 2000, identity theft cases have leveled off as more consumers have taken steps to protect themselves.

The most common types of crimes committed as identity theft include the unauthorized use of existing credit cards, savings accounts, and checking accounts; and the misuse of personal information to obtain new accounts, secure new loans, or commit other crimes. Often, identity thieves use **phishing** to find victims. Phishing emails are designed to deceive Internet users into disclosing private information such as passwords or credit card information.

In 2008 the National Crime Victimization Survey (NCVS) estimated that nearly 6.4 million American households (nearly 6% of all U.S. households) had at least one family member who had been victimized by identity theft in the past six months. The average identity theft incident resulted in $1620 in losses to the individual. The national cumulative losses are in excess of $5 billion.

When the Internet was first in use, the criminal justice system was largely unprepared for the crime of identify theft. It was not until October 1998 that Congress passed the Identity Theft and Assumption Deterrence Act (Identity Theft Act) to address the problem. Specifically, this act made identity theft a federal crime, including the action of knowingly transferring or using another person's identification with the intent to commit or abet any unlawful activity. In addition to the specific charge of identity theft, identity criminals may be in violation of several other federal laws, including those related to wire fraud, credit card fraud, bank fraud, and computer fraud.

Identity theft also is a crime at the state level. Since 1998, every state except Colorado and Vermont has enacted legislation to deal with identity theft. In 2004 Congress increased the penalties for identity theft convictions when it passed the Identity Theft Penalty Enhancement Act. This act added two years of imprisonment on two sentences for crimes involving the use of identity theft and an additional five years when the identity theft facilitated involvement with terrorists or terrorist organizations. The Enhancement Act also precludes probation as a sentence for identity theft and mandates consecutive, not concurrent, sentences.

There is no single database that captures all investigations and prosecutions of identity theft cases, and it is estimated that 75% of identity theft victims do not report the crime to police. Identity thieves often have technical knowledge about producing fraudulent documents, a working understanding of banking and other financial industries, and savvy skills that help them evade detection.

College students are especially vulnerable to identity theft. A national survey revealed that almost one-half of all college students keep personal financial information in their residence hall room, and 30% said that a room in their dormitory had been vandalized. College students are regularly sent preapproved credit applications from banks seeking relationships with potential customers.

Technology makes crimes such as identity theft possible, but technology also plays a vital role in overcoming these threats by providing ways for individuals and organizations to conduct financial transactions securely and by increasing authentication methods. Authentication helps verify the identity of the individual by using personal identification numbers or check verification processes. The next generation of authentication will most likely occur in the area of biometrics, in which an individual's unique physical attributes like fingerprints, eyes, face, and written signature are captured in electronic format.

How Cybercrime Changes the Picture

The investigation and prevention of cybercrimes moves forensic specialists into a difficult legal world. On the one hand, police want to curb such crimes to the full extent of the law; on the other hand, these very attempts often come close to violating freedoms guaranteed by the Constitution. There is, however, a new effort to curb cybercrime without venturing into murky legal ground. To fully realize it, though, would require that all countries adopt agreed-upon laws so that each system is able to prosecute cybercriminals on a level playing field. This idea is most effectively contained in the Council of Europe's Convention on Cybercrime, which presents a plan to achieve a balance:

Convinced that the present Convention is necessary to deter action directed against the confidentiality, integrity and availability of computer systems, networks and computer data as well as the misuse of such systems, networks and data by providing for the criminalisation of such conduct, as described in this Convention, and the adoption of powers sufficient for effectively combating such criminal offences, by facilitating their detection, investigation and prosecution at both the domestic and international levels and by providing arrangements for fast and reliable international co-operation;

Mindful of the need to ensure a proper balance between the interests of law enforcement and respect for fundamental human rights as enshrined in the 1950 Council of Europe Convention for the Protection of Human Rights and Fundamental Freedoms, the 1966 United Nations International Covenant on Civil and Political Rights and other applicable international human rights treaties, which reaffirm the right of everyone to hold opinions without interference, as well as the right to freedom of expression, including the freedom to seek, receive, and impart information and ideas of all kinds, regardless of frontiers, and the rights concerning the respect for privacy

There are obvious pitfalls to this plan, including the differences in legal systems and cultural standards that each country possesses, as well as the necessity of gaining most countries' cooperation with the tenets of the plan.

Another method of preventing cybercrime is to require that all computers be equipped to repel mal-intentioned software (a.k.a. **malware**). The question is: Is it reasonable to expect the average computer-literate individual between the ages of 18 and 35 to be aware of and capable of installing some type of malware protection on his or her computer? If so, the protection benefits the individual and the greater Internet community by protecting computers from malicious, remote commands that could be used to amplify criminal actions. When the possibility of mandating protective software is considered, we must also consider who is responsible for the efficacy of the software. Today, the free market is the only structure that governs the efficacy of such programs as Norton Antivirus, Symantec Antivirus, Virex, and AVG. Is it necessary for the government to play a role in our protective software?

Another legal pitfall is culpability. Let's assume that protective software is regulated at least to a point so that any brand maintains a government-sanctioned level of protection. Now let us imagine a user who lives alone and has no friends, family, or acquaintances who are aware of this technology. Is this person responsible for the absence of any government-sanctioned protective software? What if a virus begins to spread and the computer is infected? Is the person responsible for the damages that emanated from his or her computer?

These legal questions will be answered eventually, but for now they are just questions. Crime scene investigators must be aware of the legal intricacies surrounding any crime they investigate. Investigators find themselves helpless at times because laws are not able to keep up with technology. Unbiased, uninterested facts come from the investigator, and it is up to the investigator to work creatively within the bounds of our democratically created legal code to accumulate and, possibly, piece together the information that will lead to a conviction of the guilty and/or the clearing of the innocent.

Processing the Crime Scene

Processing the Physical Crime Scene

Regardless of whether a crime scene is a physical crime or a cybercrime, the methodology for processing a computer found at the location remains the same. In some cases, the crime scene that the first responder is entering will be a combination of both a physical crime scene, as described earlier in this book, with an electronic crime component. In other cases involving only cybercrime, the electronic components will be the major focus for analysis. In either case, remember that an electronic crime scene is still a crime scene. Keep your notebook, camera, and pencil as accessible as you would for any ordinary crime scene. It is imperative that the first responder does not try to turn the computer on or off because computers can be easily rigged to self-erase or do other self-damage.

The crime scene should be photographed as thoroughly as possible. Wide-angle photos should be taken first to show the overall layout of the room, the placement of the computer, the type of computer (either desktop or laptop), its peripherals, and the direction the monitor is facing. Also, make a record of any peripherals attached to the computer, such as the keyboard, the mouse, printers,

on the CRIME SCENE BTK Killer

An unknown man was responsible for a series of killings during the 1970s, 1980s, and early 1990s throughout the country in which his modus operandi was to bind, torture, and kill his victims. Because of the way he treated his victims, law enforcement officers named him the BTK (bind, torture, and kill) killer. By 2000 law enforcement investigators thought he may have died or was incarcerated for another type of crime because he had not killed since 1991. Early in 2004, however, the BTK killer began to taunt the police, first by sending a letter to a Wichita, Kansas, newspaper taking credit for a 1986 murder. Enclosed with the letter were a photocopy of the victim's driver's license and three photos of her body. Three months later, another letter surfaced that described the 1974 murder of a family. In December 2004, a package was found in a park that contained the driver's license of another victim as well as a doll in a plastic bag with its hands bound with pantyhose. The trail remained cold until February 24, 2005, when Dennis Rader was arrested as the suspected BTK killer. His arrest hinged on digital evidence. Rader had sent a floppy disk to a news station on February 17, 2005. When police analyzed the 3.5" floppy disk, they found a deleted file, which was a church newsletter. The newsletter was made using Microsoft Word and was ascribed to a man who called himself "Dennis." This information lead to an arrest of Rader and a warrant for a sample of his DNA. Rader pled guilty to ten counts of first-degree murder on June 27, 2005, and is currently serving a life sentence.

external hard drives, and any other features of the physical area.

Intermediate distance photos should show the location of evidence and its relationship to the entire scene. Close-up photos should record any Post-it notes that may contain passwords or other evidence indicating data access. Software manuals or CDs with anti-forensic software, encryption software, or Windows cleaner software also should be photographically documented.

Often the computer's mouse must be moved to disable a screen saver to capture a picture of the monitor. It is extremely important to document the time and date before moving the mouse. Photograph the taskbar area of the monitor screen, being sure that the time and date displayed on the taskbar are clearly evident relative to the actual date and time as noted by the camera. The computer's time may have been altered so that the time recorded by the computer when certain files have been accessed, or deleted, is incorrect. Perpetrators often do this to later provide them with an alibi ("I couldn't have been saving child porn then because I was at work when that happened").

After thoroughly documenting and photographing the physical scene, note whether the computer shows indications of tampering. On a laptop, check the edges of the casing for screwdriver marks that would indicate that someone has opened the machine (FIGURE 15-10). On a desktop, determine if the screws were placed by the factory and have not been removed or replaced.

After photographs have been taken, labels should be placed on all of the cords attached to the machine and its peripherals (FIGURE 15-11). If more than one computer is present at the scene, a more extensive numbering scheme should be devised.

The photographs, investigator's notebook, sketches, and labels should completely document the entire crime scene so that it will be easy to put

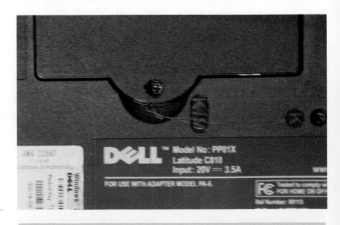

FIGURE 15-10 Scratches indicate that someone tampered with this laptop.

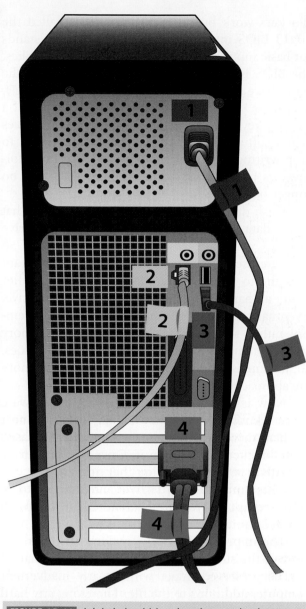

FIGURE 15-11 A label should be placed on each wire as well as a corresponding label on the port to which it was attached.

the computers back together in the laboratory just the way they were at the crime scene.

Processing the Electronic Crime Scene

Now that the physical crime scene is documented, it is time to address the electronic crime scene. Before you begin, make note of whether the computer is on. If the computer is on, *do not select the shutdown command within the operating system.* Instead, for a desktop, unplug the computer from behind the machine, and for a laptop, remove the battery. In either case, unplugging the machine or turning off any power strips may issue a command to the computer

that it is being turned off by someone other than the owner; a subsequent command to delete or destroy electronic files could be embedded in the software. After the plug is removed from behind the computer, it is safe to move it to the laboratory.

Some sophisticated cybercriminals rig battery backups to their computers that are activated if power is lost. If you encounter a battery backup, disconnect it before unplugging the computer. Be sure to record in your notebook the date and time the disconnection and unplugging occurred.

In a crime scene involving computers, knowing how to preserve the evidence is paramount. If a hard drive is exposed to a magnet or magnetic field, the data on the hard drive will be corrupted. Extra care must always be taken to shield the hard drive from magnetic fields. The powerful radio transmitters located in the trunks of most police cruisers generate magnetic fields. Seized computer evidence should, therefore, never be placed in or near the trunks of police cars.

Furthermore, it is not easy to protect or duplicate the data the HDD holds. A hard drive is always subject to change when it is in operation; every time the computer boots, it changes some information on the HDD. Therefore, a **write blocker** must be used to protect the evidence. This can be done only after the computer is secured in the crime laboratory and a write blocker is installed.

Once in custody and in the police laboratory, the computer can be opened to ensure that everything is in its proper place. Each piece of equipment inside the computer should be mounted and tight. Loose screws, missing screws, used screws, scratched metal casing, loose equipment, and any other signs of tampering should be noted.

When this examination is completed, the hard drive can be removed for either a desktop or a laptop. After the hard drive is removed, the computer should be plugged in and started up. The hard drive is removed before starting the computer up to ensure that no extraneous data is written onto it or that destroy programs are not invoked. The BIOS settings of the computer will appear on the monitor screen. The machine's date and time will be listed as will all installed **hardware**. Photograph the monitor screen and compare it with the actual date and time. Also photograph any wireless card or network interface card (NIC) that may be present. Laptops sometimes have switches to turn wireless capability on or off. Take a photograph to record whether or not the switch is in the on position. This photograph will later confirm that the computer was capable of accessing the Internet. Also note any antivirus software or firewall software. Its presence on the computer could later debunk the suspect's claim that "a virus put 1200 child porn images on my computer. . . ."

Acquiring an Electronic Image of the Crime Scene: The Hard Drive

Before viewing the hard drive, it is important to connect it to a write blocker to ensure that nothing will be changed. Write blockers come in two varieties: hardware and software. Hardware write blockers are attached to the hard drive in question. The hardware write blocker (**FIGURE 15-12**) is a simple device that filters commands that are sent to the hard drive and allows data on the HDD to be read and copied, but it ensures that the hard drive does not have any new data written onto it

Software write blockers are installed on the computer just like other software. Software write blockers work by interrupting what is called the "0x13 BIOS interface." The acronym *BIOS* stands for basic input/output system. As the name implies, the BIOS is responsible for working between the hard drive and the central processing unit (CPU). The same way the BIOS detects the presence of a mouse, keyboard, or hard drive, the BIOS realizes that a software-based write blocker is asking it to stop writing data to the hard drive. The BIOS follows its programming and the hard drive can no longer be written to.

Flash drives are the other magnetic media that store data on computer systems. Flash drive "sticks," which are easily installed through a USB port, are used to create backup copies of important files and to transfer data. Flash memory is smaller than a hard drive and is very inexpensive. Eventually all laptop computers will use flash memory rather than hard disk drives. Flash memory stores data in the same way a hard disk drive does, so, for forensic purposes, the two types of memory are approached the same way.

Now is the time either to attach a hardware device, known as a write blocker, or to connect the hard drive to a computer with a software-based write blocker installed on it. Once the hard drive is write-protected (meaning that no new data can be written onto this hard drive) and attached to an active computer, you are ready to create a copy of the electronic crime scene.

The important aspect of making an electronic copy of the suspect hard drive is to ensure that all data is preserved intact without any inadvertent computer additions to the file. Therefore, any hard drive that is used upon which the suspect hard drive is copied must be forensically clean. Forensically clean means that the hard drive reads as all zeros to the computer and that absolutely no other data exist on that drive. This task is accomplished by using a forensic tool (such as one from Encase) that wipes the disk and verifies that it is clean before any evidence is copied onto it.

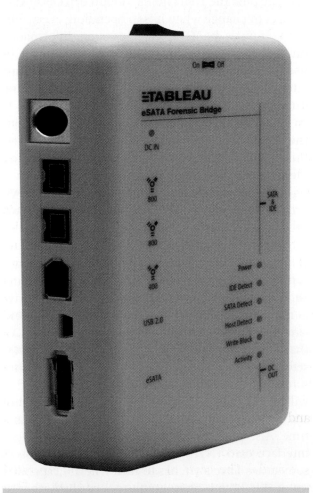

FIGURE 15-12 A hardware write blocker is a simple device.

Forensic Examination of Computer Devices

Copying the Hard Disk Drive

A forensic computer investigator will follow a protocol to gather and preserve any evidence remaining in the computer. Forensic examiners never want

After the scene has been secured, the following steps should be taken to process any and all electronic media (including computer equipment):

1. Protect all electronic evidence. Do not allow anyone to tamper with computers or other electronic devices. Isolate all electronic devices that communicate by way of infrared connectivity (cell phones, wireless devices).

2. Determine if computer equipment is turned on. If this equipment is on, then take the following steps to ensure that data is not accidentally lost.

   a. Photograph the computer screen.

   b. Make a sketch of the crime scene showing the placement of computer equipment. Document with more photographs and a video recording.

   c. Label and photograph all wires attached to the computer. As shown in Figure 15-11, a label should be placed on each wire and a corresponding label on the port to which it was attached.

   d. Do not turn off the computer in the normal way. Remove the power plug from the back of the computer. If you are dealing with a laptop computer, remove the battery. The computer may be configured to automatically overwrite sensitive information when a shutdown command is given.

   e. Gather and tag all peripherals, disks, drives, or other storage media.

3. Do not use aluminum-containing fingerprint powders near computer equipment—they can damage sensitive electronic equipment.

4. Remove cables and peripheral devices for packaging and removal.

   a. Maintain chain of custody.

   b. Package electronics in static-free containers.

   c. Seal the CD-ROM drive shut with evidence tape.

5. Transport all electronic evidence, being careful to keep it away from strong sources of magnetism. The passenger-side floorboard area is the best location in a vehicle to protect the evidence from magnetic fields and from damage by quick stops.

---

to use the original HDD for fear or altering or destroying evidence. They will therefore install a write blocker, which was discussed earlier. This device prevents the forensic examiner from writing any data onto the hard disk drive. It is like a one-way valve that allows data from the drive to be written onto a copy but no other data to enter the system and to be written onto the hard disk. After the write blocker is installed, the investigator begins by making a **bitstream copy** of the data held in the hard drive, which is an exact duplicate copy of the original disk. To ensure that any data held on the HDD is not lost, the HDD is usually removed from the seized computer and placed into a forensic computer so that a forensic image can be created. If the HDD was left in the seized computer, it is possible that that computer would alter the data held by the HDD when it was started. When this forensic image is created, the forensic computer must be in a "write blocked" state so that it will not write to the HDD.

The forensic image contains the entire contents of the HDD, not only the files that the computer's operating system can retrieve easily. Making a bit-stream or **bit-for-bit copy** of the HDD, however, is more complicated than making a copy of a computer text file. The only effective way to ensure that a bit-for-bit copy of a hard drive is acquired is to use software designed for the purpose of making bit-for-bit copies. This specialized, and expensive, software is required because ordinary copies that are made on a computer are not bit-for-bit. These copies preserve the functional aspects of programs and files, but not necessarily other aspects that are hidden and non-functional. Also, there are components of the hard drive that are not normally accessed by the operating system that need to be copied and combed through during an investigation. It is very important to recognize that the only copies admissible in court are those made by this specialized software designed to make bit-for-bit copies.

To certify that the copy of the HDD is a "true and accurate copy" of all the data on the HDD, a **message digest (MD)** or **secure hash algorithm (SHA)** is used. Both of these are accepted court standards. The copy of the HDD is fingerprinted using an encryption technique called *hashing*. This is sometimes referred to as cryptographic hash verification. Hashing generates a unique digital signature, or fingerprint, for the copied data, which is called the message digest. Comparison of the SHA produced before imaging is compared with the SHA after imaging. If even one small **bit** of data has been altered, the resulting hash value will differ greatly from the hash value of the original drive. In this way, the accuracy of the bitstream copy is verified. The same technique could be used to make a bitstream copy of the data held on floppy disks, CDs, DVDs, flash memory, or tapes.

Analyzing Digital Evidence

Most forensic software utilities begin operation with an indexing process. The **index** is automatically created by most brands of forensic utility software. The software logs the time and date of all the activities performed. The indexing process performs a number of functions on every file or file fragment. Every character, letter, and number of every file is indexed. For text or word processing documents the software indexes every word in every file, which will allow the investigator to search the entire disk drive for a particular word or phrase. Once the drive has been completely indexed, the software can easily report how many files are on the drive, what their size

and locations are, and whether or not they have been deleted. The software can also determine what kind of file each file is, without referring to the three-letter file name extension. If a criminal changes a file name extension to hide a file, the forensic software will still correctly identify the file type. For example, if a pedophile changes a pornographic image of a child from a JPG file type to a .txt file type to hide the image, then the forensic software, when indexing the file, will record that original file name but will correctly identify the file as a JPG image and will display a thumbnail image of the picture. The software also creates numerous reports that group files by type. For instance, all graphic images are grouped into one report, all video clips into another, and all email messages into still another. Forensic investigators can use these reports to quickly find files by size, date created, date modified, or type of file. Forensic examiners can also find deleted files and deleted file fragments.

When a file is deleted, nothing actually happens to the files of data at that time. As shown in **FIGURE 15-13**, executing a delete command simply tells the computer to treat that file space as blank and not to allow programs to write data to that file space until the disk is full and that space is needed later. Until new data is written to that file space, the

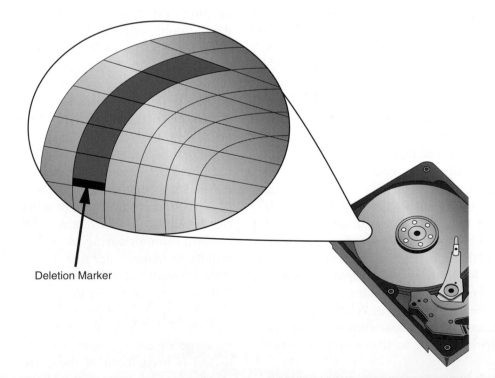

Deletion Marker

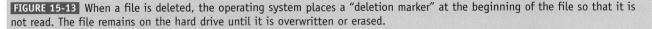

FIGURE 15-13 When a file is deleted, the operating system places a "deletion marker" at the beginning of the file so that it is not read. The file remains on the hard drive until it is overwritten or erased.

original file is still present. This is why computers have a recycle bin or a trashcan. When users are sure they no longer need the file, they can empty the recycle bin or trashcan. Many criminals believe that once the file is deleted and the recycle bin emptied, the file is gone from their hard drive and cannot be recovered. This is not true. Unless that area of the hard disk drive is overwritten by new data, forensic software can recover deleted files and any potential evidence contained therein. In fact, most forensic software calls the investigator's attention to all deleted files because these files are considered suspicious and likely to contain evidence. If the suspect computer (at the crime scene) is still turned on, there are forensic software packages, such as the Corners Toolkit and Sleuth Kit (http://www.sleuth kit.org/sleuthkit), that contain a tool called ILS, which can show deleted files that are still open.

Another useful feature of most forensic software is the ability to highlight or otherwise mark files of interest. If, for instance, an investigator locates an image that appears to be child pornography, he or she can use the software's tools to highlight that file. In such a way, numerous files or items of interest can be used to create a special report providing information such as the date files were created, the date they were modified, file names, and all other relevant data about files. This level of detail in documentation is very helpful in prosecuting child pornography cases. The FBI's Innocent Images National Initiative has made it even easier to locate child pornography images on a hard drive by routinely updating its database of known child pornography files. This database includes the MD5 hash values of files that are known to be child pornography. The National Center for Missing and Exploited Children (NCMEC) and Immigrations Customs Enforcement (ICE) also maintain databases that include SHA-1 listings of files that are known to be child pornography.

Forensic Electronic Toolkits

Most forensic software programs perform core tasks that are essential to a computer forensic laboratory. First, the software needs to have the ability to acquire digital evidence or data files by reading that information from a variety of media types and formats. Both Guidance Software's EnCase® (http://www.guidancesoftware.com) and AccessData's Forensic Toolkit® (http://www.access data.com) have the ability to read hard drives, CDs, DVDs, USBs, PDA and cell phone memory, as well a variety of other media types. After reading the data, they acquire it by creating an image file, separate and categorize data files by type (e.g., graphics, email, videos, text), compare evidence files with lists of known contraband files, recover the weeded or hidden data, crack or recover passwords to allow access to encrypted data, and systematically report findings.

E-Evidence Collection Process

Examination of a suspect computer should be able to ensure that all information is recovered. All forensic electronic toolkits should:

- Protect the computer system during forensic examination from any possible alteration or damage.
- Discover all files, including deleted files that still remain as hidden files, password-protected files, or encrypted files.
- Recover deleted files.
- Reveal the contents of hidden files as well as temporary or swap files used by application programs or the operating system.
- Access the contents of protected or encrypted files.
- Use methods to determine the existence and potential locations of any steganography, which is hidden information. Steganography generally involves hiding a message in an image.
- Analyze data found in the hard drive relating to access areas of the disk, such as unallocated space on a disk, which is space that is not currently used to store an active file but may have previously stored a file, and **slack space**, which is the remnant area at the end of a file. These spaces are discussed in detail in the next section.
- Give an overall analysis of the computer, including devices, files, and discovered file data.

File Slack

File slack is created when a file is saved if the file does not take up an entire sector on the disk. As shown in **FIGURE 15-14**, a sector is the smallest unit that can be accessed on a disk. Sectors are groups of **bytes** and can vary in size depending on the media; HDDs group sectors into 512-byte increments, while CDs allocate 2048 bytes per sector. All Microsoft operating systems read and write in

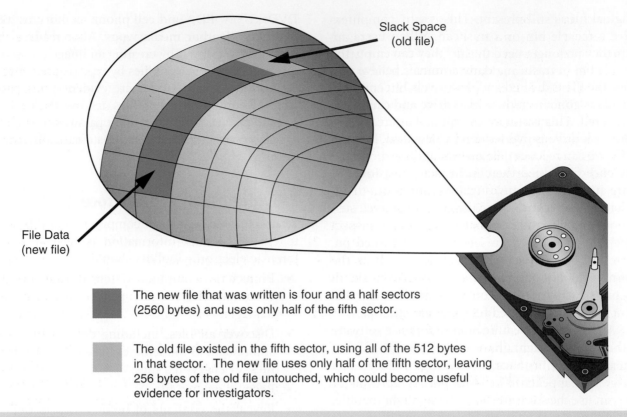

Slack Space
(old file)

File Data
(new file)

The new file that was written is four and a half sectors (2560 bytes) and uses only half of the fifth sector.

The old file existed in the fifth sector, using all of the 512 bytes in that sector. The new file uses only half of the fifth sector, leaving 256 bytes of the old file untouched, which could become useful evidence for investigators.

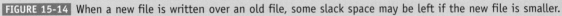

FIGURE 15-14 When a new file is written over an old file, some slack space may be left if the new file is smaller.

blocks of data called **clusters**. A cluster is a fixed block of data that consists of an even number of sectors (1024 bytes or 4096 bytes). The operating system assigns a unique number to each cluster and then keeps track of files according to which cluster they use. When a file is deleted using any Windows product, the data remain in the cluster. The number of sectors needed for a cluster depends on the type of storage device, the operating system, and the size of the storage device. The clusters that make up a deleted file are released by the operating system, which means that they remain on the disk as unallocated space until that space is overwritten with data from a new saved file.

If the minimum unit of the HDD is 512 bytes, what happens if a file contains only 200 bytes? In this example, there is 312 bytes of slack space, but it is even more complicated because there is a minimum cluster required. The minimum cluster is defined in a sector multiple of two. A cluster is a minimum of 2, 4, 6, or 8 sectors. If our 200-byte file is in an HDD that has an allocation of two sectors per cluster (1024 bytes), the operating system allocates 1024 bytes of storage space for our 200-byte file. The remaining 824 bytes would be slack space.

File slack can contain several hundred megabytes of data. It could hold data, login names, passwords, or other confidential data. File slack might also reveal historical uses of the computer, such as email messages or browsing history, which is known as legacy data.

Deleted Files

When the operator gives the command to delete a file, the computer inserts a deletion marker on the HDD at the beginning of the space where the file is stored, as shown in Figure 15-13. This makes the file no longer viewable through normal computer programs, and the computer considers that space to be empty. The data is still present on the hard drive, however. If the hard drive becomes full at a later date, the computer may choose to overwrite this file. But until that time comes, the data still reside on the hard drive and can be forensically extracted.

WRAP UP

1. Do not allow anyone to touch the computer. Isolate all electronic devices that communicate by way of infrared connectivity. If the equipment is on, photograph the screen. Make a sketch of the scene showing placement of the computer equipment, and document with more photos. Label and photograph all wires attached to the computer. Do not turn off power in the normal way. Instead, remove the power plug from the back of the computer (for a laptop, remove battery).

2. Gather and tag all peripherals, disks, drives, or other storage media. Remove cables and peripheral devices for packaging and removal. Maintain chain of custody. Package electronic devices in static-free containers. Seal the CD-ROM drive shut with tape. Transport to crime lab in passenger-side floor board area of the police vehicle.

Chapter Spotlight

- The central processing unit (CPU) is the brain of the computer. It is the CPU's job to execute stored instructions. The CPU repeats four basic steps: fetch, decode, execute and writeback.

- Read-only memory (ROM) are computer chips that hold the instructions to power up the computer and to run CDs, hard disk drives, and graphic cards. ROM is an important part of the BIOS (basic input/output system) operation.

- Random-access memory (RAM) is computer memory that holds data that can be accessed by the CPU at any time. Since the hard disk drive retrieves data slowly, RAM takes information from the hard drive that it may need to complete a task and stores it so it will be readily available.

- The hard disk drive is the computer's main storage device. It is the only component of the computer that can store data while the computer is turned off.

- The Fourth Amendment prohibits law enforcement from accessing and viewing information stored in a computer without a search warrant.

- Cybercrimes are crimes intended to damage or destroy computer networks, fraudulently obtain money, disrupt business, or commit felonies through the use of computers.

- The following steps should be taken to process any electronic media:

- Photograph the computer screen.
- Make a sketch of the crime scene.
- Label and photograph all wires attached to the computer.
- Do not turn off the computer in the normal way. Remove the power plug from the back of the computer. If it is a laptop computer, remove the battery.
- Do not use aluminum-containing fingerprint powder near the computer.
- Remove, label, and package all cables.
- Protect all electronic evidence from strong magnetic fields.

- A bit-for-bit copy of the contents of the HDD is made after a write blocker is attached to the HDD.

- Forensic analysis utilities begin by indexing the HDD copy. All graphic images are grouped into one part of the report, all video clips into another, and all email messages into still another. Examiners can use these reports to quickly find files by size, date created, or date modified.

- Forensic examiners also focus on deleted files and file slack. File slack is created when a new file does not take up an entire sector on the disk.

Key Terms

Bit A binary digit, either a 1 or a 0.

Bit-for-bit copy An exact copy of the data saved on a hard disk drive.

Bitstream copy A sequence of data in binary form.

Byte Eight bits.

Central processing unit (CPU) A microprocessor chip that carries out most of the operations of the computer.

Cluster A fixed block of data that consists of an even number of sectors (1024 bytes or 4096 bytes).

Cybercrime Disruptive activities against computers or networks, intended to cause harm.

Cyberspace The term used to describe the millions of interconnected computers, servers, routers, and switches that sustain the digital infrastructure.

Cyberterrorism Any type of attack on the computer infrastructure.

Hacking A slang term for unauthorized access to a computer or network.

Hard disk drive (HDD) A computer's main data storage device.

Hardware The physical components of a computer.

Identity theft The unauthorized use or attempted use of another person's financial records and/or identifying information without the owner's permission.

Index A list of every character, letter, number, and word in all files on a computer, created by the forensic utility software.

Malware Malicious software designed to infiltrate a computer or network without the owner's consent.

Message digest (MD) A unique digital signature for the data generated by hashing.

Phishing The use of fraudulent email sent in an attempt to trick the recipient into disclosing personal information.

Random-access memory (RAM) Computer memory that executes programs and stores temporary data. RAM data is automatically erased by the computer when it is shut off.

Read-only memory (ROM) A device that contains instructions. Computers can read the instructions, but not change or write back to ROM.

Sector The smallest unit that can be accessed on a disk. A sector is 512 bytes.

Secure hash algorithm (SHA) An algorithm developed by the National Institute of Standards and Technology (NIST) used to create or detect digital signatures.

Slack space The area from the end of the file to the end of a sector on a hard disk drive.

Smart grid The addition of computer-based technology to the U.S. power grid to manage the efficiency and reliability of electricity distribution.

Software A set of instructions that are arranged in a program that allows a computer to complete a specific task.

Spam Unsolicited bulk email messages.

Spyware A type of malware that collects information about users without their knowledge.

Swap data A file or space in the hard disk drive that swaps data with the RAM memory to free the RAM to execute other commands.

Write blocker A hardware or software device that prevents any writing to the disk drive to which it is attached.

Putting It All Together

Fill in the Blank

1. The central processing unit (CPU) repeats _____ fundamental steps.

2. When powering up the computer, the _____ enables a CPU to communicate with the HDD.

3. The program that holds the instructions to power up to the computer is held on the _____.

4. _____ is considered to be volatile memory—that is, memory that loses its information when the power is turned off.

5. The _____ is the only component of the computer that can store data while the machine is powered down.

6. A HDD is divided into _____ so that files can be organized and found more quickly.

7. A(n) _____ is a device that is attached to an HDD that prevents any new information from being written on the HDD.

8. Fingerprint powder containing _____ should not be used anywhere near a computer system.

9. An exact duplicate copy of the data held in the hard drive is known as a(n) _____.

10. To certify that the copy of the HDD is a true and accurate copy of the data a(n) _____ is used.

11. The forensic copy of the HDD is fingerprinted using an encryption technique called _____.

12. The unique digital signature for the copy data is called the _____.

13. Most forensic software utilities began by creating a(n) _____.

14. Forensic software can determine what type of file each file is without referring to the three-letter file name _____.

15. Most forensic software calls attention to all _____ files because they are considered suspicious and likely to contain evidence.

16. An attack on an Internet connection that results in an overload so large that it causes disruption of normal operation is called a(n) _____.

17. Unlawful monitoring of data traveling over a computer network is known as _____.

18. Gaining unauthorized access to a computer by sending a message from a trusted Internet address is known as _____.

19. The FBI updates a database of known child pornography files. This database includes _____ hash values of files that are known to be child pornography.

20. A remnant area at the end of a file on an HDD is known as _____.

21. A(n) _____ is the smallest unit that can be accessed on an HDD.

22. All Microsoft operating systems read and write in blocks of data called _____.

23. An HDD's sector has _____ bytes.

True or False

1. Once a file is deleted, the information it held is forever lost.

2. BIOS stands for big input/output system.

3. The first thing a forensic investigator should do when examining a computer is to turn on the computer.

4. When a file is deleted, all the digital information contained in that file is completely erased from the HDD.

5. Steganography involves hiding a message in an image.

Review Problems

1. Metadata is hidden information about computer files and folders. Go to the Microsoft support website (http://support.microsoft.com/kb/q223790), which describes types of metadata that are routinely collected on all Microsoft Word documents.

   a. Describe how you would remove your name as the author of a document.

   b. Describe how to remove network information for files you save.

   c. Define *descriptive metadata*, *structural metadata*, and *administrative metadata*.

2. You have been asked to investigate a specific file on an employee's computer. You know that he leaves his computer logged on when he goes to lunch. If you open the file, will the computer record that you have done this? Explain.

3. Go to the website of the American Institute of Certified Public Accountants at http://AICPA .org. Under the Resources tab, select Antifraud Resource Center, then select Fraud Detection, Investigation, and Prevention. Select one of the cases, then describe the case and explain how electronic evidence was used to find the fraud.

4. Windows programs use a three-letter file name extension to inform the computer of the initial program to run to open the file. For example, a file with an .xls extension will direct the computer to open it using Excel. Explain how a pedophile can hide pornography by simply changing the file extension.

5. Steganography is the science of "covered writing" and is being used by terrorists to hide secret messages. Describe what type of file is optimal for use as the carrier medium. Go to camouflage.unfiction.com to learn how to use steganography to hide information.

Further Reading

Britz, M. T. *Computer Forensics and Cyber Crime: An Introduction* (ed 2). Upper Saddle River, NJ: Prentice Hall, 2009.

Council of Europe. Convention on cybercrime, http://conventions.coe.int/Treaty/EN/Treaties/Html/185.htm

Electronic Crime Scene Investigation: A Guide for First Responders (ed 2). Washington, DC: U.S. Department of Justice, Office of Justice Programs, National Institutes of Justice, 2008.

Grabowsky, P. *Electronic Crime*. Upper Saddle River, NJ: Prentice Hall, 2007.

Knetzger, M., and Muraski, M. *Investigating High-Tech Crime*. Upper Saddle River, NJ: Prentice Hall, 2008.

Taylor, W. T., Caeti, T. J., Loper, D. K, Fritsch, E. J., and Liederbach, J. *Digital Crime and Digital Terrorism*. Upper Saddle River, NJ: Prentice Hall, 2006.

Volonio, L., Anzaldua, R., and Godwin, J. *Computer Forensics: Principles and Practices*. Upper Saddle River, NJ: Prentice Hall, 2007.

http://criminaljustice.jblearning.com/criminalistics

Answers to Review Problems

Interactive Questions

Key Term Explorer

Web Links

Explosives

OBJECTIVES

In this chapter you should gain an understanding of:

- The classification of explosives
- The composition of commercial explosives such as dynamite and ammonium nitrate/fuel oil (ANFO)
- The composition of military explosives
- How improvised explosive devices (IEDs) are made
- Field tests for explosive residue
- Confirmatory laboratory tests for explosive residue
- The use of taggants to identify the manufacturer of an explosive

FEATURES

On the Crime Scene

Back at the Crime Lab

See You in Court

WRAP UP

Chapter Spotlight

Key Terms

Putting It All Together

There are three places to look for evidence in a bombing: the target, the materials damaged by the blast (witness materials), and the area outside the bombing crime scene. Investigators' ability to find evidence at a bombing site often depends on the material that makes up the target site itself. For example, if the explosion leaves a crater, recoverable evidence may be caught in the hole. By contrast, if the bomb creates a hole through an object, only explosive residue will be left.

Materials near the target may yield clues as well. For this reason, the area immediately surrounding the crater is the first place to look for fragmented bomb components. Whether the bombing took place inside a structure or outdoors also determines how investigators look for evidence. For instance, parts of an IED may be embedded in carpet, furniture, walls, ceilings, or any object within the scene. The trajectory of bomb fragments can drop evidence in unlikely places such as rooftops, gutters, and trees.

Given the wide range of possible evidence locations, it is important to establish an outer perimeter for the bombing crime scene. Investigators begin searching at the outer perimeter and work inward toward the target. This type of search will decrease the likelihood that the search team will inadvertently destroy hidden evidence.

1. Describe how you would search for evidence in a car bombing that occurred on the basement level of a parking garage.
2. Describe how you would search for evidence in a bombing that occurred outside on the sidewalk in front of a railroad station.

Introduction

When most people hear about bombings in modern cities, their first thought may be that such acts are the work of foreign terrorists. But recall the 1995 bombing of the Murrah Federal Building in Oklahoma City, Oklahoma. It was carried out by Timothy McVeigh and Terry Nichols, two Americans who openly expressed their hatred of the U.S. federal government. In their attack, McVeigh and Nichols used fertilizer and nitromethane (race car fuel) in a truck, a simple device meant to cause massive destruction and bring maximum attention to their issue. As this case demonstrates, most bombings in the United States are actually not acts of terrorists from foreign countries, but rather the work of homegrown criminals. Explosives are an attractive weapon for members of our society who are disenchanted with our government or have some other cause to proclaim for two reasons: such weapons can generally be made at home, and they bring considerable attention to the bomber's issue thanks to the destruction they cause and the ensuing publicity.

A classic example is Ted Kaczynski, the "Unabomber," who terrorized the United States for 20 years by mailing homemade bombs to his victims. The Unabomber's manifesto, which was published in newspapers throughout the country as part of the deal made for him to turn himself in to authorities, proclaimed his anti-technology philosophy, which his bombs were intended to highlight. Kaczynski carefully built his bombs to withstand rough handling at the post office and designed them to explode only when the package was opened by the intended target. Kaczynski admitted to being the Unabomber in 1998 and accepted a life sentence in prison.

While these two high-profile cases involved explosives set off for political reasons, bombings to settle grudges between rival motorcycle gangs also are quite common. Other, less sophisticated bombers simply empty the contents of shotgun shells into cast iron pipes, seal the ends, and detonate crude pipe bombs.

Explosions

An **explosion** is a release of mechanical or chemical energy in a violent manner that generates heat (a high temperature) and the release of large quantities of gas. For the forensic examiner, explosions caused by chemical reactions are the most commonly encountered type of explosion; thus they are the primary subject of this chapter.

on the CRIME SCENE | London Bombing

On July 7, 2005, London was the target of one of the deadliest terror attacks in the United Kingdom since the explosion of Pan Am Flight 103 over Lockerbie, Scotland in 1988. On this day at 8:50 A.M., bombs exploded within seconds of each other on three trains at London Underground stations: the Liverpool Street and Edgware Road Stations on the Circle Line and the King's Cross Station on the Piccadilly Line. A fourth bomb exploded on a bus nearly an hour later in Tavistock Square. Fifty-two people died, four of whom were the perpetrators, and more than 700 people were injured as a result of the blasts.

Forensic investigators examined nearly 2500 items, including closed-circuit security tapes and forensic evidence collected at the crime scenes. Officials determined that each of the bombs contained 4.5 kg of high explosives made from homemade acetone peroxide. Using the closed-circuit television security tapes, investigators soon identified the four men involved in the bombings. Officials also recovered two unexploded devices and a mechanical timing device from the scene.

Given that the bombs on the trains exploded within 50 seconds of one another, officials suspected that cell phones were used to detonate the explosives. The unreliable cell phone service in the London Underground argued against this theory, however, and further suggested that the detonation occurred manually.

Using the security tapes, police identified four men, all British-born, as the chief suspects. They were able to trace the route these men took—from their arrival at a train station in Luton, Bedfordshire, by car to their arrival in London by train at the King's Cross Station around 8:30 A.M. Police also found a cache of explosive devices in the car that the men left behind at the train station in Luton. The identified men purchased return tickets from this station, implying that they intended to survive the explosions. It is speculated that they may have misjudged the amount of time they had to escape before the blasts occurred. Eyewitness accounts indicated that the bombs may have been left either on the train track or on the undercarriage of the trains, further disputing the suicide bomber theory.

Forensic evidence found at the bomb sites helped police connect personal belongings with the suspects. However, it was through analysis of the personal property left at the scene, bomb fragments, and explosives that ultimately led investigators to discover a bomb factory in Leeds in the north of England. Hasib Hussain, Mohammad Sidique Khan, Germaine Lindsay, and Shehzad Tanweer were found dead at the bomb sites and are believed to be the perpetrators.

Forensic evidence suggests that the terrorists intended to explode bombs in four locations in the London Underground at the point where the two train lines intersected. The bombs that did explode did so at the time when trains were entering these intersections, thereby damaging both trains with one explosion. It is believed that the fourth bomber turned away from his target in the Underground when the explosions started and then chose to detonate his bomb in a bus instead.

Source: BBC.

Any substance that is capable of producing an explosion is called an **explosive**. Such a substance must be able to participate in rapid chemical reactions that produce large quantities of gas. The explosive is usually held in a metal container, such as a sealed lead pipe. When the explosive reaction occurs, the gas produced causes the walls of the pipe to deform outward until the pressure inside the pipe becomes high enough that the walls burst. The rupture of the walls of the pipe causes it to fragment and shoot flying debris in all directions.

Bombers often add nails or glass to the pipe's contents in hopes that the bomb will cause more harm to people as these sharp objects (shrapnel) are discharged.

The gaseous products of the explosive reaction expand rapidly as they are released from confinement, and they compress layers of surrounding air as they move from the point of origin. The outward rush of gases can be greater than 5000 miles per hour, and they can be powerful enough to move walls, lift roofs, and flip over automobiles. If the explosive charge is large enough, more lethal damage will be inflicted by the blast effect of the gases than by fragmentation debris emitted from the walls of the device. Sometimes the most lethal effects are caused by collapse of a structure or by large objects that become airborne missiles.

Types of Explosives

Explosives are classified as high or low explosives based on the speed at which they produce gas. **Deflagration** is a chemical explosion in which the reaction moves through the explosive at less than the speed of sound. Low explosives cause a deflagration. A **detonation** is a chemical explosion in which the reaction front moves though the explosive at greater than the speed of sound. High explosives detonate.

Low Explosives

Low explosives (also known as deflagrating explosives) include black powder and smokeless powder. In Chapter 8, we learned that black powder (also known as gunpowder) is a mixture of charcoal (which is a crude form of carbon, C), sulfur (S), and potassium nitrate (also known as saltpeter, KNO_3). When ignited, carbon and sulfur act as fuel, and potassium nitrate acts as an oxidizer. No oxygen (or air) is required for this ignition, because potassium nitrate provides the necessary oxygen:

$$2 \ KNO_3 + C + S \rightarrow K_2SO_4 + CO_2 + N_2$$

In this reaction, the solid carbon is oxidized (gains oxygen) and becomes gaseous carbon dioxide. The potassium nitrate is reduced (loses oxygen) and is converted into gaseous nitrogen. The conversion of the solid reactants into gaseous products ($CO_2 + N_2$) provides the pressure for the expulsion of the bullet from the barrel of the gun. The optimal ratio of these components for gunpowder is 15 parts potassium nitrate to three parts charcoal to two parts sulfur, or 15:3:2. Although black powder is no longer used in the manufacture of ammunition for guns, it still is the explosive most commonly included in fireworks. The composition of black powder manufactured for pyrotechnic use today is 75% potassium nitrate, 15% softwood charcoal, and 10% sulfur.

Smokeless powder is the name given to any propellants used in firearms that produce little smoke when fired, unlike the older black powder (which it replaced in this application). Smokeless powder consists of two components: cordite and ballistite. Cordite and ballistite are made by mixing together two high explosives, nitrocellulose and nitroglycerine. Nitrocellulose (also known as guncotton) is a highly flammable compound that is formed by nitrating cellulose with nitric acid. Nitroglycerine (NG) is an explosive liquid obtained by nitrating glycerol. In its pure form, NG is shock-sensitive (physical shock can cause it to explode), and it can degrade over time to even more unstable forms. As a result, the pure form of NG is highly dangerous. To make NG less shock-sensitive so that it will be safe for use in modern ammunition, manufacturers add 10% to 30% ethanol, acetone, or dinitrotoluene to the NG, which stabilizes it.

The explosive power of NG derives from its ability to detonate. This reaction sends a shock wave through the smokeless powder at a supersonic speed, which in turn generates a self-sustained cascade of pressure-induced combustion that grows exponentially. An explosion is essentially a very fast combustion, and combustion requires two components: a fuel and an oxidant. Nitroglycerine ($C_3H_5N_3O_9$), as can be seen from its composition and structure (**FIGURE 16-1**), contains carbon atoms that are ready to be oxidized and nitrogen that is ready to be reduced. One of the decomposition reactions of NG is

$$4 \ C_3H_5N_3O_9 \rightarrow 12 \ CO_2 + 6 \ N_2 + O_2 + 10 \ H_2O$$

Thus, from 4 molecules of solid or liquid NG, 29 molecules of gas are formed. If NG is detonated in a closed container, it explodes to produce hundreds of times its original volume in hot gas. This reaction increases the pressure in the container until it ruptures and explodes. The production of nitrogen, N_2, and carbon dioxide, CO_2, both of which are very stable, is highly exothermic; that is why nitrogen is a main constituent of most explosives.

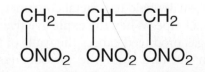

FIGURE 16-1 The chemical structure of nitroglycerin ($C_3H_5N_3O_9$).

Low explosives will burn rather than explode if they are not confined in some type of container. To create a fuse for a bomb, one simply wraps a fabric, such as cotton, around black powder. When ignited, the fuse burns at a slow, steady rate that allows the bomber enough time to escape. When low explosives are held in a container and do explode, the result resembles a pushing action rather than a shattering, smashing action.

Condensed explosives may be either a solid or a liquid. **Dispersed explosives**, by contrast, may be either a gas or an aerosol. Grain elevators, for example, have been destroyed by deflagrating explosions of dust that is suspended as an aerosol in the air. Other deflagrations occur when natural gas escapes into a confined area and mixes with air. Mixtures of air and natural gas will explode or burn only when the natural gas concentration in air reaches 5.5% to 19%. This mixture, when ignited in a home, produces so much gas and heat that it can force walls outward, causing the roof to collapse on the structure.

The most commonly encountered illegal explosive device in the United States is the pipe bomb (**FIGURE 16-2**). Pipe bombs usually contain black or smokeless powder or an improvised explosive mixture. Regardless of the type of explosive, any such material confined in a closed pipe that is fitted with a detonating device (fuse) is considered to be an **improvised explosive device (IED)**.

High Explosives

High explosives are classified into two groups: primary explosives and secondary explosives. Primary explosives are extremely sensitive to shock and heat and will detonate powerfully. As a consequence, they are rarely used as the major components in pipe bombs. Their shock sensitivity makes them useful for detonating other explosives; for this reason, they are often referred to as primers. Lead azide, lead styphnate, and diazodinitrophenol are the primary explosives used in the primers in bullets and in blasting caps.

Compared with primary explosives, secondary explosives are more stable—that is, they are not as sensitive to heat, shock, or friction. Like unconfined low explosives, these materials usually burn rather than detonate if ignited in an unconfined space (i.e., not in a container). For this reason, primers are often used to set off a bomb that contains a charge of secondary explosives.

Commercial Explosives

The first high explosive that was widely used in commercial application (in the 1850s) was NG. It is an oily liquid that is extremely sensitive to shock, thus making it difficult to handle and transport. Although users learned they could freeze this material to reduce its sensitivity, accidents were common as the NG was being thawed. In 1866 Alfred Nobel discovered that when NG was adsorbed onto an inert material, such as clay, its sensitivity was greatly reduced. The resulting high explosive, which Nobel called dynamite, could be shipped and handled without danger.

Dynamite

Nobel's invention made large-scale blasting available to the road-building and mining industry. Later it was learned that replacing the clay filler with a porous combustible material, such as rice hulls or sawdust, improved dynamite's gas production and, as a result, its explosive power.

If dynamite is used in cold climates, the NG can freeze, which makes its ignition more unreliable. This problem was solved by adding ethylene glycol dinitrate (EGDN) to the mixture, which reduces the explosive's freezing point and improves its reliability. The most common dynamite composition found in North America is an 80/20 mixture of EGDN and NG. **FIGURE 16-3** shows the chemical structure of EGDN.

FIGURE 16-2 The pipe bomb is the most commonly encountered illegal explosive device in the United States.

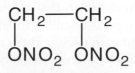

$$CH_2\!-\!CH_2$$
$$\;|\qquad\quad|$$
$$ONO_2\;\;ONO_2$$

FIGURE 16-3 The chemical structure of ethylene glycol dinitrate (EGDN).

Ammonium Nitrate/Fuel Oil

Ammonium nitrate (AN; chemical formula = NH_4NO_3) is a fertilizer that is widely used by farmers. In the early part of the twentieth century, commercial AN was sold as a white crystalline material that would stick together in clumps. In the 1940s, a new production method, called **prilling**, was discovered. It produced free-flowing, adsorbent, porous AN spheres (prills) that were easily handled and stored.

Mining companies, which used dynamite to blast away rock, found that if the AN was mixed with a source of carbon, it made a very cheap and effective explosive. At first these firms mixed AN with powdered coal, but these two solids proved difficult to mix into a uniform material. Miners eventually found that fuel oil (home heating oil or diesel fuel) was easier to mix than powdered coal and that the pores in the AN soaked up the oil to make a slurry. Later it was found that other fuels, such as nitrobenzene and nitromethane, were also effective in combination with AN.

ANFO, a mixture containing 94% AN and 6% fuel oil, is probably the most commonly used explosive material in the world today. Because its use is so widespread, dynamite production in the United States has been greatly reduced in recent decades. In 1959 there were 34 dynamite plants in this country; today, there is only one. Because ANFO requires a primer charge to initiate a blast, ANFO is classified as a blasting agent. Blasting agents do not perform well when used in small quantities, so ANFO is rarely encountered in small, improvised bombs. Large amounts of ANFO, however, have been used in terrorist bombings.

Military Explosives

Military explosives are manufactured for specific purposes. To determine the suitability of an explosive substance for military use, its physical properties must be thoroughly investigated. In view of the enormous quantity of explosives that are used in modern warfare, these explosives must be produced from cheap raw materials that are not strategic and are available in great quantity. In addition, the operations used to manufacture them must be reasonably simple, cheap, and safe. Finally, the density of military explosives needs to be as high as possible. Grenades, for example, are small military devices that can produce large explosions.

TNT

During World War I, 2,4,6-trinitrotoluene (TNT) became the dominant military explosive. TNT has a reasonably low melting point 178°F (81°C), which means it can be easily melted and poured into shells and bombs (**FIGURE 16-4**). TNT is not sensitive to shock, nor will it spontaneously explode. In fact, cast TNT is so stable that terrorists in some regions of the world routinely recover and use TNT from unexploded munitions. To initiate an explosion, a detonator is placed in the TNT to produce a pressure wave from another, more easily induced explosion involving another explosive in the detonator. Some detonators contain lead azide, $Pb(N_3)_2$, which explodes when it is struck or if an electric discharge is passed through it.

TNT is a powerful explosive because it can very quickly change from a solid into hot, expanding gases. Upon detonation, two molecules of solid TNT decompose to produce 15 molecules of hot gases plus some powdered carbon, which gives a dark sooty appearance to the explosion:

$$2\ \underset{\text{TNT}}{C_7H_5N_3O_6} \rightarrow 3\ N_2 + 7\ CO + 5\ H_2O + 7\ \underset{\text{Soot}}{C}$$

Although TNT has less potential energy than gasoline, the high velocity at which this energy is released from TNT produces its enormous blast pressure. TNT has a detonation velocity of 6825 meters/second (m/s), compared with 1700 m/s for the detonation of gasoline in air.

During World War II, TNT was conserved by mixing it with various amounts of AN to produce amatol, the explosive used by the German army in the V-2 rockets that were shot at London.

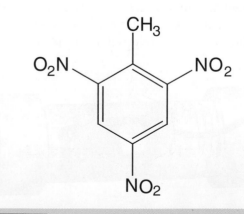

FIGURE 16-4 The chemical structure of 2,4,6-trinitrotoluene (TNT).

RDX

Cyclotrimethylenetrinitramine, also known as RDX, is one of the most widely used military explosives. It is believed that its RDX acronym is derived from "Research and Development Explosive." RDX is stable in storage and is considered the most powerful and brisant of the military high explosives. In a **brisant** explosive, the maximum pressure is reached so rapidly during the combustion reaction that a shock wave forms; this shock wave then shatters any material in its path.

RDX is a colorless solid with a density of 1.82 g/cm^3. It is obtained by reacting concentrated nitric acid with hexamine:

$$(CH_2)_6N_4 + 4\ HNO_3 \rightarrow (CH_2-N-NO_2)_3 + \\ 3\ HCHO + NH_4^+ + NO_3^-$$

RDX is a cyclic molecule (FIGURE 16-5). Its structural formula is hexahydro-1,3,5-trinitro-1,3,5-triazine, or $(CH_2-N-NO_2)_3$; it is also called cyclonite hexogen. RDX starts to decompose at a temperature of approximately 338°F (170°C) and melts at 399°F (204°C); thus it is stable at room temperature. This lack of heat and shock sensitivity means that RDX must be used in conjunction with a detonator to produce an explosion—otherwise, it just burns. Even small arms fire will not initiate an RDX explosion. Thus RDX-based explosives usually contain a variety of materials—for example, other explosives, plasticizers, and desensitizers—to maximize their effectiveness, rather than just RDX alone.

Several common military explosives contain RDX, including composition A, composition A5, composition B, composition C, composition D, and HBX. The well-known military plastic explosive C-4 contains about 90% RDX; the remainder consists of a plastic binder material and oil. In such so-called plastic bonded explosives, the plastic binder fulfills two functions: (1) It coats the explosive material, making it less sensitive to shock and heat and therefore safer to handle, and (2) it makes the explosive material easy to mold into different shapes. During the explosive reaction of C-4, the RDX, plastic binder, and oil in the C-4 react to release nitrogen and carbon oxides. These gases rush outward at a speed of 8050 m/s, exerting a huge amount of pressure on everything in the surrounding area. At this expansion rate, the explosion is nearly instantaneous. Given that only a small amount of C-4 is required to produce a huge explosion, it is a popular choice among terrorists.

HMX

High-melting-point explosive (HMX), also known as Octogen, is said to be about 30% more powerful than TNT. HMX is a powerful, yet shock-sensitive, nitroamine-based high explosive that is chemically related to RDX (FIGURE 16-6). This military explosive has a density of 1.84 g/cm^3 and produces a blast velocity of 9124 m/s. It may be blended with TNT or some other material to reduce its shock sensitivity.

Improvised Explosives

From 1990 to 1994, 64% of all bombings reported in the United States involved low explosives. A low-explosive IED has two necessary components: a container to confine the explosive and a fuse or primer to detonate it. By contrast, a high explosive does not need confinement to explode. There are two types of improvised low explosives: commercially available products that are modified to act as explosives and combinations of chemicals.

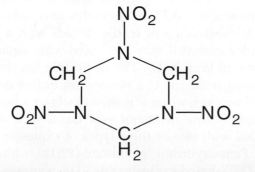

FIGURE 16-5 The chemical structure of cyclotrimethylenetrinitramine (RDX).

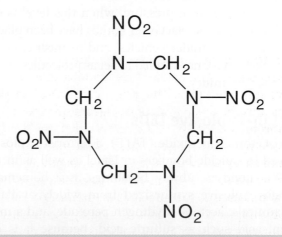

FIGURE 16-6 The chemical structure of high-melting-point explosive (HMX).

Low-Explosive IEDs

Many commercial products have been modified to act as explosives. For example, road safety flares contain a pyrotechnic mixture in a cardboard tube and are usually ignited by twisting or scratching the end. Bombs have been made by emptying the contents of road flares into pipes and then sealing the ends. Match heads contain chlorate/sulfur mixtures; they, too, have been carefully packed into pipe bombs. Smokeless powder from shotgun shells and chemicals from fireworks are sometimes used to fill pipe bombs as well.

Homemade mixtures of (1) potassium nitrate with sugar and aluminum, (2) sulfur/charcoal, potassium chlorate, and sugar, and (3) potassium perchlorate with sugar and aluminum have all been used to construct pipe bombs. Swimming pool chlorinators often contain calcium hypochlorite; when this chemical is mixed with automobile brake fluid, the mixture spontaneously ignites in a few minutes. Fast-moving arsonists have used this combination to start fires.

Juveniles who want to make IEDs have found numerous inexpensive methods for assembling simple bombs. Although most are made with materials found in the home, the danger posed by these devices should not be underestimated. One such device is a chemical reaction bomb called the "MacGyver bomb" (named after the television character who was a master at improvising devices from common household items). Its main ingredients—toilet bowl cleaner and aluminum foil—are placed in a two-liter plastic soda bottle. When the aluminum and liquid react, they produce large amounts of oxygen gas, which quickly pressurizes the bottle. The plastic bottle can withstand only 80 pounds per square inch of pressure; when this level is exceeded, it bursts. MacGyver bombs have been placed in mailboxes, under vehicles, and in trash containers. Some have nails added to act as projectiles, causing severe injury.

High-Explosive IEDs

Triacetone triperoxide (TATP), a terrorist explosive used by suicide bombers in Israel as well as in the 2005 London subway bombings, is a homemade high explosive synthesized from widely available chemicals: acetone, hydrogen peroxide, and a mineral acid such as sulfuric acid. Because it is extremely sensitive to shock, temperature change, and friction, it is very easy to detonate accidentally while being made in makeshift facilities such as garages and basements. This tendency has earned TATP the title "Mother of Satan" in the Middle East. When TATP explodes, each molecule of TATP decomposes to produce one ozone molecule and three acetone molecules. Just 10 ounces of the white solid material produce hundreds of liters of gas in a fraction of a second. The explosion of TATP is similar to an automobile airbag deployment.

TATP is the explosive of choice for terrorists because it does not have a nitrogen atom in its structure. Since most explosive compounds contain nitrogen, presumptive tests for explosives are designed to respond to a reactive nitrogen atom in the residue being tested. The fact that TATP does not contain nitrogen (even though constructing bombs using TATP is a dangerous activity in itself), makes it easier to sneak past airport and other bomb detectors.

Initiators

The term **initiator** is used to describe any device that is used to start a detonation or deflagration. The most common IED found in the United States is the pipe bomb that is filled with a low-explosive charge; the most popular initiators used for these bombs are the safety fuse (hobby fuse) and the electric match. **Detonators**, which are commonly known as blasting caps, are devices that are used to set off high explosives. Although detonators can be used to initiate a low-explosive blast, they are more challenging for a bomber to obtain and, as a result, are seldom used in low-explosive IEDs.

Safety Fuses

The safety fuse is a cord that, when ignited at one end, is intended to burn uniformly and to transmit the flame from the point of ignition to the IED. Approximately 0.25 inches in diameter, this initiator is constructed of textile threads with a black powder core that may be coated with asphalt or plastic to keep it dry. The safety fuse fits into the opening at the end of a pyrotechnic or fuse detonator (fuse cap), where it is used to trigger a fuse detonator. A fuse detonator is a metal shell that is loaded with two or three types of explosive powder. Pentaerythritol tetranitrate (PETN, which is a SEMTEX-based explosive) or RDX is pressed into the bottom of the shell, with an ignition mixture of lead azide or lead styphnate being pressed into the top. When the flame from the fuse reaches the fuse detonator, it ignites this mixture and starts a deflagration.

Electric Matches

An electric match triggers an IED when it is set off by an electric current. The electric match is made by attaching a pair of insulated copper wires to a thin resistance wire that heats when current is applied. The resistance wire is coated with a pyrotechnic composition that ignites when the wire is heated, and it provides enough energy to set off the main explosive charge. You may have experience with electric matches if you have ever set off a model rocket, because many of these rockets' motors are ignited by these types of initiators.

In the past, bombers typically attached the electric match to a timer that was rigged to close a switch in an electrical circuit at a specific time. The timer gave the bomber a chance to flee the scene and establish an alibi. More recently, bombers have attached the electric match to the ringer on a cell phone: When the phone is called, it sets off the electric match and the IED. Using the cell phone, the bomber can set off the charge remotely at any time he or she desires.

Detonators

A detonator is a device used to trigger high explosives. Most detonators are mechanical or electrical. Military mines use mechanical detonators, which are activated when someone steps or drives on them (mechanical motion). Commercial and military high explosives, by contrast, use electrical detonators. The electric detonator is similar in some ways to the electric match (FIGURE 16-7). Insulated wires are attached to a high-resistance bridge wire (match head). A pyrophoric material, which is a mixture of lead azide, lead styphnate, and aluminum, is pressed into place above the match head. When electric current from a blasting machine (located far away) passes through the wires, the bridge wire heats very quickly to ignite the pyrophoric material, which in turn ignites the high explosive.

Collection of Explosive-Related Evidence

Personnel who are involved in bomb scene or bomb threat searches should be instructed that their pri-

FIGURE 16-7 In an electric detonator, insulated wires are attached to a high-resistance bridge wire. Pyrophoric material is pressed into place above this bridge wire. When the electric current from a blasting machine passes through the wires, the bridge wire heats very quickly to ignite the pyrophoric material, which in turn ignites the high explosive.

mary mission is to search for and report suspicious objects. Under no circumstances should anything suspicious be moved or touched. People should be moved away from the bomb, rather than the bomb being moved away from people.

In a post-blast situation, bomb disposal experts should first conduct a search to determine whether a secondary bomb has been set to entrap crime scene investigators. Next, investigators should attempt to determine the point of detonation and the types of blast effects exhibited by the explosive. This search should begin at the site of a crater and proceed outward in ever-widening circles. Sift, sort, and collect samples of the rubble, keeping in mind that most bombs leave some parts behind. Only a few, extremely powerful bombs will completely destroy all of the components used in their construction. This sort of evidence can provide powerful clues to the bomber's identity. Pieces of wire may display tool marks, pieces of the triggering mechanism may be traced to a manufacturer or retail dealer, and pieces of an explosive wrapper may permit a trace through past records to the last known purchaser. Individuals who construct pipe bombs also like to use duct tape, and they may leave their fingerprints in the tape's glue.

Wire mesh screens may be used to sift through debris to search for parts of the explosive device. When pipe bombs are used, residue from the explosive is often found adhering to pieces of the pipe or

The Crash of TWA Flight 800

Trans World Airlines (TWA) Flight 800 left New York's Kennedy International Airport at 10:31 P.M. on July 17, 1996, headed to Paris. Shortly after takeoff, the Boeing 747 exploded in mid-air and crashed into the Atlantic Ocean off the coast of Long Island, New York. All 230 people on board were killed. Investigators from the National Transportation Safety Board (NTSB) and the Federal Bureau of Investigation (FBI), because of the possibility of terrorism, initiated parallel investigations into the crash.

Pieces of wreckage were recovered from the ocean, and the plane was reconstructed in a hangar in Calverton, New York. Investigators looked at the distribution pattern of the scattered debris and analyzed data from the flight recorder ("black box"), the intact parts of the aircraft, and soot patterns to understand the sequence of events that occurred during the breakup of the aircraft. The NTSB and FBI considered the possibility that this had been a terrorist attack based on eyewitness reports of a streak of light leading from the ground to the airplane and unexplained radar images picked up at a nearby airport in Islip, New York. It was possible that there had been a bomb on board or a missile fired from a shoulder-launched weapon. The bomb theory gained credibility when it was found that the seat backs of several of the passenger chairs had an unknown red/brown substance that could not be readily explained and when FBI investigators found trace levels of explosive residue on separate pieces of recovered wreckage. Although the damage pattern did not follow the typical pattern of a high-powered bomb or missile, these findings raised many questions.

To rule out the terrorist bomb theory, the NTSB analyzed material from within the aircraft. Bombs leave signature residues that can be found through high-level chemical analysis. James Girard, the author of this text, was asked to provide more extensive analysis of fragments from the aircraft to help determine the cause of the explosion. His laboratory tested fuel line samples with GC–MS analysis. These tests showed that the residues taken from the aircraft debris were consistent with a jet fuel fire and not an explosive device. However, the red/brown chair substance still needed an explanation.

NTSB personnel supplied Girard with samples from the exploded aircraft's seat backs and fibers found in the air ducts above the passenger compartment. This particular 747 had different carpet and seat coverings in the first, business, and economy classes. The distribution of fibers found in this 747 was compared with that of a retired 747 (of the same vintage) that was in storage in the desert of Arizona.

Girard's analysis of the red/brown substance showed that it came from an adhesive used in the manufacture of the fabric and was not the result of bomb residue. Furthermore, the distribution of fibers in the exploded aircraft and the retired one was found to be the same. Therefore, Girard concluded that there was no evidence of a bomb or missile explosion.

The NTSB concluded that the accident began with an explosion that originated in the center fuel tank. To save space in the wing, it was common to have electrical cables run inside the fuel tanks of Boeing 747s. This 747 was more than 20 years old; it was speculated that faulty electrical wiring caused a spark that ignited the fuel tank and the ensuing structural failure and decompression of the aircraft. The careful analysis of the fragments and debris of the aircraft prompted the Federal Aviation Administration (FAA) to change its rules, no longer allowing aircraft manufacturers to run cables through the fuel tanks. New airplanes have nitrogen systems that displace air from fuel tanks and replace it with nitrogen, making the headspace over the fuel more inert and not able to support combustion. Careful analysis in the face of emotional theories led investigators to the correct conclusion, which further resulted in safer aircraft. The wreckage is permanently stored in an NTSB facility near Dulles International Airport in Washington, D.C., and is used to train accident investigators.

in the pipe threads. If a suspect is present at the scene, test his or her hands by wiping them with a swab that has been moistened with acetone.

All materials that will be sent to the laboratory must be placed in sealed, labeled containers. Debris from different areas should be packaged in separate containers. Soil and other loose debris are best stored in metal containers or plastic bags. Since some explosive residue can diffuse through plastic, it is important to keep these containers separated. Sharp objects should be placed in metal containers, not plastic.

An explosives detection canine should also search the scene. These dogs can detect even minute quantities of explosives, and they have successfully detected bombs hidden in luggage, stowed in vehicles, and buried underground.

All manufacturers that sell and distribute explosives must identify them with a date, shift, and place of manufacture. These markings are known as the date/shift code (**FIGURE 16-8**). Explosive manufacturers, distributors, and users are required to maintain records of these explosives by amount, type, and date/shift code. This information can be used to trace these products in the same way that serial numbers are used to trace firearms.

The Bureau of Alcohol, Tobacco, Firearms, and Explosives operates the National Explosives Tracing Center to trace commercial and military explosives recovered by law enforcement. The center traces recovered, stolen, or abandoned explosives and provides information about the manufacturer and distributor.

Not all explosives come from the United States, of course. For instance, although the U.S. military is the major supplier of C-4 and tightly controls its supply, Iran has become a well-known source of this explosive in recent years—a fact exploited by international terrorists.

Field Tests for Explosive Residue

Investigators can screen objects found at the scene for the presence of explosive residues with an **ion mobility spectrometer (IMS)**, shown in **FIGURE 16-9** . This portable instrument generates a small vacuum that will collect residues from surfaces. Alternatively, the surface to be investigated may be wiped with a special paper disk that is inserted into the instrument.

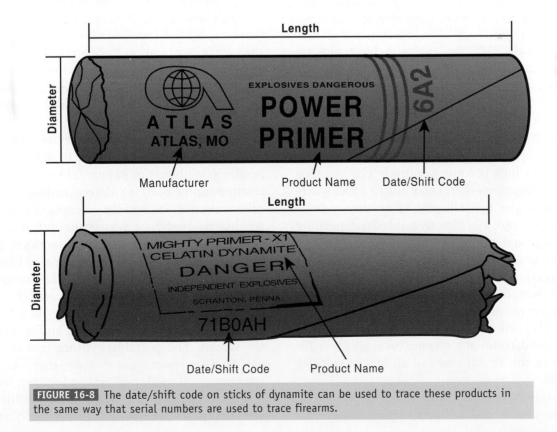

FIGURE 16-8 The date/shift code on sticks of dynamite can be used to trace these products in the same way that serial numbers are used to trace firearms.

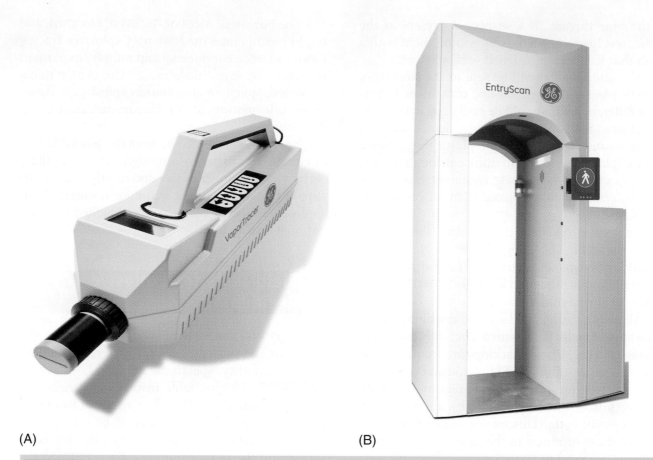

(A)

(B)

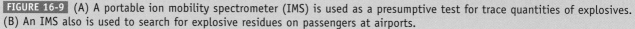

FIGURE 16-9 (A) A portable ion mobility spectrometer (IMS) is used as a presumptive test for trace quantities of explosives. (B) An IMS also is used to search for explosive residues on passengers at airports.

Once in the IMS, the explosive molecules are vaporized into the gas phase and converted into electrically charged species (ions). When the molecules enter the IMS, they are bombarded with beta rays from a radioactive source (**FIGURE 16-10**). Beta rays, which are similar to electrons, are negatively charged. When they hit explosive molecules, those molecules in turn become negatively charged ions (anions). These ions then enter a drift tube, where they come under the influence of an electric field. The ions will move at different speeds, depending on their size and structure—that is, small, compact ions move faster. The IMS measures the time it takes for the residue sample to move through the drift tube and then, based on the measured time, identifies which explosive is present. For example, molecules of TNT will move through the drift tube and arrive at the detector at a specific time, which is different from the arrival times of all other explosive molecules.

The IMS can detect a wide range of explosives, including plastic and military explosives, even if they are present only at trace levels. It is a presumptive test, meaning that it is used only as a screening tool. The definitive identification of all residues tentatively made by IMS must later be verified by confirmatory tests.

A portable hydrocarbon detector (discussed in Chapter 10) also can be used to detect explosive residues on objects or people. This highly sensitive instrument is used to detect hydrocarbons and other organic molecules present in accelerant residue by analyzing vapors present at the crime scene. The hydrocarbon detector works by sucking in air and then passing it over a heated filament. If a flammable vapor is present in the air sample, it is oxidized immediately and increases the temperature of the filament. The ensuing rise in temperature is recorded by the detector and displayed on the screen. The portable hydrocarbon detector is yet another presumptive test—that is, it is a screening tool used to identify a location for future testing. It can detect even minute quantities of accelerant (parts per million). Additionally, specially

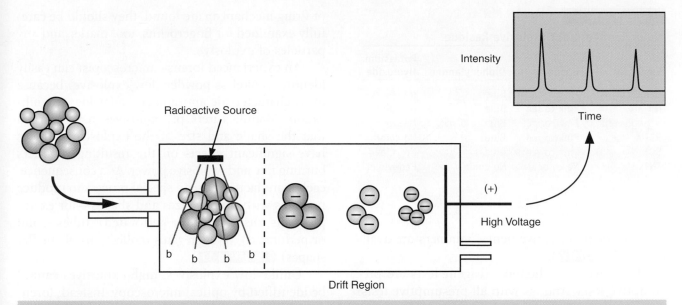

FIGURE 16-10 In an IMS, the ions from different explosives drift at different speeds. The IMS measures the time that it takes for the residue sample to move through the drift tube and then, based on the measured time, identifies which explosive is present.

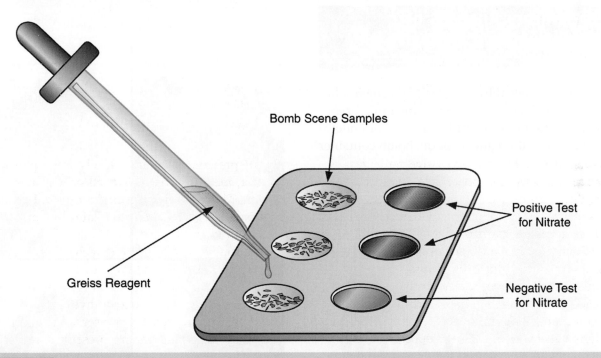

FIGURE 16-11 Some explosive residues induce a change in color of a specific reagent.

trained dogs may be used to detect hydrocarbon residue.

Chemical reagents may also be applied in the field in an effort to detect explosive residue. These chemicals change color when they come in contact with explosive residues. In such a test, the suspect material is placed in the well of a sample tray, and a few drops of a test solution are added to it (**FIGURE 16-11**). If a specific explosive residue is present, it will cause the reagent to change colors. **TABLE 16-1** lists some common color test reagents. As we learned in Chapter 9, the Modified Greiss Test is used as a test for nitrate spatter from the discharge of a gun. Commercial kits that provide

TABLE 16-1

Color Spot Tests for Explosive Residue

Explosive	Greiss	Diphenylamine	Potassium Hydroxide
Chlorate		Blue	No color
Nitrate	Pink to red	Blue	No color
Nitroglycerin	Pink to red	Blue to black	No color
PETN	Pink to red	Blue	No color
RDX	Pink to red	Blue	No color
TNT	No color	No color	Red–violet

these reagents in convenient containers are available (FIGURE 16-12).

Like the IMS, chemical reagent tests are presumptive tests. And, as with all presumptive tests, their results must be verified through confirmatory tests.

Laboratory Analysis of Explosives and Explosive Residues

Once debris collected from the crime scene reaches the lab, this evidence is first examined microscopically in an attempt to identify particles of undetonated explosive. If fragments of the bomb container or firing mechanism are found, they should be carefully examined for fingerprints, tool marks, and any particles of explosive.

An experienced forensic microscopist can easily identify smokeless powder (low explosive) because of its characteristic appearance under low magnification. Manufacturers of explosives have learned that the shape and size of the explosive particles have significant effects on the resulting product's burning rate and explosive power. As a consequence, each manufacturer uses a special process to produce particles with unique sizes and shapes—for example, balls, disks (solid or perforated), tubes (solid or perforated), or aggregates (collections of smaller shapes) (FIGURE 16-13).

Unlike low explosives, high explosives cannot be identified by optical microscopy. Instead, forensic examiners use chemical tests to establish their composition.

After the microscopic examination is complete, the examiner rinses the recovered debris with ace-

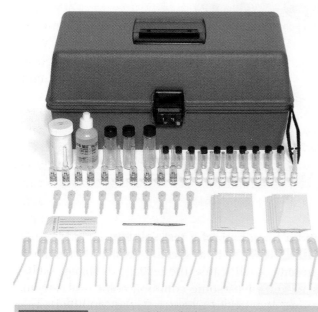

FIGURE 16-12 A commercial explosives residue test kit provides chemical reagents in convenient containers.

FIGURE 16-13 Each manufacturer of low explosives uses a special process to produce particles with unique sizes and shapes.

tone. Acetone is a common solvent (often found in fingernail polish remover) that will dissolve a wide variety of organic materials, including most organic explosive residues. Acetone also has a very low boiling point, so it evaporates very quickly. Thus the washed-off organic residue is easily concentrated by evaporating the acetone.

Thin-Layer Chromatography

Thin-layer chromatography (TLC) is another presumptive chemical test that can be used to screen for the presence of an explosive. As we learned in Chapter 11, this separation technique takes advantage of the solubility and physical properties of the controlled substance to separate and distinguish compounds in a mixture. Compounds are separated based on their size, shape, solubility in a solvent, and interaction with the thin-layer plate. A separation occurs as a result of the interaction taking place between the mobile and stationary phases and the compounds being separated. The more tightly a compound binds to the silica surface of the thin-layer plate, the more slowly it moves up the thin-layer plate.

For a specific solvent, a given compound will always travel a fixed distance relative to the distance traveled by the solvent front. This ratio of distances, which is called the R_f, is expressed as a decimal fraction:

$$R_f = \frac{\text{distance traveled by compound}}{\text{distance traveled by solvent front}}$$

The R_f value for a specific compound depends on its structure. Its value is a physical characteristic of the compound, just as its melting point is a physical characteristic of the compound.

Multiple samples and standards can be spotted on the same TLC plate and the results of the separation compared. FIGURE 16-14 is an example of a separation of two bomb scene residue samples, which are being compared to a TNT standard and an RDX standard. In this case, the TLC screening test indicates that sample 2 is RDX, and sample 1 is neither TNT nor RDX.

Gas Chromatography

Gas chromatography (GC) is a dynamic method for separating and detecting volatile organic compounds that is used as a confirmatory test for organic explosive residue. As we learned in Chapter 10, the GC separates the gaseous components of a mixture by partitioning them between an inert gas mobile phase and a stationary phase (FIGURE 16-15). Chromatography functions on the same principle as extraction, except that one phase is held in place while the other moves past it in a carrier gas (usually helium). If one component of the gaseous mixture is more strongly absorbed to the stationary phase (usually silicone), then it will move through the system more slowly than less strongly absorbed components.

The GC is operated with helium flowing at a constant rate (F). Each component of the sample requires a specific volume of gas, the retention volume (V_r), to push it through the GC. The time needed to push a specific component through the GC is the retention time, t_r.

$$V_r = t_r F$$

The retention time, t_r, is directly proportional to the retention volume, V_r. It is generally more convenient to describe the elution of components in terms of their retention times than in terms of their retention volumes. Each component in the sample will have a unique retention time.

GCs that are used for explosive analysis use a detector that is called a thermal energy analyzer (TEA). This detector contains a small furnace that decomposes the separated substance. If the compound contains nitrogen, then nitrogen oxides are produced from the decomposition reaction. Ozone is then added to the decomposition products. If nitrogen oxides are present, a chemiluminescence reaction takes place and light at a specific wavelength is emitted. This detector is extremely sensitive and responds only to nitrogen-containing compounds.

Once the forensic scientist has set the flow rate and temperature of the TEA's furnace for the separation, a reference sample is injected into the GC, and the GC then determines its composition. When a standard mixture of explosives is injected, a complex pattern of peaks is observed. FIGURE 16-16 shows the pattern observed for 10 different explosives that might be found at a bomb scene, where each peak represents a different explosive compound. The time it takes to exit from the GC is characteristic of the specific compound, and the area under each peak represents its concentration. TNT, for example, can be identified in debris samples if the residue sample produces a peak that exits the

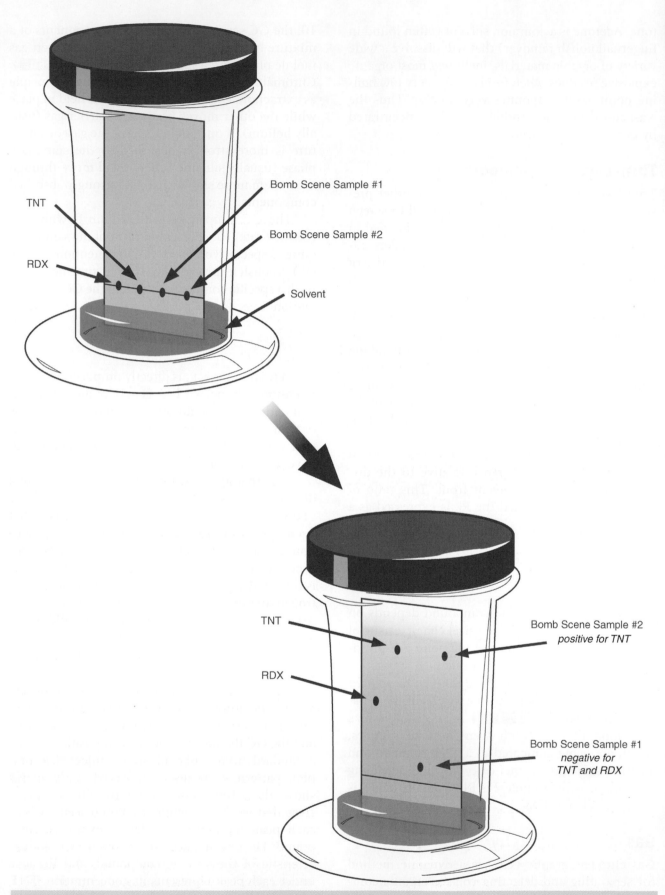

FIGURE 16-14 Thin-layer chromatography of TNT and RDX.

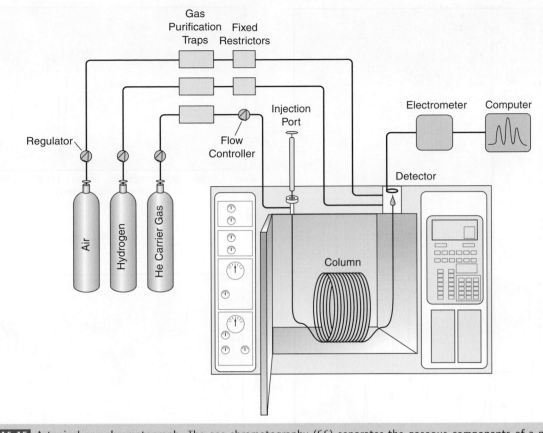

FIGURE 16-15 A typical gas chromatograph. The gas chromotography (GC) separates the gaseous components of a mixture by partitioning them between an inert gas mobile phase and a stationary phase.

Source: © Agilent Technologies, Inc. 2010. Reproduced with permission, courtesy of Agilent Technologies, Inc.

GC in 6.9 minutes; RDX residue, by comparison, exits the GC in 8.2 minutes.

Infrared Tests

If sufficient quantities of unexploded organic explosives are recovered, confirmatory infrared tests (IR) can be carried out on this evidence. The IR spectrum not only identifies each organic functional group by its characteristic absorption, but each explosive gives a characteristic IR spectrum—that is, an IR "fingerprint." To provide definitive confirmation of the material's identity, the IR spectrum of the residue found at the bomb scene is then compared to the standard IR spectrum of each explosive.

FIGURE 16-17 shows the IR spectra of NG, TNT, and RDX. Because these molecules have different functional groups, they absorb IR energy at different wavelengths. TNT, for example, has two major peaks at 1534 and 1354 cm^{-1}. NG's major peaks are at 1663, 1268, 1004, and 843 nm, and RDX's major peaks are at 1592, 1573, 1267, 1039, 924, and 783 cm^{-1}. If you were to overlay these spectra, you would see that the major IR peaks are unique for each compound. The fact that the IR spectrum of a residue collected at a bomb site matches that of a TNT standard will be accepted by a court as proof that TNT was used at the bomb scene.

Unfortunately, IEDs often are assembled by placing several different chemicals into a pipe. As a consequence, the material recovered at the bomb scene may be a mixture of unexploded explosive molecules. The IR spectrum of this mixture, although it may give very useful probative information, will not match the spectrum of any pure explosive.

To overcome this problem, forensic technicians often chemically separate the components of the debris before IR analysis. The separation is usually based on the different solubilities of the various components in one or more solvents. After separation of the components, the IR spectrum of each relevant component is recorded, and the forensic chemist decides whether the quality of the spectrum permits an unambiguous identification of the compound separated from the debris.

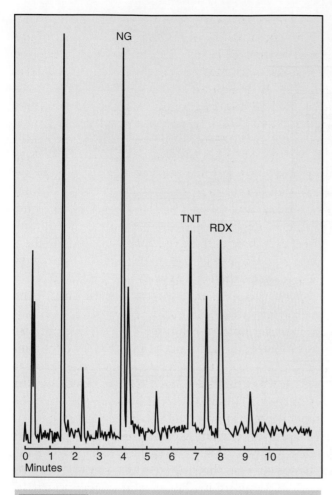

FIGURE 16-16 GC separation of a standard explosive mixture.

Source: © Agilent Technologies, Inc. 2010. Reproduced with permission, courtesy of Agilent Technologies, Inc.

Analysis of Inorganic Explosive Residues

In many bombings, materials such as black powder, AN, and homemade mixtures are used in IEDs such as pipe bombs. These crude bombs typically contain an oxidizer, such as potassium nitrate, mixed with a fuel. When detonated, these homemade devices do not consume all of their charge; indeed, as much as 60% of the original weight of the explosive might be left behind in the debris. Analysis of the inorganic cations and anions in the debris may, therefore, help investigators determine the composition of the explosive used. **TABLE 16-2** lists the ions that will be left in debris from various explosives.

When it is suspected that an explosive contains AN (NH_4NO_3) or potassium perchlorate ($KClO_4$), the residue can be washed with water so that water-soluble inorganic substances will be removed. The

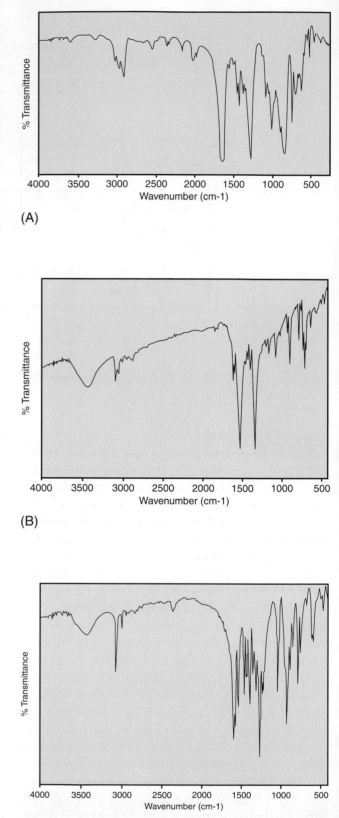

(A)

(B)

(C)

FIGURE 16-17 The IR fingerprint spectra of (A) nitroglycerin, (B) TNT, and (C) RDX.

Source: © Agilent Technologies, Inc. 2010. Reproduced with permission, courtesy of Agilent Technologies, Inc.

TABLE 16-2

Ions Produced by Explosives

Explosive	Components	Ions Present
Black powder	KNO_3, S, C	K^+, NO_3^-, HS^-, SO_4^{2-}, OCN^-, SCN^-
Flash powder	$KClO_4$, S, Al	K^+, HS^-, SO_4^{2-}, Cl^-, ClO_4^-, ClO_3^-
Smokeless powder	Nitrocellulose, NG, KNO_3, K_2SO_4	K^+, Cl^-, NO_3^-, SO_4^{2-}
ANFO	NH_4NO_3, Fuel oil	NH_4^+, NO_3^-, HCO_3^-

Source: Alexander Beveridge. *Forensic Investigation of Explosions.* Boca Raton, FL: CRC Press, 1998.

spot tests described earlier can then be carried out to screen for the presence of inorganic residue. If these spot tests indicate that an inorganic substance such as nitrate or chlorate is present, then a confirmatory **ion chromatography (IC)** test should be performed. For analysis of the inorganic ions left behind following an explosion, IC is the most specific and sensitive method available.

Ion Chromatography

FIGURE 16-18 shows the components of an IC instrument—namely, a solvent reservoir, solvent pump, sample injector, separation column, detector, and data system. Given that all ions in solution conduct electricity, a detector that measures conductance is most commonly used. The IC instrument must also be set up to handle anion separations because a specific separation column and solvent are used for the separation of anions.

The separation column used for anion separations contains an ion-exchange resin; the anions pass over small beads that have positively charged ion-exchange sites on their surfaces that attract anions. Although all anions in the sample are attracted to the surfaces of the beads, some anions are held more tightly by the beads than others. A weakly held anion will flow through the IC to the detector more quickly than an anion that is held tightly. Because of this difference in affinity, a separation of anions in a sample can be completed in 15 minutes. The anions expected to be found in bomb scene evidence are nicely separated by IC, and their identity is determined by the time it takes them to reach the detector as indicated on the x-axis, which shows their retention time (FIGURE 16-19).

A separate IC analysis can be performed using the cations in a bomb scene sample. IC is an extremely important technique in this sense because it can detect the ammonium ion (NH_4^+). Other analytical methods that detect metal cations, such as atomic absorption and inductively coupled plasma spectroscopy, cannot detect ammonium ion. The separation column used for cation separations contains an ion-exchange resin; the cations pass over small beads that have negatively charged ion exchange sites on their surfaces that attract them. As with the anions in the previously mentioned IC techniques, some cations are held more tightly by the beads than others. A weakly held cation will flow through the IC to the detector more quickly than a cation that is held tightly. The identities of the cations are determined by the time it takes each one to reach the detector (FIGURE 16-20).

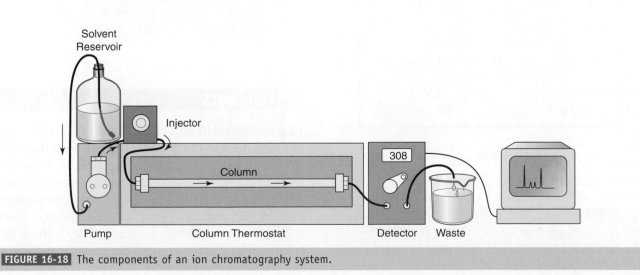

FIGURE 16-18 The components of an ion chromatography system.

Relative to fire, an explosion burns its substrate much more rapidly, almost instantly turning a small volume of a solid or a liquid into a gaseous phase that takes up much more space.

The power of an explosion can be measured using the ideal gas law, which states that the pressure and volume of a gas are proportional to the number of molecules present multiplied by the temperature ($PV \propto nT$). If the volume remains constant but the number of gas molecules and the temperature increase dramatically, then the pressure within the volume has to increase as well. If we take the example of PETN, then 13 gas molecules are formed from each molecule of PETN and the detonation temperature is 3400 K, about 12 times the background temperature of 290 K ($17°C/62.5°F$). Thus the change in pressure and therefore the explosive force depend on the amount of explosive used, the initial gas pressure and temperature, and the volume of the bomb created.

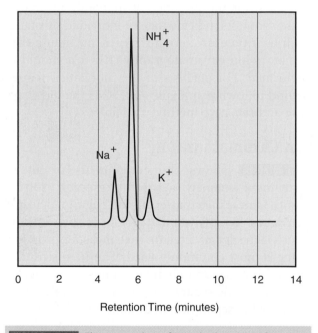

FIGURE 16-20 The separation of standard cations that can be present in explosive debris.

Source: © Agilent Technologies, Inc. 2010. Reproduced with permission, courtesy of Agilent Technologies, Inc.

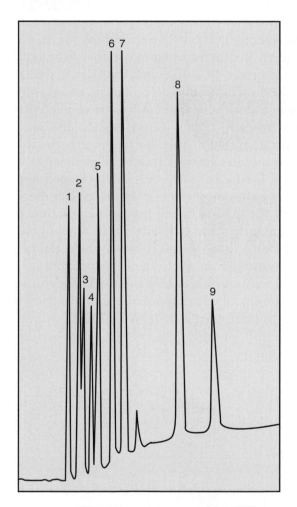

(1) Cl^- , (2) NO_2^- , (3) OCN^- , (4) ClO_3^- , (5) NO_3^- ,

(6) SO_4^{2-} , (7) HS^- , (8) SCN^- , (9) ClO_4^-

FIGURE 16-19 The separation of standard anions that can be present in explosive debris.

Source: © Agilent Technologies, Inc. 2010. Reproduced with permission, courtesy of Agilent Technologies, Inc.

Taggants

Law enforcement personnel would like to have an easy way to identify which company manufactured the explosive used for a bombing. Because explosive manufacturers have to keep detailed records of their sales, identifying the manufacturer will help establish the sale and movement of the explosive. Since the bombing in Oklahoma City in 1995, there has been renewed interest in putting identifiers, called **taggants**, into each batch of explosive produced.

Two types of taggants are being evaluated to achieve this goal.

One type of taggant technology uses tiny multicolored chips of plastic (FIGURE 16-21). With this type of taggant, each manufacturer would produce chips with different colors and identifying patterns. To the naked eye, the taggant looks like black specks. When examined under the microscope, however, it becomes apparent that the taggant contains as many as 10 mini-slabs of colored melamine plastic bonded to a magnetic material. The different colors make up a unique code, like a bar code. After an explosion, investigators could gather these taggants out of the debris with a magnet, look at them under a microscope, and recognize and identify the tag. When the chips are added to an explosive, they will uniquely identify which factory made the explosive, in which batch, on which date, who distributed it, and which shops sold it.

The most widely used taggants were invented in the 1970s by Richard G. Livesay, who worked for the 3M Company. Because there was little interest in taggants at that time, Livesay bought back the license for his invention from 3M and set up his

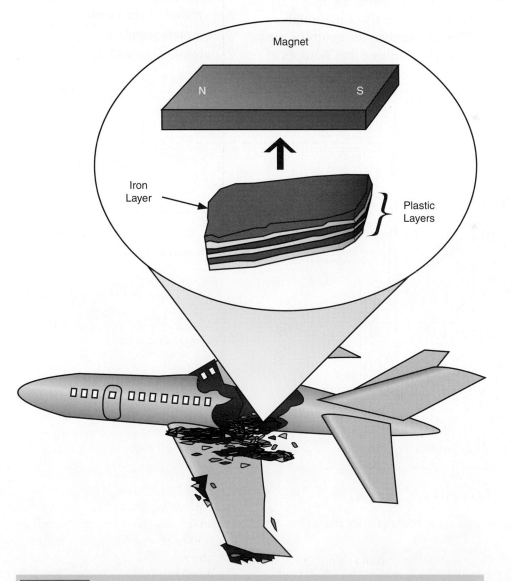

FIGURE 16-21 The color sequence within a taggant forms a type of barcode that is unique to a specific manufacturer and a specific date of manufacture.

own company, Microtrace. The Swiss government has mandated that all explosives manufactured in Switzerland contain Microtrace taggants.

At one point, explosives manufacturers in the United States thought that adding taggants was a good idea. Today, however, most explosive and fertilizer manufacturers are opposed to adding these materials to their explosives. The National Rifle Association also opposes the addition of taggants to gunpowder, stating that the plastic-tagged gunpowder might degrade more rapidly than normal.

The second technology that is being studied as a taggant is the addition of isotopes. Understanding this technology requires understanding the basic structure of the atom. The numbers of protons and electrons in the atoms of a given element are always constant, but the number of neutrons may vary. Atoms of an element that have the same numbers of protons but different numbers of neutrons are called **isotopes**. It therefore follows that isotopes of a particular element all have the same atomic number (i.e., the number of protons) but different mass numbers (because they have different numbers of neutrons).

Naturally occurring elements are mixtures of isotopes in which one isotope usually predominates. For example, hydrogen (atomic number 1) is found as a mixture of three isotopes: hydrogen, deuterium, and tritium. Just a few atoms in every thousand hydrogen atoms have one neutron or two neutrons. Oxygen (atomic number 8) is a mixture of three isotopes, more than 99.7% of which have eight neutrons; the remaining 0.3% of oxygen atoms have either 9 or 10 neutrons.

The following notation is used to specify a particular isotope of an element:

X = symbol of the element

$^{A}_{Z}X$ A = mass number

Z = atomic number

Thus the three isotopes of hydrogen (H) are written as follows: ^1_1H (hydrogen), ^2_1H (deuterium), ^3_1H (tritium). The three isotopes of oxygen (O) are written this way: $^{16}_8O$, $^{17}_8O$, $^{18}_8O$.

Although an isotope of oxygen has a different number of neutrons, it is otherwise chemically identical to the other isotopes of oxygen. Thus an isotope does not change the physical properties of compounds in which it is included—a characteristic that could be exploited when using isotopes as taggants in explosives. For example, TNT manufacturers might substitute a different isotope of oxygen or hydrogen for the normal isotope, thereby "tagging" the TNT molecules without changing their explosive behavior. Gas chromatography–mass spectrometry can easily detect TNT molecules that contain these isotopes in concentrations as low as a few parts per billion. The TNT molecules with the heavier isotopes would have a larger mass than regular TNT molecules.

The National Research Council (NRC) was commissioned to study taggants in explosives following the 1995 Oklahoma City bombing. In its 1998 report, *Containing the Threat from Illegal Bombings*, the NRC stated that none of the proposed isotopic taggants has been shown to survive a severe blast or to be amenable to standard collection and analysis procedures.

The Federal Bureau of Investigation (FBI) has determined that the current technologies cannot support the requirements of an identification system based on taggants. Additionally, crime statistics show that less than 2% of criminal bombings in the United States involve commercial explosives; thus the vast majority of bomb scene investigations would not be facilitated by use of taggants. As a result, the FBI has determined that taggants will not serve as a significant deterrent to criminals.

WRAP UP

1. Bomb disposal experts should first conduct a search to determine whether a secondary bomb has been set to entrap crime scene investigators. An explosives-detection canine should search the scene.

   Next, investigators should attempt to determine the point of detonation and the types of blast effects exhibited. Their search should begin at the site of a crater and proceed outward in ever-widening circles. They should sift, sort, and collect samples of the rubble, keeping in mind that most bombs leave some parts behind. Wire mesh screens may be used to sift through debris to search for parts of the explosive device. When pipe bombs are used, residue from the explosive are often found adhering to pieces of the pipe or in the pipe threads.

   All materials to be sent to the laboratory must be placed in sealed, labeled containers, with debris from different areas being packaged in separate containers. Soil and other loose debris are best stored in metal containers or plastic bags; sharp objects should be placed in metal containers.

2. Proceed in the same way as in the parking garage investigation. In this case, because the bombing occurred outside, the area to be searched is much larger. This search will be very difficult because fragments from the bombing might have been blown blocks away and may have landed in obscure places such as the roofs of buildings and storm drains.

Chapter Spotlight

- An explosion is a release of mechanical or chemical energy. High temperature and large quantities of gas are generated in this reaction. Explosives are classified as high explosives or low explosives based on the amount of energy released.

- The most commonly used low explosives are black powder and smokeless powder. Black powder is a mixture of charcoal, sulfur, and potassium nitrate. Smokeless powder is made from cordite and ballistite.

- High explosives are classified into two groups: primary explosives and secondary explosives. Primary explosives are extremely sensitive to shock and heat and will detonate violently. Secondary explosives are relatively insensitive to heat, shock, or friction.

- The first high explosive that was widely used commercially was NG, an oily liquid that is extremely dangerous to handle and transport. Alfred Nobel invented dynamite, a much safer alternative, when he discovered that NG could be absorbed onto an inert material, such as clay. ANFO is a combination of a common commercial ammonia fertilizer and fuel oil.

- TNT is a powerful explosive that can very quickly change from a solid into hot expanding gases.

- RDX is typically used in mixtures with other explosives and plasticizers or desensitizers.

- Regardless of the type of explosive, an explosive confined in a closed pipe that is fitted with a detonating device is considered to be an IED.

- To collect evidence from a bombing, investigators sift, sort, and collect samples of the rubble. They use wire mesh screens to sift through debris. All materials collected in this way must be placed in sealed containers or plastic bags before transport to the lab for further analysis.

- Ion mobility spectrometry can test objects for explosive residues at the scene.

- TLC is a chemical test in which compounds are separated by their size, shape, solubility in a solvent, and interaction with a thin-layer plate.

- GC is a dynamic method of separation and detection of volatile organic compounds. It is used as a confirmatory test for organic explosive residue.

- IR tests are used to identify unexploded organic explosives.

- IC can detect inorganic substances such as nitrate or chlorate.

- Taggants are microscopic pieces of multilayered colored plastic that are added to an explosive to indicate its source of manufacture.

Key Terms

Brisant Of or relating to the shattering effect from the sudden release of energy in an explosion.

Condensed explosive An explosive made of a solid or a liquid.

Deflagration An explosion in which the reaction moves through the explosive at less than the speed of sound.

Detonation An explosion in which the reaction moves through the explosive at greater than the speed of sound.

Detonator A device that is used to set off a high explosive.

Dispersed explosive An explosive made of a gas or an aerosol.

Explosion A sudden release of mechanical, chemical, or nuclear energy, generating high temperatures and usually releasing gases.

Explosive A substance, especially a prepared chemical, that explodes or causes an explosion.

High explosive An explosive that when detonated produces a shock wave that travels at a speed greater than 1000 meters per second (e.g., military explosives).

Improvised explosive device (IED) A homemade bomb.

Initiator Any device that is used to start a detonation or a deflagration.

Ion chromatography (IC) A chromatography method for separating material based on ions.

Ion mobility spectrometer (IMS) A device that identifies ions by measuring their speed.

Isotope One of two or more atoms with the same number of protons but differing numbers of neutrons.

Low explosive An explosive that when detonated produces a shock wave that travels at a speed less than 1000 meters per second (e.g., black powder).

Prilling A production method that produces free-flowing, adsorbent, porous ammonium nitrate spheres (prills) that are easily handled and stored.

Taggant A substance, such as microscopic pieces of multilayered colored plastic, that is added to an explosive to indicate its source of manufacture.

Putting It All Together

Fill in the Blank

1. A(n) _____ is a chemical explosion in which the reaction front moves through the explosive at less than the speed of sound.

2. A(n) _____ is a chemical explosion in which the reaction front moves though the explosive at greater than the speed of sound.

3. The chemicals in smokeless powder are _____ and _____.

4. Low explosives will burn, rather than explode, if they are not _____.

5. Condensed explosives are either a(n) _____ or a(n) _____.

6. Dispersed explosives are either a(n) _____ or a(n) _____.

7. The most commonly encountered illegal explosive device in the United States is the _____ bomb.

8. High explosives are categorized into two groups: _____ explosives and _____ explosives.

9. Explosives that are insensitive to shock, heat, and friction are called _____ high explosives.

10. Small, porous ammonium nitrate spheres are called _____.

11. The most commonly used explosive in the world is a mixture of _____ and _____.

12. The most widely used military explosive is _____.

13. A(n) _____ explosive is one in which a shock wave is formed that shatters the material surrounding it.

14. The military explosive C-4 contains _____.

15. The search of a bomb scene should begin at the site of a(n) _____.

16. Soil and loose debris are best stored in _____ containers.

17. If a suspect is present at the scene, the investigator should test his or her hands by wiping them with a swab that has been moistened with _____.

18. The _____ operates the National Explosives Tracing Center.

19. Once in the lab, debris from a bomb scene is first examined using a(n) _____.

20. The solvent _____ is used to remove organic material from recovered bomb scene debris.

21. Gas chromatographs that are used for explosive analysis use a(n) _____ detector.

22. The GC detector used by crime labs for explosive detection detects only molecules that contain the element _____.

23. If a mixture of chemicals is used in an explosive device, the substances must be _____ prior to IR analysis.

24. If inorganic chemicals were used to construct a bomb, the inorganic residue can be removed from debris by washing it with _____.

25. Materials removed from debris by washing with water can be tested for inorganic residues by means of _____ tests.

26. A confirmatory test for inorganic residues uses _____ chromatography.

27. An ion chromatograph detects the inorganic ions with a(n) _____ detector.

28. Ion chromatography can detect the _____ cation, which is not detected by atomic absorption spectroscopy and is commonly found in bomb debris.

29. The identifiers that are put into manufactured explosives are called _____.

30. The country that requires taggants in all explosives is _____.

31. Atoms of an element that have the same numbers of protons but different numbers of neutrons are called _____.

32. A molecule of TNT that contained a deuterium isotope would have a _____ (higher/lower) mass than regular TNT.

True or False

1. Natural gas and air will explode when mixed in any ratio.

2. Only a small quantity of C-4 is needed to produce a large explosion.

3. A TATP explosion produces large quantities of gas and heat.

4. The searchers of a bomb scene should first determine whether a secondary bomb is present.

5. The ion mobility spectrometer is used as a confirmatory test for explosive residues.

6. Portable hydrocarbon detectors are used to conduct presumptive tests for explosive residues.

7. Color spot tests for explosive residue are confirmatory tests.

8. IR is a sensitive technique that is used to analyze trace samples.

Review Problems

1. List two presumptive field tests for explosive residues. List two confirmatory tests.

2. Residue from a bombing crime scene is placed in three wells of a sample tray. When a few drops of Modified Greiss Test solution are added to the first well, no color change is observed. When a few drops of diphenylamine are added to the second well, no color change is observed. When a few drops of potassium hydroxide are added to the third well, a red color is observed. Which explosive residue is suggested by these presumptive tests? When this residue is analyzed by gas chromatography, a peak at 6.8 minutes (refer to Figure 16-16) is observed. Which explosive does this test confirm?

3. The residue from a bombing is collected and is placed in two wells of a sample tray. When a few drops of diphenylamine are added to the first well, a blue color is observed. When a few drops of potassium hydroxide are added to the second well, no color change is observed. Which explosive residue is suggested by these presumptive tests? When this residue is analyzed by ion chromatography to identify the anions present, a peak with a retention time identical to that of peak 9 in Figure 16-19 is observed. Which explosive residue does this test confirm?

4. Explain how the ion mobility spectrometer works and how it identifies which explosive residue might be present in a bomb residue sample.

5. You have reason to expect that a bomber used TNT to blow up a factory. Should you order a gas chromatography confirmatory test or an ion chromatography test? Why?

Further Reading

Beveridge, A. *Forensic Investigation of Explosions*. Boca Raton, FL: CRC Press, 2002.

Marshall, M., and Oxley, J. C. *Aspects of Explosive Detection*. Amsterdam: Elsevier, 2008.

Midkiff, C. R. Arson and explosive investigations. In R. Saferstein (Ed.), *Forensic Science Handbook*, vol. 1, pp. 479–524. Upper Saddle River, NJ: Prentice-Hall, 2002.

Thurman, J. T. *Practical Bomb Scene Investigation*. Boca Raton, FL: CRC Press, 2006.

Yinon, J. *Counterterrorist Detection Techniques of Explosives*. Amsterdam: Elsevier, 2008.

http://criminaljustice.jblearning.com/criminalistics

Answers to Review Problems

Interactive Questions

Key Term Explorer

Web Links

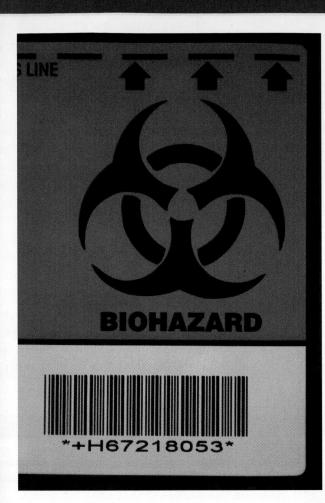

Detecting Weapons of Mass Destruction

The need to detect weapons of mass destruction (WMD) without violating individuals' basic constitutional rights presents an interesting dilemma for law enforcement professionals. In 2001 in the case of *Kyllo v. United States*, the U.S. Supreme Court ruled that "[W]here, as here, the Government uses a device that is not in general public use to explore details of the home that would previously have been unknowable without physical intrusion, the surveillance is a 'search' and is presumptively unreasonable without a warrant." In this case, police used an infrared camera, without a search warrant, from the outside of the Kyllo residence to determine whether heat emanating from the house indicated that high-intensity lights were being used to cultivate marijuana plants. The scan revealed that the garage roof and a wall of the Kyllo home were comparatively hotter than the rest of the home and neighboring houses. Based on the information gleaned from the infrared scan, police officers were able to obtain a search warrant for the inside of the premises, resulting in Kyllo's arrest. The Supreme Court ruled that the use of the infrared camera without a search warrant violated the Fourth Amendment, which prohibits unreasonable search and seizure.

Shortly after this ruling, terrorists attacked New York City and the Pentagon in Washington, D.C. These events were followed by mailings of anthrax spores to members of Congress and several media outlets. In response to these terrorist acts, Congress quickly passed the USA PATRIOT Act, commonly known as the Patriot Act, which greatly expanded the Federal Bureau of Investigation's (FBI) authority to investigate Americans and significantly reduced Fourth Amendment checks and balances, such as a citizen's right to challenge government searches in court.

The scientific and technical community then was called upon to design and deploy advanced sensors to detect WMD. For national safety purposes, these detectors must be able to be used in a nonintrusive way. Equally important, these sensors must be developed such that they can readily detect WMD but no other information that would be considered personal and private.

To maintain national security while still protecting civil liberties, it will be vital that the scientific and technical communities working on these sensors understand the intricacies of the Fourth Amendment and the Patriot Act. Such understanding will ensure they create devices that provide law enforcement personnel with accurate intelligence without violating an individual's civil rights.

1. Using the Department of Justice website (http://www.justice.gov), list the new powers granted to the FBI by Section 215 of the Patriot Act.
2. Would the "plain view" exception to the Fourth Amendment give a police officer the right to search a suspicious backpack that was left next to the entrance of a government building?

Introduction

It has become clear that some terrorist organizations are seeking to develop **weapons of mass destruction (WMD)** that can inflict massive casualties by means of chemical, biological, and radiological or nuclear devices. The bombings of the World Trade Center in New York City in February 1993 and the Murrah Federal Office Building in Oklahoma City in April 1995 are just two examples of terrorist acts on the United States (**FIGURE 17-1**) in which the attackers used conventional explosives. Even though these terrorists used less technologically advanced weapons, the Oklahoma City bombing nevertheless resulted in hundreds of casualties and produced considerable destruction. If a terrorist group uses a biological, chemical, or nuclear weapon in the future, the consequences will be much more severe than the earlier terrorist attacks and will quickly overwhelm local and state emergency responses.

This chapter explores the chemical nature of these threats and considers how they inflict bodily harm. It also discusses the analytical tests that are used in the field by emergency responders to detect WMD. Emergency personnel must be able to quickly determine whether a WMD was used and, if so, which one. Only then can they determine

FIGURE 17-1 The Murrah Federal Office Building in Oklahoma City, Oklahoma, was the target of terrorists in 1995.

whether it is safe to enter the "hot zone" and bring appropriate emergency aid to the victims.

Chemical Warfare Agents

Of the three types of WMD (chemical, biological, and radiological/nuclear), chemical weapons have the greatest potential for terrorist use. **Chemical warfare agents (CWAs)** are easy to obtain and make, and only small amounts of chemicals are needed to immobilize an entire city. Originally, these chemicals were created by makers of toxic industrial chemicals (TICs) and had nonmilitary applications, but eventually they evolved into compounds intended primarily for use by the military. The mil-

itary has given these agents two-letter designations (given in parentheses after the chemical name in this section) to make it easier for troops to describe the agent they have encountered.

CWAs are classified into six categories, where each category contains chemicals that generally act on the body in the same way. These categories are choking agents, blister agents, blood agents, irritating agents, incapacitating agents, and nerve agents.

Choking Agents

Choking agents, such as chlorine (CL), chloropicrin (PS), phosgene (CG), and diphosgene (DP), stress the respiratory tract and lead to a buildup of fluid in the lungs (a condition known as edema). If the edema is severe enough, the person may suffer from asphyxiation—essentially drowning from his or her own bodily fluids.

Choking agents or asphyxiants were some of the first chemicals used for warfare. During World War I, the U.S. army was caught off guard when the German army first used "mustard gas." The term "mustard" refers to the yellow–green color of the clouds of chlorine gas (Cl_2) that were the first chemical agents used in modern warfare. Chlorine is a corrosive poison that forms hydrochloric acid when it comes in contact with the wet mucous membranes of the nose and lungs. The hydrochloric acid then destroys protein in the lung and produces edema.

Blister Agents

Blister agents (vesicants) are corrosive poisons that burn and blister the skin or any other part of the body with which they come in contact. They have a delayed effect: The person does not experience any symptoms immediately after coming in contact with the blister agent, but cell damage nevertheless begins and becomes manifest some 2 to 24 hours after the initial exposure. These colorless gases and liquids are likely to be used to produce casualties rather than to kill, although exposure to such agents can be fatal.

Two major groups of blister agents (vesicants) are distinguished: (1) sulfur mustard (HD) and nitrogen mustard (HN), and (2) the arsenical vesicants (L). The arsenic-containing agents—diphenylchloroarsine (DA), diphenylcyanoarsine (DC), lewisite (L), and ethyldichloroarsine (ED)—

cause blistering of the skin as well as blistering inside the lungs.

Extensive research by the United States during World War I to find a more effective CWA led to the production of lewisite, an arsenic-containing gas that was named for its discoverer, W. Lee Lewis. Although this blister agent was never used in warfare, the threat of lewisite is credited with bringing an end to World War I. Like other compounds containing heavy metals, arsenical vesicants inactivate enzymes by reacting with sulfhydryl groups in enzyme systems.

Blood Agents

Blood agents, such as hydrogen cyanide (AC), arsine (SA), and cyanogen chloride (CK), interfere with the ability of the blood to transfer oxygen, resulting in asphyxiation. These metabolic poisons are more specific in their action than corrosive poisons. They cause harm—and frequently death—by interfering with an essential metabolic process in the body.

Many of us have watched a movie in which a captured spy being interrogated suddenly pops something into his or her mouth and dies instantly. The spy probably bit down on a cyanide capsule. A dose of sodium cyanide as small as 90 mg (0.003 oz) can be fatal. When cyanide enters the body, it inhibits cytochrome oxidase, an enzyme that plays an essential role in the process that allows oxygen to be used by the cells in the body. Cyanide binds to the iron in the enzyme, thereby preventing its reduction. If this reduction does not occur, oxygen cannot be utilized. Cellular reparation ceases and death results within minutes. An antidote for cyanide poisoning is sodium thiosulfate ($Na_2S_2O_3$). Because cyanide acts so rapidly, however, thiosulfate must be administered almost immediately after the poison is ingested to be effective.

Irritating Agents

Irritating agents, such as tear gas (CS), Mace (CN), pepper spray (OC), and dibenzoxazepine (CR), are collectively known as riot-control agents (or just "tear gas"). These chemicals are designed to incapacitate their targets temporarily by causing the person to cough and making his or her eyes water. Tear gas belongs to a group of harassing agents called lachrymators that cause the eye to protect itself by releasing tears to wash the chemical away.

Incapacitating Agents

Incapacitating agents, such as lysergic acid diethylamide (LSD), 3-quinucildinyl benzilate (BZ), and Agent 15, are designed to cause hallucinations, confusion, and motor coordination problems.

Nerve Agents

Nerve agents, such as tabun (GA), sarin (GB), Soman (GD), and Soviet nerve agent (VR), disrupt nerve-impulse transmission or cause peripheral nervous system effects. These organophosphates, which all contain a phosphate (PO_2) group, are absorbed through both the skin and the lungs. Victims lose muscle control and die within minutes from suffocation.

Nerve impulses travel along nerve fibers by electrical impulses. To pass from the end of one nerve fiber to receptors on the next nerve fiber, the impulse must cross a small gap called the synapse (FIGURE 17-2). When an electrical impulse reaches the end of a nerve fiber, chemicals called neurotransmitters are released; these chemicals allow the impulse to cross the gap and travel to receptor cells on the receiving nerve fiber. Each neurotransmitter must fit into a specific receptor to bring about the transfer of the message. Once the nerve impulse has been received, the neurotransmitter is destroyed; the synapse is cleared and is ready to receive the next electric signal.

One important neurotransmitter is acetylcholine. Once it has mediated the passage of an impulse across the synapse, acetylcholine is broken down to acetic acid and choline in a reaction catalyzed by the enzyme cholinesterase (FIGURE 17-3). Other enzymes convert acetic acid and choline back to acetylcholine, which can then transmit another impulse across the synapse.

Different neurotoxins disrupt the acetylcholine cycle at different points. For example, they may block the receptor sites, block the synthesis of acetylcholine, or inhibit the enzyme cholinesterase. Organophosphate nerve agents are anticholinesterase poisons that prevent the breakdown of acetylcholine by inactivating (binding to) the enzyme cholinesterase. As the acetylcholine builds up, nerve impulses are transmitted in quick succession, and nerves, muscles, and other organs become overstimulated. The heart begins to beat erratically, causing convulsions and death.

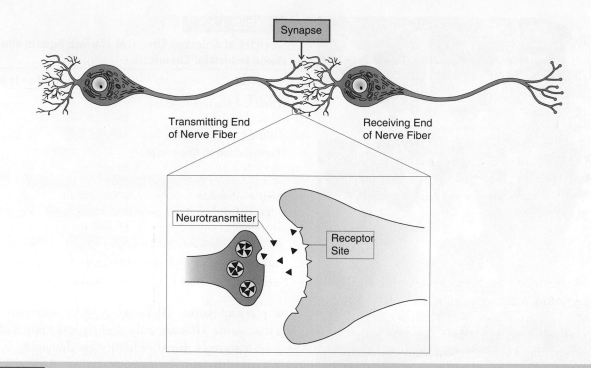

FIGURE 17-2 Transmission of a nerve impulse from one nerve fiber to another occurs when an electrical impulse stimulates the release of neurotransmitter molecules from the transmitter end of one nerve fiber. The neurotransmitter molecule crosses the synapse and fits into receptor sites on the receiving end of another nerve fiber.

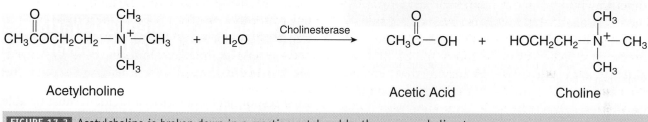

FIGURE 17-3 Acetylcholine is broken down in a reaction catalyzed by the enzyme cholinesterase.

German chemists originally developed the organophosphate nerve gases tabun and sarin during World War II (FIGURE 17-4); after the war, the U.S. army also began manufacturing these nerve agents. Tabun has a fruity odor, but sarin, which is four times as toxic as tabun, is odorless and, therefore, more difficult to detect.

In 1995 the Aum Shinrikyo, a Japanese religious cult obsessed with the apocalypse, released sarin into the Tokyo subway system (FIGURE 17-5). The attack came at the peak of the Monday morning rush hour in one of the busiest commuter systems in the world. Witnesses said that subway entrances resembled battlefields, because injured commuters lay gasping on the ground with blood gushing from their noses or mouths. The attack killed 12 people and sent more than 5000 others to hospitals. As a result of this attack and the September 11 terrorist attacks, chemical detection systems have been installed in the subway systems in Washington, D.C., and New York City.

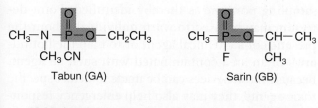

18.08.chem

FIGURE 17-4 Nerve agents such as tabun and sarin are organophosphates; they contain a PO$_2$ functional group.

TABLE 17-1

Toxicity of Selected Chemical Warfare Agents and Toxic Industrial Chemicals

Chemical	Lethal Concentration (ppm)
Chemical Warfare Agents	
Sarin (GB)	36
Hydrogen cyanide (AC)	120
Toxic Industrial Chemicals	
Chlorine (CL)	293
Hydrogen chloride	3000
Carbon monoxide	4000
Ammonia	16,000
Chloroform	20,000
Vinyl chloride	100,000

Source: Data from the National Academies.

FIGURE 17-5 In 1995 a religious cult released sarin into the Tokyo subway system, killing 12 people and injuring more than 5000 others.

once) exposures. Although CWAs are extremely lethal, some TICs are only slightly less toxic and can pose a serious threat, whether by damaging a person's health (e.g., by causing cancer, damaging the person's reproductive capabilities, producing corrosive effects) or by destroying the surroundings (e.g., flammable or explosive chemicals).

Toxic Industrial Chemicals and Materials

Toxic industrial materials (TIMs) take many forms—gases, liquids, and solids—and are used throughout the world for a variety of legitimate purposes. Unfortunately, they may also be used for more nefarious purposes. The chemical industry continues to make and use TICs that also have served as CWAs through the years. For example, chlorine gas has been used as a CWA, yet it is found in large quantities at many water treatment facilities. Phosgene gas was used as a CWA in World War I, yet it remains an essential element in many of today's chemical processes.

The more toxic a chemical, the smaller the amount required to cause harm. **TABLE 17-1** compares the lethal concentrations of some CWAs and TICs in parts per million (ppm) for acute (all at

Detection of Chemical Warfare Agents and Toxic Industrial Materials

First responders to a terrorist incident must be able to test the air for any chemicals that may have been released during the attack. Three major categories of technologies are used for this purpose: point detection technologies, analytical instruments, and standoff detectors.

Point Detection Technologies

A **point detector** is a sensor that samples the environment around it—that is, the air. If that air contains a CWA or TIM, it enters the detector's sampling port and is thereby identified. Point detectors may be used to warn individuals about the presence of a chemical agent or to map the boundaries of an area contaminated with such an agent. Because these devices can be made to detect specific toxic agents, they may also help emergency responders choose appropriate protective clothing when responding to an attack. In addition, handheld point detectors might potentially be used to distinguish who has been contaminated (and how extensive the contamination is) and who has avoided contamination following a terrorist attack.

The U.S. Department of Homeland Security has placed point detectors in crucial places to detect the presence of CWAs or TIMs. For example, these devices are currently being used in subways of major cities, where the detectors are placed in underground tunnels, upwind of a suspected CWA or TIM entry point. In a chemical attack on the subway system, the toxic chemical would be carried by the movement of air in the tunnel to this location, where its encounter with the detector would set off an alarm signaling that immediate evacuation should begin. Unfortunately, some chemical agents may be so toxic or present in high enough concentrations that these warnings might not come quickly enough to enable all subway passengers to exit before encountering hazardous amounts of the toxin.

Several technologies are used for point detectors, including photo-ionization detectors, surface acoustic wave detectors, and colorimetric tubes.

Photo-Ionization Detector

A photo-ionization detector (PID) uses an ultraviolet (UV) light as a source of energy to dislodge electrons from any molecules entering the detector (FIGURE 17-6). In such a device, the molecule is first bombarded with high-energy ultraviolet (UV) light. The absorption of this light excites an electron, causing a temporary loss of an electron, which leads to the formation of a cation (a positively charged ion). This process is called ionization. The gaseous sample then enters the detector cell, and the cations (which now conduct electricity) produce a current that is then amplified and converted by a computer to a ppm concentration; this concentration indicates how much of the toxin is present in the environment. After exiting the detector, the ions quickly recombine with free electrons to reform the original molecule. The PID is a nondestructive detector; it does not cause a permanent change in the sample (or neutralize it).

All chemicals can be ionized. Some are easier to ionize than others, such that they require a smaller amount of energy to boost electrons into an excited state. A compound's ionization potential (IP)—that is, the amount of energy required to excite an electron in a molecule of the compound—and the energy in the UV light used in the PID to bombard the sample molecules are both measured in units of electron-volts (eV). If the IP of the sample gas is less than the eV output of the UV lamp, then the sample gas will be ionized—that is, its electrons will be temporarily excited. Using a UV lamp with a smaller energy output will make the PID sense only those molecules that are easily ionized; using a more powerful UV lamp will lead to the detection of almost all organic molecules. Because the ionized molecules will conduct electricity, they act as a switch that closes a circuit and hence sounds an alarm if any chemical agent is found in the air.

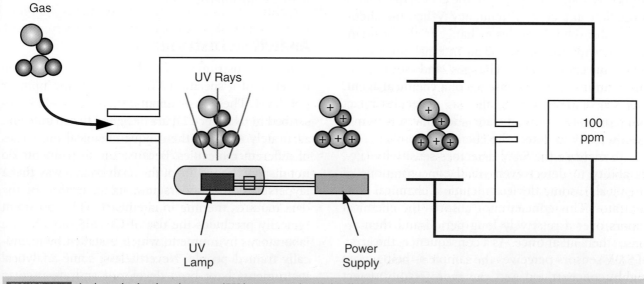

FIGURE 17-6 A photo-ionization detector (PID) uses an ultraviolet light as a source of energy to dislodge electrons (ionization) from any molecules entering the detector.

A PID is very sensitive: It can measure vapors in the low ppm or even parts per billion levels. It is not a very selective monitor, however: It has little ability to differentiate between chemicals. The PID will sound an alert if it detects a chemical in the air, but more sophisticated detectors must be used to identify precisely which chemical is present and to determine what kind of a threat it poses. An air sample may be taken and sent to the forensic laboratory for gas chromatography–mass spectrometry (GC–MS) analysis to determine what chemicals are present. GC-MS analysis was described in Chapter 11.

Surface Acoustic Wave Detectors

Compared with PIDs, surface acoustic wave (SAW) detectors demonstrate superior selectivity when it comes to detecting specific chemical vapors. A SAW sensor is coated with a specially designed polymer film, whose unique physical properties allow a reversible absorption of the chemical it was designed to trap. Each polymer film is designed to detect only certain types of molecules (e.g., just hydrocarbons, just amines). A variety of these films may be collected in an array in the detector, thereby enabling the SAW device to identify more than one type of chemical. For example, selection of the appropriate polymer film may enable detection of both nerve and blister agents. Note, however, that a SAW detector cannot provide definitive identification of precisely which agent is present—just its general class.

SAW sensors use piezoelectric crystals that sense the chemical vapor absorbed onto the sensor surface. A voltage is applied to the crystal that makes it oscillate at a certain frequency. When the chemical is absorbed, it causes a change in the resonant frequency of the sensor. The internal microcomputer measures these changes and sounds an alarm, indicating the presence of a chemical agent. The sensor then releases the gas molecules intact and returns to its initial state, such that it is immediately ready to detect the chemical again.

To increase the SAW detector's sensitivity (i.e., its ability to detect even small concentrations of chemicals), some devices include a chemical concentrator. This concentrator absorbs the chemical vapors over a relatively long period and then releases them all at once. As a consequence, the array of SAW sensors perceives the sample as being more highly concentrated and can more readily detect any chemical agents present in the sample.

Colorimetric Tubes

Colorimetric tubes are a relatively old technology (dating back to the 1930s) that can detect a broad range of chemical vapors. Despite their long history, these tubes, which change color when they are exposed to low concentrations of specific chemical vapors, are actually better at detecting CWAs than PIDs are.

To use this kind of detector, the end of the colorimetric tube is broken off and air is sucked through the tube. The pump draws air through the tube at a constant rate. As the reaction with the indicator in the tube proceeds, the material in the tube changes colors farther into the tube. This process takes 4 to 5 minutes. The side of the tube is calibrated so the concentration of the CWA can be directly read from the side of the tube (**FIGURE 17-7**).

Each tube is designed to test for only one CWA, and more than 100 different tubes are available (**FIGURE 17-8**). Kits have also been developed for the North Atlantic Treaty Organization that hold five different tubes that can detect five CWAs at once. Responders can activate these tubes before entering the "hot zone" and manipulate them while wearing protective suits. Colorimetric tubes are relatively inexpensive: Individual tubes start at around $10, and a five-tube CWA set costs $200, although it must be thrown away after one use.

Unfortunately, colorimetric tubes have two weaknesses. First, responders must guess which CWA agent or TIC is present so that they can pick which tubes to use for the test. Second, first responders have to be in the hot zone to sample the air with a colorimetric tube.

Analytical Instruments

Analytical instruments can be used to confirm the presence of a specific CWA or TIC at the ppm or ppb level. These instruments, such as GC–MS (described in Chapter 11), are designed to separate and accurately measure the unique chemical properties of different molecules. Because the instruments do not display the results of the analysis in a way that a layperson could understand, interpretation of the data requires training in chemistry. This constraint typically precludes the use of GC–MS outside of a laboratory environment, which is staffed by technically trained people. Nevertheless, some analytical instruments have been developed and are used in mobile laboratories for field applications.

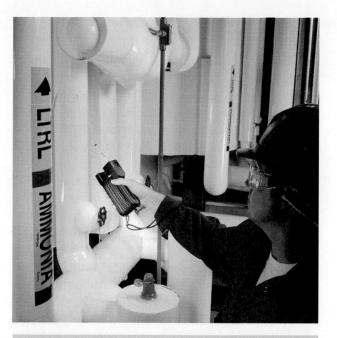

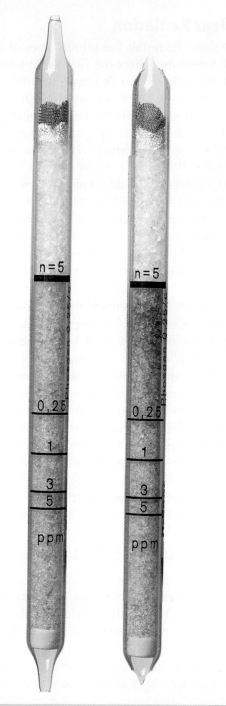

FIGURE 17-7 The colorimetric tube is calibrated so the concentration of chemical vapor in the air can be determined. The material inside the tube changes color to indicate its concentration.

Standoff Detectors

A **standoff detector** reacts to distant events or hazards and can be used to warn of approaching clouds of CWAs or TIMs. These detectors measure infrared radiation (IR) emitted or absorbed from the atmosphere and can detect CWA or TIM clouds at distances as great as 5 km (**FIGURE 17-9**). The IR spectra of CWAs and common TIMs are stored in the detector's database, and the detector compares each measured spectrum that may contain CWAs with these known spectra. Some standoff detectors are placed in strategic positions so as to monitor the atmosphere from a distance, whereas others are mounted on top of vehicles. To date, these devices have proven less dependable and more difficult to operate than point detectors.

FIGURE 17-9 A standoff detector can measure clouds of chemical warfare agents or toxic industrial chemicals from as far as 5 km away.

Nuclear Weapons

A 2003 Central Intelligence Agency (CIA) report warned that Al Qaeda's "end goal" is to use WMD. According to the CIA's report, Al Qaeda has "openly expressed its desire to produce nuclear weapons" and Al Qaeda sketches and documents recovered in Afghanistan included plans for a crude nuclear device.

Although the technical expertise required to construct a nuclear device might be beyond the capabilities of some terrorists, a more realistic pathway to use of a nuclear device would involve its theft or purchase. The collapse of the Soviet Union led to loose internal control over the nuclear weapons and weapons-grade fissionable materials held in the former Soviet republics. In 2003, for example, a cab driver in Georgia was arrested while transporting containers of cesium-137 and strontium-90, fissionable materials that could be used to make a dirty bomb. A "dirty bomb," also known as a radiological dispersal device (RDD), would use conventional explosives to spread nuclear contamination. Later in 2003, a traveler in Bangkok, Thailand, was also arrested while carrying a canister of cesium-137; Thai authorities traced the cesium back to its origin in Russia.

A 2003 study found that the United States and other countries were providing too little assistance to Russia and other former Soviet republics in their efforts to destroy the poorly protected nuclear material and warheads left from the Cold War. Given that the amount of plutonium needed to make a bomb is so small that it will fit into the pocket of an overcoat, it can be easily smuggled. The CIA study also warned that most nuclear reactors in Eastern Europe are "dangerously insecure," suggesting that they might also be sources of material for would-be nuclear terrorists.

Nuclear Radiation

Radioactive materials emit three types of radiation, which are named after the first three letters in the Greek alphabet: alpha (α) rays, beta (β) rays, and gamma (γ) rays. All three types of radiation have sufficient energy to break chemical bonds and disrupt living and nonliving materials upon contact. Hence they are collectively referred to as "ionizing radiation."

Alpha rays are made up of particles, where each particle consists of two protons and two neutrons. An **alpha particle**, therefore, has a mass of 4 atomic mass units (amu), has a 2+ charge, and is identical to a helium nucleus (a helium atom minus its two electrons). Alpha particles are represented as $_2^4\alpha$, or $_2^4\text{He}$.

Beta rays are also made up of particles. A **beta particle** is identical to an electron; thus it has negligible mass and a charge of 1−. It is usually represented as $_{-1}^0\beta$ (or $_{-1}^0\text{e}$). Although a beta particle is identical to an electron, it does not come from the electron cloud surrounding an atomic nucleus, as might be expected. Instead, it is produced from inside the atomic nucleus and then ejected.

Gamma rays, unlike alpha and beta rays, are not made up of particles and therefore have no mass. They are a form of high-energy electromagnetic radiation with very short wavelengths, similar to X-rays, and are usually represented as $_0^0\gamma$.

Penetrating Power and Speed of Radiation

Alpha particles, beta particles, and gamma rays are emitted from radioactive nuclei at different speeds and have different penetrating powers (**TABLE 17-2**). Alpha particles are the slowest, traveling at speeds approximately equal to 1/10 the speed of light; they can be stopped by a sheet of paper or by the outer

---

TABLE 17-2

Properties of Three Types of Radiation Emitted by Radioactive Elements

Name	Symbol	Identity	Charge	Mass (amu)	Velocity	Penetrating Power
Alpha	$_2^4\alpha$, $_2^4\text{He}$	Helium nucleus	2+	4	1/10 the speed of light	Low, stopped by paper
Beta	$_{-1}^0\beta$	Electron	1−	0	Close to the speed of light	Moderate, stopped by aluminum foil
Gamma	$_0^0\gamma$	High-energy electromagnetic radiation	0	0	Speed of light (3×10^{10} cm/s)	High, stopped by several centimeters of lead

layer of skin (FIGURE 17-10). Beta particles are emitted at speeds almost equal to the speed of light. Because of their greater velocity and smaller size, their penetrating power is approximately 100 times greater than that of alpha particles. Beta particles can pass through paper and several millimeters of skin but are stopped by aluminum foil. Gamma rays are released from nuclei at the speed of light and are even more penetrating than X-rays. They pass easily into the human body and can be stopped only by several centimeters of lead or several meters of concrete.

Nuclear Reactions

When a radioactive isotope of an element emits an alpha or a beta particle, a nuclear reaction occurs, and the nucleus of that isotope is changed. The changes that occur during a nuclear reaction can be represented by a nuclear equation.

The element uranium has several radioactive isotopes, including $^{238}_{92}U$ (also written uranium-238), which spontaneously emit alpha particles. The notation $^{238}_{92}U$ tells us that the atomic number of uranium is 92, and the mass number of this par-

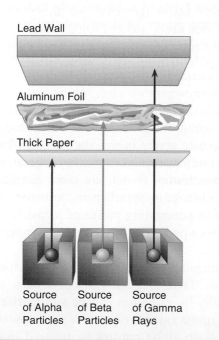

Lead Wall

Aluminum Foil

Thick Paper

Source of Alpha Particles   Source of Beta Particles   Source of Gamma Rays

FIGURE 17-10 The penetrating abilities of alpha particles, beta particles, and gamma rays differ. Alpha particles are the least penetrating and are stopped by a thick sheet of paper or the outer layer of the skin. Beta particles pass through paper but are stopped by aluminum foil or a block of wood. Gamma rays can be stopped only by a lead wall that is several centimeters thick or a concrete wall that is several meters thick.

ticular isotope is 238. Thus $^{238}_{92}U$ has 92 protons and 146 (238 − 92) neutrons in its nucleus.

When a uranium-238 nucleus emits an alpha particle, $^{4}_{2}\alpha$, it loses four atomic mass units (two protons and two neutrons). The resulting atomic nucleus has an atomic mass number of 234 (238 − 4) and, because it has lost two protons, an atomic number of 90 (92 − 2). An atom with an atomic number of 90 is no longer an atom of uranium. If you refer to the periodic table, you will see that the element with atomic number 90 is thorium (Th). Thus, the spontaneous emission of an alpha particle from a uranium atom results in the formation of a completely different element. This **transmutation** of uranium to thorium is represented by the following nuclear equation:

$$\text{mass number} = 238 \text{ sum of mass numbers}$$
$$= 234 + 4 = 238$$

$$^{238}_{92}U \rightarrow\ ^{234}_{90}Th + ^{4}_{2}\alpha$$

$$\text{atomic number} = 92 \text{ sum of atomic numbers}$$
$$= 90 + 2 = 92$$

For the equation to be properly balanced, the mass number on the left side of the equation must equal the sum of the mass numbers on the right side of the equation. Similarly, the atomic number on the left must equal the sum of the atomic numbers on the right.

Radioisotope Half-life

Different radioisotopes—whether they occur naturally or are produced artificially—decay at characteristic rates. The more unstable the isotope is, the more rapidly it will emit alpha or beta particles and change into a new element. The rate at which a particular radioisotope decays is expressed in terms of its **half-life**, or the time required for one-half of any given quantity of the isotope to decay. There is no way of knowing precisely when any one nucleus will disintegrate, but after one half-life, half of the nuclei in the original sample will have disintegrated.

Half-lives range from billionths of a second to billions of years. For example, the half-life of boron-9 is only 8×10^{219} seconds. The half-life of thorium-234 is 24 days, while that of uranium-238 is 4.5 billion years. Note that many of the radioisotopes in the radioactive wastes generated by the production of nuclear weapons and by nuclear power plants have long half-lives.

We can construct a decay curve to show graphically how much of a given radioisotope will remain after a specific length of time (FIGURE 17-11). Assume that we start with 16 g of $^{32}_{15}P$, a radioisotope that has a half-life of 14 days and decays by beta emission to form $^{32}_{16}S$. After one half-life (14 days), one-half of the original 16 g of $^{32}_{15}P$ will have decayed and been converted to $^{32}_{16}S$; 8 g of $^{32}_{15}P$ will therefore remain (and 8 g of $^{32}_{16}S$ will have been formed). At the end of two half-lives (28 days), one-half of the 8 g of $^{32}_{15}P$ that was present at the end of one half-life will have decayed; 4 g will remain. At the end of three half-lives (42 days), 2 g will remain, and so on. After six half-lives (84 days), just 0.25 g of the original isotope will remain; the rest of the sample will be the new element $^{32}_{16}S$. Even after many half-lives, a minute fraction of the original radioisotope will persist.

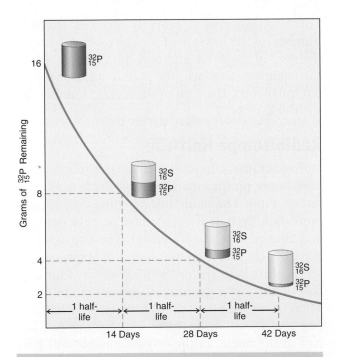

FIGURE 17-11 The decay curve for a 16 g sample of 3215P (half-life of 14 days) to 3216S by emission of beta particles. After one half-life, half of the original 16 g of 3215P will be converted to 3216S; that is, 8 g will remain. After another 14 days, half of the 8 g will remain. At the end of 42 days (three half-lives), 2 g of 3215P will remain, and 14 g of 3216S will be formed. *Source:* Data from National Council on Radiation Protection and Measurement (NCRP87b), Washington, D.C., National Academies Press, 1990.

The Harmful Effects of Radiation on Humans

Why Is Radiation Harmful?

Radiation emitted by radioisotopes is harmful because it has sufficient energy to knock electrons from atoms, thereby forming positively charged ions. For this reason, it is called **ionizing radiation**. Some of the ions formed in this way are highly reactive. By disrupting the normal workings of cells in living tissues, these ions can produce abnormalities in the genetic material DNA and increase a person's risk of developing cancer.

Factors Influencing Radiation Damage

The degree of damage caused by ionizing radiation depends on many factors, of which four are particularly important:

- The type and penetrating power of the radiation
- The location of the radiation—that is, whether it is inside or outside the body
- The type of tissue exposed to the radiation
- The amount and frequency of exposure

The major types of radiation vary in terms of their penetrating power, as described earlier. In general, their penetrating power—that is, their ability to enter the body from the outside environment—can be summarized in this way: alpha particles<beta particles<X-rays<gamma rays. Alpha particles cannot penetrate the skin and enter the body. Beta particles can penetrate the outer layers of skin and clothing, and produce severe burns; they can also cause skin cancer and cataracts. Gamma rays, because of their great penetrating power, are more dangerous than alpha or beta particles when encountered outside the body. The penetrating power of X-rays lies between that of beta particles and gamma rays (refer to Figure 17-10).

Things are reversed when the radiation enters the body. That is, the amount of damage produced internally by radiation can be summarized in this way: gamma rays<X-rays<beta particles<alpha particles. If an alpha emitter is ingested in contaminated food or is inhaled in air containing radon gas, for example, it can cause severe damage to tissues in the body; this kind of ionizing radiation readily strips electrons from molecules in the surrounding tissues and produces numerous harmful ions. Alpha particles have a highly localized effect, however. They do most of their damage in a small area, rather

than affecting large swaths of the body. By comparison, beta particles and gamma rays travel farther than alpha particles inside the body and transfer their energy over a wider area of tissue. Because less radiation is directed at each area, beta particles and gamma rays form fewer disruptive ions.

Different tissues within the body have different sensitivities to all types of ionizing radiation. Cells that undergo rapid replication—such as the bone marrow cells (which manufacture red blood cells), the cells lining the gastrointestinal tract, and the cells in the lymph glands, spleen, and reproductive organs—are readily damaged by radiation, which disrupts the mechanism by which they are reproduced. This phenomenon also explains why cancer cells, which are characterized by uncontrolled growth, may be killed by treatment with radiation. (Paradoxically, radiation-related damage to healthy cells may induce genetic mutations that *produce* cancer.) Likewise, embryonic tissue (i.e., tissue within a fetus) is also vulnerable to the effects of ionizing radiation, which explains why pregnant women are advised to avoid X-rays.

Units of Radiation

Nuclear disintegrations are measured in **curies** (Ci); 1 curie = 3.7×10^{10} disintegrations per second. A curie represents a very high dose of radiation—natural background radiation amounts to only about 2 disintegrations per second. The damage caused by radiation depends not only on the number of disintegrations per second but also on the radiation's energy and penetrating power (as discussed earlier). Another unit of radioactivity, the **rad** (radiation absorbed dose), measures the amount of energy released in tissue when it is struck by radiation. A single medical X-ray is equivalent to 1 rad. The rad is currently being replaced by a new international unit, the **gray** (Gy); 1 gray = 100 rad.

A more useful unit for measuring radiation is the **rem**, which takes into account the potential damage to living tissues caused by the various types of ionizing radiation (TABLE 17-3). For X-rays, gamma rays, and beta particles, 1 rad is essentially equivalent to 1 rem. For alpha particles, which have a much greater ionizing ability, 1 rad is equivalent to 10 rem. The new international unit to replace the rem is the **sievert** (Sv); 1 sievert = 100 rem. A typical dental X-ray exposes a person to approximately 0.0005 rem. An exposure of 600 rem will be fatal to most humans.

TABLE 17-3

Effects of Short-Term, Whole-Body Exposure to Radiation

Dose (rem)	Effect
0–25	No detectable effect
25–50	Temporary decrease in white blood cell count
100–200	Nausea; significant decrease in white blood cells
500+	Death in 50% of the exposed population within 30 days

Source: National Council on Radiation Protection and Measurement (NCRP87b), Washington, D.C.: National Academies Press, 1990.

Detection of Radiation

To discuss radiation exposure, we need to be able to measure it. The instrument that is most commonly used for detecting and measuring radioactivity is the **Geiger counter**, which is essentially a modified cathode-ray tube (FIGURE 17-12). In such a device, argon gas is contained in a metal cylinder, which acts as the cathode; a wire anode runs down the axis of the tube (Figure 17-12A). Radiation from a radioactive source enters the tube through a thin mica window. When it does so, it causes ionization of the argon gas. This process causes pulses of electric current to flow between the electrodes. These pulses are then amplified and converted into a series of clicks that are counted automatically. The Geiger tube can be used to search for radioactive materials (Figure 17-12B).

If the radioisotope is a gamma emitter, such as cesium-137, a gamma spectrometer can be used for its detection. The gamma spectrometer includes multiple solid-state detectors that are able to measure the wavelength of the emitted gamma ray. Each gamma-emitting element releases a gamma ray with a unique wavelength. By using a gamma spectrometer, one can not only determine how much gamma radiation is present, but also identify which radioisotopes may be present.

Biological Weapons

Biological weapons (BWs) were banned by the 1975 International Biological Weapons Convention. Nevertheless, biological terrorism, which seeks to disperse disease-producing biological agents within the civilian population, is of great concern today. Bioterrorism is defined by the Centers for Disease

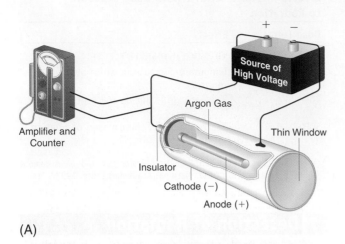

(A)

(B)

FIGURE 17-12 (A) In a Geiger counter, the metal cylinder, which is filled with a gas (usually argon) acts as the cathode; the metal wire projecting into the cylinder acts as the anode. The window in the cylinder is permeable to alpha, beta, and gamma radiation. When the radiation enters the cylinder, the gas is ionized and small pulses of electric current flow between the wire and the metal cylinder. The electrical pulses are amplified and counted. The number of pulses per unit of time is a measure of the amount of radiation. *Source:* Adapted from Brown, T. E., LeMary, H. E., and Bursten, B. E. *Chemistry: The Central Science,* 5th ed. Upper Saddle River, NJ: Prentice Hall, 1991. (B) The Geiger counter can be used to check a scene of an explosion for radiation.

Control and Prevention (CDC) as the "intentional or threatened use of viruses, bacteria, fungi, or toxins from living organisms to produce death or disease in humans, animals, or plants." The 2001 anthrax letters exemplify the use of bioterrorism to create widespread fear among the civilian population. Other possible bioterrorism agents include botulism toxin, brucellosis, cholera, glanders, plague, smallpox, and tularemia Q fever. Viral agents, such as hemorrhagic fevers and severe acute respiratory syndrome, might also potentially be used for this purpose.

BWs present some serious challenges for law enforcement personnel: They are odorless, colorless, and tasteless; they can be readily acquired (and often cultured, if necessary); they are easily transported; and relatively small amounts are needed to produce major damage to human life. These infectious microorganisms can also be released in a variety of ways—for example, through contamination of food or water, use of aerosol sprays, and use of explosives. The last technique is not very effective,

however, because BW agents are living creatures that may be killed by the combustion reaction.

Of course, BWs also present stiff challenges to would-be terrorists. For instance, it is difficult to control the direction in which the pathogens spread, because their dispersal may be affected by the vagaries of wind or water currents. Once an agent is released, many people other than those targeted may be infected, including the terrorists themselves. If contamination of the water supply is the goal, massive quantities of microorganisms would need to be added to water after its exit from a regional treatment facility; in addition to the difficulty in obtaining an adequate supply of BWs to carry out this plan, terrorists might find it difficult to escape detection while accessing the site.

If the goal is to disseminate biological agents through the air, terrorists must ensure that the particles are small enough to be absorbed through the respiratory system—typically, less than 5 mm in size. Nevertheless, this route of administration likely poses the greatest threat to the general public, be-

on the CRIME SCENE — Anthrax in the Mail

Shortly after terrorists attacked the World Trade Center in New York City and the Pentagon in Washington, D.C., on September 11, 2001, by crashing airplanes into the buildings, another form of terrorist attack was carried out in the United States: the use of the postal system as a means to deliver biological WMD. On September 18, 2001, letters containing anthrax bacteria arrived at ABC News, CBS News, NBC News, the *New York Post*, and the *National Enquirer* in New York; and at the American Media, Inc. (AMI), in Boca Raton, Florida. The letters were postmarked from Trenton, New Jersey. On October 9, two more of the anthrax-contaminated letters were found in the offices of South Dakota Senator Tom Daschle and Vermont Senator Patrick Leahy.

The anthrax found in the first set of letters sent to the New York-based media outlets appeared as a brown granular substance, somewhat like ground dog food. This form of anthrax, called cutaneous anthrax, only causes skin infections. The anthrax sent to the AMI office in Florida and to the senators' offices, however, contained a more dangerous version known as inhalation anthrax. Both forms of anthrax were derivative of the Ames strain, a bacterial strain that was first researched at the U.S. Army Medical Research Institute of Infectious Diseases in Fort Detrick, Maryland. Robert Stevens, who worked for AMI, subsequently died of the disease.

The letter sent to Senator Daschle was opened on October 15 by one of his aides, who found a highly refined dry powder containing nearly 1 g of pure spores. Consequently, the government mail service was closed and secured. The letter sent to Senator Leahy was actually discovered on November 16 in Sterling, Virginia, at a state department mail annex because of a misread ZIP code. This facility also was shut down to be disinfected but not until a postal worker, David Hose, had contracted inhalation anthrax. Hose was lucky—he survived.

A total of 22 people developed inhalation anthrax as a result of the mailings. Ultimately, five people died of the disease, including two postal workers from the Brentwood mail facility in Washington, D.C., who likely handled the letters sent to Senators Daschle and Leahy. Thousands of others took an antibiotic, Cipro, to prevent infection in case they had inadvertently come in direct contact with the contaminated mail.

Radiocarbon dating of the anthrax found in the letters determined that it had been produced no more than two years earlier. In August 2002, using special detectors, investigators found a post office box near Princeton University in New Jersey that was likely used to mail all or some of the contaminated letters.

The affected post offices and Senate offices were closed for months while decontamination efforts took place. Special detectors now have been strategically placed in post offices around the United States to detect biologically contaminated mail before it is sent. However, the individual(s) who created the anthrax that was placed in the letters sent in 2001 remained at large at the time of this text's publication.

cause infectious organisms might be delivered through a variety of readily available technologies—via airplane (e.g., agricultural crop dusters), by use of aerosol generators on motor vehicles or boats, or through personal devices, such as a sprayer concealed in a backpack, briefcase, or perfume atomizer.

Government Efforts to Thwart Bioterrorism

In recent years, as concerns about the risk of bioterrorism have increased, government agencies have been undertaking a more coordinated effort to detect and respond to such threats. For example, the CDC has fine-tuned its system for reporting suspected terrorism-related acts. Working in collaboration with state and local health agencies, the CDC and myriad other agencies have begun systematically recording and evaluating suspicious events, with the twin goals of identifying existing BW-related problems and minimizing their damage.

Part of the challenge in dealing with biological agents is determining how much of the BW agent is present and what level of threat it poses. A credible threat is defined as the presence of a biowarfare agent in sufficient quantity to kill people who come into contact with the substance. To date, large-scale systems have focused on distant threats, with the goal of allowing people to escape the area when a credible threat, such as a large cloud of biological agents, is detected. The U.S. Department of Defense has placed a high priority on research and development of systems to detect these threats. For example, the military has developed the Long Range Biological Standoff Detection System that uses an IR detector to detect biological aerosols at distances of 30 km. The Short Range Biological Standoff Detection System uses UV and laser-induced fluorescence to detect biological aerosols at distances of 5 km. Other strategies to detect a biological aerosol cloud include using an airborne pulsed laser system to scan the lower altitudes upwind from a possible target area. The U.S. military is also developing a detection system that can be mounted on a vehicle for mobile use.

But what should we do about biological agents that pose a significant danger even when they are present in minute quantities and would be difficult to detect at much closer range? For example, approximately 10,000 spores of anthrax are needed to infect one person. But the anthrax bacterium is tiny: 10,000 spores, which weigh about 10 ng, is about 1/100th the size of a single speck of dust. How, then, might detectors be expected to identify BW agents that are present in low concentrations? To date, terrorists (and perpetrators of bioterrorism hoaxes) have tended to use much larger quantities in their attacks. For example, the letter sent to Senator Daschle's office in October 2001 contained approximately 2 g of material, the equivalent of about 200,000 infectious doses.

Detection of Biological Agents

Detection of biological agents involves either finding the agent in the environment or making a medical diagnosis of the agent's effects on human or animal victims. Early detection of a biological agent in the environment allows for early specific treatment and time during which medical treatment would be effective.

Unfortunately, current biological detection systems are not as reliable as are chemical detection systems. Chemical detectors can provide information about a CWA within seconds or minutes in the field. BW detectors, by comparison, are slow to recognize the presence of a pathogen, have difficulty discriminating between pathogenic and nonpathogenic organisms in the environment, and lack adequate sensitivity. Two detection techniques do show considerable promise—immunoassays and polymerase chain reaction (PCR)-based technologies. As yet, the only reliable test for BW threats targets anthrax.

Immunoassays

Immunoassays use antibodies to detect the organism of interest. When a human ingests an antigen, such as anthrax, his or her body responds by producing antibodies (proteins) that bind to the invading antigen, thereby inactivating it. The antigen usually has several different sites to which an antibody may bind. When the person is inoculated with the antigen, he or she will produce a series of different antibodies, all of which are synthesized by the immune system to attack a particular site on the antigen of interest. The antibodies that are produced can be easily isolated from blood. The antibodies can then be used in an **enzyme-linked immunosorbent assay (ELISA)**.

To understand how an ELISA works, consider the test developed for anthrax. Anthrax is a serious disease caused by infection with the bacterium *Bacillus anthracis*. It is typically seen in farm animals, such as cattle, sheep, and other herbivores. Humans can contract anthrax by handling products from infected animals or by breathing in or coming in close contact with *B. anthracis* spores from infected animal products, such as unprocessed hides and bones. In 2006, for example, a New York City man developed this disease after exposure to anthrax-infected animal skins used as the coverings for drums. Anthrax also can be transmitted to humans when anthrax spores are used as a bioterrorist weapon.

When a person comes in contact with *B. anthracis*, his or her body reacts by developing antibodies to the infectious invader. This fact has been exploited to develop a new test for anthrax—rather than testing for the microorganism itself, the test looks for the antibodies indicating the person has been exposed to this pathogen. The first point test approved for detecting antibodies to anthrax, called the Anthrax Quick ELISA test, became available to state and private labs in June 2004. This test is quick (results are available in 1 hour, rather than the 4 hours required for older tests), and the results are easy to interpret. Prior to its development, few laboratories other than those operated by the CDC and the U.S. army could test blood for antibodies to anthrax.

In the ELISA test for anthrax, antibody 1, which is specific for anthrax, is bound to a plastic test strip. The suspect sample is then incubated with the plastic-bound antibody. If anthrax is actually present in the sample, it forms a complex with the antibody (FIGURE 17-13). The fraction of antibody sites

1. Add Sample Suspected to be Anthrax
2. Wash to Remove Unbound Molecules

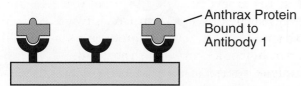

Anthrax Protein Bound to Antibody 1

3. Add Enzyme-Labeled Anthrax Antibody 2
4. Wash to Remove Unbound Antibody

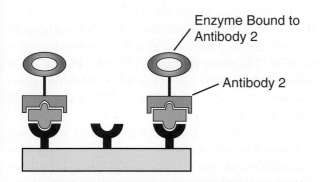

Enzyme Bound to Antibody 2

Antibody 2

FIGURE 17-13 An enzyme-linked immunosorbent assay (ELISA). Antibody 1, which is specific for the protein of interest, is bound to a plastic strip and treated with the sample. The protein binds with the antibody. The bound protein is then treated with antibody 2, which recognizes a different site on the protein and to which an enzyme is covalently attached.

that bind anthrax is proportional to the concentration of anthrax in the sample.

After washing the plastic strip, the antigen–antibody complex is incubated with another anthrax antibody (antibody 2), which recognizes a different region on the surface of the anthrax antigen. An enzyme is attached to antibody 2 before it is incubated with the plastic strip. This enzyme is necessary for the final step in the analysis, in which a special reagent is added that reacts with the enzyme to produce a colored product (**FIGURE 17-14**). Because one enzyme molecule can catalyze the reaction many times, many molecules of colored product are formed for each anthrax molecule. The enzyme amplifies the signal that is captured on the surface of the strip. The higher the concentration of anthrax in the sample, the more enzyme bound and the darker the color produced. This test kit is similar to

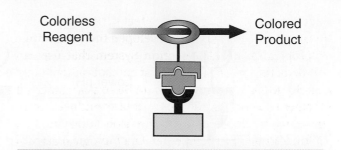

FIGURE 17-14 The enzyme bound to antibody 2 can catalyze reactions that produce colored or fluorescent products. Each molecule of bound protein in the immunoassay produces many molecules of colored or fluorescent product, which are easily measured.

home pregnancy tests, which are based on an ELISA that detects placental protein in the mother's urine (**FIGURE 17-15**).

The sensitivity of an immunoassay can be enhanced by a factor of more than 1000 times with the use of a time-resolved fluorescence immunoassay. When a europium ion (Eu^{3+}, element 63) is excited with a laser, it produces a strong luminescence at 615 nm that has a relatively long lifetime—about 700 microseconds (μs). In such an immunoassay, a group that binds a europium ion is attached to antibody 2 of the assay (**FIGURE 17-16**). While bound to the antibody, Eu^{3+} has only weak luminescence. When the pH of the solution is lowered, the Eu^{3+} ion is released, it becomes strongly luminescent, and a time-resolved measurement is taken to detect its presence.

In a time-resolved fluorescence measurement, a UV laser pulse is generated at 340 nm; the luminescence is measured approximately 600 μs later and

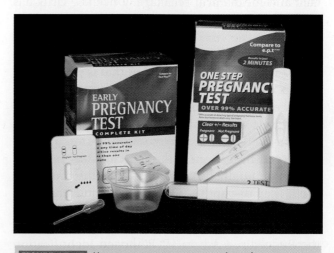

FIGURE 17-15 Home pregnancy tests are based on an ELISA that detects placental protein in a woman's urine.

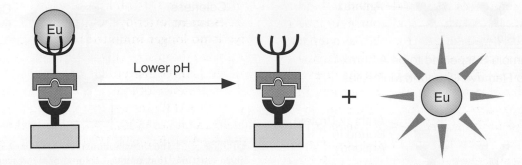

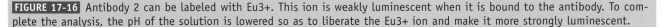

FIGURE 17-16 Antibody 2 can be labeled with Eu3+. This ion is weakly luminescent when it is bound to the antibody. To complete the analysis, the pH of the solution is lowered so as to liberate the Eu3+ ion and make it more strongly luminescent.

compared with the europium luminescence. If the laser-generated emission is with 200 μs of the europium emission, it is rejected—it is deemed to be part of the background fluorescence. The next laser pulse is flashed at 1000 μs and again compared with the europium luminescence. This cycle is repeated approximately 1000 times per second. The result generated from such an immunoassay is more sensitive because it has discarded any background fluorescence that might otherwise interfere with the emissions generated by the europium.

DNA PCR-Based Technologies

Viruses are packets of infectious nucleic acid surrounded by protective coats. These pathogens are very efficient, self-reproducing, intracellular parasites. The complete extracellular form of a virus is called a virion or virus particle. In a virion, the nucleic acid of the virus is surrounded by a protein layer, which protects the DNA from enzymatic attack and mechanical breakage. Once the virus has entered a susceptible host, the protein layer is dissolved and the viral genes multiply rapidly.

The PCR (discussed in depth in Chapter 14) can be used to amplify the DNA that is part of each BW. Because the PCR reagents must contact the pathogen's DNA, many of the successful PCR methods include a mechanical step to open the protective protein shell, such as ultrasonic vibration or high-speed centrifugation. Although chemical methods could potentially be used to dissolve the protective shell, they disrupt the PCR process and hence are avoided.

One PCR-based pathogen detection system is the Ruggedized Advanced Pathogen Identification Device (RAPID), which is designed to be used in the field (FIGURE 17-17). RAPID has already been used in

several practical circumstances—for example, to screen attendees at the Super Bowl and President George W. Bush's inauguration. Developed originally for military applications, it relies on two components: (1) use of a thermal cycler to amplify DNA probes followed by (2) use of a very sensitive fluorescence detector to determine whether the sample contains a BW.

In this device, the collected samples are first shaken with zirconia beads to break the virus's protein shell. Next, the samples are filtered and placed in long tubes. Freeze-dried reagents for a specific test are then added to the tube, which is centrifuged to mix the reagents and to concentrate the sample at the bottom of the tube. RAPID uses hydrolysis, with fluorescent dyes attached, to test for the presence of specific organisms. The probes consist of

FIGURE 17-17 The Ruggedized Advanced Pathogen Identification Device (RAPID) is designed to be used in the field. It uses PCR amplification to enhance the pathogen of interest in the sample.

short DNA sequences that are designed to bind to an organism's DNA (which is, of course, unique to that particular organism). Each probe has a reporter dye (R) and a quencher (Q) molecule attached to each of its ends (FIGURE 17-18). When the quencher is near to the reporter dye, it snuffs out any fluorescence signal that the reporter might emit, so the hybridization probe itself does not release a fluorescent signal. When a probe finds a complementary site on a strand of viral DNA, however, it binds to it. During this binding process, the probe releases the reporter dye. Once released, the reporter dye is no longer inhibited by the quencher, and it begins to fluoresce. This fluorescence signal is read by the RAPID instrument as a positive indication of the presence of that specific pathogen.

The RAPID instrument has a carousel that can take 32 samples (FIGURE 17-19). Probe kits are available for 10 different BWs, but unfortunately all 10 tests cannot be carried out simultaneously.

FIGURE 17-19 The RAPID instrument can accommodate multiple samples, though it carries out tests one at a time.

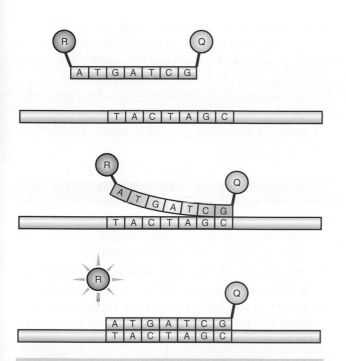

FIGURE 17-18 When the hydrolysis probe binds to a complementary strand of DNA. it releases the reporter dye (R). The RAPID instrument then measures the fluorescence of the released reporter dye.

WRAP UP

1. Section 215 of the USA Patriot Act allows the FBI to order any person or entity to turn over "any tangible things," so long as the FBI "specifies" that the order is "for an authorized investigation . . . to protect against international terrorism or clandestine intelligence activities." Section 215 vastly expands the FBI's power to spy on ordinary people living in the United States, including U.S. citizens and permanent residents.
   a. The FBI need not show probable cause, nor even reasonable grounds to believe, that the person whose records it seeks is engaged in criminal activity.
   b. The FBI need not have any suspicion that the subject of the investigation is a foreign power or an agent of a foreign power.
   c. The FBI can investigate U.S. citizens based in part on their exercise of First Amendment rights, and it can investigate non-U.S. citizens based solely on their exercise of First Amendment rights. For example, the FBI could spy on a person because it doesn't like the books she reads, or because it doesn't like the websites she visits. It could spy on her because she wrote a letter to the editor that criticized government policy.
   d. Anyone served with Section 215 orders is prohibited from disclosing that fact to anyone else. The people who are the subjects of the surveillance are never notified that their privacy has been compromised.
2. Yes, the police officer would be able to search the backpack under the emergency exceptions, plain view doctrine, and open fields exceptions.

Chapter Spotlight

- There are three types of WMDs: chemical, biological, and nuclear weapons.
- Chemical weapons, which have the greatest potential for terrorist use, are classified into six categories: choking agents, blister agents, blood agents, irritating agents, incapacitating agents, and nerve agents.
- TIMs, which are commonly used by the chemical industry, can be used as WMDs.
- A point detector is a sensor that samples the air around it.
- The PID, SAW, and colorimetric tubes are all used as point detectors for sensing chemicals present in the air.
- Standoff detectors are used to warn of clouds of CWA or TIM approaching from a distance.
- A "dirty bomb," which is also known as an RDD, uses conventional explosives to spread nuclear contamination.

- Radioactive materials can emit three types of radiation: alpha, beta, and gamma rays.
- The amount of damage caused by radioactive materials depends on the type and penetration power of the radiation, the location of the radiation, the type of tissue exposed, and the amount or frequency of exposure.
- Radioactive contamination is detected by a Geiger counter.
- BWs disseminated through the air pose a threat to the general public.
- Current biological detection systems that use immunoassays or DNA-based technologies are not as reliable as are chemical detection systems.

http://criminaljustice.jblearning.com/criminalistics

Key Terms

Alpha particle A radioactive particle that is a helium nucleus.

Beta particle A radioactive particle that has properties similar to those of an electron.

Biological weapon (BW) Disease-producing microorganisms, toxic biological products, or organic biocides used to cause death or injury.

Blister agent A chemical that injures the eyes and lungs and burns or blisters the skin.

Chemical warfare agent (CWA) A chemical used to cause disease or death.

Choking agent A chemical agent that attacks the lung tissue, causing the lung to fill with fluid.

Cholinesterase An enzyme found at nerve terminals.

Curie A unit of radioactivity.

Enzyme-linked immunosorbent assay (ELISA) A sensitive immunoassay that uses an enzyme linked to an antibody or antigen as a marker for the detection of a specific protein, especially an antigen or antibody. It is often used as a diagnostic test to determine exposure to a particular infectious agent.

Gamma ray A high-energy photon emitted by radioactive substances.

Geiger counter An instrument that detects and measures the intensity of radiation.

Gray The new international unit that is intended to replace the rad (l Gy = 100 rad).

Half-life The time required for half of the atoms originally in a radioactive sample to decay.

Immunoassay A test that makes use of the binding between an antigen and its antibody to identify and quantify the specific antigen or antibody in a sample.

Incapacitating agent A chemical that disables but does not kill immediately.

Ionizing radiation Radiation capable of dislodging an electron from an atom, thereby damaging living tissue.

Nerve agent A chemical that incapacitates its target by attacking the nerves.

Neurotransmitter A chemical that carries nerve impulses across the synapse between nerve cells.

Point detector A sensor that samples the environment wherever it is located.

Rad Radiation absorbed dose; the basic unit of measure for expressing absorbed radiant energy per unit mass of material.

Rem Roentgen equivalent for man; a dose of ionizing radiation.

Sarin A nerve gas.

Sievert The new international unit intended to replace the rem (l Sv = 100 rem).

Standoff detector A sensor that reacts to distant events or hazards and can be used to warn of approaching chemicals.

Synapse A narrow gap between nerve cells across which an electrical impulse is carried.

Tabun A nerve gas.

Transmutation Conversion of one kind of atomic nucleus to another.

Weapon of mass destruction (WMD) A weapon that kills or injures civilians as well as military personnel. WMDs include nuclear, chemical, and biological weapons.

Putting It All Together

Fill in the Blank

1. Of the three types of WMDs, _____ weapons have the greatest potential for use by terrorists.

2. The chemical warfare agents (CWAs) that cause fluid to accumulate in the lungs and can result in asphyxiation are known as _____ agents.

3. One of the first "mustard gases" to be used was _____ gas.

4. The CWAs that produce no immediate symptoms on contact but produce pain 2 to 24 hours later are known as _____ agents.

5. The CWAs that interfere with the ability of the blood to transfer oxygen and result in asphyxiation are known as _____ agents.

6. If cyanide is taken, it inhibits cytochrome oxidase, which is a(n) _____ that is essential for oxygen metabolism.

7. Tear gas belongs to a group of harassing agents called _____, which cause the eye to protect itself by releasing tears.

8. Chemical agents called _____ carry nerve impulses across a synapse.

9. Nerve agents inhibit the enzyme _____.

10. _____ _____ _____ are industrial chemicals that are manufactured, stored, and used throughout the world.

11. The ion mobility spectrometer can be used as a(n) _____ (point/standoff/analytical) detector.

12. The photo-ionization detector (PID) can be used as a(n) _____ (point/standoff/analytical) detector.

13. The surface acoustic wave (SAW) detector can be used as a(n) _____ (point/standoff/analytical) detector.

14. The PID uses _____ light as a source of energy.

15. The SAW uses a(n) _____ coating on the detector to trap the chemical vapor of interest.

16. Standoff detectors measure _____ radiation to determine whether a CWA agent is in the air.

17. A radiological dispersal device is also known as a(n) _____ _____.

18. A(n) _____ particle has a mass of 4 amu and a 2+ charge.

19. A(n) _____ particle has a negligible mass and a charge of 1−.

20. A(n) _____ ray has no mass and no charge.

21. The _____ particle can be stopped by a piece of paper.

22. The _____ ray is stopped only by several centimeters of lead or several meters of concrete.

23. When an alpha particle or a beta particle is emitted from an isotope, a process called _____ takes place and a new element is formed.

24. The _____ life of a radioisotope is the time required for one-half of any given quantity of that isotope to decay.

25. Radiation is detected using a(n) _____ counter.

26. A(n) _____ spectrometer, when used to measure emitted radiation, can determine which radioisotope is present.

27. A dose of ___ rem causes nausea and a decrease in white blood cells.

28. A dose of _____ rem causes death in 50% of the exposed population within 30 days.

29. When a human ingests an antigen, such as anthrax, his or her body will respond by producing _____.

True or False

1. Chemicals originally developed by the civilian chemical industry that can be used as chemical weapons are known as toxic industrial chemicals.

2. An antidote for cyanide poisoning is sodium thiosulfate.

3. LSD can be used as an irritating agent.

4. An acute chemical exposure is an all-at-once exposure.

5. Point detectors determine the presence of CWAs at specific locations.

6. A photo-ionization detector is not very sensitive, but it is very selective.

7. A surface acoustic wave detector can identify which nerve gas is in the air.

8. Analytical instruments are typically used by emergency response personnel.

9. Beta particles can be stopped by a sheet of aluminum foil.

10. BW agents are difficult to detect or protect against.

11. Contamination of a municipal water supply with a BW is a likely scenario.

12. Biological agent detectors are not as reliable as chemical detection systems.

Review Problems

1. List three point detection methods that can be used to detect CWAs. What are the major limitations of point detectors?

2. List two standoff detection methods that can be used to detect CWAs.

3. Which type of protective clothing should emergency personnel wear if a dirty bomb has been detonated that contained a radioisotope that emits:

   a. Alpha particles?

   b. Beta particles?

   c. Gamma rays?

4. A dirty bomb containing the radioactive element cesium-137 was detonated in New York's subway system. The half-life of cesium-137 is 30 years. How long will it take for the cesium to naturally decay to a concentration that is 25% of its initial concentration? Which type of detector would be used to sense its presence?

5. You are sent to a post office mail sorting center that has reported a white powder leaking from a torn envelope. Describe how you would test this powder to determine if it is anthrax.

Further Reading

Cocciardi, J. A. *Weapons of Mass Destruction and Terrorism Response Field Guide*. Sudbury, MA: Jones and Bartlett, 2004.

Harris, D. C. *Quantitative Chemical Analysis* (ed 7). New York: Freeman, 2006.

Peruski, A. H., and Peruski, L. F. Immunological methods for detection and identification of infectious disease and biological warfare agents. *Clinical and Diagnostic Laboratory Immunology*, 2003, 506–513.

Stewart, C. *Weapons of Mass Casualties and Terrorism Response Handbook*. Sudbury, MA: Jones and Bartlett, 2006.

http://criminaljustice.jblearning.com/criminalistics

Answers to Review Problems

Interactive Questions

Key Term Explorer

Web Links

Forensic Science Resources

Professional Organizations

American Academy of Forensic Sciences
410 North 21st Street
Colorado Springs, CO 80904
(719) 636-1100
http://www.aafs.org

American Board of Criminalistics
PO Box 1123
Wausau, WI 54402
http://www.criminalistics.com

American Board of Forensic Toxicology
410 North 21st Street
Colorado Springs, CO 80904
(719) 636-1100
http://www.abft.org

American Board of Questioned Document
  Examiners
PO Box 18298
Long Beach, CA 90807
(562) 901-3378
http://www.asqde.org

American Chemical Society
1155 16th Street NW
Washington, DC 20036
(800) 227-5558
http://www.acs.org

American Society of Crime Laboratory Directors
139K Technology Drive
Garner, NC 27529
(919) 773-2044
http://www.ascld.org

Association for Crime Scene Reconstruction
PO Box 51376
Phoenix, AZ 85076
http://www.acsr.org

California Association of Criminalists
CAC Membership Secretary
California Department of Justice—Riverside
1500 Castellano Road
Riverside, CA 92509
(909) 782-4170
http://www.cacnews.org

Canadian Society of Forensic Sciences
PO Box 37040
3332 McCarthy Road
Ottawa, Ontario K1V 0W0
Canada
(613) 738-0001
http://www.csfs.ca

College of American Pathologists
325 Waukegan Road
Northfield, IL 60093
(800) 323-4040
http://www.cap.org

Forensic Sciences Society
Clarke House
18A Mount Parade
Harrogate, North Yorkshire HG1 1BX
United Kingdom
44 (0)1423 506 068
http://www.forensic-science-society.org.uk

International Association of Arson Investigators
12770 Boenker Road
Bridgeton, MO 63044
(314) 739-4224
http://www.firearson.com

International Association of Bloodstain Pattern
 Analysts
12139 East Makohoh Trail
Tuscon, AZ 85749
http://www.iabpa.org

National Association of Criminal Defense Lawyers
1150 18th Street NW, Suite 950
Washington, DC 20036
(202) 872-8600
http://www.nacdl.org

National Association of Medical Examiners
430 Pryor Street SW
Atlanta, GA 30312
(404) 730-4781
http://www.thename.org

National District Attorneys Association
99 Canal Center Plaza, Suite 510
Alexandria, VA 22314
(703) 549-9222
http://www.ndaa.org

Society for Forensic Toxicologists
One MacDonald Center
1 North MacDonald Street, Suite 15
Mesa, AZ 85201
(888) 866-7638
http://www.soft-tox.org

Federal Forensic Laboratories

Bureau of Alcohol, Tobacco, Firearms, and
 Explosives (ATF)
http://www.atf.gov

Department of Homeland Security
245 Murray Lane SW, Building 410
Washington, DC 20528
(202) 354-1000
http://www.dhs.gov

Drug Enforcement Administration (DEA)
Office of Diversion Control
2401 Jefferson Davis Highway
Alexandria, VA 22301
(202) 305-8500
http://www.dea.gov

Federal Bureau of Investigation (FBI)
Washington Metropolitan Field Office
601 4th Street NW
Washington, DC 20535
(202) 278-2000
http://www.fbi.gov

Federal Emergency Management Agency (FEMA)
500 C Street SW
Washington, DC 20472
(800) 621-3362
http://www.fema.gov

National Institute of Justice
810 7th Street NW
Washington, DC 20531
(202) 307-2942
http://www.ojp.usdoj.gov/nij

Non-Government Forensic Testing Laboratories

DNA

Bode Technology
7364 Steel Mill Drive
Springfield, VA 22150
(770) 752-7730
http://www.bodetech.com

GeneTree
2495 South West Temple
Salt Lake, UT 84115
(888) 404-GENE
http://www.genetree.com

Orchid Cellmark
13988 Diplomat Drive, Suite 100
Farmers Branch, TX 75234
(214) 271-8400
http://www.orchidcellmark.com

Mitotyping Technologies
2565 Park Center Boulevard, Suite 200
State College, PA 16801
(814) 861-0676
http://www.mitotyping.com

Fingerprints

Lynn Peavey Company
10749 West 84th Terrace
Lenexa, KS 66214
(800) 255-6499
http://www.lynnpeavey.com

Sirchie
100 Hunter Place
Youngsville, NC 27596
(800) 356-7311
http://www.sirchie.com

Measurement and the International System of Units (SI)

In science and in many other fields it is very important to make accurate measurements. For example, the establishment of the Law of Conservation of Matter was dependent on accurate weight measurements; medical diagnosis relies on accurate measurements of factors such as temperature, blood pressure, and blood glucose concentration; a carpenter building a deck must make careful measurements before cutting his wood.

All measurements are made relative to some reference standard. For example, if you measure your height using a ruler marked in meters, you are comparing your height to an internationally recognized reference standard of length called the meter.

Most people in the United States use the English system of measurement and think in terms of English units: feet and inches, pounds and ounces, gallons and quarts, and so forth. If a man is described as being 6 foot 4 inches tall and weighing 300 pounds, we immediately visualize a large individual. Similarly, we know how much to expect if we buy a half-gallon of milk or a pound of hamburger at the grocery store. However, most of us have a far less clear idea of the meaning of meters, liters, and kilograms—common metric units of measurement that are used by the scientific community and by every other major nation in the world. Although the United States is committed to changing to the metric system, the pace of change, so far, has been extremely slow.

The International System of Units (SI)

The International System of Units, or SI (from the French "Système International"), was adopted by the International Bureau of Weights and Measures in 1960. The SI is an updated version of the metric system of units that was developed in France in the 1790s following the revolution.

The standard unit of length in the SI is the meter. Originally, the meter was defined as one ten millionth (0.0000001) of the distance from the North Pole to the equator measured along a meridian. However, this distance was difficult to measure accurately, and for many years the meter was defined as the distance between two lines etched on a platinum-iridium bar kept at 32°F (0°C) in the International Bureau of Weights and Measures at Sèvres, France. Today, the meter is defined even more precisely as equal to 1,650,763.73 times the wavelength of the orange-red spectrograph line of $^{86}_{36}$ Kr.

The standard unit of mass is the kilogram. It is defined as the mass of a platinum-iridium alloy bar that, like the original meter standard, is kept at the International Bureau of Weights and Measures.

Note: The difference between mass and weight is explained at the end of Appendix B.

Base Units

There are seven base units of measurement in the SI. They are shown in **TABLE B-1**.

TABLE B-1
SI Base Units

Quantity Measured	Name of Unit	SI Symbol
Length	Meter	m
Mass	Kilogram	kg
Time	Second	s
Electric current	Ampere	A
Thermodynamic temperature	Kelvin	K
Amount of a substance	Mole	m
Luminous intensity	Candela	Cd

Prefixes

The SI base units are often inconveniently large (or small) for many measurements. Smaller (or larger) units, defined by the use of prefixes, are used instead. Multiple and submultiple SI prefixes are given in **TABLE B-2**. Those most commonly used prefixes in general chemistry are underlined.

TABLE B-2
SI Prefixes

Factor		Prefix	SI Symbol
Exponential Form	Decimal Form		
10^6	1,000,000	mega	M
10^3	1000	kilo	k
10^2	100	hecto	h
10	10	deka	da
10^{-1}	0.1	deci	d
10^{-2}	0.01	centi	c
10^{-3}	0.001	milli	m
10^{-6}	0.0000001	micro	μ
10^{-9}	0.0000000001	nano	n
10^{-12}	0.000000000001	pico	p

Thus, for example:

1 kilogram equals 1000 grams (or 1/1000 kg or 0.001 kg equals 1 g).

1 centimeter equals 1/100 or 0.01 meter (or 100 cm equals 1 m).

Derived SI Units

In addition to the seven SI base units, many other units are needed to represent physical quantities. All are derived from the seven base units. For example, volume is measured in cubic meters (m^3), area is measured in square meters (m^2), and density is measured in mass per unit volume (kg/m^3). Derived units commonly used in general chemistry are listed in **TABLE B-3**.

TABLE B-3
SI Derived Units

Physical Quantity	Unit	SI Symbol	Definition
Area	square meter	m^2	m^2
Volume	cubic meter	m^3	m^3
Density	kilogram/meter^3	kg/m^3	kg/m^3
Force	newton	N	$kg \cdot m/s$
Pressure	pascal	Pa	$N/m^2 = kg/m \cdot s^2$
Energy (quantity of heat)	joule	J	$N \cdot m = kg \cdot m^2/s^2$
Quantity of electricity	coulomb	C	$A \cdot s$
Power	watt	W	$J/s = kg \cdot m^2/s^3$
Electric potential difference	volt	V	$J/A \cdot s = W/A$

Note: In 1964, the liter (L) was adopted as a special name for the cubic decimeter (dm^3).

Probably the least familiar of the derived units are the ones used to represent force (force = mass × acceleration), pressure (pressure = force/area), and energy, which is defined as the ability to do work (work = force × distance).

The SI unit of force, the newton (N), is defined as the force that when applied for 1 second will give a 1-kilogram mass a speed of 1 meter per second.

The SI unit of pressure, the pascal (Pa), is defined as the pressure exerted by a force of 1 newton acting on an area of 1 square meter. Often, it is more convenient to express a pressure in kilopascals (1 kPa = 1000 Pa). For example, atmospheric pressure at sea level is approximately equal to 100 kPa.

The SI unit of energy (or quantity of heat), the joule (J), is defined as the work done by a force of 1 newton acting through a distance of 1 meter. Often it is more convenient to express energy in kilojoules (1 kJ = 1000 J).

Note: The relationship between the joule and the more familiar calorie will be discussed later in Appendix B.

Conversions Within the SI

Since the SI is based on the decimal system, conversions within it are much easier than conversions within the English system. Subunits and multiple units in the SI always differ by factors of 10 (refer to Table B-2). Thus, conversions from one unit to another are made by moving the decimal point the appropriate number of places. This procedure is best explained by some examples.

Example 1

Convert 0.0583 kilograms to grams.

a. Obtain the relationship between kg and g from Table B-2.

$$1 \text{ kg} = 1000 \text{ g}$$

b. From the relationship, determine the factor by which the given quantity (0.0583 kg) must be multiplied to obtain the answer in the required unit (g).

$$\text{Factor} = \frac{1000 \text{ g}}{1 \text{ kg}}$$

c. Multiply 0.0583 kg by the factor to obtain the answer.

$$0.0583 \text{ kg} \times \frac{1000 \text{ g}}{1 \text{ kg}}$$

Answer: 58.3 g

Example 2

Convert 72,600 grams to kilograms.

a. 1 kg = 1000 g
b. The answer is required in kg.

Therefore, the factor is:

$$\frac{1 \text{ kg}}{1000 \text{ g}}$$

(not 1000 g/1 kg as in Example 1, where the answer was required in g)

c. $72{,}600 \text{ g} \times \dfrac{1 \text{ kg}}{1000 \text{ g}}$

Answer: 72.6 kg

Example 3

Change 4236 millimeters to kilometers.

a. Table B-2 does not give a direct relationship between mm and km, but it does give relationships between mm and m between m and km:

$$1000 \text{ mm} = 1 \text{ m} \quad 1000 \text{ m} = \text{km}$$

b. The given quantity (4236 mm) must be multiplied by *two* factors to obtain the answer in the required unit (km). In order for the proper terms to cancel out, the two factors must be:

$$\frac{1 \text{ m}}{1000 \text{ mm}} \quad \text{and} \quad \frac{1 \text{ km}}{1000 \text{ m}}$$

c. $4236 \text{ mm} \times \dfrac{1 \text{ m}}{1000 \text{ mm}} \times \dfrac{1 \text{ km}}{1000 \text{ m}}$

$$= \frac{4236 \text{ km}}{1{,}000{,}000}$$

Answer: 0.004236 km

Conversions from the SI to the English System and Vice Versa

Units commonly used in the English system of measurement for length, mass, and volume are given in TABLE B-4.

TABLE B-4

Units of Measurement in the English System

Length
12 inches (in.) = 1 foot (ft)
3 feet = 1 yard (yd)
1760 yards = 1 mile (mi)
Mass
16 ounces (oz) = 1 pound (lb)
2000 pounds = 1 ton
Volume
16 fluid ounces (fl oz) = 1 pint (pt)
2 pints = 1 quart (qt)
4 quarts = 1 gallon (gal)

Relationships that must be used to convert from SI units to English units, and vice versa, are given in TABLE B-5.

TABLE B-5

Conversion Factors: Common SI and English Units

Length
1 inch (in.) = 2.54 centimeters (cm)
1 yard (yd) = 0.914 meter (m)
1 mile (mi) = 1.61 kilometers (km)
Mass
1 ounce (oz) = 28.4 grams (g)
1 pound (lb) = 454 grams (g)
1 pound (lb) = 0.454 kilogram
Volume
1 fluid ounce (fl oz) = 29.6 milliliters (mL)
1 U.S. pint (pt) = 0.473 liter (L)
1 U.S. quart = 0.946 liter (L)
1 gallon (gal) = 3.78 liters (L)

Example 4

Convert 25 inches to centimeters.

a. Table B-5 gives the relationship between inches and centimeters.

$$1 \text{ in.} = 2.54 \text{ cm}$$

b. The answer is required in cm.

Therefore, the factor is:

$$\frac{2.54 \text{ cm}}{1 \text{ in.}}$$

c. $25 \text{ in.} \times \dfrac{2.524 \text{ cm}}{1 \text{ in.}}$

Answer: 63.5 cm

Example 5

Convert 60 pounds to kilograms.

a. From Table B-5: 1 lb = 0.454 kg
b. The answer is required in kg.

Therefore, the factor is:

$$\frac{0.454 \text{ kg}}{1 \text{ lb}}$$

c. $60 \text{ lb} \times \dfrac{0.454 \text{ kg}}{1 \text{ lb}}$

Answer: 27.24 kg

Example 6

Convert 3.5 liters to quarts.

a. From Table B-5: 1 qt = 0.946 L
b. The answer is required in qt.

Therefore, the factor is:

$$\frac{1 \text{ qt}}{0.946 \text{ L}}$$

c. $3.5 \text{ L} \times \dfrac{1 \text{ qt}}{0.946 \text{ L}}$

Answer: 3.7 qt

Example 7

Convert 4.83 m to feet.

a. From Table B-5: 1 yd = 0.914 m and 1 yd = 3 ft
b. The answer is required in ft.

Therefore, the two factors needed are:

$$\frac{1 \text{ yd}}{0.914 \text{ m}} \text{ and } \frac{3 \text{ ft}}{1 \text{ yd}}$$

c. $4.83 \text{ m} \times \dfrac{1 \text{ yd}}{0.914 \text{ m}} \times \dfrac{3 \text{ ft}}{1 \text{ yd}}$

Answer: 15.8 ft

Useful Approximations Between SI and English Units

To make rough estimates of English units in terms of SI units, and to get a feel for the meaning of SI units, it is useful to memorize the following approximations:

$$1 \text{ kg} = 2 \text{ lb}$$

$$1 \text{ m} = 1 \text{ yd}$$

$$1 \text{ L} = 1 \text{ qt}$$

$$1 \text{ km} = 2/3 \text{ mi}$$

Other Commonly Used Units of Measurement

For certain measurements, including temperature and energy, scientists continue to use units that are not SI units.

Temperature

The SI unit for temperature is the kelvin (K) (Table B-1), but for many measurements scientists use the Celsius scale. On this scale the unit of temperature is the degree (°); 0°C corresponds to the freezing point of water, and 100°C corresponds to its boiling point at atmospheric pressure. The 100 degrees between the two reference points are of equal size.

On the Kelvin scale, zero temperature corresponds to the lowest temperature it is possible to attain, or absolute zero. Absolute zero, as determined theoretically and confirmed experimentally, is equal to −273°C (more accurately −273.15°C). The unit of temperature on the Kelvin scale is the same size as a °C. Therefore, to convert from °C to K it is only necessary to add 273.

$$K = °C + 273 \text{ or } °C = K − 273$$

Thus, the boiling point of water is 373 K, and its freezing point is 273 K.

In the United States, most temperatures, including those given in weather reports and cooking recipes, are measured on the Fahrenheit scale. On this scale, the freezing point of water is 32°F, and its boiling point is 212°F. Thus, there are 180 (212 − 32) degrees between the two reference points, compared to 100 degrees on the Celsius scale.

Conversions Between °C and °F

$$100 \text{ divisions } °C = 180 \text{ divisions } °F$$

or, dividing both sides of the equation by 20:

$$5 \text{ divisions } °C = 9 \text{ divisions } °F$$

To convert from °C to °F, the following equation is used:

$$°F = \frac{9}{5}(°C) + 32$$

32 is added because the freezing point of water is 32° on the Fahrenheit scale, compared to 0° on the Celsius scale. To convert from °F to °C, the following equation is used:

$$°C = \frac{5}{9}(°F - 32)$$

In this case, 32 must be subtracted from the °F before multiplying by 5/9.

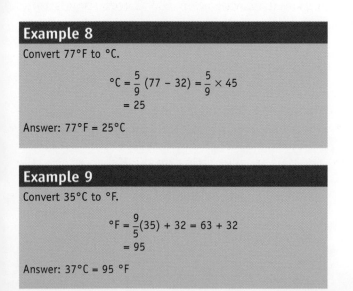

Example 8

Convert 77°F to °C.

$$°C = \frac{5}{9}(77 - 32) = \frac{5}{9} \times 45$$
$$= 25$$

Answer: 77°F = 25°C

Example 9

Convert 35°C to °F.

$$°F = \frac{9}{5}(35) + 32 = 63 + 32$$
$$= 95$$

Answer: 37°C = 95 °F

Energy

The SI unit for measuring heat, or any form of energy, is the joule (J) (Table B-3), but scientists often use the more familiar calorie (cal). For measuring the energy content of food, the kilocalorie (kcal) or Calorie (Cal) is used.

A calorie is the amount of heat required to raise the temperature of 1 g of water 1°C.

$$1 \text{ cal} = 4.18 \text{ J}$$

The Difference Between Mass and Weight

Although the terms *mass* and *weight* are frequently used interchangeably, they have different meanings.

Mass is a measure of the quantity of matter in an object. Weight is a measure of the force exerted on an object by the pull of gravity.

The difference between mass and weight was dramatically demonstrated when astronauts began to travel in space. An astronaut's mass does not change as he is rocketed into space, but once free of the gravitational pull of the Earth, the astronaut becomes weightless. On the moon, an astronaut's weight is approximately one-sixth that on Earth because of the moon's much weaker gravitational pull.

Problems

1. Convert 825 mL to L. (Answer: 0.825 L)
2. Convert 153,000 mg to kg. (Answer: 0.153 kg)
3. Convert 0.00061 cm to nm. (Answer: 61,000 nm)
4. Convert 56,800 mg to oz. (Answer: 2 oz)
5. Convert 1.7 m to feet and inches. (Answer: 5 ft 7 in.)
6. Convert 3 lb 8 oz to kg. (Answer: 1.6 kg)
7. A road sign informs you that you are 60 km from Calais. How many minutes will it take you to arrive there if you travel at 60 miles per hour? (Answer: 37 minutes)
8. A doctor orders a one fluid ounce dose of medicine for a patient. How much is this in mL? (Answer: 29.6 mL)

9. You need to purchase approximately 2 lb of meat at an Italian grocery store. Should you ask for 1/2, 1, or 2 kg? (Answer: 1 kg)

10. Normal body temperature is 98.6°F. A patient has a temperature of 40°C. Is this above or below normal? By how many (a) °F and (b) °C? (Answer: above by (a) 5.4°F and (b) 3°C)

11. The temperature is 15°C. Should you turn on the air conditioning or the heat? (Answer: 59°F; heat)

12. A piece of pie contains 200 Cal. This is equivalent to how many kilojoules? (Answer: 836 kJ)

Glossary

Accelerant Any material that is used to start a fire, but usually an ignitable liquid.

Acid etching method A method in which strong acid is applied to a firearm to reveal the serial number.

Activation energy The amount of energy that must be applied to reactants to overcome the energy barrier to their reaction.

Active bloodstain A bloodstain caused by blood that traveled by application of force, not gravity.

Adenine One of the two double-ring bases found in nucleic acids. It belongs to the class of compounds called purines.

Agglutination The clumping of blood cells in response to an antibody.

Alcohol The common name for ethyl alcohol, a central nervous system depressant.

Alkaloids Organic compounds that normally have basic chemical properties and usually contain at least one nitrogen atom in a ring, occurring in many vascular plants and some fungi. Many alkaloids, such as nicotine, quinine, cocaine, and morphine, are known for their poisonous or medicinal attributes.

Alkane A hydrocarbon containing only carbon–carbon single bonds; the general formula is C_nH_{2n+2}.

Allele The sequence of nucleotides on DNA that constitutes the form of a gene at a specific spot or a chromosome. There can be several variations of this sequence, each of which is called an allele. For example, in the gene determining a person's blood type, one allele may code for type A, whereas the other allele may code for type B.

Alpha particle A radioactive particle that is a helium nucleus.

Alveoli Tiny air sacs within the lungs where the exchange of oxygen and carbon dioxide takes place.

Amorphous material A material without order in the arrangement of its atoms.

Amorphous solid A material in which the atoms have a random, disordered arrangement.

Amphetamines Drugs that have a stimulant effect on the central nervous system and can be both physically and psychologically addictive when overused. The street term "speed" refers to stimulant drugs such as amphetamines.

Amplicon A DNA sequence that has been amplified by PCR.

Anabolic steroids Molecules that promote the storage of protein and the growth of tissue; sometimes used by athletes to increase muscle size and strength.

Anagenic phase The initial phase of hair growth, when the hair follicle is producing hair.

Analgesic A medicine used to relieve pain.

Anisotropic material Material that appears different when the direction of observation is changed.

Anneal When two complementary strands of DNA bind together.

Annealing Heat treatment that produces tempered glass.

Anthropometry A method of identification devised by Alphonse Bertillon in the nineteenth century that used a set of body measurements to form a personal profile.

Antibody A protein that inactivates a specific antigen.

Anticoagulant A chemical that inhibits the coagulation of blood.

Anticodon The three-base sequence carried by the tRNA molecule that determines which specific amino acid it will deliver.

Antigen Any substance (usually a protein) capable of triggering a response from the immune system.

Antiserum Blood serum containing antibodies against specific antigens.

Arch A fingerprint pattern in which ridges enter on one side of the print, form a wave, and flow out the other side.

Arrest warrant A judicial order requiring that a person be arrested and brought before a court to answer a criminal charge.

Associative evidence Evidence that associates individuals with a crime scene.

Atomic absorption spectrophotometry (AAS) A quantitative analysis technique that measures absorption of light by vaporized elements in a sample.

Atomic mass number The sum of the number of protons and the number of neutrons in the nucleus of an atom.

Atomic number The number of protons in the nucleus of an atom of an element.

Backdraft An event in which an oxygen-starved fire suddenly receives oxygen. The sudden rush of oxygen causes the smoldering materials present to reignite at the same time, which causes an explosion.

Barbiturates A group of barbituric acid derivatives that act as central nervous system depressants and are used as sedatives.

Basecoat The layer of automotive paint that contains the colored pigment.

Baseline method A technique used to record measurements for crime scene sketches. A line is drawn between the fixed points (A, B) and the distance to the evidence (X, Y) is measured at a right angle from this line.

Becke line When glass fragments are observed in oil under a microscope, the Becke line is a bright halo that appears around the glass fragment.

Beta particle A radioactive particle that has properties similar to those of an electron.

Binder The material that hardens as the paint dries forming a continuous film.

Biological weapon (BW) Disease-producing microorganisms, toxic biological products, or organic biocides used to cause death or injury.

Biometrics A technology using features of the human body for identification.

Birefringent material An anisotropic material.

Bit A binary digit, either a 1 or a 0.

Bit-for-bit copy An exact copy of the data saved on a computer's hard disk drive.

Bitstream copy A sequence of computer data in binary form.

Black powder The oldest type of gunpowder, which was composed of a mixture of potassium nitrate (saltpeter), charcoal, and sulfur.

Blister agent A chemical that injures the eyes and lungs and burns or blisters the skin.

Block capitals Uppercase, unjoined letters.

Blood alcohol concentration (BAC) The amount of alcohol in a person's blood.

Blood serum Blood plasma with its protein content removed.

Bore The interior of a gun barrel.

Breechblock The metal block at the back end of a gun barrel.

Brisant Of or relating to the shattering effect from the sudden release of energy in an explosion.

Broach cutter A tool that is pushed through the gun barrel to form the rifling.

Bulk samples Samples of drugs that are large enough to be weighed.

Byte Eight bits (binary digits).

Caliber The diameter of the bore of a firearm (other than a shotgun). Caliber is usually expressed in hundredths of an inch (.38 caliber) or in millimeters (9 mm).

Capillary electrophoresis (CE) A method of separating DNA samples based on the rate of movement of each component through a gel-filled capillary while under the influence of an electric field.

Cartridge Ammunition enclosed in a cylindrical casing containing an explosive charge and a bullet that is fired from a rifle or handgun.

Catagenic phase The intermediate stage of hair growth, which occurs between the anagenic and telogenic phases.

Catalyst A substance that increases the rate of a chemical reaction without being consumed during that reaction.

Cellulosic fibers Fibers that are produced from cellulose-containing raw materials, such as trees or other plants.

Celsius scale A temperature scale on which water freezes at 0°C and boils at 100°C at sea level.

Central nervous system (CNS) The vertebrate nervous system, consisting of the brain and the spinal cord.

Central processing unit (CPU) A microprocessor chip that carries out most of the operations of the computer.

Chain of custody The chronological record of each person who had an item of evidence in his or her possession and when they had it.

Chemical property A characteristic of a substance that describes how it reacts with another substance.

Chemical warfare agent (CWA) A chemical used to cause disease or death.

Choke A device placed on the end of a shotgun barrel to change the dispersion of pellets.

Choking agent A chemical agent that attacks the lung tissue, causing the lung to fill with fluid.

Cholinesterase An enzyme found at nerve terminals.

Chromosomes A strand of DNA in the nucleus of eukaryotic cells that carries the genes and functions in the transmission of hereditary information.

Class characteristic A feature common to a group of items.

Clearcoat Outermost layer of automobile paint that contains no pigment.

Cluster A fixed block of computer data that consists of an even number of sectors (1024 bytes or 4096 bytes).

Cocaine A white crystalline alkaloid that is extracted from coca leaves and is widely used as an illicit drug for its euphoric and stimulating effects.

Codeine An alkaloid narcotic that is derived from opium and is used as a cough suppressant and analgesic.

Coding region A section of DNA that contains the code for protein synthesis.

Codons Three-base sequences along the mRNA strand that determine the order in which tRNA molecules bring the amino acids to the mRNA.

Combined DNA Index System (CODIS) The national database containing the DNA profiles of individuals convicted of sexual and violent crimes.

Combustible liquids Liquids with flash points of 100°F (38°C) or higher, such as kerosene and fuel oil.

Comparison microscope Two microscopes linked by an optical bridge.

Compound A substance formed by the chemical combination of two or more elements in fixed proportions.

Compound microscope A microscope with one body tube that is used for magnification in the range of 25 to 1200 times.

Compressive force A force that squeezes glass.

Concentric cracks Cracks in glass that appear as an imperfect circle around the point of fracture.

Condensed explosive An explosive made of a solid or a liquid.

Condenser A lens under the microscope stage that focuses light on the specimen.

Confirmation test A test that identifies the presence of an illegal drug. Its results may be used as confirmatory evidence in court proceedings.

Connecting stroke A line joining two adjacent letters in handwriting.

Continuous spectrum A spectrum that contains all wavelengths in a specified region of the electromagnetic spectrum.

Corpus delicti The "body of the crime."

Cortex The layer of a hair that lies between the medulla and the cuticle.

Crystalline solid A solid in which the atoms are arranged in a regular order.

Curie A unit of radioactivity.

Cursive A type of handwriting in which the letters are joined and the pen is not lifted after writing each letter.

Cuticle A scale structure covering the exterior of the hair.

Cybercrime Disruptive activities against computers or networks, intended to cause harm.

Cyberspace The term used to describe the millions of interconnected computers, servers, routers, and switches that sustain the digital infrastructure.

Cyberterrorism Any type of attack on the computer infrastructure.

Cytoplasm The part of the cell surrounding the nucleus.

Cytosine One of the three single-ring bases found in nucleic acids. It belongs to the class of compounds called pyrimidines.

Deflagration An explosion in which the reaction moves through the explosive at less than the speed of sound.

Denature The separation of two complementary strands of DNA upon heating.

Density The mass of an object per unit volume.

Density gradient column A glass tube filled (from bottom to top) with liquids of sequentially lighter densities.

Density gradient tube A tube filled with liquids of successively higher density.

Deoxyribonucleic acid (DNA) A double-stranded helix of nucleotides that carries the genetic information of a cell.

Deoxyribose A sugar that is a component of DNA.

Depressant A drug that decreases the rate of vital physiological activities.

Depth of focus The depth of the area of the specimen that is in focus.

Dermis The second layer of skin, which contains blood vessels, nerves, and hair follicles.

Detonation An explosion in which the reaction moves through the explosive at greater than the speed of sound.

Detonator A device that is used to set off a high explosive.

Digital imaging The recording of images with a digital camera or scanner.

Dispersed explosive An explosive made of a gas or an aerosol.

DNA polymerase An enzyme that is a catalyst in the polymerase chain reaction.

Double helix A DNA structure that can be compared to a spiral staircase, consisting of two polynucleotide chains wound around each other.

DUI Driving under the influence of alcohol.

DUID Driving under the influence of drugs.

Ejector The mechanism in a semiautomatic weapon that ejects the spent cartridge from the gun after firing.

Electrocoat primer First layer of paint applied to the steel body of an automobile.

Electromagnetic radiation (EMR) A general term used to describe energy that is encountered daily.

Electron A subatomic particle with a negative charge of one (−1) and negligible mass.

Electrostatic detection apparatus (ESDA) An instrument used to visualize writing impressions.

Element A substance that cannot be broken down by chemical means. Elements are differentiated by the number of protons they possess.

Endothermic reaction A chemical reaction that absorbs heat from its surroundings.

Energy The ability to do work. Energy has many forms—for example, heat, chemical, electrical, and mechanical.

Energy level Any of several regions surrounding the nucleus of an atom in which the electrons move.

Energy-dispersive X-ray A technique in which the energy of X-rays emitted in the scanning electron microscope is measured.

Entrance side The side of an object where a projectile enters; the load side.

Enzyme-linked immunosorbent assay (ELISA) A sensitive immunoassay that uses an enzyme linked to an antibody or antigen as a marker for the detection of a specific protein, especially an antigen or antibody. It is often used as a diagnostic test to determine exposure to a particular infectious agent.

Epidermis The tough outer layer of skin.

Erasure The removal of writing or printing from a document.

Evidence Information about a crime that meets the state or federal rules of evidence.

Excited state The state of an atom in which an electron has acquired sufficient energy to move to a higher energy level.

Excretion The act or process of discharging waste matter from the blood, tissues, or organs.

Exemplar Representative (standard) item to which evidence can be compared.

Exit side The side of an object where a projectile exits; the unloaded side.

Exothermic reaction A chemical process that releases heat to the surroundings.

Explainable differences Differences between evidence and reference material that have a sound scientific explanation.

Explosion A sudden release of mechanical, chemical, or nuclear energy, generating high temperatures and usually releasing gases.

Explosive A substance, especially a prepared chemical, that explodes or causes an explosion.

Extensive physical property A property that is dependent on the amount of material present.

External ballistics The events that occur after a bullet leaves the barrel of the gun but before it strikes its target.

Extractor The device that extracts a spent cartridge from a gun's chamber.

Fahrenheit scale A temperature scale on which water freezes at 32°C and boils at 212°C at sea level.

Federal Rules of Evidence (FRE) Rules that govern the admissibility of all evidence, including expert testimony.

Field of view The area that can be seen through the microscope lenses.

Finished sketch A drawing made by a professional artist that shows the crime scene in proper perspective and that can be presented in court.

Flammable liquids Liquids with flash points of 100°F (38°C) or lower.

Flammable range The range of vapor concentrations in air that is capable of burning.

Flashover The temperature at which a fire begins unrestrained growth and can cause complete destruction.

Flash point The minimum temperature at which a liquid fuel will produce enough vapor to burn.

Follicular tag Tissue surrounding the hair shaft that adheres to hair when it is pulled out.

Forgery When a person creates a document (or other object) or alters an existing document (or other object) in an attempt to deceive other individuals, whether for profit or for other nefarious purposes, such as entering the country illegally.

Fourth Amendment This amendment in the U.S. Constitution gives citizens the right to be secure in their persons, houses, papers, and effects against unreasonable searches and seizures.

Fractional distillation The separation of the components of petroleum by boiling, followed by condensation into fractions with similar boiling ranges. Small molecules with low boiling points emerge first, followed by larger molecules with higher boiling points.

Fracture match The alignment of the edges of two or more pieces of glass, indicating that at one time the pieces were part of one sheet of glass.

Freehand method A method of forging a signature by simulation rather than by tracing.

Frequency The number of crests that pass a fixed point in one second. A frequency of one cycle (passage of one complete wave) per second is equal to one hertz (Hz).

Friable Easily broken into small particles or dust.

Friction ridge characteristics Skin on the soles of the feet, palms of the hands, and fingers in humans that forms ridges and valleys.

Fuel cell A device that converts chemical energy directly into electrical energy through chemical reactions.

Gamma ray A high-energy photon emitted by radioactive substances.

Gas chromatography (GC) A technology used to separate complex mixtures of hydrocarbons, alcohols, ethers, and ketones.

Gas chromatography–mass spectrometry (GC-MS) A technology that makes use of a gas chromatograph, which separates components of a mixture, in conjunction with a mass spectrometer, which identifies each component by measuring its mass.

Gauge The unit used to designate size of a shotgun barrel. The smaller the gauge, the larger the diameter of the shotgun barrel.

Geiger counter An instrument that detects and measures the intensity of radiation.

Gel electrophoresis A method of separating DNA samples based on the rate of movement of each component in a gel while under the influence of an electric field.

Genetic code The sequence of nucleotides on DNA that determines the sequence of amino acids in the synthesis of proteins.

Gray The absorption of one joule of energy by one kilogram of matter (1 Gy = 100 rad).

Grooves The cutout sections of a rifled barrel.

Ground state The state of an atom in which all of its electrons are in their lowest possible energy levels.

Group One of the vertical columns of elements in the periodic table.

Guanine One of the two double-ring bases found in nucleic acids. It belongs to the class of compounds called purines.

Gunshot residue (GSR) Material discharged from a firearm other than the bullet.

Hacking A slang term for unauthorized access to a computer or network.

Half-life The time required for half of the atoms originally in a radioactive sample to decay.

Hallucinogen A drug that induces hallucinations or altered sensory experiences.

Hard disk drive (HDD) A computer's main data storage device.

Hardware The physical components of a computer.

Headspace The space above fire debris that has been stored in a sealed container.

Heroin Derived from morphine, heroin is a white, odorless, bitter crystalline compound that is a highly addictive narcotic.

Heterozygous Having different alleles at one or more corresponding chromosomal loci.

High explosive An explosive that when detonated produces a shock wave that travels at a speed greater than 1000 meters per second (e.g., military explosives).

Homozygous Having identical alleles at corresponding chromosomal loci.

Hydrocarbon A chemical that contains only two elements, carbon and hydrogen.

Identification The process of matching a set of qualities or characteristics to uniquely identify an object.

Identity theft The unauthorized use or attempted use of another person's financial records and/or identifying information without the owner's permission.

Illuminator The part of a microscope that illuminates the specimen for viewing.

Immunoassay A test that makes use of the binding between an antigen and its antibody to identify and quantify the specific antigen or antibody in a sample.

Improvised explosive device (IED) A homemade bomb.

Incapacitating agent A chemical that disables but does not kill immediately.

Incipient stage The growth phase of a fire, which begins at ignition.

Indexing A list of every character, letter, number, and word in all files on a computer, created by the forensic utility software.

Individual characteristic A feature that is unique to one specific item.

Individualization The process of proving that a particular unknown sample is unique, even among members of the same class, or proving that a known sample and a questioned sample share a unique common origin.

Inductively coupled plasma (ICP) A high-energy argon plasma that uses radio-frequency energy.

Infrared luminescence Light given off in the infrared region when ink is irradiated with visible light.

Infrared spectrophotometry A technique that measures the absorption of infrared radiation by a drug sample at different wavelengths.

Initiator Any device that is used to start a detonation or a deflagration.

Intaglio A printing method in which a metal printing plate is engraved, producing a raised image on the document. Intaglio is used for security printing, such as for money, passports, and identity documents.

Intensive physical property A property that is independent of the amount of material present.

Internal ballistics A description of the events that transpire within a firearm.

International System of Units (SI) The system of measurement (metric units) used by most scientists.

Iodine fuming The use of iodine sublimation to develop latent prints on porous and nonporous surfaces. Iodine vapors react with lipids in the oil from the latent print to form a visible image.

Ion chromatography (IC) A technique that separates and measures ions.

Ion mobility spectrometer (IMS) A technique capable of detecting and identifying low concentrations of chemicals: molecules are ionized and are identified by the speed with which they move through a magnetic field.

Ionizing radiation Radiation capable of dislodging an electron from an atom, thereby damaging living tissue.

Isotope One of two or more atoms with the same number of protons but differing numbers of neutrons.

Isotropic materials Materials that have the same optical properties when observed from any direction.

Keratin The primary protein that forms hair and nails.

Kerosene A petroleum fraction that boils at temperatures between 300°F (149°C) and 550°F (288°C). Kerosene is used in space heaters, cook stoves, and wick lamps.

Known sample Standard or reference samples from verifiable sources.

Laminated glass Two sheets of glass bonded together with a plastic sheet between them.

Lands The raised section of a rifled barrel.

Latent fingerprint A friction ridge impression that is not readily visible to the naked eye.

Letterpress printing A printing method in which detail is raised off the printing plate, resulting in a corresponding indentation of the characters of the printed document.

Line spectrum A spectrum produced by an element that appears as a series of bright lines at specific wavelengths, separated by dark bands.

Lithography A printing method in which an image is transferred from a printing plate to an offset sheet and then onto the document. The resulting image is not raised or indented.

Locard's exchange principle Whenever two objects come into contact with each other, there is an exchange of materials between them.

Locus A specific location on the DNA chain.

Loop A fingerprint pattern in which ridges enter on one side of the print, form a wave, and flow out the same side.

Low explosive An explosive that when detonated produces a shock wave that travels at a speed less than 1000 meters per second (e.g., black powder).

Lower explosive limit (LEL) The lowest concentration of vapor in air that will burn.

Lysergic acid diethylamide (LSD) A crystalline compound derived from lysergic acid and used as a powerful hallucinogenic drug.

Magnaflux method A method of restoring a gun's serial number.

Malware Malicious software designed to infiltrate a computer or network without the owner's consent.

Marijuana A hallucinogenic drug made from the dried flower clusters and leaves of the cannabis plant. It is usually smoked or eaten to induce euphoria.

Mass A measure of the quantity of matter.

Mass spectrometry A method used to identify chemicals in a substance by their mass and charge.

MDMA (3,4-methylenedioxymethamphetamine) A drug that is chemically related to amphetamine and mescaline and is used illicitly for its euphoric and hallucinogenic effects. MDMA reduces inhibitions and was formerly used in psychotherapy but is now banned in the United States.

Medulla A column of cells running down the center of the hair.

Mescaline An alkaloid drug, obtained from mescal buttons, that produces hallucinations. Also called peyote.

Message digest (MD) A unique digital signature for computer data generated by hashing.

Metabolites The metabolic breakdown products of drugs.

Metals Malleable elements with a lustrous appearance that are good conductors of heat and electricity and that tend to lose electrons to form positively charged ions.

Methadone A synthetic narcotic drug that is less addictive than morphine or heroin and is used as a substitute for morphine and heroin in addiction treatment programs.

Microcrystalline tests Tests used to identify drugs based on the formation of crystals during a chemical reaction. When certain illegal drugs are mixed with testing reagents, they produce colored crystals with specific geometry that can then be used to identify the drug.

Micrometer One-millionth of a meter (μm).

Microspectrophotometer A microscope that measures the interaction of infrared or ultraviolet radiation with a sample.

Mineral A naturally occurring inorganic substance found in the Earth's crust as a solid.

Minutiae Bifurcations, ending ridges, and dots in the ridge patterns of fingerprints.

Modified Greiss Test A test for the presence of nitrites in gunshot residue.

Modus operandi The "method of operation"; also known as the MO.

Mohs scale A scale that measures the hardness of minerals and other solids.

Molecule The smallest unit of a compound that retains the characteristics of that compound.

Morphine A bitter crystalline alkaloid, extracted from opium, that is used in medicine as an analgesic, a light anesthetic, or a sedative.

Multiplexing A technique in which multiple STR loci are amplified and analyzed simultaneously.

Narcotic An addictive drug that reduces pain, alters mood and behavior, and usually induces sleep or stupor.

National Institute on Drug Abuse (NIDA) A U.S. government agency that implements the drug testing program that has become a condition of federal employment.

Nerve agent A chemical that incapacitates its target by attacking the nerves.

Neurotransmitter A chemical that carries nerve impulses across the synapse between nerve cells.

Neutron An electrically neutral subatomic particle found in the nuclei of atoms.

Neutron activation analysis (NAA) An analysis method that bombards the sample with neutrons and then measures the isotopes produced.

Ninhydrin A chemical used to visualize latent fingerprints. Ninhydrin reacts with amino acids in latent fingerprints to form a blue-purple compound called Ruhemann's purple.

Nitrocellulose A cotton-like material produced when cellulose is treated with sulfuric acid and nitric acid; also known as "guncotton." Nitrocellulose is used in the manufacture of explosives.

Nitroglycerin An explosive chemical compound obtained by reacting glycerol with nitric acid. Nitroglycerin is used in the manufacture of gunpowder and dynamite.

Noncoding region A section of DNA that does not code for protein synthesis.

Nonmetals Elements that lack the properties of metals.

Nucleic acids The most common nucleic acids are deoxyribonucleic acid (DNA) and ribonucleic acid (RNA).

Nucleotides Nitrogen-containing compounds that link together to form DNA and RNA.

Nucleus The small, dense, positively charged central core of an atom, composed of protons and neutrons.

Objective lens The lower lens of a microscope; the lens closest to the specimen.

Ocular lens The upper lens of a microscope; the lens nearest to the eye.

Orbital The sublevels of principal energy levels, identified by the letters s, p, d, and f.

Oxidation A chemical reaction in which oxygen is combined with other substances.

OxyContin A prescription painkiller that has become a popular and dangerous recreational drug. Known as "oxy" on the street, this time-released morphine-like narcotic is prescribed by physicians to relieve chronic pain.

Passive bloodstain A pattern of blood formed by the force of gravity.

Patent fingerprint A fingerprint that is readily visible to the eye.

Period One horizontal row of elements in the periodic table.

Periodic table The elements arranged in order of atomic number (i.e., number of protons). Each element has a unique atomic number.

Phencyclidine (PCP) An anesthetic used by veterinarians; an illicit hallucinogen known as "angel dust."

Phenotype The observable characteristics of an organism (e.g., blue eyes, blond hair).

Phishing The use of fraudulent email sent in an attempt to trick the recipient into disclosing personal information.

Physical dependence The situation in which regular use of a drug causes a physiological need for it. When the drug is stopped, withdrawal sickness begins.

Physical evidence Any object that provides a connection between a crime and its victim or between a crime and its perpetrator.

Physical property A property that can be measured without changing the composition of a substance.

Pigment Added to paint to give it color.

Plane-polarized light Light that oscillates in only one plane.

Plasma (blood) The colorless liquid in which the red and white blood cells are suspended in blood.

Plasma (high energy) A high temperature gas that has a portion of its atoms ionized which makes the gas capable of conducting electricity.

Plastic fingerprint A fingerprint indentation left by pressing a finger into a soft surface.

Pleochroism A property of a substance in which it shows different colors when exposed to polarized light from different directions.

Point detector A sensor that samples the environment wherever it is located.

Point of convergence The most likely point of origin of blood that produced a blood stain.

Polar coordinate method A technique used to record measurements for crime scene sketches, using a transit or a compass to measure the angle from the north and the distance to the evidence (X). This method is most commonly used in a large area crime scene (outside or in a warehouse) when a wall or side of a building is used to establish the fixed points (A, B).

Polarizer A lens that passes light waves that are oscillating only in one plane.

Polarizing microscope A microscope that illuminates the specimen with polarized light.

Polymer A large organic molecule made up of repeating units of smaller molecules (monomers).

Polymerase chain reaction (PCR) A reaction that is used to make millions of copies of a section of DNA.

Postmortem interval Time elapsed since death.

Precipitin An antibody that reacts with a specific soluble antigen to produce a precipitate.

Presumptive test A test that provides a reasonable basis for belief.

Prilling A production method that produces free-flowing, adsorbent, porous ammonium nitrate spheres (prills) that are easily handled and stored.

Primer (DNA) A short sequence of DNA that is specifically designed and synthesized to be complementary to the ends of the DNA region to be amplified.

Primer (igniter) An igniter that is used to initiate the burning of gunpowder.

Primer surfacer A layer of automobile paint that slows corrosion of the underlying steel.

Principal energy level One of the energy levels to which electrons are limited.

Projectile A fired or thrown object, such as a bullet.

Proton A subatomic particle with a relative positive charge of one (+1) and a mass of 1; a hydrogen ion.

Psychological dependence A pattern of compulsive drug use caused by continual craving.

Pyrolysis The decomposition of a substance by the application of heat.

Quantum The smallest increment of radiant energy that may be absorbed or emitted.

Questioned document An object that contains typewritten or handwritten information but whose authenticity has not yet been definitively proven.

Questioned sample A sample of unknown origin.

Rad Radiation absorbed dose; the basic unit of measure for expressing absorbed radiant energy per unit mass of material.

Radial cracks Cracks in glass that radiate in many directions away from the point of fracture.

Radial loop A fingerprint loop pattern that flows in the direction of the thumb.

Radio frequency identification (RFID) chip A computer chip that uses communication via a radio frequency to uniquely identify an object, such as a visa.

Random-access memory (RAM) Computer memory that executes programs and stores temporary data. RAM data is automatically erased by the computer when it is shut off.

Read-only memory (ROM) A device that contains instructions. Computers can read the instructions, but not change or write back to ROM.

"Real" image An actual nonmagnified image.

Reference standard Physical evidence whose origin is known and that can be compared to evidence collected at a crime scene.

Refraction The bending of light waves.

Refractive index Ratio of the speed of light in air to the speed of light in another material (such as glass).

Regenerated fibers Fibers made by treating cotton or wood with a variety of chemicals.

Rem Roentgen equivalent for man; a dose of ionizing radiation.

Restriction fragment length polymorphism (RFLP) A DNA typing technique in which sections of the DNA strand are cut out by restriction enzymes and differentiated based on their length.

Retention time The amount of time it takes a compound to be separated via a gas chromatograph.

Reticle A network of fine lines in the eyepiece of a microscope that allows the examiner to measure the distance between magnified objects.

Rib marks The set of curved lines that is visible on the edges of broken glass.

Ribonucleic acid (RNA) A nucleic acid that is an essential component of all living cells composed of a long chain of nucleotides that contain the sugar ribose.

Ribose A sugar that is found in ribonucleic acids (RNA).

Ricochet Deviation of a bullet's trajectory because of collision with another object.

Rifling The spiral grooves on the inside surface of a gun barrel that make the bullet spin.

Rough sketch A drawing made at a crime scene that indicates accurate dimensions and distances.

Ruhemann's purple The blue-purple compound formed in latent fingerprints when they are developed with ninhydrin.

Sarin A nerve gas.

Scanning electron microscope (SEM) A microscope that illuminates the specimen with a beam of electrons.

Scanning electron microscopy with energy-dispersive X-ray (SEM-EDX) The preferred method for the detection of GSR. The investigator applies an adhesive tape directly to the suspect's hands (or other surface) to lift the GSR.

Screen printing A printing method in which ink is passed through a screen.

Screening test A preliminary test performed to detect illegal drugs.

Script A writing style characterized by lowercase unjoined letters.

Search warrant A court order that gives the police permission to enter private property and search for evidence of a crime.

Secondary crime scenes Sites where subsequent criminal activity took place.

Secretor An individual who has significant concentrations of antigens not only in his or her blood but also in other body fluids, such as perspiration, saliva, semen, and urine.

Sector The smallest unit that can be accessed on a computer disk. A sector is 512 bytes.

Secure hash algorithm (SHA) An algorithm developed by the National Institute of Standards and Technology (NIST) used to create or detect digital signatures.

Selectively neutral loci Inherited characteristics that confer neither a benefit nor any harm to the individual's ability to reproduce successively.

Semiautomatic pistol A firearm that reloads itself after firing.

Semimetal (metalloid) Refers to elements with properties that lie between those of metals and nonmetals.

Serology The science that deals with the properties and reactions of blood serum.

Shear force Force that moves one part of a material in one direction while another part is moving in a different direction.

Short tandem repeat (STR) A DNA locus where sequences of 3 to 7 bases repeat over and over.

Sievert The new international unit intended to replace the rem (l Sv = 100 rem).

Single-nucleotide polymorphism (SNP) A DNA sequence variation that occurs when a single nucleotide (A, T, C, or G) in the genome sequence is altered.

Slack space The area from the end of a computer file to the end of a sector on a hard disk drive.

Smart grid The addition of computer-based technology to the U.S. power grid to manage the efficiency and reliability of electricity distribution.

Smokeless powder An explosive charge composed of nitrocellulose or nitrocellulose and nitroglycerin (double-base powders).

Software A set of instructions that are arranged in a program that allows a computer to complete a specific task.

Solvent The liquid in which the components of paint are suspended.

Spam Unsolicited bulk email messages.

Spectroscopy Measurement of the absorption of light by different materials.

Spyware A type of computer malware that collects information about users without their knowledge.

Standoff detector A sensor that reacts to distant chemical or biological events or hazards and can be used to warn of these approaching substances.

Stereoscopic microscope A microscope with two separate body tubes that allow both eyes to observe the specimen at low or medium magnification.

Stimulant A drug that temporarily increases alertness and quickens some vital processes.

Striations Fine scratches left on bullets, formed from contact of the bullet with imperfections inside the gun barrel.

Structural isomers Organic compounds that have the same formula but different molecular structures.

Subsoil Soil lying beneath the topsoil that is compacted and contains little or no humus.

Super Glue™ fuming A technique for visualizing latent fingerprints in which fumes from heated Super Glue™ (cyanoacrylate glue) react with the latent print.

Swap data A file or space in the hard disk drive that swaps data with the RAM memory to free the RAM to execute other commands.

Synapse A narrow gap between nerve cells across which an electrical impulse is carried.

Synthetic fibers Fibers produced from chemicals made from refined petroleum or natural gas.

Tabun A nerve gas.

Taggant A substance, such as a mixture of microscopic pieces of multilayered colored plastic, that is added to an explosive to indicate its source of manufacture.

Tangential cracks Cracks in glass that appear as an imperfect circle around the point of fracture.

Telogenic phase The final phase of hair growth, during which hair falls out of the follicle.

Tempered glass Glass that has been heat-treated to give it strength.

Tensile force Force that expands material.

Terminal ballistics A description of what happens when the bullet strikes its target.

Terminal stroke In handwriting, the final stroke trailing away from a letter, word, or signature.

Tetrahydrocannabinol (THC) The psychoactive substance present in marijuana.

Thermocycle The heat–cool cycle used in a PCR reaction.

Thin-layer chromatography (TLC) A technique for separating components in a mixture. TLC is used to separate the components of different inks.

Thrombocytes Platelets; substances found in blood that are essential for blood clotting.

Thymine One of the three single-ring bases found in nucleic acids. It belongs to the class of compounds called pyrimidines.

Tool mark Any impression, cut, scratch, or abrasion caused when a tool comes in contact with another surface.

Topsoil The surface layer of soil, which is rich in humus.

Trace evidence Evidence that is extremely small or is present in limited amounts that results from the transfer from the victim or crime scene to the suspect.

Trace-over method The copying of a genuine signature by tracing over it.

Transfer bloodstain A bloodstain deposited on a surface as a result of direct contact with an object with wet blood on it.

Transmutation Conversion of one kind of atomic nucleus to another.

Triangulation method A technique used to record measurements for crime scene sketches, measuring the location of the evidence (X, Y) from fixed points (A, B).

Typica Points in a fingerprint where ridge lines end or where one ridge splits into two.

Ulnar loop A fingerprint loop pattern that flows in the direction of the little finger.

Upper explosive limit (UEL) The highest concentration of vapor in air that will burn.

Uracil One of the three single-ring bases found in nucleic acids. It belongs to the class of compounds called pyrimidines.

Virtual image An image that is seen only by looking through a lens.

Visible fingerprint A fingerprint that can be seen with the naked eye.

Vitreous humor Liquid inside the eye.

Wad A plastic, cardboard, or fiber disk that is placed between the powder and the shot in a shotgun shell.

Warp Lengthwise strand of yarn on a loom.

Watermark A translucent design impressed into more expensive paper during its manufacture.

Wavelength The distance between adjacent wave crests, usually measured in nanometers (nm).

Weapon of mass destruction (WMD) A weapon that kills or injures civilian as well as military personnel. WMDs include nuclear, chemical, and biological weapons.

Weft Crosswise strands of yarn on a loom.

Weight A measure of the force of attraction of the Earth for an object.

Whorl A fingerprint pattern forming concentric circles in the center of the finger pad.

Working distance The distance between an object being investigated and the objective lens of a microscope.

Write blocker A hardware or software device that prevents any writing to the disk drive to which it is attached.

X-ray fluorescence spectrometry (XRF) A technique that measures the emission of an X-ray when a sample is exposed to higher-energy X-rays. The energy of the emitted X-ray indicates which element released it.

Author Index

Subject Index

Ammonium nitrate/fuel oil (ANFO), 420
Ammunition evidence. *See also* Bullets; Firearms
    cartridges, 192, 196–198, 196*f*, 204, 205*f*
    characteristics of, 195
    collection of, 16, 201
    preservation of, 201
    primers and, 197–198, 223
    shotguns, 198, 198*f*
Amobarbital, 285
Amorphous materials, 86
Amorphous solids, 112
Amphetamines, 285–287
Amplicon, 363
Amyl nitrites, 288
Anabolic steroids, 288–289
Anagenic phase of hair growth, 92, 92*f*
Analgesics, 277
Analytical detection instruments, 450
Anderson sequence, 382
Andrews, Tommy Lee, 349
Anions, 433, 434*f*
Anisotropic materials, 86
Annealing
    DNA strands, 360
    glass, 122–123
Anthrax bacterium, 456, 457, 458
Anthrax letter attacks (2001), 456, 457
Anthrax Quick ELISA test, 458–459, 459*f*
Anthropometry, 134–135
Antibodies, 296–297, 333–334, 458–459, 459*f*
Anticoagulants, 321
Anticodons, 346–347, 348*f*
Antigens, 333
Antimony (Sb), 196
Antiquities, 229
Antiserum, 335
Anti-spam laws, 399
Antivirus software, 398, 401, 404
Arches, fingerprints, 139–140, 139–140*f*
Arcus 94 9-mm handgun, 200, 200*f*
Arrest, search incident to, 20–21, 397
Arrest warrants, 20, 21
Arsenical vesicants, 445–446
Arsine (SA), 446
Arson, 246–263. *See also* Accelerants
    burn patterns, 255–258, 256–257*f*
    charring of floor surfaces, 258–259
    chemical reactions, 247–248
    collection of evidence, 15*t*, 16–17, 260, 262
    concentration effects, 249
    crude petroleum, 252
    determination of cause, 246, 254–258
    flammable residue analysis, 260–263
    free-burning stage, 250
    gas chromatography, 260, 261–263, 262–263*f*
    heated headspace sampling, 260, 261*f*
    hydrocarbons, 250–252, 251–252*f*, 253*t*
    incipient stage, 249–250, 250*f*

liquid containers, 259
melting of materials, 258, 258*t*
odors, 259, 259*f*, 260–261
oxidation, 246–247
passive headspace diffusion, 260–261
petroleum refining, 253–254, 254*f*, 254*t*
point of origin, 246, 254–258
preservation of evidence, 260
smoldering stage, 250
solid-phase microextraction, 261
temperature effects, 248–249, 248*t*
V pattern on vertical surfaces, 249–250, 250*f*, 257*f*
wildfire arson cases, 259
Asphyxiants, 445
Association of Official Analytical Chemists, 122
Associative evidence, 39–41, 41–43, 112
ATF. *See* Alcohol, Tobacco, Firearms and Explosives Bureau
Atomic absorption spectrophotometry (AAS)
    functions of, 228
    gunshot residue, 234, 236, 236*f*, 237
    limitations of, 433
Atomic mass number, 219, 436, 453
Atomic number, 219, 219*f*, 436
Atomic spectroscopy
    elements of glass and, 231–232
    inductively coupled plasma emission spectroscopy, 229–231, 230–231*f*, 309, 433
    inorganic analysis, 229–233
    X-ray fluorescence spectrometry, 232–233, 232–233*f*, 237
Audio recordings, 9, 13
Aum Shinrikyo sarin attack, 447, 448*f*
Authentication methods, 178, 178*f*, 400
Automated Fingerprint Identification System (AFIS), 150–151, 152
Automobile paints, 101–102, 101*f*
Automobiles. *See* Car accidents; Hit-and-run homicides
Autopsies, 17, 306–307, 307*t*. *See also* Forensic toxicology
Azoospermia, 337

B
BAC. *See* Blood alcohol concentration
Backdraft, 250
Backwards theory, 254–255
BACTrack, 319, 320*f*
Baden, Michael, 286
Balco Labs, 289
Ballistics, 44, 198–199, 199*f*. *See also* Firearms
Bank of Canada, 176–177
Barbiturates, 284–285, 285*f*
Barton, Terry, 259
Basecoats, 101–102
Baseline method, of sketching, 12, 12*f*
Battelle Corporation, 34
Bayer Company, 277, 277*f*
Becke line, 100–101, 121, 121*f*, 122
Beer's law, 318
Bell, Everett, 84
Belushi, John, 286

Benjamin N. Cardozo School of Law Innocence Project, 34, 362
Bertillon, Alphonse, 134–135
Betadine, 321
Beta particles, 452–455, 452t, 453f
Binders in paints, 101
Biological agents, 458–461, 459–461f
Biological fluids, 330–351. *See also* Blood tests
    blood components, 331–332
    collection of, 16, 338–340, 339f
    DNA inheritance, 341–349, 342–348f, 349t
    environmental factors and, 37
    mitochondrial DNA, 350–351, 351f
    nuclear DNA and laws, 349–350
    paternity and, 340–341, 340f
    saliva, 337, 339
    seminal stains, 16, 337–338
Biological weapons (BWs), 455–461, 459–461f. *See also* Weapons of mass destruction
Biometrics
    facial scans, 152, 178–179
    fingerprint identification, 151–153
    finger scanning, 151–153, 151f, 178–179
    identity authentication, 178, 178f, 400
    security issues, 153
    US-VISIT program and, 178–179, 178f
BIOS operation, 395, 404
Bioterrorism, 455–458, 459–461f
Birefringent materials, 86
Bit-for-bit copy of HDD, 405
Bit of data, 406
Bitstream copy of HDD, 405–406
Black powder, 196–197, 418
Blassie, Michael, 383
Blasting caps, 422
Blister agents, 445–446
Block capitals handwriting, 164
Blood, 16, 18, 37, 47, 331
Blood agents, 446
Blood alcohol concentration (BAC), 310–320
    absorption-elimination data, measurement of, 315–316
    alcohol in circulatory system, measurement of, 316–317, 317f
    alcohol metabolism and, 313–315, 314f
    driving risks and, 310, 312–313, 312f
    effects on human body, 311, 311t
    handheld devices and, 319, 320f
    putrefactive blisters and, 316
    zero tolerance levels, by state, 313, 313f
Blood serum, 331
Bloodstain patterns, 45–49
    active bloodstains, 45–47
    in crime scene reconstruction, 45–49
    passive bloodstains, 47, 48f
    point of convergence, 47–49
    transfer bloodstains, 47
Blood tests
    alcohol in blood, 320–322, 321–322f

    collection of blood, 321
    gas chromatography, 321–322, 321–322f
    preservation of blood and, 321
    presumptive tests, 332, 333f
    serological blood typing, 335–337, 336–337t, 336f
    serological tests, 332–335, 334–335f
Blood types, 335–337, 336–337t
Body fluids. *See* Biological fluids
Body temperature, 65–66, 306
Boggs Bill (1951), 272
Bomb evidence. *See also* Explosives
    collection of, 16–17
    dirty bombs, 452
    MacGyver bombs, 422
    pipe bombs, 419, 419f, 422, 423, 425, 432
    primers and, 419
Bombings. *See* Terrorism and terrorist attacks
Border searches, 20
Bores in gun barrels, 188–189
Boundaries of crime scenes, 6–8
Brannon, United States v. (1998), 323
Brass, 229
Breathalyzers, 311–312, 317–318, 323
Breechblocks, 204
Brisant explosives, 421
Broach cutters, 189–190, 190f
Browning, John, 192
Buckland, Richard, 358–359
Buell, Robert Anthony, 100
Bulk samples, 289
Bullets. *See also* Ammunition evidence; Firearms
    bullet force, 191
    examination of fired bullets, 202–203, 203f
    glass fractures and, 117, 117f
    recovery of, 15
    ricochets, 45
    trajectory estimates, 44–45, 45f
Burglary, 15t, 214
Burn patterns, 255–258, 256–257f. *See also* Arson
Bush, George W., 163
Butane, 251, 252f
Butyl nitrates, 288
BWs. *See* Biological weapons
Bytes, 407, 409

C

CAD (computer-aided design) software, 12, 13f
Caliber, 189
California Proposition 215, 281
Camarena Salazar, Enrique "Kiki," 61
Cameras, 10
Canadian currency, 176–177, 177f
Cancer cells, 455
Canines, 259, 425, 427
Cannabis, 280
Capacitive scanners, 152
Capillary array electrophoresis (CAE) analyzers, 372, 373f
Capillary electrophoresis (CE), 368–370, 368–370f

Explosives (*cont.*)
  low explosives, 418–419, 428*f*
  military explosives, 420–421, 420–421*f*
  RDX, 421, 421*f*, 430*f*, 431
  taggants, 434–437, 435*f*
  thin-layer chromatography and, 429, 430*f*
  TNT, 420, 420*f*, 429, 430*f*, 431, 432*f*
Extensive physical properties, 60
External ballistics, 198
Extractors in cartridges, 204

F

FAAS (flameless atomic absorption spectrophotometry), 125
Facial scans, 152, 178–179
Fahrenheit scale, 65
Fast Fourier transform (FFT), 149
Federal Aviation Administration (FAA), 424
Federal Bureau of Investigation (FBI). *See also* CODIS
  on Clinton's RFLP match, 359
  crime laboratories, 35, 35*t*, 80
  fingerprint database, 137
  on glass density samples, 119, 123–124*f*
  guidelines on glass evidence, 124, 125
  on hair comparison errors, 93
  ICP-OEP use by, 229
  Innocent Images National Initiative, 407
  Integrated AFIS, 150
  on Madrid terrorist bombings, 135
  on metabolites, 308
  Patriot Act, investigations under, 444
  search pattern recommendations, 13, 14*t*
  SWGs and, 38
  on taggants, 437
  on TWA Flight 800 crash, 424
  on US-VISIT program, 179
Federal Bureau of Narcotics. *See Now* Drug Enforcement Agency
Federal crime laboratories, 34–35, 35*t*
Federal Firearms License (FFL), 18
Federal Rules of Evidence (FRE), 33, 33*t*, 138, 179
  Rule 401, 49
  Rule 702, 49–50, 138
FedEx, 18
Fentanyl patches, 307
FFL (Federal Firearms License), 18
FFT (Fast Fourier transform), 149
Fibers, 94–101
  natural fibers, 94–97, 95*f*
  synthetic fibers, 90*f*, 97–101, 98*t*, 99–100*f*
Field of view, 85
Field sobriety testing, 310–311
Fifth Amendment, 166
Filament yarns, 95
File slack, 407–409, 408*f*
Fingerprints
  AFIS and, 150–151
  biometric identification, 151–153, 151*f*

classification of, 149–150, 149*f*, 150*t*
collection of, 16, 37, 134
development of, 140–144
digital imaging of, 148–149
finger scanning, 151–153, 151*f*, 178–179
genetics and, 136
on hard/nonabsorbent surfaces, 142–143, 142–143*f*
HDD, 405–406
history of fingerprinting, 134–137
identification points, 138, 140, 141*f*
individualization and, 15
latent fingerprints, 16, 134, 141
lifting tapes for, 142, 142*f*
patent fingerprints, 141
patterns of, 139–140, 139*f*
permanence of, 138–139, 138*f*, 139*f*
photography and, 145–148, 146–148*f*
plastic fingerprints, 141
powders, 142, 142*f*
preservation of, 145–149
on soft/porous surfaces, 143–144, 143*f*
uniqueness of, 137–138
Finger scanning, 151–153, 151*f*, 178–179
Finished sketches, 12, 13*f*
Firearms, 188–205. *See also* Ammunition evidence; Gunshot residue
  accuracy of, 188–191, 189–190*f*, 191
  ammunition, 195–198, 196*f*, 198*f*, 201
  ballistics, 44, 198–199, 199*f*
  bullet force, 191
  cause of death and, 188
  collection of evidence, 16, 199–200
  court testimony, 202
  firing rate, 191–192
  firing reliability, 191–192
  forensic examination of, 201–205, 202–203*f*, 205*f*
  gun barrels, 188–191, 189–190*f*
  handguns, 192–194, 193–194*f*, 205*f*
  preservation of evidence, 199–200
  recovery of bullets, 15
  rifles, 191–192, 191*f*, 194
  serial number restoration, 200–201, 200–201*f*
  shipping regulations, 18
  shotguns, 195–196*f*, 198, 198*f*
Fires. *See* Arson
Firing rate and reliability, 191–192
First responders, 6–7, 8
Flameless atomic absorption spectrophotometry (FAAS), 125
Flammable liquids, 248–249, 259
Flammable range, 249
Flash drives, 404
Flash memory, 394–395
Flashover, 250
Flash points, 248
Flintlock rifles, 191–192, 191*f*
Follicular tags, 92, 92*f*
Food and Drug Administration, 281, 307, 308

collection of evidence and, 93, 93f, 102, 102f
hair analysis and, 90–94
lenses, 81, 82f
magnification, 80, 82f
microspectrophotometers, 88–89
natural fibers and, 94–97, 95f
paints and, 101–103, 101f, 103f
refraction, 80–81, 81f
synthetic fibers and, 90f, 97–101, 98t, 99–100f
types of, 81–90, 83f
Microsoft software, 407, 409
Microspectrophotometers, 88–89
Microtrace, 435–436
Military
explosives and, 420–421, 420–421f
identification of soldier remains, 380, 382–383, 383t
Long or Short Range Biological Standoff Detection System, 458
RAPID devices, 460
Mincey v. Arizona (1978), 21–22
Mincey, Rufus, 21–22
Minerals, 69, 86
Minimal administered dose, 310
Minutiae, fingerprints, 138–139, 141f, 146f, 147, 153
Miranda v. Arizona (1966), 34
Miranda warnings, 34
Missing persons identification, 380
Mitchell, Byron, 149
Mitchell, United States v. (1999), 138
Mitochondrial DNA
DNA typing analysis, 380–383, 381f, 383t
in human cell, 340, 340f
maternal inheritance of, 350–351, 351f
nuclear DNA and, 350–351
Modified Greiss Test, 239, 239f, 427
Modus operandi (MO), 36–37
Moeller, Donald, 72
Mohs scale, 115–116
Molecules, 216–217, 216–217t, 216f. *See also* Inorganic analysis
Monomers, 98
Monson, Diane, 281
Morphine, 270, 275, 277
Morphology, 80, 90–92, 91f
Mothers Against Drunk Driving (MADD), 312
mtDNA. *See* Mitochondrial DNA
Multiplex DNA analysis
by dye color, 371–372, 373f
interpretations of, 378, 378t
on multiple capillaries, 372, 373f
multiplexing, 371
by size, 371, 372f
Municipal crime laboratories, 35
Munsell color system, 68, 69f
Murrah Federal Building bombing (1995), 416, 444, 445f
Museums, 229
Mustard gas, 445

N

NAA (neutron activation analysis), 229, 234
Narcotic Drug Act (1956), 272
Narcotics
codeine, 275, 277
heroin, 274, 275, 277–278, 278f, 294f
methadone, 278–279, 279f
morphine, 270, 275, 277
opiates, 270, 272, 274–277, 276f
OxyContin, 279, 279f
Narcotics Act (1946), 272
National Board of Forensic Odontology, 35
National Center for Missing and Exploited Children (NCMEC), 407
National Crime Information Center, 200
National Crime Victimization Survey (NCVS), 400
National DNA Index System. *See* CODIS
National Explosives Tracing Center (ATF), 425
National Fire Protection Association, 246
National Forensic Science Technology Center, 35
National Institute on Drug Abuse (NIDA), 296, 296t
National Integrated Ballistic Information Network, 203
National Research Council (NRC), 350, 437
National Rifle Association, 436
National security. *See* Homeland Security Department (DHS); Terrorism and terrorist attacks
National Strategy to Secure Cyberspace (DHS), 398
National Transportation Safety Board (NTSB), 424
Native American Church, 280, 282
NATO (North Atlantic Treaty Organization), 450
Natural disaster relief, 152
Natural fibers, 94–97, 95–97f
Natural gas, 419
Nature on Jefferson's descendants, 340
Naval Criminal Investigative Service, 255
NCMEC (National Center for Missing and Exploited Children), 407
NCVS (National Crime Victimization Survey), 400
Near-infrared reflectance spectroscopy, 322–323
Nebulizers, 230, 230f
Nerve agents, 446, 447–448f
Neurotransmitters, 446, 447f
Neutron activation analysis (NAA), 229, 234
Neutrons, 218. *See also* Inorganic analysis
New York City, Civil Service Commission, 137
New York State Prison System, 137
NG. *See* Nitroglycerine (NG), and explosive residue
Nicholas II, Tsar, 350–351, 382
Nichols, Terry, 416
NIDA (National Institute on Drug Abuse), 296, 296t
Ninhydrin, 143–144, 144f, 147, 148f, 171
Nitrate, 433
Nitrocellulose, 197, 418
Nitroglycerine (NG), and explosive residue, 197, 197f, 418, 418f, 431, 432f
Noncoding regions, DNA, 340, 364–365, 380–381
Nonmetals, 221

PIDs (photo-ionization detectors), 449–450, 449f
Pigments, 101. *See also* Paint evidence
Pipe bombs, 419, 419f, 422, 423, 425, 432. *See also* Explosives
Pitchfork, Colin, 359
Pixels, 148
Plain view doctrine, 21, 397
Plain weave, 96, 97f
Plane-polarized light, 87, 87f
Plasma, 229, 331
Plasma-optical emission spectrometer (ICP-OES)
    ammonium ion and, 433
    functions of, 228
    glass evidence, 113, 125, 231–232
    multichannel instrument, 230–231, 231f
    trace metals, 229–231, 230–231f, 309
Plastic fingerprints, 141
Plastics industry, 98
Pleochroism, 87
Point detection technologies. *See also* Weapons of mass destruction
    colorimetric tubes, 450, 451f
    handheld devices, 448–449
    photo-ionization detectors, 449–450, 449f, 451f
    surface acoustic wave detectors, 450
Point detectors, 448–449
Point of convergence, blood, 47–49, 48f
Polar coordinate method, of sketching, 12, 12f
Polarized light microscopy, 87, 87f
Polarizer lenses, 87
Polarizing microscopes, 86–87, 87f
Polymerase chain reaction (PCR), 360–361f, 360–363, 383, 460
Polymers, 98, 99f
Polyvinyl butyral (PVB), 113–114
Poppy plant (*Papaver somniferium*), 274–275, 275f. *See also* Opiates
Population genetics, 376–378, 377–378t
Postmortem interval, 65–66, 306
Postmortem toxicology, 306–310
    collection of specimens, 306–308, 307t
    interpretations of, 309–310
    specimen analysis, 308–309, 308f
Post offices, 457
Precipitin serological tests, 332–334, 333–334f
Pregnancy tests, 459, 459f
Presumptive tests
    on blood, 332, 333f
    color tests, 290–291, 290f
    on drugs of abuse, 289–293, 308
    microcrystalline tests, 291
    on semen, 337–338
    thin-layer chromatography, 170, 291–293, 291–293f, 296, 429, 430f
Prilling, 420
Primary explosives, 419
Primers
    in ammunitions, 197–198, 233
    in bombs, 419
    in DNA typing, 360, 361f
Primer surfacer, 101
Principal energy levels, 225–226
Printers and printing
    commercial printing, 175
    printer drums, 172–173, 172–173f
    security printing of currency, 175–177, 176–177f
    types of printers, 171–173
Privacy, 397. *See also* Fourth Amendment
Probable cause, 20, 21
Proficiency testing, 36
Prohibition era, 271–272
Projectiles, 117, 117f, 191. *See also* Bullets
Propane, 251, 251f
Proposition 215 (California), 281
Protein synthesis, 345–347, 347–348f. *See also* DNA
Protocols
    in computer forensics, 405
    with hazardous materials, 289
    in lab testing, 35–36, 38
Protons, 218. *See also* Inorganic analysis
Pseudoephedrine, 286, 286f
Psychological dependence on drugs, 274
p30 tests, 338
Punch cards, 393, 393f
Pure Food and Drug Act (1906), 271
Putrefactive blisters, 316
Pyrex, 112–113
Pyrolysis, 249, 256
Pyrometers, 65, 65f

Q

Quadrant search pattern, 13–14, 14f, 14t
Quality assurance manuals, 35
Quality control manuals, 35
Quantum of energy, 225
Questioned documents. *See* Documentary evidence
Questioned samples, 15, 38
Quinine, 278
3-quinucildinyl benzilate (BZ), 446

R

Racial groups, 94, 94t
Rad, 455
Rader, Dennis, 402
Radial cracks, 116–117, 116f, 118, 118f
Radial loops, 140
Radiation
    detection of, 455, 456f
    effects on human body, 454–455, 455t
    measurement of, 455
    from nuclear weapons, 452–454
Radioactive materials, 367
Radio frequency identification (RFID) chips, 178
Radioisotope half-life, 453–454, 454f
Radiological dispersal devices (RDDs), 452

Photo Credits

Chapter 6

Chapter Opener, Courtesy of Forensics Source © 2010; 6-1, © Ed Reschke/Peter Arnold, Inc.; 6-3, 6-4, Courtesy of Sirchie Finger Print Laboratories, Youngsville, NC; 6-5, © coko/ShutterStock, Inc.; 6-8, 6-9, 6-10, 6-11, Courtesy of Sirchie Finger Print Laboratories, Youngsville, NC; 6-13, © Mauro Fermaiello/Photo Researchers, Inc.; 6-14, Courtesy of Sirchie Finger Print Laboratories, Youngsville, NC; 6-21, © Krill Roslyakov/ShutterStock, Inc.; Page 152, Courtesy of Cross Match Technologies, Inc.; 6-22, 6-23, 6-24, Courtesy of Sirchie Finger Print Laboratories, Youngsville, NC

Chapter 7

Chapter Opener, © basel101658/ShutterStock, Inc.; 7-3, 7-4, Courtesy of Sirchie Finger Print Laboratories, Youngsville, NC; 7-5, 7-6, 7-7, Courtesy of Jon Girard; 7-8, 7-9, Courtesy of Foster & Freeman Ltd.; 7-10, Courtesy of James Girard; 7-11, Courtesy of Foster & Freeman Ltd.; 7-12, 7-16, Courtesy of James Girard; 7-17 (top), © Robert H. Creigh/ShutterStock, Inc.; 7-17 (bottom), © Feng Yu/ShutterStock, Inc.; 7-19, © Bank of Canada/Banque du Canada; 7-20, © Sukree Sukplang/Reuters/Landov; 7-21, Courtesy of Jim Tourtellotte/U.S. Customs and Border Protection

Chapter 8

Chapter Opener, © LiquidLibrary; 8-1, © Janko Stermecki/ShutterStock, Inc., 8-2, Courtesy of Pioneer Broach Company, Los Angeles, CA; 8-5, © Travis Klein/ShutterStock, Inc.; 8-8, © vadim kozlovsky/ShutterStock, Inc.; 8-9A, © LockStockBob/ShutterStock, Inc.; 8-9B, © Erik De Castro/Reuters/Landov; 8-10A, © Gualberto Becerra/ShutterStock, Inc.; 8-10B, © Mariusz Lubkowski/Dreamstime.com; 8-12, © Elena Kalistratova/ShutterStock, Inc.; 8-16, Photo courtesy of Corbin Manufacturing & Supply, Inc.; 8-17, 8-19, Courtesy of William Sherlock; 8-20, 8-21, 8-22, Courtesy of Forensic Technology Inc.; 8-23, Courtesy of Maine State Police Crime Laboratory; 8-25, Courtesy of William Sherlock

Chapter 9

Chapter Opener, © Corbis; 9-17, Courtesy of William Sherlock and the Illinois State Police, Division of Forensic Services, Forensic Sciences Command; 9-19, Courtesy of Sirchie Finger Print Laboratories, Youngsville, NC; 9-20, 9-22, Courtesy of McCrone Associates; 9-23, Courtesy of Jack Dillon; 9-24, Courtesy of Jeffrey Scott Doyle, www.firearmsid.com

Chapter 10

Chapter Opener, © Bernd Jürgens/ShutterStock, Inc.; 10-7, © Vladimir Zanadvorov/ShutterStock, Inc.; 10-10, 10-11, 10-12, Courtesy of Sirchie Finger Print Laboratories, Youngsville, NC; 10-14, 10-15, © Agilent Technologies, Inc. 2010. Reproduced with Permission, Courtesy of Agilent Technologies, Inc.

Chapter 11

Chapter Opener, © Andrew Burns/ShutterStock, Inc.; 11-1, © Philip Scalia/Alamy Images; 11-4, © DEA/AP Photos; 11-8, © Mitchell Brothers 21st Century Film Group/ShutterStock, Inc.; 11-10, Courtesy of Orange County Police Department, Florida; 11-13, Courtesy of DEA; 11-22, Courtesy of Smiths Detection; 11-25, © David Karp/AP Photos; 11-26, © Carolina K. Smith, M.D./ShutterStock, Inc.

Chapter 12

Chapter Opener, Courtesy of Axygen, Inc.; 12-8, Courtesy of CMI, Inc.; 12-10, BACtrack Breathalyzer, courtesy of KHN Solutions LLC; 12-12, 12-13, 12-14, © Agilent Technologies, Inc. 2010. Reproduced with Permission, Courtesy of Agilent Technologies, Inc.

Chapter 13

Chapter Opener, © Jan Kliciak/ShutterStock, Inc.; 13-1, Courtesy of Sirchie Finger Print Laboratories, Youngsville, NC; 13-2, Courtesy of Maine State Police Crime Laboratory; 13-6, © John Walsh/ Photo Researchers, Inc.; 13-7, Courtesy of Sirchie Finger Print Laboratories, Youngsville, NC; 13-14C, © Jan Kliciak/ShutterStock, Inc.

Chapter 14

Chapter Opener, Courtesy of Forensics Source © 2010; 14-2, © Reuters/HO/Landov; 14-3, Courtesy of Techne Inc.; 14-14, Courtesy of Life Technologies, Carlsbad, CA

Chapter 15

Chapter Opener, © Stillfx/ShutterStock, Inc.; 15-1, © c./ShutterStock, Inc.; 15-2, Courtesy of Brett Robinson; 15-3, Courtesy of Erik Garcia; 15-4, © Stillfx/ShutterStock, Inc.; 15-5, Courtesy of Erik Garcia; 15-6, © ExaMedia Photography/ ShutterStock, Inc.; 15-7, Courtesy of Erik Garcia; 15-9, © LeahKat/ShutterStock, Inc.; 15-10, Courtesy of Jon Girard; 15-11, Courtesy of Erik Garcia; 15-12, Courtesy of Digital Intelligence; 15-13, 15-14, Courtesy of Erik Garcia

Chapter 16

Chapter Opener, © Chas/ShutterStock, Inc.; 16-7, Courtesy of Timo Halén; 16-9, Courtesy of Morpho Detection, Inc.; 16-12, Courtesy of Sirchie Finger Print Laboratories, Youngsville, NC; 16-13, Courtesy

of Western Powders, Inc.; 16-15, 16-16, 16-17, 16-19, 16-20, © Agilent Technologies, Inc. 2010. Reproduced with Permission, Courtesy of Agilent Technologies, Inc.

Chapter 17

Chapter Opener (left), © Richard C. Bennett/ ShutterStock, Inc.; Chapter Opener (right), © WilleeCole/ShutterStock, Inc.; 17-1, Courtesy of FEMA; 17-5, © Kyodo/Landov; 17-7, 17-8, Courtesy of Draeger Technology; 17-9, Courtesy of Jon Girard; 17-12B, Courtesy of Draeger Technology; 17-15, Courtesy of Jones & Bartlett Learning. Photographed by Kimberly Potvin; 17-17, 17-19, Courtesy of Idaho Technology, Inc.

Unless otherwise indicated, all photographs and illustrations are under copyright of Jones & Bartlett Learning.